AF333789

Information and Collaboration Models of Integration

NATO ASI Series

Advanced Science Institutes Series

A Series presenting the results of activities sponsored by the NATO Science Committee, which aims at the dissemination of advanced scientific and technological knowledge, with a view to strengthening links between scientific communities.

The Series is published by an international board of publishers in conjunction with the NATO Scientific Affairs Division

A	**Life Sciences**	Plenum Publishing Corporation
B	**Physics**	London and New York
C	**Mathematical and Physical Sciences**	Kluwer Academic Publishers
		Dordrecht, Boston and London
D	**Behavioural and Social Sciences**	
E	**Applied Sciences**	
F	**Computer and Systems Sciences**	Springer-Verlag
G	**Ecological Sciences**	Berlin, Heidelberg, New York, London,
H	**Cell Biology**	Paris and Tokyo
I	**Global Environmental Change**	

NATO-PCO-DATA BASE

The electronic index to the NATO ASI Series provides full bibliographical references (with keywords and/or abstracts) to more than 30000 contributions from international scientists published in all sections of the NATO ASI Series.
Access to the NATO-PCO-DATA BASE is possible in two ways:

– via online FILE 128 (NATO-PCO-DATA BASE) hosted by ESRIN,
Via Galileo Galilei, I-00044 Frascati, Italy.

– via CD-ROM "NATO-PCO-DATA BASE" with user-friendly retrieval software in English, French and German (© WTV GmbH and DATAWARE Technologies Inc. 1989).

The CD-ROM can be ordered through any member of the Board of Publishers or through NATO-PCO, Overijse, Belgium.

Series E: Applied Sciences - Vol. 259

Information and Collaboration Models of Integration

edited by

Shimon Y. Nof

School of Industrial Engineering,
Purdue University,
West Lafayette, IN, U.S.A.

Kluwer Academic Publishers

Dordrecht / Boston / London

Published in cooperation with NATO Scientific Affairs Division

Based on the NATO Advanced Research Workshop on
Integration: Information and Collaboration Models
Il Crocco, Italy
June 6–11, 1993

A C.I.P. Catalogue record for this book is available from the Library of Congress

ISBN 0-7923-2753-5

Published by Kluwer Academic Publishers,
P.O. Box 17, 3300 AA Dordrecht, The Netherlands.

Kluwer Academic Publishers incorporates the publishing programmes of
D. Reidel, Martinus Nijhoff, Dr W. Junk and MTP Press.

Sold and distributed in the U.S.A. and Canada
by Kluwer Academic Publishers,
101 Philip Drive, Norwell, MA 02061, U.S.A.

In all other countries, sold and distributed
by Kluwer Academic Publishers Group,
P.O. Box 322, 3300 AH Dordrecht, The Netherlands.

Printed on acid-free paper

TABLE OF CONTENTS

PREFACE

The objective of this book is to bring together contributions by eminent researchers from industry and academia who specialize in the currently separate study and application of the key aspects of integration. The state of knowledge on integration and collaboration models and methods is reviewed, followed by an agenda for needed research that has been generated by the participants. The book is the result of a NATO Advanced Research Workshop on "Integration: Information and Collaboration Models" that took place at Il Ciocco, Italy, during June 1993.

Significant developments and research projects have been occurring internationally in a major effort to integrate increasingly complex systems. On one hand, advancements in computer technology and computing theories provide better, more timely, information. On the other hand, the geographic and organizational distribution of users and clients, and the proliferation of computers and communication, lead to an explosion of information and to the demand for integration.

Two important examples of interest are computer integrated manufacturing and enterprises (CIM/E) and concurrent engineering (CE). CIM/E is the collection of computer technologies such as CNC, CAD, CAM, robotics and computer integrated engineering that integrate all the enterprise activities for competitiveness and timely response to changes. Concurrent engineering is the complete life-cycle approach to engineering of products, systems, and processes including customer requirements, design, planning, costing, service and recycling. In CIM/E and in CE, computer based information is the key to integration. Integration can be viewed in three main aspects: (1) *vertical*, between organizational layers of authority or detail; (2) *horizontal*, between a variety of related functions at the same level; (3) along *progression in time*. Therefore, integration is more vital when an organization is more complex structurally and functionally, and is under more pressure to produce timely responses to its customers' demands.

The process of effective integration, if properly design and controlled, offers substantial advantages for reliable, high quality and timely results. Are there generic theories and models of integration?

Four important recent directions of research have evolved separately with a common relation to integration: (a) Computer supported collaborative work (CSCW), and coordination theory and collaboration technology (CTCT); (b) Integrated cognitive systems, including distributed AI and distributed problem solving; (c) Digital multimedia technology and prototype systems; (d) Research in CIM/E and in CE. Useful results and developments have been presented already in each of these areas. All these issues are fundamental to the future development of integration.

The purpose of the advanced research workshop and this resulting book is to juxtapose the developments and research in the above related yet separate directions in an effort to understand, compare and merge them. It is felt that it is now timely and necessary to do so in order to leap forward in the multidisciplinary modeling and design of integration processes. The output will benefit the whole research community in understanding how the different approaches to integration can complement and enrich each other. The research agenda is proposed for future international cooperation in this area.

I wish to express my gratitude and appreciation to the NATO Division of Scientific Affairs which sponsored the workshop, particularly, Dr. Luigi Sertorio, the ARW Programme Director, for his invaluable support in the organization of the workshop; the co-sponsors of the workshop: NSF Division of Design and Manufacturing, especially its director, Dr. Thom Hodgson; NSF Program on Integration Engineering and its director, Dr. Stan Settles; and the School of Industrial Engineering, Purdue University.

I would also like to express my thanks to my distinguished colleagues and friends who participated in the workshop and helped create this book; the resourceful members of our Advisory Board: Asbjorn Rolstadas, Agostino Villa, and Richard Weston; Frank Fallside, the Head of the Information Engineering Division at Cambridge University, U.K., who was on the Advisory Board in the beginning but, sadly, passed away and left a big gap in our community.

Special thanks to Bruno Giannasi from the Il Ciocco International Center, who helped in the local arrangements and also taught us about beautiful Tuscany and its native Carlo Lorencini Collodi who created Pinocchio; Nava Nof, Moriah Nof, and Jasmin Nof, for organizational support; Joyce Hinds and Brenda Thomas, for administrative support, and Barbara Kester from the NATO Publication Coordination Office for helping in the publication of this book.

The challenges and significance of integration and collaboration research are great. It is hoped that this book will serve as a contribution to progress in modeling and design of integration methods and systems. After all, collaborative integration of research results and of researchers' work is one of the targets of this endeavor!

West Lafayette, Indiana
October 1993

Shimon Y. Nof

INTEGRATION AND COLLABORATION MODELS

SHIMON Y. NOF
School of Industrial Engineering
Purdue University
West Lafayette, IN 47907-1287   U.S.A.

> "If we understand each other,
> That's communication.
> ...
> If we help each other home,
> That's cooperation.
> And all these ations added up
> Make civilization."
> Shell Silverstein, "*A Light in the Attic*"

1. INTRODUCTION

The purpose of this introductory chapter is to define several important types of integration and collaboration issues and describe briefly some research efforts to model them.

Integration is an intelligent response to an increasing fragmentation of distributed organizations. Information integration is necessary to gain better control over the different units or functions of an organization, leading to improved performance and, potentially, better results. With the advent of computer and communication techniques, information integration has become more feasible, but at the same time, also more necessary. Increasing amounts of information are generated and handled over complex networks of users, clients, and servers. As a result, successful integration is viewed as a useful competitive ability.

Typical examples of integration are on-going design revisions and process improvements that are integrated with previous product information to produce a more desirable product model; integration of sensory information from multiple sensors; integration of several expert-systems embedded in microprocessors; integration of distributed human and machine function in a factory or a service organization.

INTEGRATION: Integration is a process by which sub-systems share or combine physical or logical tasks and resources so that the whole system can produce better (synergistic) results. *Internal integration* occurs among sub-systems of the same system, and *external integration* is with sub-systems of other systems (as in customers-supplier relations). However, integration always depends on a cooperative behavior.

COOPERATION: The success of any integration is a function of the degree of cooperation among the integrated sub-systems. Without cooperation, integration is impossible. Cooperation is defined as the willingness and readiness of sub-systems to share or combine their tasks and resources as in "open systems".

1

S. Y. Nof (ed.), Information and Collaboration Models of Integration, 1–6.
© 1994 *Kluwer Academic Publishers. Printed in the Netherlands.*

2

COLLABORATION: Even with cooperation, there is still the issue of how to perform the process of integration effectively. Collaboration is defined as the active participation and work of the sub-systems towards accomplishing collaborative integration. Collaboration can also be characterized as *internal* or *external*. Usually, collaboration is the opposite of competition among sub-systems, although there can be situations of competitive integration. An important function of collaborative integration is to overcome conflicts among the sub-systems.

2. TYPES OF INTEGRATION AND COLLABORATION

An attempt to provide an integration taxonomy is shown in Figure 1. The taxonomy is along three main axes: integration, interaction, and collaboration. Examples are also given to illustrate the different types of integration and collaboration. Recent research has focused on two main types of integration: (1) coordination and collaboration among distributed, often remote, teams, facilities, suppliers, and clients; (2) integration of software tools. By coordination and collaboration, distributed workers or processors can integrate information via a network of workstations (Eberts and Nof, 1993). In software tool integration, information and programs are combined to provide a smooth transition of information between them to increase overall computational effectiveness (Weston, 1993).

Several attempts to develop a theory of integration have been reported, e.g., Papastavrou and Nof (1992), in the area of decision integration. We can define the integration process with six elements of integration as follows:

$$\text{Integration: } (I, \ IP, \ I/O, \ S_i \ (i=1,...n), \ t, \ IM) \tag{1}$$

where

(1) I is the *Integration Operator*, e.g., $I(Y_1, Y_2) \rightarrow u$ means that two input streams, Y_1 and Y_2, are integrated to produce output u; $I(Y_1, Y_2) \rightarrow u = Y_1$ or Y_2 means that one of the inputs is ignored and the other accepted, etc.

(2) IP is the *Integration Procedure* or algorithm which specifies *how* the integration process is realized, e.g., the collaboration type and the interaction protocols that are applied.

(3) The *I/O streams* indicate the information inputs and outputs.

(4) S_i is a *sub-system* of a distributed organization defined by $G = (N,A)$, a directed graph of the message flow diagram specifying the A arcs architecture and communication among the N S_is.

(5) t is the *Integration Time*, a time function specifying the durations and timing of integration tasks and events.

(6) IM is the *Integration Model*, the model of the integration which can be used to analyze, design, test, evaluate, or operate the integration process.

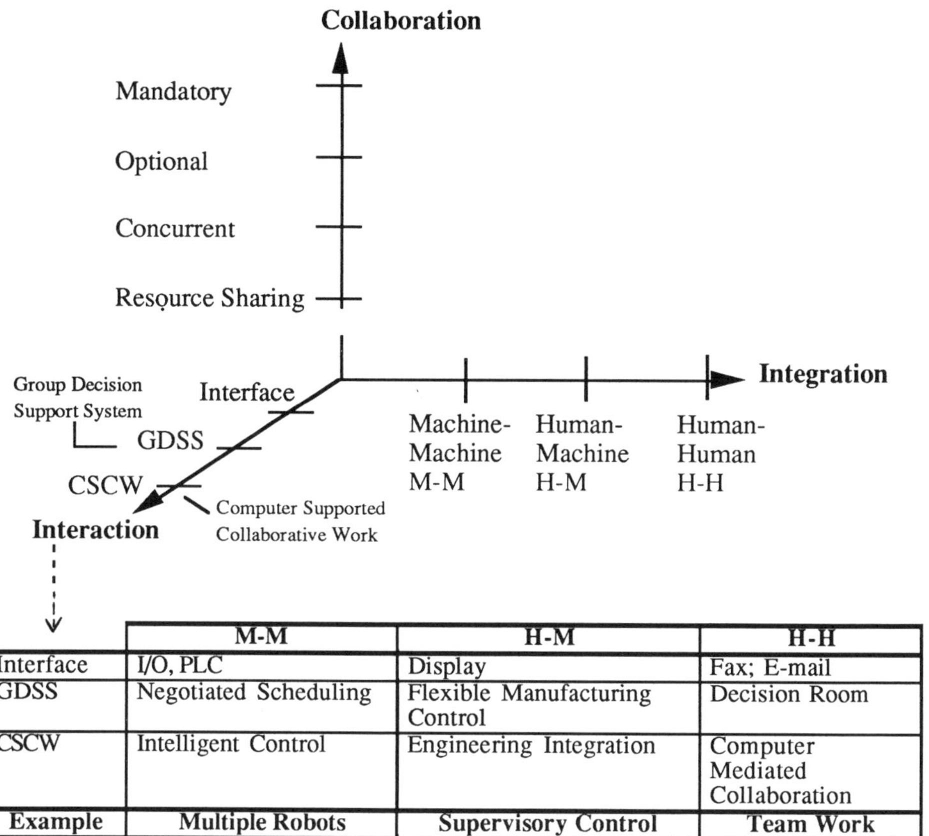

	M-M	H-M	H-H
Interface	I/O, PLC	Display	Fax; E-mail
GDSS	Negotiated Scheduling	Flexible Manufacturing Control	Decision Room
CSCW	Intelligent Control	Engineering Integration	Computer Mediated Collaboration
Example	**Multiple Robots**	**Supervisory Control**	**Team Work**

	Integration Problem	Integration Example *		Collaboration Type
1	Processing of task without splitting and with sequential processing	Concept design followed by physical design	H-H	Mandatory, sequential
2	Processing of task with splitting and parallel processing	Multi-robot assembly	M-M	Mandatory, parallel
3	Processing of task without splitting. Very specific operation.	Single human planner (particular)	H-M	Optional, similar processors
4	Processing of task without splitting. General task.	Single human planner (out of a team)	H-M	Optional, different processor types
5	Processing of task, can have splitting	Engineering team design	H-H	Concurrent
6	Resource allocation	Job-machine assignment	M-M	Competitive
7	Machie substitution	Data base backups	M-M	Cooperative

* For every problem, examples exist for H-H, H-M, and M-M; only one was arbitrarily given.

Figure 1 Integration taxonomy

4

3. MODELING RESEARCH

A variety of modeling approaches have been researched and developed for integration and collaboration, depending on the application area. *Planning and design models* have included enterprise information flow models, software tool integration models, operational models such as CRP (Collaboration Requirement Planning), game-theoretic models, and reasoning models such as multi-agent models and DPS (Distributed Problem Solving) models. *Enabling models* have included interface models and interface generators, GDSS/CSCW models, and object-oriented integration models.
Some specific modeling examples:

- *Organizational integration:* e.g., to evaluate the impact of integration on organizational models (Sproull and Kiesler, 1991).

- *Database integration:* e.g., by federated databases with loose coupling and database schema integration models (Sheth and Larson, 1990).

- *Semantic integration:* Here, the focus is on semantic models for unifying and agreeing on terms, and resolving semantic conflicts (Sciore et al., 1993).

- *Task integration:* e.g., integration and collaboration models for multiple machines and FMS based on game theory and decision models (Nof, 1992); distributed hypothesis testing models for decision task integration (Papastavrou and Nof, 1992); computer mediation models for decision integration by large, distributed, loose teams (Sudweek and Rafaeli, 1994).

- *Computer integrated engineering:* e.g., software design tool integration in a simulation/emulation workstation (Nof, 1994); parallel simulation and parallel computing models of engineering task integration, (Nof and Fortes, 1994); (see Figure 2).

- *Computer integrated manufacturing:* e.g., use of neural networks and object-oriented models to integrate horizontal and vertical information flows in distributed manufacturing organizations (Eberts and Nof, 1993).

- *Enterprise integration:* e.g., information models are used to design integration of information flows for functional integration throughout the whole enterprise (Petrie, 1992).

Measures and limits of integration and collaboration
An essential objective of the above models is to provide measures or metrics of integration and collaboration and identify their theoretical limits. Examples of some initial efforts in this area include the MRC, multi-robot cooperation measures, and CCC, Collaborative Coordination Control performance measures (Nof, 1992).
Current and future work on the PIE (Parallel Integrated Engineering) communication-driven simulator addresses the evaluation of alternative integration organizations by measuring speedup, number of errors, and quality of the integrated results (Nof, 1994).

4. CONCLUSIONS

The competitive benefits of integration and collaboration are significant: *Flexibility* by utilizing the right information and resources at the right time and place and responding correctly to change; *Reliability* by mutual backup, substitution, and recovery, by conflict resolution, and by look-ahead, predictive planning; *Quality* by responding correctly to evolving client needs, and by creative, synergistic solutions to new problems; *Quality of worker life* by well-managed, computer-supported teamwork.

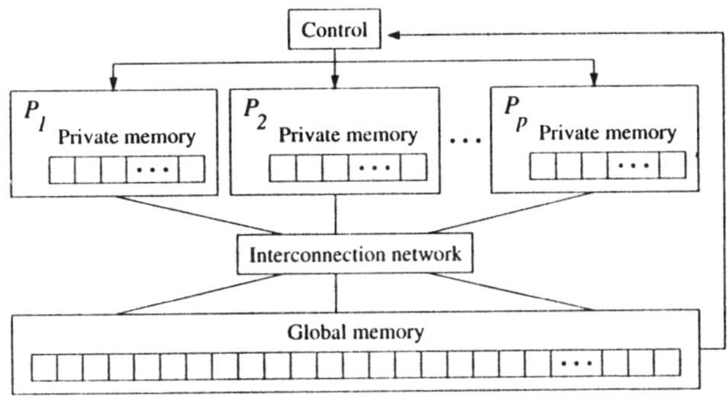

a) The Parallel Random Access Memory model of parallel computation (Quinn, 1994). The PIE Simulator applies the Pi's to parallel workers.

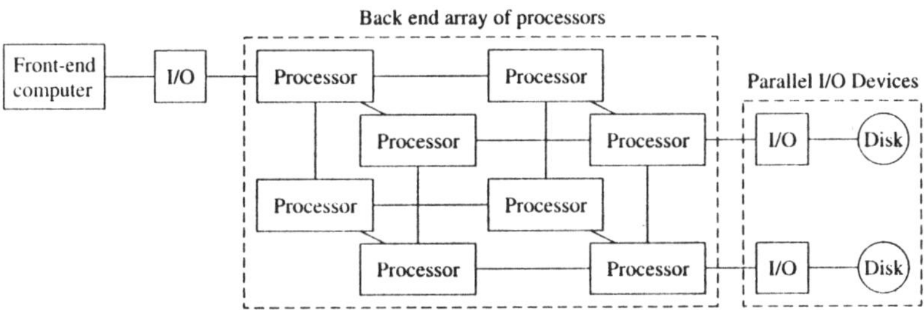

b) Block diagram of an nCUBE 2 multicomputer, for communication-based simulation of concurrent work .

Figure 2 Parallel Computing Models of Engineering Tasks Integration.

We further observe that computer-based integration and collaboration can support the development of Asimov's Gaia (in Italian, happiness), an organism living by combining many collaborating individuals. But we also observe that integration and collaboration occur as a looping phenomenon of organizations, as follows:

$$\textbf{Conflict} \rightarrow \textbf{Negotiations} \rightarrow \textbf{Compromise} \rightarrow \textit{New}\ \textbf{Conflict} \dots \qquad (2)$$

Hence, we face the following challenging issues:

- How many individuals processors, databases, machines can/should be included?

- How much effort and time should be devoted for negotiations as part of the integration?

- How shall we negotiate and mediate?

- How good is the compromise, and how long will it last?

- How can information gaps and conflicts be minimized in number and in severity?

These issues and related interesting questions about integration and collaboration, as intelligent features of civilization, merit additional investigation.

REFERENCES

Eberts, R.E. and Nof, S.Y., (1993) "Distributed Planning of Collaborative Production," *Int. J. Adv. Mfg. Tech.*, Vol. 8, pp. 258-268.

Nof, S.Y., (1992) "Collaborative Coordination Control (CCC) of Distributed Multi-Machine Manufacturing," *Annals of the CIRP*, Vol. 41, No. 1, pp. 441-445.

Nof, S.Y., (1994) "Recent Developments in Simulation of Integrated Engineering Environments," *Proc. of the SCS Symposium on Computer Simulation and A.I.*, Mexico City, Mexico, February.

Nof, S.Y. and Fortes, J.A.B., (1994) "Parallel Computing Models for the Design of Engineering Tasks Integration," *Proc. NSF Design and Mfg. Conf.*, Boston, MA, January.

Papastavrou, J.D. and Nof, S.Y., (1992) "Decision Integration Fundamentals in Distributed Manufacturing Topologies," *IIE Trans.*, Vol. 24, No. 3, July, pp. 27-42.

Petrie, C.J., (1992) "Introduction," *Proc. of the 1st Int. Conf. on Enterprise Integration Modeling*, MIT Press, pp. 1-14.

Quinn, M.J., (1994) *Parallel Computing, Theory and Practice*, McGraw-Hill.

Sciore, E., Siegel, M., and Rosenthal, A., (1993) "Using Semantic Values to Facilitate Interoperability among Heterogeneous Database Systems," *ACM Trans. on Database Syst.*

Sheth, A.P. and Larson, J.H., (1990) "Federated Database Systems for Managing Distributed, Heterogeneous, and Autonomous Databases," *ACM Computing Surveys*, Vol. 22, No. 3, September, pp. 183-236.

Sproull, L. and Kiesler, S., (1991) *Connections: New Ways of Working in the Networked Organization*, MIT Press.

Sudweeks, F. and Rafaeli, S., (1994) "How Do You Get a Hundred Strangers to Agree: Computer Mediated Communication and Collaboration," in *Computer Networking and Scholarship in the 21st Century*, Harrison, T.M. and Stephen, T.D. (Eds.), SUNY Press.

Weston, R.H., (1993) "Steps Towards Enterprise-wide Integration: A Definition of Needs and First-Generation Open Solutions," *Int. J. Prod. Res.*, No. 8, August.

I. Integration in Concurrent Design

PLANNING AND CONTROL OF CONCURRENT ENGINEERING ACTIVITIES UNDER UNCERTAINTY

PROFESSOR ASBJØRN ROLSTADÅS
University of Trondheim
Division of Production Engineering
Trondheim, Norway

KEYWORDS. Concurrent engineering, project planning and control, scheduling, cost estimating, uncertainty, engineering process, modeling.

ABSTRACT. The engineering process provides technical documents. It can be modeled by extending the Walrasian model. Concurrent engineering involves shortening of lead time and life cycle engineering. Shortening of lead time can be done by choosing an uncertainty level and using an iterative algorithm to determine optimal concurrency. Cost uncertainty can be included by adding contingency. Cost and time uncertainty in a project can be catered for by providing risk buffers.

1. The Engineering Task in Large Projects

The engineering function in a project or in one of a kind production is responsible for all technical specifications and documents needed for fabrication or manufacturing.

The engineering task may be split on two phases:

- Preengineering
- Detailed engineering

Preengineering is the conceptual engineering. The overall framework is determined. Detailed engineering will provide the ultimate result, i.e. fabrication and contracting documents. The distribution between preengineering and detailed engineering is vague. The two phases overlap and interact considerable. The need to maintain two phases are due to need to:

- Split the engineering task in manageable units.
- Provide possibilities for sub-contracting detailed engineering
- Allow possibilities for including detailed engineering in fabrication contracts

In the North Sea, a number of large projects has been run in order to design, fabricate and install oil drilling and production facilities off shore. The usual structure is a platform consisting of a base structure in steel or concrete and a topside comprising a deck frame and a number of modules.

Engineering, fabrication and commissioning of such a platform run typically over a number of stages and phases as shown in figure 1. A typical schedule is shown in Figure 2. The engineering task is usually split on disciplines. Table 1 shows typical drawing statistics pr. discipline.

S. Y. Nof (ed.), Information and Collaboration Models of Integration, 9–25.
© 1994 *Kluwer Academic Publishers. Printed in the Netherlands.*

	Project Development							
Project Stage	Project Identification			Project Definition		Project Execution		
Project Phase	Con-cession	Explora-tion	Feasib. Study	Concept. Study	Basic Eng.	Engineer-ing	Construc-tion	Commis-sioning
	Steps & Deliverables							

Figure 1. Project Stages and Phases.

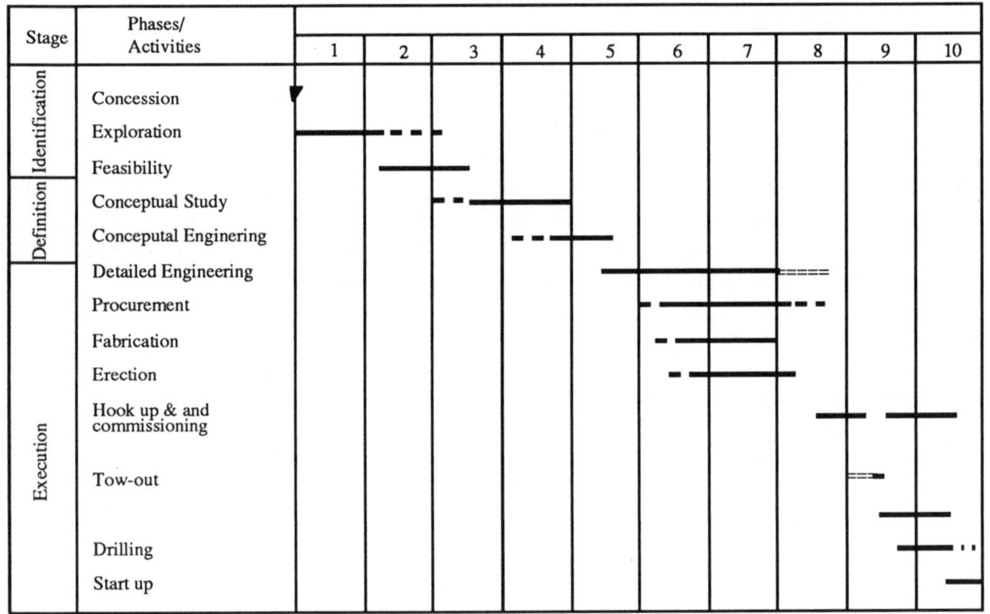

Figure 2: Typical Schedule, major Offshore Oil/Gas Project.

Discipline	No. of Drawings	%	Man-hours/Drawing
Flow diagrams	200	2	500
Structural steel	3 200	29	140
Piping	1 100	10	600
Piping isometrics	2 500	22	15
Architectural	200	2	200
HVAC	250	2	240
Electrical	1 100	10	150
Instrumentation	2 500	22	70
Loss prevention	100	1	700
Total	11 150	100	250

Table 1: Planning Statistics pr. Discipline for a GBS Platform

The output of an engineering organization are documents. In many respects the flow and control of documents through an engineering company are similar to the flow of materials through a workshop. There are some difference that create special problems for engineering. The most important are:

- Insufficient accuracy level of data. Decisions have to be taken based on incomplete data
- Change orders and variations are likely to occur
- Limited possibility of physical consistency check of the results (with the exception of scale models) and collision tests on special software

The "material flow" in engineering will be referred to as a flow of object documents. These documents describe the results of the engineering function, and comprise mainly:

- Engineering drawings
- Specifications
- Technical calculations

The object documents are produced in accordance with and constrained by a set of rules, regulations and limitations, such as:

- The operator's (client's) product specifications
- Governmental rules and regulations
- Design and fabrication standards (including company and client standards)

To control the object document flow through the company another set of documents is used, referred to as control documents. These documents are:

- Schedules and cost estimates for the engineering work (plan documents)
- Progress reports
- Procedures

The plan documents and the progress reports are similar to those used for physical material flow. The procedures are documents describing the processing of documents in detail. These can be compared to process and operation sheets for physical material flow. However, they tend to be more detailed and comprehensive, since the final result is not so stringent defined as the drawing of a physical part would be.

Furthermore, procedures tend to be developed specifically for each project, at least for very large projects. Examples of procedures are:

- Drawing approval
- Change order
- Purchasing
- Inter discipline check
- Etc.

A model for an engineering company will contain a description of the previously mentioned type of documents:

- Object documents
- Control documents
- Rules, regulations and restrictions

Major problems arise from the fact that many engineering tasks are large and complex and have to be performed on a tight schedule. In order to check consistency and consequences for other disciplines, each development produced in one discipline will have to be sent on an inter discipline check before final approval.

The processing of changes for design and fabrication is an important task in all engineering. Changes may be motivated to correct errors, from safety viewpoint to improve or simplify design or to reduce costs. The consequences of each change as to cost, schedule and design will have to be computed for decision support. Implementation of changes will have to be run on an inter discipline check. Contracts will have to be revised.

2. Modeling of the engineering process

It is possible to regard a manufacturing company as consisting of three interdependent processes as visualized in Figure 3. A process is defined as a set of related operations performed on, or in connection with, a flow of concrete or abstract items.

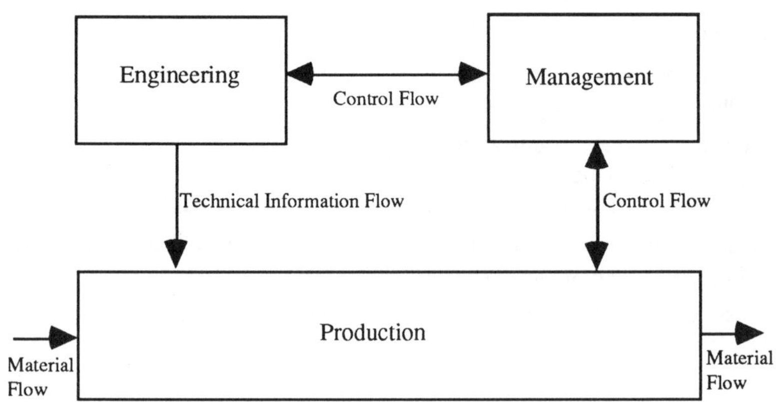

Figure 3. Processes in an OKP System.

In Figure 3 the processes are:

- Production
- Engineering
- Management

Production is connected to a flow of materials. The purpose of production is to transform raw material to finished products.

Engineering is connected to a flow of technical information, usually represented as drawings or other documents. The purpose of engineering is to provide technical specifications on what products to produce and how to produce the products.

Management is connected to a flow of operational information, usually represented in the form of work orders or planning or status documents. The purpose of management is to release and monitor work orders for production and engineering.

In a manufacturing system there are two basic data structures. In accordance with the Walrasian model these are:

- Product data (P-graph)
- Resource data (R-graph)

The interconnection of these two can be seen as management of production. The P- and R-graph and these interconnections represent three generalizations of the Walrasian model as described in the succeeding.

The Walrasian model (Figure 4) depicts the transformation process of production factors into finished products. Basically, the abstract structure of this model can be conceived as a network. The system considered as the production process or production function itself is represented by the network in the first quadrant. The system can be considered from two viewpoints. Horizontally, corresponding to each production factor, it defines a stage or department as the parallel connection of the consumption of all the production factors involved in producing the single product. The relationship, i.e., production function, or ratio of transformation between the amount of a certain productive service and the amount produced of a given product, is known as a technical coefficient. In Figure 4 a technical coefficient is symbolized by a cross surrounded by a circle.

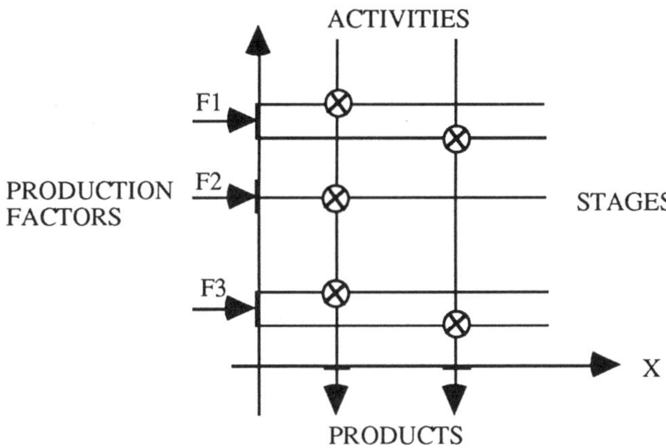

Figure 4. The Walrasian Production Model.

The Walrasian production model as described above does not consider ordering of products (assemblies and subassemblies, i.e., bill of materials) and production operations (routing) as required for real production situations in both repetitive production and one of a kind production. The first generalization of Walras is to include ordering of products represented by the so-called Product graph (P-graph), see Figure 5. For a given product the graph represents the number of components and subassemblies needed to produce one unit of the product. To some extent this represents a requirement for resources that has to be satisfied by the availability of these resources. However, in general this requirement is defined by production, assembly and other operations performed. Consequently, the graph has to be extended to cover that. This means that every node in the graph represents a sequence of operations.

The P-graph as described here, represents the primary flow in the production process. In fact, it defines the work to be done. The purpose of the engineering process is to establish the P-graph as defined here. In fact, the engineering process can be regarded as a production process, however, with a different primary flow. In this case a similar graph may be used to describe the work flow of the engineering process. Again the same is also valid for the management process.

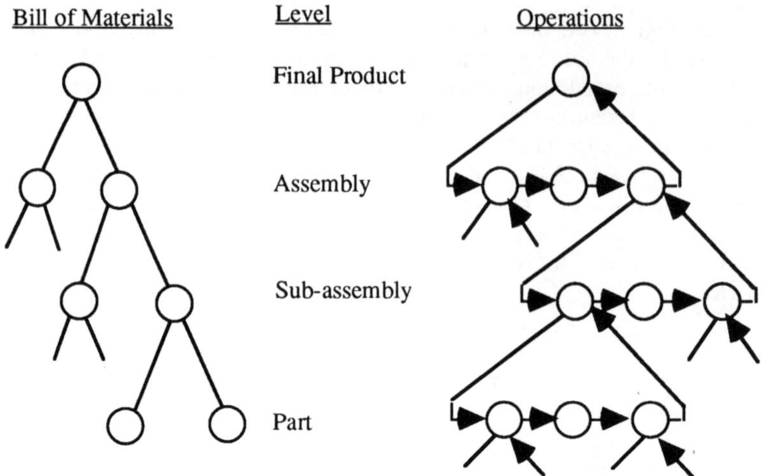

Figure 5. P-graph

The P-graph identifies the requirement for resources. This requirement is defined by capability and capacity.

The second generalization of Walras arranges resources in a similar way. The resources may be depitched by a graph. The two generalizations can be summarized in one figure as shown in Figure 6.

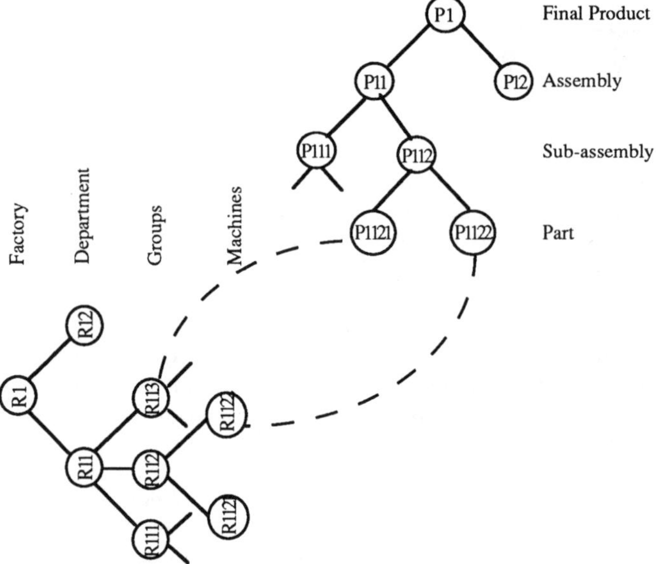

Figure 6. From Walras to Ordered Resources and Products.

The P-graph represents operations to be performed and defines the requirement for resources. The R-graph defines in a similar way the supply of resources, however, still independent of time. Requirement and availability of resources are defined by two parameters:

- Capacity
- Capability

The capacity indicates how much of the resource is required or available. This is a quantitative number. The capability defines more precisely the type of resource needed or available. Dependent on the detailing level, the capability will be organized in a hierarchical graph structure.

In an actual situation a demand for products is defined, and availability of resources is given in form of working hours. This means that the P-graph will define the actual requirement of components, subassemblies, etc., and subsequently operations in a time phased manner, i.e., the requirements will be defined within predefined time intervals. Having multiplied the requirement figures of the P-graph with the demand, time phased it, and aggregated require-ments for the same operations in the same time interval, an activity is established.

Quite parallel, the inclusion of the time dimensions to the resources is established from knowing their basic capacity (units per time unit) and the working hours. These resources are allocated on the activities, i.e., an element of the R-graph is allocated to an element of the P-graph in a given time interval. This process is usually referred to as scheduling and loading. This connectance represents information on:

- Which task to perform
- What quantity
- Which resource
- When and how much (of the resource)

This is the type of information we refer to as a job order.

Consequently, when P- and R-graphs are connected and delivery time and quantity are supplied, the job orders are created.

The Walrasian model is a static model of the primary process of resources and products with a fixed production function in terms of technical coefficients. Turning it into a dynamic control model as the same time and putting emphasis on the decision functions and their structure gives the Grai Grid in Figure 7.

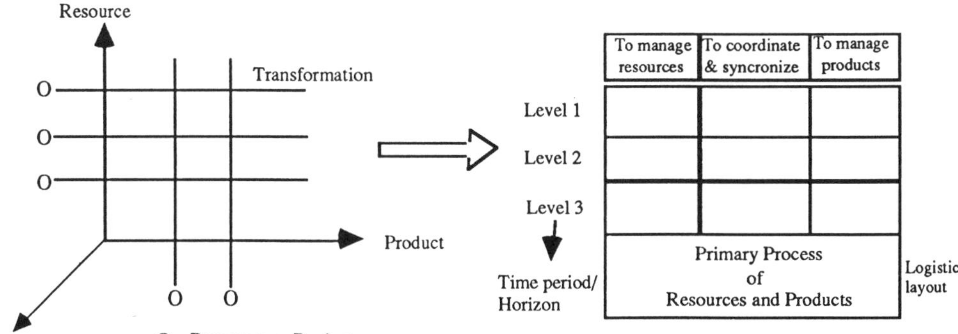

Figure 7. From Walras to Levels of Control (Grai Grid).

In the left part of Figure 7, the Walras model is shown with its resources and products. In addition, a time dimension is included, making the model 3-dimensional. The right part of the figure shows how this is transformed into a GRAI grid. The columns of the grid represent the resources and the products of the Walras model as well as the connection of these (i.e., allocation of resource to products). This connection is referred to as "to coordinate and synchronize". This corresponds with the GRAI method which defines management of production as "to synchronize in time". The time dimension is included in the GRAI grid as several levels. The GRAI method consequently provides a way of representing the 3-dimensional model in the left part of Figure 7 as a 2-dimensional grid as shown in the right part of Figure 7.

The Grai Grid represents the control system. The controlled system, i.e., the primary process is usually not depicted with the Grid. The concept of synchronization and coordination in the Grid corresponds to the production function in Walras.

We refer to this generalization of Walras as the 3rd generalization

3.   Concurrent engineering issues

There is yet no common understanding of what concurrent engineering means. The first international workshop on this topic held in Tokyo during July 1992, showed quite a diversity of opinions on the definition of concurrent engineering.

From the position papers a sample of definitions are:

> *"Concurrent engineering is a strategic know-how to improve product quality at lower costs and shortened product development time through the integration of product and process design as well as manufacturing and construction planning."*

> *"Concurrent engineering is an effort at teamwork to correct the flaws of the functional organization."*

> *"Concurrent engineering characterizes simultaneous and massive interactions between product design and design evaluations regarding producibility, functionality, cost estimation and the like."*

> *"Concurrent engineering is parallel execution of different development tasks and the replacement of traditional department specific division by cooperated enterprise-wide problem solving."*

> *"Concurrent engineering is a systematic approach to the integration of design, production and related processes which considers all aspects of a product life-cycle."*

> *"Concurrent engineering is an effective strategy to design robust and durable products that are environmentally safe to manufacture, use and recycle or dispose."*

In spite of this diversity of opinions, there are some common denominators in this. These may probably be summarized in the following definition which we will use in this text:

> **Concurrent engineering is parallel execution of different development tasks in multidisciplinary teams with the aim of obtaining in minimum time and with minimum costs an optimal product with respect to functionality, quality and producibility.**

The need for concurrent engineering is market driven. The market conditions for mechanical products both in the consumer and durable goods market, is gradually changing into a larger degree of diversification and shorter product life.

Concurrent engineering involves two basic aspects:

- Time to market
- Life cycle engineering

Time to market involves shortening the time from an idea is created until it is presented as a product on the market for the customer to buy. Time to market has in the later years proved to be a factor of crucial importance to the competitiveness of a company.

The total time from idea to market spans a number of phases, of which the following may be regarded as the most important:

- Product design and development
- Process and operations planning
- Production
- Distribution and installation

The focus here will be on design, development and process planning, i.e. the activities from decision on a new product until it is ready for (mass) production. We shall refer to this period as lead time for design and engineering.

The total lead time for design and engineering is influenced by a number of factors. Some of the more common ones are:

- Lead time of the various activities
- Quality of the design and engineering work
- Productivity in design and engineering
- Change orders to specifications
- Risk and uncertainty

We shall focus on lead time and uncertainty in chapter 5.

Life Cycle Engineering implies the parallel consideration of product, function, design, material, production process and cost with regard to later stages such as testing, service, maintenance, operation, repair, destruction and further product development. In this way aspects of the whole life cycle can be used already at the stage of product development.

Concurrent engineering provides in this respect a market oriented and systematic technique for integrated and parallel development of products and the associated processes, including subcontracting, manufacturing, maintenance and customer service. The purpose is to "optimize" all elements in the product life cycle, from idea to recycling, inclusive user needs, quality, time and costs. In this way "non-optimal" things are discovered and acknowledged by people who can affect them in the "best" way.

This requires a closer co-operation and a more effective information flow than usual in the traditional sequential system i.e. interdisciplinary knowledge and communication across the traditional borders, both internally and externally.

There are two possible ways to implement Life Cycle Engineering in practice. The first one is to start at the beginning by forming interdisciplinary groups who try to solve tasks right first time (Top Down approach). The other one is to start at the end by identifying the "non optimal" aspects, and try to resolve them by involving those causing this solution (Bottom Up approach).

4.   Project management - state of the art.

As described in chapter 1, a project could be split in three stages:
- Project identification
- Project definition
- Project execution

The first stage will result in a field development plan, and the purpose is to take a decision on whether to develop a field or not and to provide the first overall plans solutions and estimates. The second stage will give the conceptual design and the necessary plans to run the project, i.e. the stage is aiming at defining what to do. It will result in a project execution plan and design specifications. The third stage is to execute the project in accordance with the approved plan.

Each stage is usually subdivided in a number of phases. For each phase a schedule and a cost estimate is provided.

Project management involves:
- Project planning (stages 1 and 2)
- Project control (stage 3)

Project planning will result in a project execution plan that will form the baseline for project control. This project control baseline will contain three elements:

- Work breakdown (MCWBS)
- Cost estimate (MCE)
- Schedule (MCS)

The abbreviation MC here refers to "master control" which really means that it is the first original plan. Master control plans can be revised if the scope of work is changed. Otherwise plans are revised on a continuous basis all the time predicting final results. These plan are referred to as "current control" (CC) plans, i.e. CCWBS, CCE and CCS.

In any project, control of costs are considered extremely important. Some tools that explain in a simple diagram to management how the project is performing, is necessary. As a matter of fact such a diagram can easily be developed and partly exist in some companies. However, to understand such a diagram requires some effort. An example is given in Figure 8.

The difference between MCE and CCE is a forecasted total difference between original estimate (MCE) for current work scope and the current forecasted costs. The difference expresses a possible total deviation (negative figures represent overrun) and would mainly be due to other market prices and different work productivity than estimated. In Figure 8 this total difference is:

$$MCE - CCE = -1.1$$

The curves are shown in the upper part of the diagram.

In the bottom part three curves are shown. ECWS shows the estimated costs over time for the current work scope. It contains both estimate and schedule information. At current date the planned amount of work moneywise is 6.0. However, the accounts show that only 5.0 have been spent (ACWP curve). Note that this "underusage" does not imply cost saving. On the contrary, the last curve (ECWP) shows that the physical amount of work actually performed in the original estimating terms only represents 4.3, i.e. 4.3 is the earned value. The difference 0.7 represents a cost overrun that has already occurred:

$$OCD = ECWP - ACWP = -0.7$$

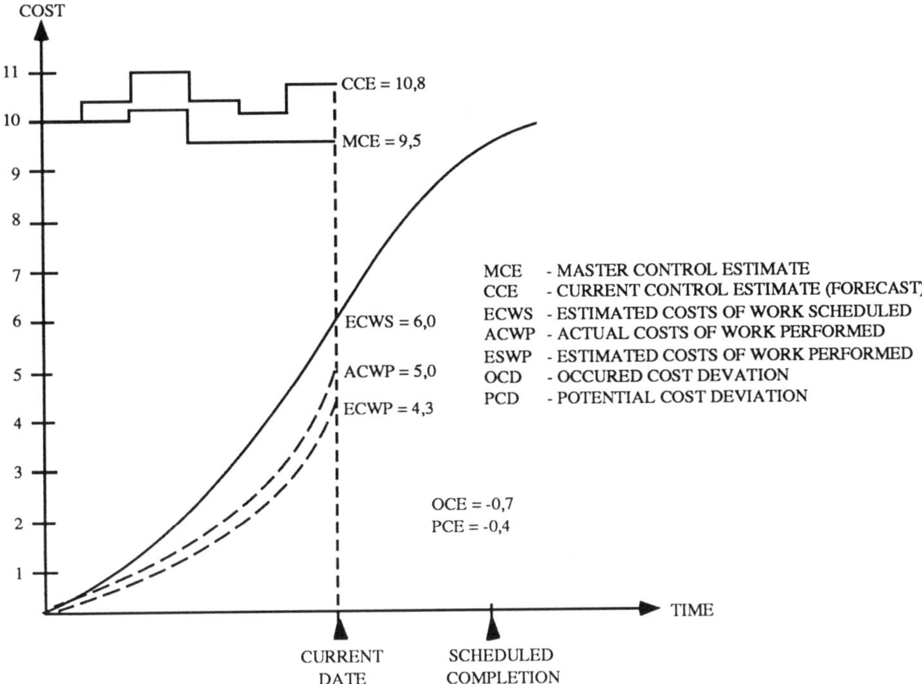

Figure 8. Cost Control Diagram

However, the forecast says that the total cost overrun will be 1.1, which means that there is still a potential of a further cost overrun of 0,4:

$$PCD = MCE - CCE - OCD = -0,4$$

The difference ECWS - ECWP shows the delay of the project in work scope, which is here 1,7. The reason may be insufficient availability of resources or less productivity that estimated. The latter can be expressed by the ratio: ACWP/ECWP = 1,16.

Although the diagram in Figure 8 is considered sufficient for top management monitoring of a project, is does not very clearly spell out schedule and progress information. A second diagram, shown in Figure 9, could be useful in this respect. It depitches cost and schedule indexes defined as:

$$CE = ACWP/ECWP$$
$$SE = ECWS/ECWP$$

The indexes should both be 1, if the project follows the plan with respect on cost and schedule. Figures above 1 indicate that the project performance is behind plans.

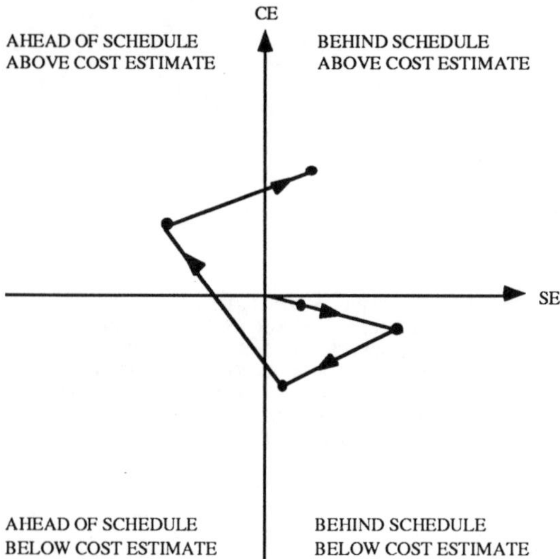

Figure 9. Project Performance Diagram

Diagrams like the one in Figure 9 clearly show how the project has developed up till now. However, to see the forecast, a diagram like Figure 8 must be used.

State of the art in project control today is that the necessary tools exist and are well developed. It may, however, still take some time before they are widely adopted and applied.

Concerning state of the art in project planning, the situation is not equally encouraging. As mentioned, three elements are to be planned: WBS, schedule and estimate.

Concerning WBS very little techniques or guidelines exist except for the American DOE standards. WBS influence project performance considerably, and more effect should be placed in developing more firm techniques, and better skill.

Scheduling is mainly done by network analysis. The traditional CPM and PERT techniques are well developed. Various refinements exist serving different purposes:

- Precedence networks allowing activity overlapping
- Resource allocation
- Time/cost tradeoff
- Gert networks allowing conditional branching
- Stochastic networks enabling risk analysis

In cost estimation, the major problem today is to find the appropriate estimation models and to develop the necessary databases.

5.   Scheduling concurrent activities under uncertainty

Lead time and quantity are probably two of the most important planning parameters. The third is of course cost. It is important that risk is also considered as a planning parameter. Risk is a management decision variable that will heavily influence the outline of an optimal schedule (and budget). We shall try to treat both these parameters simultaneously in order to obtain minimal time to market.

An important question is how lead time can be reduced. In production, there are three possibilities:

- Overlap activities
- Split activities
- Shrink transfer time between activities

In a project oriented environment as we consider, there are no transfer time between activities, so we are left with overlap and splitting. We have illustrated this in figure 10, where a) show the original schedule as a bar chart. The dashed line indicate the deadline. b) show the overlap situation and c) show splitting. Splitting is really the same as extending the available resources, f.ex. applying a second machine or increasing the number of men. In c) we have split activity A. However there is little difference to b) where A overlap B. Actually, we can regard both options as an overlap. In the first case, the overlap is provided by running A and B partly simultaneously. In the second case, the overlap is provided by adding resources and "running A in parallel with itself".

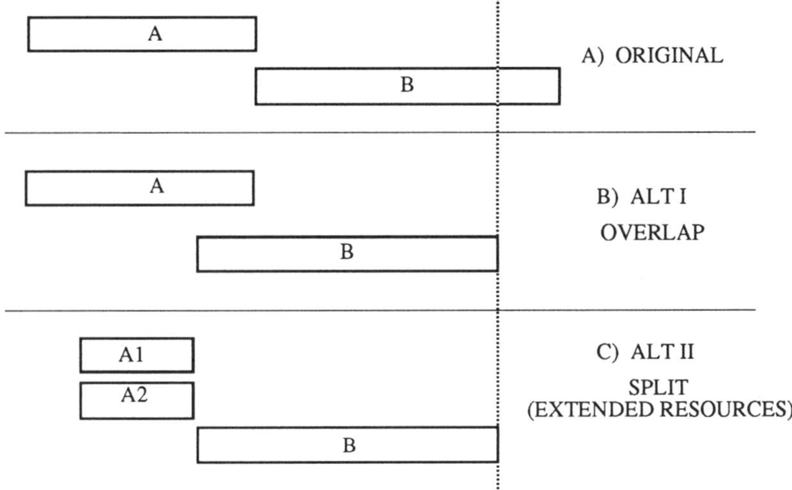

Figure 10. Reducing lead time of two consecutive activities

We shall now describe an algorithm for scheduling of design and engineering activities under consideration of uncertainty. The algorithm is tailored to deal with a limited time approach, as it realizes that there is a time window within which the design and engineering must be performed. It will also take into consideration overlapping of activities. As a matter of fact it will vary the degree of overlapping in order to obtain acceptable risk level within the feasible time window.

The algorithm will be based on a number of **decision points**. These must be determined by the user or planner. Such decision points represent stages or milestones in the design and engineering work.

The algorithm operate in six steps:

1. Define all decision points. Estimate a **confidence level** for each of them.
2. Define **main activities** based on precedence relationships between the decision points.
3. Detail each main activity by providing a network plan. Define uncertainty in each activity in the network, and estimate maximum overlap between any two activities.
4. Compute statistical distribution on the time for each decision point.
5. Compute a **nominal duration** for each main activity.

6. Determine overlap between each pair of activities in each main activity to obtain the nominal duration of the main activity.

As pointed out, the basis of the planning is initially the definition of the decision points. The user must decide on these and give them as input. They are defined as a point in time. These decision points may be given as fixed or they may be of a type that allow a some deviation. The user can indicate this by defining a confidence level for each decision point as indicated in step 1 in the algorithm. This confidence level is defined as the probability of not being delayed compared to the defined decision point.

The next step is to define the precedence relationships between the decision points. This means indicating which decisions are dependent on other decisions to be taken. Such relations actually indicate some type of activity between the decision points. Will denote this a **main activity**. We then have a network of main activities defined by the precedence relationships.

Step 3 is repeated for each main activity. The purpose here is to make a detailed plan including all activities needed to reach the succeeding decision point(s). These activities will be referred to as **activities** in contrast to the main activities.

In order to deal with uncertainty, we must introduce statistical distributions for the duration of each activity. The traditional way of doing this is by providing triple estimates for all durations.

Assuming stochastic independent activities, we can perform a linear sum to obtain expected value and variance for the total project time (i.e. main activity) along the critical path. By use of the central limit theorem, we can approximate the distribution for the total project time by a normal distribution. We shall then assume that this distribution is valid for the time of the decision point. That is we will place a normal distribution around the decision point with a variance equal to what we have outlined in the preceding. This is what is included in step 4 in the algorithm.

There is one more operation associated with the activities. We need to define overlap possibilities. This can in principle be done for any pair of adjacent activities. In practice, we need to pick only those activities where a overlap is really possible. We will define a window of overlap where x would represent maximum overlap.

Step 5 involves computing a nominal duration for each main activity. **The nominal duration** is the duration we will schedule the main activity with. This must be selected so that we meet the decision point by the desired probability. We have computed a duration based on the underlying network. Probably this is too long. We shall then fix the duration based on the statistical distributions on the decision points. The situation is illustrated in figure 11. We denote the duration T. Assume that the main activity starts on the exact time of the decision point (indicated by expected value). Based on our confidence level we can calculate an interval Δt_2 that need to be subtracted from the duration based only on expected values. This is the security margin to compensate for risk of delay at the decision point. If we keep the previous decision point fixed, this T will be the nominal duration. This is the case when we will wait to execute succeeding activities if we finish early at a decision point.

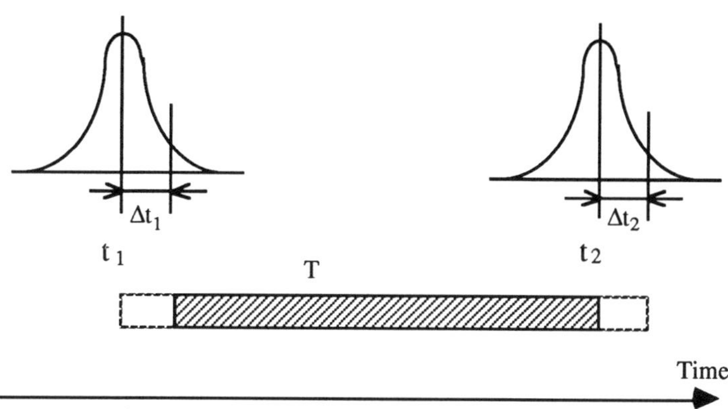

Figure 11: Main activity duration

If the decision point is not such that we need to wait, there might be another Δt_1 that will be added to T to take advantage for the risk of being finished early. In this case however, the distribution of the finish decision point need to add variance from both the start and finish distribution.

Now when the nominal duration of the main activity is given, we need to shorten the network of the main activity to this value. This is done by systematic increasing overlap for the activities that allow for that. The question is however, which activities to overlap and how much, since we probably have several options. We will now adopt a policy of shortening most the activities with the least uncertainty. We will consequently use the uncertainty of each pair of overlapping activities to calculate the size of each overlap. To deal with such problems in a network is similar to dealing with time/cost-relationships is necessary.

6.   Integrating cost in the planning cycle

As soon as the activities duration is fixed and the resource consumption is known, the costs can be easily derived. However, costs are also associated with uncertainty, and the method for dealing with this is required.

Today's networks are optimized trying to reduce costs. A better approach is to optimize the net present value, since this is the parameter usually applied in profitability analysis of a project. Using net present value (or rate of return) simultaneously takes both time and costs into consideration when optimizing.

An efficient planning tool should have the capability to work with several alternatives and options at the same time and all the time calculate net present value and cost and schedule (milestone) risk. In this way the project management can test various technical options, various schedules, contract options etc.

The plans should include risk buffers, preventing unexpected project performance to affect final result. A delay or a cost overrun should at some stage (milestone or node) in the project be stopped. At such a node a risk buffer will exist.

A risk buffer can be provided by one or several of the following options:

- A cost contingency
- A time buffer
- Acceleration possibilities of other activities
- Contractual options that can be released

24

-     Alternative activities

All these can and should be quantified. Risk buffers should be purposely placed at strategic points in a project and dimensioned to pick up unexpected performance at a risk level decided by management on a cost and profitability basis.

A trained project staff would probably be able to forecast possible trouble areas in a future project. with a proper tool such possible trouble areas could be tested by simulating a project execution based on an existing plan. In this way the plan can be modified and tuned to be nonsensible to these effects.

The basis of all cost estimating is bill of quantities. This is also basic input for scheduling. Therefor focusing on bill of quantities will allow the simultaneous estimating of uncertainty in both cost and schedule.

Uncertainty in cost is usually covered by including a contingency item in the cost estimate.

The calculation of project contingency is therefore a matter of great concern. A theory has been developed based on the assumption that the estimate provided by an estimator is the most likely value. For control purposes, however, a fair estimate should have equal chances of overrun and underrun.

In the statsistical terms this means the median . In this context the median can be approximated by the expected value which is more convenient to compute.

Estimates are based on bill of quantities picked off drawings. In early stages these drawing are uncomplete. Detailed design will uncover something. A contingency shall cover this something which is likely to occur, but cannot be identified.

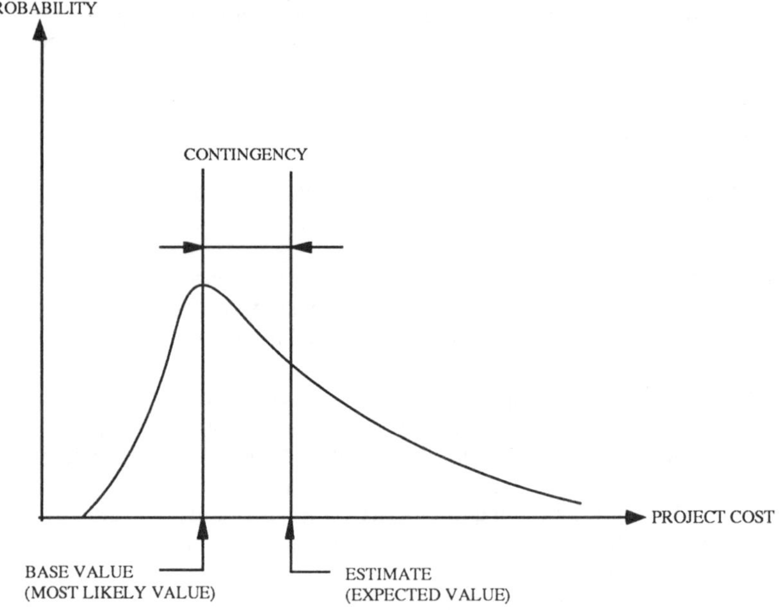

Figure 12. Base Estimate and contingency.

Assuming that the estimated cost follows a skewed distribution as depitched in Figure 12 the contingency can be defined as the amount that must be added to the base estimate to obtain a

50/50 probability. The base estimate is then defined as the most likely value provided by the estimator based on the existing drawings.

7. References

1. Granli, Hetland, Rolstadås: Applied Project Management. Tapir, 1986
2. Rolstadås: Praktisk prosjektstyring. Tapir, 1988
3. Rolstadås: Estimate Classification and Risk Evaluation. SINTEF, 1986
4. Rolstadås: Cost Effectiveness and Consciousness in Exploration Development and Production - Project Management and Execution, paper for the ONS conference, August 1988
5. Rolstadås. A., Moseng B, Blankenburg D: Concurrent engineering as a tool for improved competitiveness, SINTEF, February 1993.
6. Rolstadås, A: Engineering for one of a kind production, paper for the IFIP WG 5.7 conference, Bremen, November 1991.
7. Rolstadås, A: A Conceptual reference model Seen from the Functional view. FOF/SINTEF, August 1990.
8. Falster, P: The Conceptual model and related topics, FOF/DTH, December 1989.

INTEGRATION OF RECYCLING CONSIDERATIONS INTO PRODUCT DESIGN - A SYSTEM APPROACH

G. Seliger, E. Zussman, A. Kriwet
Technical University Berlin
Institute for Machine Tools and Production Technology
Department of Assembly Technology
Pascalstraße 8-9
D - 1000 Berlin 10

ABSTRACT. This paper presents an integrative approach for design-for-recycling of products. A system approach is suggested integrating the product's recycling features, the recycling process and the product logistic support during the product's life cycle. Design-for-recycling is defined as a design for ease of product recycling and maximum output. Rules dedicated for design-for-recycling are given particularly to the disassembly process of a product during the recycling stage. The design approach is demonstrated by investigating a washing machine as a representative of a "white" household machines family. The recyclability of the machine is evaluated, where different design-for-recycling rules are applied in order to improve the machine recycling characteristics.

1.   Introduction

1.1   OVERVIEW

One of the most disturbing by-products of industrialization and the subsequent rise of living standards is the ongoing destruction of our environment. This has lead to increased public awareness of the need for environmental protection. Industry is confronted with new challenges. In addition to the need for cleaner production processes and increased energy efficiency during usage, strong emphasis is currently put on what happens to the products after usage. Complex consumer products like cars or household appliances pose a specific problem due to their high production volume and the diversity of their materials. In Europe alone, more than 10 million cars have to be scrapped every year, and with the economic recovery of eastern European countries, this figure is likely to rise steeply (Table 1).

The vast amount of used consumer products have until now, at best, been shredded, allowing to regain most ferrous and some non-ferrous metals, while having to dispose the rest. However, in many cases the products are disposed of altogether, i.e. because harmful substances included in the product might contaminate the whole product during a shredding process. Both the lack of natural resources, raw materials and energy, and the shortage of landfill or waste burning capacities force the industry to consider ways to increase the amount of components and materials that can be reused for a "second life".

S. Y. Nof (ed.), Information and Collaboration Models of Integration, 27–41.

	Stock of used cars (millions)	Yearly scrapping (millions)	Tons of non metallic materials
Germany	35,53	2,4	500.000
West. Europe	147,06	10,6	2,2 million
USA	144,46	10,4	2,2 million
Japan	34,95	2,4	500.000
World	420	28	5,9 million

Table 1: Stock of used cars and yearly scrapping, 1990 /1/

1.2   REASONS FOR RECYCLING

1.2.1   *Motivations for the Industry.* The pressure on the industry to produce economically sound and recyclable products comes from three directions:
- *The legislation,*
- *The need for cost reduction and*
- *The marketplace.*

1.2.2   *Legislation.* Several European countries are tightening their legislation for environmental protection. The German Bundestag is currently discussing a proposal for an act regarding "The Avoidance of Residues, the Exploitation of Used Materials and the Disposal of Waste". It provides the legal frame for the German government to force the industry and importing companies to recollect their products after usage at no cost. The proposal is due to come into effect by the end of 1993, but the industry will be granted a transitional period of two years before the law will be enforced. The act will force the industry to take financial as well as organizational responsibility for recycling their used products in an economically sound way. Concepts for recycling plants and regional recycling networks are already developing in the car industry, while the producers of electronic consumer goods (TVs, VCRs) and household appliances are still far behind. Most European car manufacturers have set up prototypical disassembly plants /2,3,4/ , and many participate in regional or nationwide groups to exchange experiences and organize logistical collection and distribution systems /5/.

1.2.3   *Cost reduction.* The main cost benefit in recycling is currently usually the cost reduction for the disposal. The serious shortage of landfill capacity, combined with increased concern about "garbage tourism", have sent landfill dumping fees to skyrocketing heights. However, also the value of regained materials and components itself could be increased, if they could be regained intact and pure enough for further utilisation at the same "level" for tasks as demanding as their original ones. Copper can serve as an example to demonstrate the price gap between pure (=raw) and polluted (=used) materials. While raw copper costs approx. DM 3000,-/ ton, copper from shredding residues and cables can be sold only for approx. DM 300,- / ton, hardly covering separation and transportation costs /6/ . A major key to cost reduction in recycling lies in enhanced purification of the regained materials and a higher degree of regained reusable components.

1.2.4   *Marketplace.* Increased public awareness of environmental issues is also reflected in consumer demands for environmentally friendly products. Many companies have already responded to that development by offering to collect and recycle their used products and by stressing the ecological orientation of their company in their marketing strategy (i.e. IBM, Rank

Xerox) /7/. Other companies have suffered serious setbacks that had until now been unimaginable, because their products were seen as an offence to the rule of sparingly using natural resources (i.e. Mercedes Benz with their new S series). Offering products that are economically sound and easy to recycle becomes a necessity in market competition.

1.3   LIFE CYCLE OPTIMIZATION

All this leads towards extending the segment of the products life cycle that companies take into consideration when optimizing the design of their products. Until now, products were designed for ease of production, delivery and maintenance. The disposal was excluded from the optimisation because it was paid for by somebody else, the customer, or more likely, the state. In future, with companies having to organize and pay for the recycling, they will put the same emphasis on designing their products for ease of recycling. A lot of questions still remain unsolved: What are the recycling options ? How can one deal with conflicts between optimal design for recycling and other design goals, i.e. design for manufacturing ? How strong should the emphasis on recycling aspects be ? How can we create the infrastructure needed for reusing materials and possibly components ? Research in these areas is already well under way, but fast results and simple answers are unlikely due to the complexity of the problem.

Two paths have to be followed simultaneously: While on one hand developing more sophisticated recycling technologies, more advanced separating and purifying methods, the industry must on the other hand think about how to produce future product generations in a recycling-friendly matter. This paper focuses on the second aspect, the "Design-For-Recycling" (DFR) or how to integrate recycling aspects into product design. We suggest a system approach, in which product, production and support life cycles are planned simultaneously in order to observe all relevant recycling parameters. The following chapter will deal with the current understanding of recycling and the developments in recycling technologies. In the third chapter, we describe our system approach and how it effects the design activities. The fourth chapter details the aspects of DFR with a strong emphasis on the necessity of disassembly prior to shredding. A case study is given describing the current recyclability features of a washing machine. Design-for-recycling rules were employed to improve the machine recycling features in future construction.

2   Recycling: Definitions and State of the Art

2.1   TYPES OF RECYCLING

Recycling aims at "closing the loop" of materials or components after usage by reusing them for new products. Three loops can be distinguished, during which recycling activities can take place /8/ :

* *Recycling of production scrap*
  The renewed use of production scrap is by far the most developed form of recycling. This is due to the fact that production scrap usually occurs in large quantities of pure materials (i.e. rest parts from sheet metal cutting, plastic residues from injection moulding). Most companies stating that their products "contain xx % of recycled materials" refer to this form of recycling.
* *Recycling during product usage*
  The reuse of a product for the same or a different purpose using its original shape is called recycling during usage.
* *Recycling after product usage*

Since recycling of production scrap is already far developed, and recycling during usage is limited by the obvious restraints of the products shape, the highest potential for future development lies in the area of recycling after usage.

During all loops, different forms of recycling are possible. Keeping the shape of the original product for future tasks is defined "using", while making use of the material after dissolving the original shape is called "utilization". If the function of the recycled product is the same as the one of the original product, we speak of "reusing" or "reutilization". If the functions of original and recycled products differ, we define the process as "using on" or "utilizing on" /9/. Examples for the different forms of recycling are given in <u>Table 2</u>.

Loop	Form of recycling	Original product	Recycled product
Recycling during product usage	Reusing	Bottle	Refilled bottle
		TV set	Repaired TV set
		Car tire	Remoulded car tire
	Using on	Milk bottle	Flower vase
		Shopping bag	Waste bag
		Old tire	Ship fender
Recycling after product usage / Recycling of production scrap	Reutilization	Glass bottle	Bottle from recycled glass
		Aluminium can	Can from recycled aluminium
		Sheet metal scrap	Sheet metal
	Utilizing on	Glass window	Bottle from recycled glass
		Aluminium cans	Aluminium window frame
		Sheet metal scrap	Wires

<u>Table 2:</u> Examples for different forms of recycling

As written above complex consumer goods pose specific recycling problems that have to be dealt separately from recycling of normal household waste. In this paper, emphasis is put on complex consumer goods including cars, household appliances, TV sets, personal computers and the like. For some of them, mainly cars and kitchen appliances, recycling has already in the past been a profitable business due to their high amount of valuable metals. In coming product generations however, the amount of plastic and other non-metals is likely to rise due to their superior and still improving material characteristics. This adds to the strong need for research in all aspects of recycling.

2.2 AIMS OF RECYCLING

Recycling aims at facilitating secondary use of products, that are not considered suitable to further perform their original task. Like in all production processes, the principle optimization goals of recycling are reducing the effort necessary while at the same time optimizing the output of the recycling process, which can be products, components, materials or energy. To pursue these goals, there are two areas of improvement, which are closely related to each other: To enhance the productivity and effectiveness of the recycling processes and to design future products for ease of recycling.

In the case of complex consumer products, recycling usually requires reducing the "complexity" (product -> component -> material -> energy) of the used product. In many cases, the optimal recycling form for the old products is reached, when most of the product is reused at the highest level of complexity /9/. However, this is not the decisive factor. What determines the optimal way to get most out of a used product is its "substitution value". All parts of the products should be reused in such a way, that the value of the new products, new components, raw material or energy they substitute reaches its highest value.

The difference can be illustrated by a simple example: Consider two ways of recycling a used car rim. One can either scrap and melt it to produce steel, or reuse it as a component for a stand of a temporary traffic sign. The rule "component recycling before material recycling" would decide in favour of the sign stand, because it allows to save the shape of the rim. However, that does not lead to the overall optimum, because the value of the recycled steel from the rim is higher than the cost of casting a sign stand from concrete. This shows, that evaluating the optimal solution for recycling, which is necessary for DFR, can be a difficult task and must be developed further.

2.2   OVERVIEW OF RECYCLING PROCESSES

2.2.1   *Preparation.* Before identifying the appropriate recycling technology, collection of information about the product is necessary. This information regards the existence and location of reusable components, valuable or harmful materials as well as process information, i.e. hints for the most effective disassembly path. The information can be supplied with the product (e.g. warning signs for harmful materials), presented by the producing company upon request (e.g. by using network accessible data banks) or be regained from former experiences by the recycling company. Preparation also includes collection and storage of similar products to be processed together, cleaning the product (to allow sensor identification of joining elements or to enhance the purity of reclaimed materials), drying.

2.2.2   *Resolving original shape.* Recycling after usage requires a reduction of complexity of the used product. This can be done by several ways, including disassembly, shredding, chemical dissolving, melting or burning (see the following Section 2.3).

2.2.3   *Separation.* In order to reuse components and materials, they have to be identified and sorted. That can either be accomplished by individual identification of parts or fragments, as done in manual sorting, laser identification and microwave sorting, or by separating material streams by means of process engineering using physical or chemical material characteristics like specific weight, magnetism, electrical conductivity etc..

2.2.4   *Exploitation.* In order to channel components and materials back into the production process, additional processes may be necessary: Components need testing and possibly repair, materials might need to be processed to enhance their purity.

2.3   STATE OF THE ART OF RECYCLING TECHNOLOGY

2.3.1   *The Shredding Approach.* Currently, industrial scale recycling activities are limited to few areas. Secondary utilization of materials exists only for ferrous and some non ferrous metals, while large scale secondary use of components is even rarer: Examples exist only in the area of capital goods, especially regarding goods from short term leasing contracts. In Germany, recycling as the dominant form of phase out is limited to cars (approx. 95 %) /10/ and kitchen appliances (> 80 %) due to their significant amount of valuable metals.
Current recycling technologies for these consumer products mainly rely on a shredding process followed by several stages of sorting. Shredders are built in a large variety of sizes and with hourly capacities of 10 to 120 tons /11/. All are based on the same operating principle: The input material is condensed by hydraulic presses and fed into a drum, where it is ripped apart by a set

of rotating hammers, until it is small enough to drop out of an output grid. Subsequently, lightweight materials like textiles and some plastics are separated in an air tunnel. They form the "Light fraction", which currently has to be disposed. Usually, the next step is a magnetic separation, regaining the high amount of steel and other ferrous metals. What is left over are heavy, non magnetic materials like glass, rubber and some plastics. The magnetic fraction is often subject to hand picking of valuable non ferrous metals, that were connected to ferrous material and thus transported to the ferrous fraction. The material flow trough a car shredding facility is shown in Fig. 1:

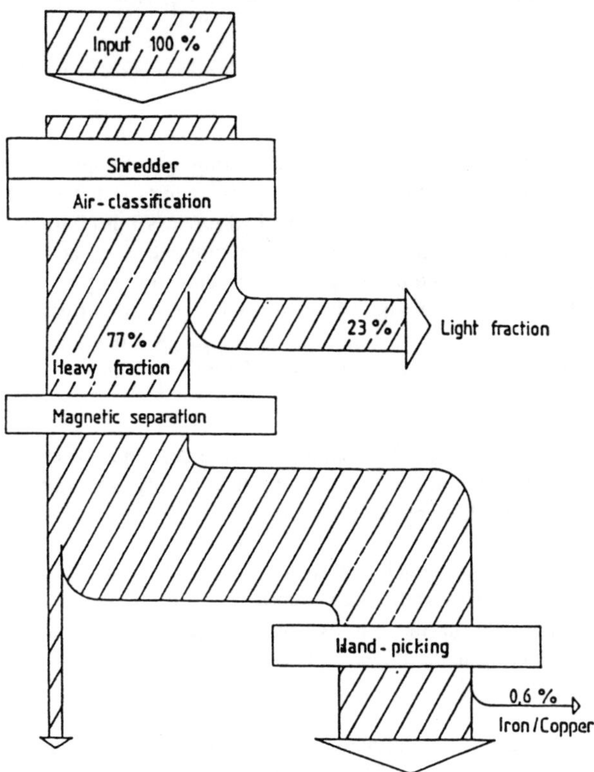

Fig. 1: Material flow through a car shredding facility

The light fraction as well as the non magnetic heavy fraction can not be separated further and have to be deposited. Due to the rising importance of these fractions including plastics, aluminium and ceramics in modern cars, it becomes more and more important to find different methods of dealing with these materials. A specific problem is posed by the different types of fluids in cars, including motor oil, cooling fluid, hydraulic oil and the like. If even small portions of these liquids are burned in waste burning facilities, highly poisonous dioxins can be generated. This has prompted the German government to declare shredder residues as hazardous waste, which has to be dealt with using extra care and making disposal very expensive.

One of the key problem of the shredding process is that it cuts the product at random lines, which always lead to fractions that contain more than one material. This is important for the metals, which are the only materials anyway that can be reused after a shredding process. If the mixed metal fractions can not be identified and sorted out, they tend to produce unwanted alloys during later metallurgical processes. These alloys can show greatly reduced material characteristics and thus be of very little value, if they can not be removed. Reducing the grain size of the shredding process will lead to a smaller fraction of mixed materials, but requires a bigger shredding effort. Additionally, the effort for later individual part identification and separation rises steeply with the reduction of grain size.

2.3.2   *The Disassembly approach.* The currently favoured approach to this problem is to regain all fluid components before shredding and disassemble all materials and components, that can not be regained in sufficient purity after shredding and that justify the effort of disassembly because of their value or reduced disposal cost. The rest of the car is subsequently shredded and processed as described. The disassembly of used complex consumer products is, however, confronted with numerous difficulties. First of all, it is difficult to gain all the information necessary to plan the disassembly. Parts of the product might have been exchanged during repair, adding to the uncertainties. Then there are changes of the product during usage, i.e. joining elements are often difficult to disassemble due to corrosion or wear and tear. Many consumer products are not designed for disassembly, they contain rivets, welding spots or glued connections that may have to be destroyed.
All this leads to the fact that, until now, disassembly takes place in a repair shop atmosphere, where highly skilled workers using only hand tools take the products apart. For large scale disassembly, this is not economically feasible. Most car manufacturers, that have already set up pilot disassembly plants, run them to gain information about disassembly processes and future product design for ease of disassembly /2/.

2.3.3   *The Metallurgical Approach.* A different method to make use of the non metal fraction of consumer products is to use them as fuel to melt and regain the metallic components. The German car manufacturer Mercedes Benz favours this method and claims that the overall energetically balance of this method is better than that of a disassembly /12/. It also includes the regaining of fluids and the disassembly of valuable materials. The rest is compacted and forwarded to a melting reactor. Mercedes claims that the remaining textiles and plastic materials substitute for as much as 40 % of the fuel that would normally be necessary for melting. However, critics of the concept fear, that non ferrous metals like copper could form alloys with the melted steel and thus reduce its value significantly and permanently, because copper is very difficult to remove from an alloy it has formed with steel. They see the concept only as a camouflaged waste burning facility.

3 Recycling Considerations in the System Life-Cycle

3.1 CONCEPT OF THE SYSTEM LIFE-CYCLE

When companies take an ecological approach, they may change the nature of their business. Ecological orientation is not only limited to production and marketing. The entire product life cycle must be considered. The traditional view on the product's life cycle started with the recognition of the consumers needs and ended with the phase out after usage. In future, however this view is widened to include also the recycling phase /13/.
In addition to extending the time horizon that companies take into account while designing new generation of products, there must also be a broadened approach in another dimension: Besides the environmentally friendly composition of their <u>products,</u> companies have to consider the environmental impact of the <u>processes</u> and of the <u>logistics / support</u> organisation. Therefore, an integrative approach is desired which considers these three elements during the whole life cycle. Such an approach is denoted here as a system approach and is represented in this section with strong emphasis on its recycling aspects.

34

A system is defined in general as an assemblage or combination of elements or parts forming a complex /14/. All relevant interdependencies and interactions must be enclosed. A product in general cannot function properly without an operator, a production system, a support capability, and so on. Therefore, when companies adopt the system approach described above, not only the product must be considered, but its production process, way of use, its maintenance capabilities, support, and recycling options. A system approach must be a strategic approach of a company when introducing new products. Proper functioning and competitiveness of such a system can not be achieved through efforts applied largely after the product's design stage. Especially, the ecological aspects are the ones that must be planned in advance and considered in order to form an ecologically sound system.

The system approach presented here includes the life-cycles of three elements: the product, its related processes, and its logistic support (Fig. 2). These three life cycles should be considered simultaneously while following the system life-cycle during the acquisition, utilization, and recycling phases.

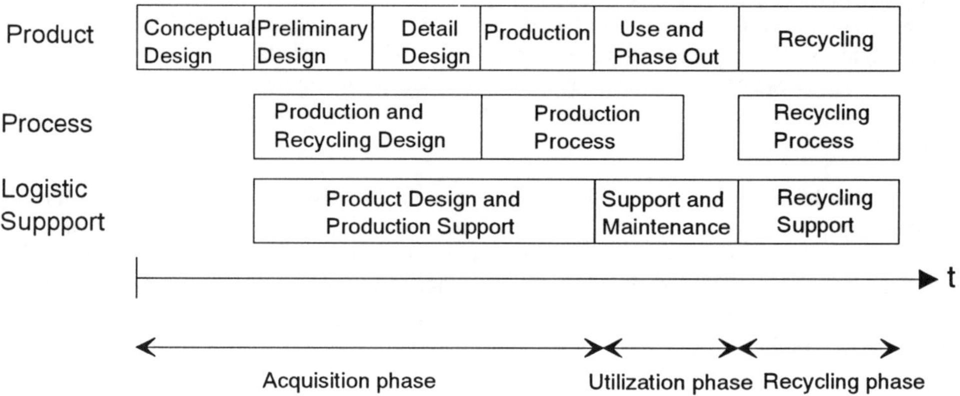

Fig. 2: System life-Cycles

3.2 PRODUCT LIFE CYCLE

The product life cycle begins with the identification of the needs and extends through design, planning, production, assembly, usage, phase out and recycling stages. For the recycling, the most important and difficult phase is the design. It determines the recycling options of the product and influences the options for recycling processes, as well as the logistic-support during phase out. Its difficulty evolves from the time gap between design and recycling, which can amount to 10 to 15 years for cars and household appliances. Additionally, the design must not only consider the product recycling aspects, but also the requirements of all other system life cycle phases.

3.3 PROCESS LIFE CYCLE

The process life cycle begins with the definition of the production task by the product design. It encloses the design of the production and recycling systems and processes. With regard to recycling, production process planning has to deal with minimizing production waste and finding ways to recycle it. As to the recycling system, the aim is to find the processes which lead to maximum output, namely components, materials and energy for further use while minimizing recycling effort.

3.4 LOGISTIC SUPPORT LIFE CYCLE

The logistic support life-cycle encloses the support during the design and production stages, the consumer support and maintenance during the product usage, and the support for the product recycling. Of interest to the recycling are the collection and transport of used products, providing product information about used materials etc. for the recycling industry and possibly transferring used materials and components back to the production of future products.

3.5 SYSTEM MEASURES

A system as defined above is measured in general by: performance, effectiveness, ease of production, reliability, maintainability, and cost. Central to our approach of integrating recycling considerations to the system life cycle is the design of the products. It influences both the selection of recycling processes and their logistic support and thus allows for an ecologically sound phase out of the product.

4 Design-for-Recycling

4.1 AIMS

This section deals with design-for-recycling at the conceptual design stage. For simplicity, the different steps in the conceptual design stage are considered as one block (see <u>Fig. 3</u>). The input to this stage includes the clarified design task comprising functional product requirements as well as the necessary informations about relevant processes, tools and optimization goals.

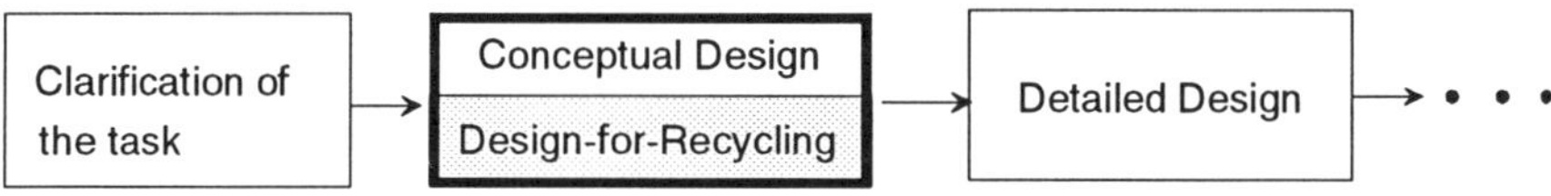

<u>Fig. 3</u>: Design-for-recycling at the conceptual design stage

One of the optimization goals is the ease of recycling. The importance of design for both the reduction of effort and the increase of output of the recycling process has already been stated in chapter 2. When designing a product for ease of recycling, the designer has to take into account several areas, that influence the optimal choice for DFR:

- *Future ways of collecting, transporting and storing the product after usage*
- *Current and future developments of recycling methods*
- *Possibilities to reuse components in future products*
- *Existence of technologies to reprocess the materials*
- *Existence or future development of markets for the recycled materials*

Recycling can be represented as a black box that processes used products and outputs regained products, components, materials, and energy (<u>Fig. 4</u>). The black box includes processes like disassembly, shredding, sorting, metallurgical processing and the like. The prime task in DFR is to plan the future recycling method of a product simultaneously with the product design, so that the product design can be oriented towards easing the recycling processes. DFR is thus a part of concurrent engineering.

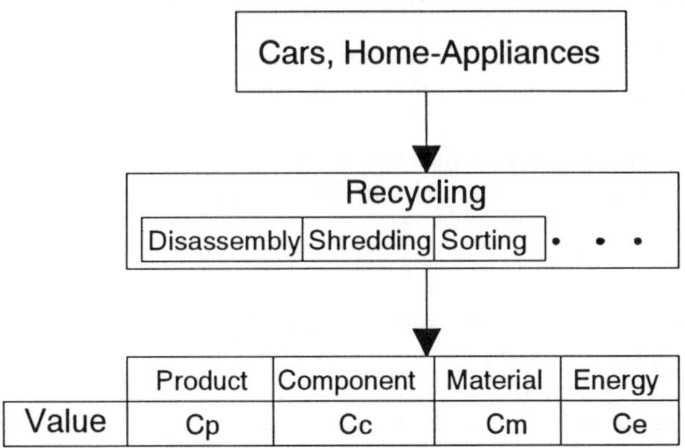

Fig. 4: Black Box approach to Recycling

The total value of the recycling output is the sum of the value of all output components $C_p+C_c+C_m+C_e$. The aim of recycling has already been described as maximizing total output substitution value while minimizing the recycling effort /15/. Thus, DFR aims at <u>designing a product for ease of recycling and maximal output</u>.

Where:

    Ease of recycling regards the effort for the recycling process and its logistic support

        Process:

            The different recycling processes as described in Section 2.3

        Logistic support:

            The services a company has to provide to facilitate the phaseout of their products

    Output:

        The output of the recycling process is determined by the difference of the total output substitution value minus the residue disposal cost.

This shows, that DFR is strongly dependant on methods to evaluate costs and benefits of operations, components and materials. This is especially difficult in areas, where due to the long life span of a the product design and recycling of the product may be years apart. The area of evaluation must thus be dealt with special care in future research work.

4.2   RULES FOR DFR

4.2.1   *Structuring DFR.* DFR is still subject to research. A specific problem lies in the complexity of the problem, as most design decisions are influenced by many factors and do themselves influence many following decisions /16,17,18/. The aim of the research must be to provide the designer with a set of guidelines, that are simple, easy to apply and easy to evaluate. The current state of knowledge, however, is a more or less unstructured collection of specific rules, that are hard to apply to realistic design problems. The possibility to structure these rules that is presented here is to form groups of rules relating to the same design aspects.

4.2.2   *Product recycling.* The first priority for recycling is extending the products life span by recycling during usage. It aims at ensuring that the product can fulfil its function for a longer period using minimal resources. This can be achieved by the following aims:
- *A modular structure of the product allows to modernize components who are outdated without having to change the whole product (Example: HiFi-Tower, Personal Computer)*
- *To design those sections of a product which are subject to heavy wear and tear as separate elements allow to have them exchanged easily and thus extend the products life.*

4.2.3   *Ease of Disassembly.* Significant to the DFR rules is the ease of disassembly, since the disassembly process allows to regain components intact and materials of higher purity than the alternative shredding process. Design for disassembly can also be of advantage for other stages of the products life cycle, namely:
- *for easiest packaging and transportation during the distribution phase*
- *for repair and maintenance during the usage phase*
- *during the recycling phase*

In this paper we focus on the disassembly during the recycling phase. Disassembly should not be seen as just the opposite operation of the assembly since the time horizon of the operation, the task requirements, and the product condition are different /19,20/. However, while taking a system approach the design-for-assembly and the design-for-recycling activities should be done concurrently in order to achieve a global optimum of the system. However, conflicts can not always be avoided. For example, design-for-assembly rules prescribe reducing part counts by combining functions /21/. Doing so cuts assembly times, but complicates disassembly efforts and recycling in case that different materials are involved. As another example, design-for-assembly rules recommend snap or press fit when possible, whereas from a disassembly point of view metal inserts are not recommended because they are difficult to remove by a worker. Design for disassembly rules include:
- *Form subassemblies from harmful materials (Cluster components) in order to cut the disassembly time.*
- *Design for easiest access to harmful materials al well as valuable materials and reuseable components.*
- *Consider the optimal disassembly sequence enabling to disassemble first the harmful and reused components.*
- *Use plastic fasteners or multi usage fasteners such as thumb screw.*
- *Use quick disconnecting electrical connectors.*

4.2.4   *Choice of material.* An issue of big importance for recycling is the usage of the right materials. If recycling of the whole product or product components is impossible, the combination of materials has to be separated and regained. Rules regarding this aspects are /22/:
- *Selecting environmentally compatible and recyclable materials for components.*
- *Reducing the volume of plastic and composite materials used, since now a days most recycled plastic find their way into less-demanding applications.*
- *Avoid secondary finishing operations such as painting, plating, coating and so forth.*
- *For ease of sorting: use similar materials and colors in any part or assembly. Dissimilar materials must be identified and separated, often a labour-intensive operation. In case that different material must be used marking the parts should be done.*
- *For ease of shredding avoid using not shreddable material , i.e. avoid using concrete as a counterweight in a washing machine*

4.2.5   *Design for logistics.* Besides designing the product for ease of recycling, an important factor is to make sure the product is actually fed back to the recycling process by the last user. This can be assisted by following these rules:
- *Design the product in a way that it can be transported easily after usage, i.e. by allowing pre-disassembly.*
- *Develop a simple and efficient system support approach which will encourage the consumers to start the recycling process, and will be cost effective.*

4.2.6   *Conclusion.* All the given rules effect the efficiency and effectiveness of the recycling process. They must be integrated in the conceptual design stage and evaluated simultaneously to the manufacturing, maintainability, reliability and other design rules of a product. The following case study demonstrates the ideas of design-for-recycling using a washing machine.

5. A Case Study

Currently, design-for-recycling aspects are not considered widely enough in white household machines such as: washing machines, dryers and ovens. Although white household machines include a complex composition of valuable materials, just like car wrecks, the regaining of these materials is less well developed than in the case of the car wrecks. The following section describes a case study on the options of recycling washing machines. Different recycling aspects are considered with an emphasis on disassembly aspects. This case study aims at:
- *evaluating the current recycling situation of washing machines,*
- *analysing the options of reused components and materials and*
- *generating recomendations for future DFR for washing machines.*

5.1 PRESENT RECYCLING SITUATION

Almost 84% of Germans households own a washing machine /23/. After an average life cycle of 8 - 12 years these machines are reaching their phase out stage. Currently, the recycling approach is limited, usually they are shred as a whole product at local shredding plants, where the aim is to regain only the valuable metals. The rising amount of plastics and rubber as well as glass fractions are not reused and have to be deposited.

In the area of logistics, there is also room for improvement: The old machines can be deposited at the roadside at certain dates, where they are collected by the community. They are than stored on uncovered community disposal sites, until a shredding company collects them for further processing. A small percentage of machines is collected by local dealers upon delivery of new machines and forwarded directly to shredding companies. The high cost for collection and transport of the washing machines, relative to the scrap price, is the main obstacle for the limited scale of reusing this source of materials and components. Small sized household appliances are just thrown to the garbage.

Due to rising ecological awareness, the product recyclability aspects have slightly gained importance in some companies. Harmful materials like PCB in the capacitors or mercury from switches are not used anymore. In order to enable efficient shredding, some companies have changed the concrete counterweight mass into a cast iron part /24/.

5.2 ANALYSIS OF THE PRODUCT'S OPTIONS FOR RECYCLING

The reasons for recycling a washing machine can be both ecological and economical. Investigating the current design of washing machines shows, that there are valuable materials that can be used further. An example of the material composition of a typical washing machine is shown in Table 3.

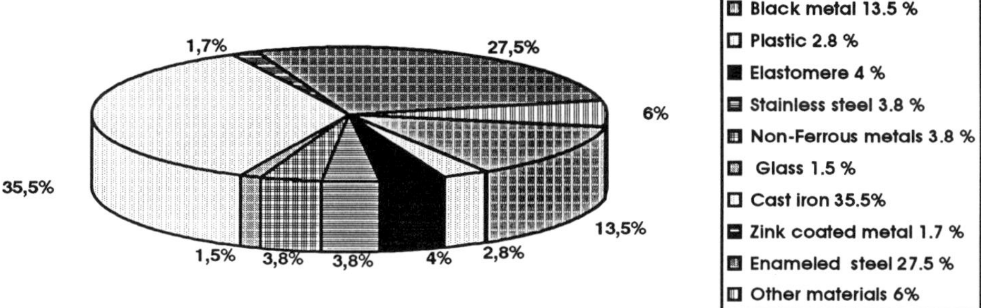

Table 3:: Material composition of a washing machine /24/

In the current recycling technology, the plastic, rubber, glass, and composite materials cannot be reused. These materials are make up for 11% of the washing machine materials. By employing a different recycling approach these materials could be regained. Besides the valuable materials the washing machine contains several components that can be considered for recycling. While it is not likely that components of one series of washing machines can be reused in an industrial scale for future series of machines, it is sensible to reuse them as spare parts for the repair of other machines of the same series /25/.

5.3 RECOMMENDATION FOR FUTURE DESIGN

Improving the recyclability of washing machines can be done in several aspects. The following suggestions represent options of new construction and using new materials.

Materials
- Use cast iron instead of concrete for the counterweight mass to prevent breakage of the shredder and to increase the percentage of recyclable material
- Avoid use of cooper as far as possible since it is difficult to separate and forms inferior alloys with steel
- Avoid the application of galvanized plate work for easier steel extraction.
- Reduce the variety of plastics in the machine in order to make the separation easier.
- Mark all materials for easier identification

Disassembly
- Locate all of the machine's electronic components closely together, so easy disassembly is possible.
- Locate all harmful materials (Cadmium, Asbestos) in a location with good accessibility.
- Use joining methods that are easy to disassemble

Logistics
- Organize collection system to regain machines after usage. Collection of the machine at the phase out stage can be done by one of three options:
  - Collection from the consumer house by the community,
  - Collection by the local dealer

- The consumer itself will bring the machine to community collection place, or to a private organization.
- Provide technical details about the product composition, the harmful materials , and the disassembly options to third companies that recycle your products

6. Conclusions and Summary

Due to the continuing destruction of our environment, industry is more and more expected to make an important contribution to solving the ecological problems created by their products. Reducing the burden from the ecological system while maintaining the current standards of living requires the development of innovative technologies. Such a technology is the recycling, aiming to reuse materials and components. In this paper recycling is considered as part of the system life cycle, which includes three elements: the product, the processes, and logistic support. These elements must be considered simultaneously in order to achieve an optimal system. Different interfaces were considered in the system: Between the production system and the market, the market and the recycling system, and an interface of the recycling system to the production system as recycled material, or reused components to be assembled in the new products.

Rules regarding design-for-recycling are given aiming at the minimal effort for the recycling process and the maximum output value from the recycling process. Disassembly is represented as an important element of recycling processes, because it allows to regain components and subassemblies whereas shredding only results in materials.

A case study of a washing machine as a representative of the 'white' household appliances is investigated. Current recycling activities are reviewed and their potential for recycling is checked. Different aspects for improving the recyclability of the washing machine are considered.

Following the above recycling approach via a system concept can be an opportunity for companies to comply with legislation and consumers demands. Taking such an approach requires changing the traditional design, production, and phase out phases of the life cycle. In a long term companies adopting this approach while combining innovative technologies may make the recycling activities profitable.

References

[1] J. Schmidt. Im Schrittempo - In zahlreichen Pilotprojekten werden Erfahrungen mit der Altautoverwertung gesammelt. *MÜLLMAGAZIN*, 1:34-42, 1993.

[2] P. Mast and K. Strobel. *Demontagefreundliche Gestaltung von Automobilen.* Forschungsvereinigung Automobiltechnik e.V. (FAT), 1989.

[3] *Was Übrigbleibt, Wenn Mercedes einen Mercedes Wiederverwertet.* Mercedes-Benz AG.

[4] *Unternehmensaufgabe Umweltschutz,* VOLKSWAGEN AG.

[5] M. Simms. Driving away scrap. *ENGINEERING*, 39, 39-40, 1993.

[6] P. Börkey. Recycling von Haushaltsgroßgeräten - Rahmenbedingungen und Bewertung technischer Alternativen. Interner Bericht-IWF, TU-Berlin, 1993.

[7] Die Zeche zahlt der Anwender. *PC Magazin*, 13:53-56, 1991.

[8] VDI-Richtlinie 2243E. *Konstruieren recylinggerechter technischer Produkte*, VDI-Verlag, Düsseldorf, 1991.

[9] G. Pahl and W. Beitz. *Konstruktionslehre Methoden und Anwendung.* Springer-Verlag Berlin Heidelberg, 1993.

[10] H. Trapp. Aufgaben und Chancen der Schrott-Recycling-Wirtschaft. *Recycling*, 3:7-12, 1991

[11] F. Schmieg. Die Shreddertechnologie und das stoffliche Recycling. *VDI Berichte*, 7, 1992.

[12] L.P. Schreiber. Metallurgische Auto-Verwertung braucht zukunftssicheren Stahlstandort. *VDI Nachrichten*,1, 1993.

[13] L. Alting. Life Cycle design of industrial products. *Concurrent Engineering*, 1, 1991.

[14] B.S. Blanchard and W.J. Fabrycky. *Systems engineering and analysis*. PrenticeHall Inc., 1990.

[15] W. Beitz. Ein ökologisch und ökonomisch leistungsfähiges Recycling beginnt bei der Produktgestaltung. *VDI -V*, 47:17-19, 1993.

[16] H. Franze. Konstruieren fürs Recyceln. *Automobil-Produktion*, 88-90, 1992.

[17] E. Garbe and H. Salomon. Recyclinggerechtes Konstruieren-Erfordernis moderner Produktgestaltung. *VDI-Z*, 131(4):79-83, 1989.

[18] R. Hoffmann-Kroll. Recycling-eine Aufgabe für Konstrukteure. *UMWELT Bd.* 22(5): 311-313.

[19] M. Kahmeyer and T. Leicht. Demontieren leicht gemacht. *Kunststoffe*, 81(12):1135-7, 1991.

[20] W. Eversheim, M. Hartmann and M. Linnhoff. Zukunftsperspektive Demontage, *VDI-Z*, 134(6):83-86, 1992.

[21] G. Boothroyd and L. Alting. Design for Assembly and Disassembly. *Annals of the CIRP*, 41(2), 1992.

[22] K.J. Thome-Kozmiensky. *Materialrecycling durch Abfallaufbereitung*. EV-Verlag für Energie und Umwelttechnik GmbH, 1992.

[23] *Der Markt für Elektro-Grossgeräte Elektro-Kleingeräte in Deutschland*, Gfk Handels-Forschung, 1991.

[24] *Stoffliche Verwertung von Elektro-Hausgeräten*, Miele Umweltreferat-Produkte, 1993.

[25] *Vom Haushalten - Schritte zur neuen Harmonie*. AEG Hausgeräte AG, 1992.

SCHEDULING DESIGN ACTIVITIES

Andrew Kusiak
and
Upendra Belhe
Intelligent Systems Laboratory
Department of Industrial Engineering
The University of Iowa
Iowa City, IA 52242 - 1527

ABSTRACT. The design process can be divided into a number of activities that are performed according to precedence constraints among them. The objective of concurrent engineering approach is to minimize the total design time without violating any of the constraints. A limited number of available resources constrain the simultaneous execution of design activities. The execution of succeeding design activities depends on the available information provided by the preceding activities. The smooth flow of information from the beginning to the end of the design process is a key factor for successful and timely completion of the design process. In this paper, a pull system approach for management of activities in the design process is proposed. In this approach, the dynamic scheduling policies are adopted. A heuristic for scheduling of activities is developed.

1. Introduction

The design process involves a number of activities and relationships among them. The basic input information to the design process are the customer requirements. At each stage of the design process decisions are made based on the data or information available at that stage which is provided by the 'preceding' stages. Hence, in a network of design activities data dependency relationships can be established among the activities. An arrow can be drawn from design activity 'i' to activity 'j', indicating that the activity 'j' depends upon the activity 'i' (Figure 1).

It is assumed that every design activity is indivisible, and hence must be treated as one entity with its 'start' and 'finish' times.

A design stage involves set of resources to perform a group of design activities that share some similarity. For example, in testing, the design activities are breadboard testing, electrical testing of a device, thermal testing, and so on. Design activities are performed with the available resources and according to the precedence and some pre defined scheduling rules. The information generated is passed on to the stages where the succeeding design activities are performed.

The supply of information to any particular design stage can be considered in two ways.

S. Y. Nof (ed.), Information and Collaboration Models of Integration, 43–60.

1) At each design stage the execution of design activities results in generation of some data or information. This information is provided to the succeeding stages. In a way, the information is 'pushed' down to the succeeding design stages.
2) Another strategy is to provide information when needed. A design stage pulls down the information from a preceding stage.

The specific problem addressed in this paper is the problem of scheduling design activities (as per information requests) at different design stages. Limited amount of resources are available and the objective is to schedule simultaneously as many activities as possible with the minimum delay. It is not possible to schedule all activities concurrently due to precedence and resource constraints.

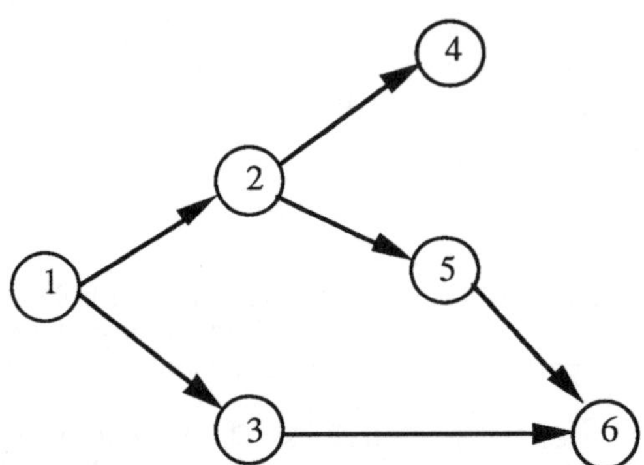

Figure 1. Data dependency relationship among design activities

The problem of minimizing the weighted number of late jobs was studied by Karp (1972) and Pinedo (1983). Karp (1972) showed that this problem is NP-hard. When all jobs have a common due-date, the problem is equivalent to the knapsack problem and remains NP-hard. The problem of scheduling the design activities is still more complex in nature because of multiple resource constraints. Since the duration of design activities are not known in advance, in this paper the dynamic scheduling policies are adopted (see Pinedo 1983). In dynamic policy, decisions can be made at any moment of time based on the information available.

The analogy of the problem of scheduling design activities to the project scheduling problem with multiple resource constraints is studied. The latter problem has been considered by Kurtulus and Davis (1982) and Kurtulus and Narula (1985). Numerous procedures for solving certain variants of the problem were developed by Christofides et al. (1987), and Bell and Park (1990). The scheduling process at lower level translates the global schedule into the short term schedule. Bel et al. (1988) developed a software tool for short term scheduling, however, their approach does not deal with problem of assigning activities to resources.

2. Pull System

A proper flow of information is required for timely execution of design activities. In a pull system the consumption of information rules the flow throughout the design process rather than top-down schedules and releases. The 'pull' strategy provides a set of rules that controls the issuing of information to the design stages. The simple pull system rule considered in this paper is that, *send required information when needed.* This means that the information is transferred whenever demanded. Excess generation of information is considered as waste.

2.1. CONGESTION IN PULL SYSTEM

To be general, it is not assumed that the information demand at any particular design stage is stationary. The number of requests per time unit coming to a design stage varies for different projects or products. Whenever it appears that the design stage will soon catch up with the information requests, the scheduling system can reduce the resources used at that stage. It is believed that, in most cases the scheduling system will be able to maintain a comfortable level of information demand in the entire design process. For this reason, it is assumed that there is always demand at each design stage for at least some type of information.

In the pull system, the decision variable is how much 'work in process' is to be maintained between various stations. The information throughput of a pull system is a result of this 'work in process' configuration. Hence, although the arrival rate of design projects to the design process is non stationary, the scheduling system will be able to operate the design process in steady state.

2.2. INFORMATION REQUESTS AND ANALOGY TO KANBAN SYSTEM

The information can be requested from the preceding stages in a specified format. The Kanban principle is useful in this case. A number of requests are queued at a particular design stage from the succeeding design stages. The number of requests made by a design stage to another is decided dynamically and it usually varies for different projects. An individual stage can also build its own strategy to process (serve) the requests. Similar to the Kanban system, the request flag is attached to the information sent back to the stage and in this way the receiving design stage can keep track of the incoming information and detect discrepancies, if any.

The rules for issuing the information to the following design stages can be depicted as follows:
1) The requests should be processed only according to the predetermined strategy (by managers or project planners).
2) If any discrepancies with the information provided to the succeeding stages is detected then the same information should not be sent to the other succeeding stages (before a proper review).
3) The information should not be sent if the request is not fully satisfied.

3. Scheduling System

The scheduling system should be efficient enough so that, the number of waiting information requests at various stages remains low. The organizational conditions dictate that higher level of service (quality of information) is required. The reduction in the duration of the design process depends on the efficient utilization of resources, that in turn are controlled by the system. The scheduling system is considered as a cooperative hierarchical decision making system.

3.1. STRUCTURE OF THE DESIGN ORGANIZATION

The push system causes some difficulties when planners (managers) request results within certain date without considering the availability of information required. The amount and nature of information needed at different design stages may vary for different products. Pull system works well in self-adjusting such design process variations. The individual design stages need not be aware of the global schedule in order to perform the assigned activities.

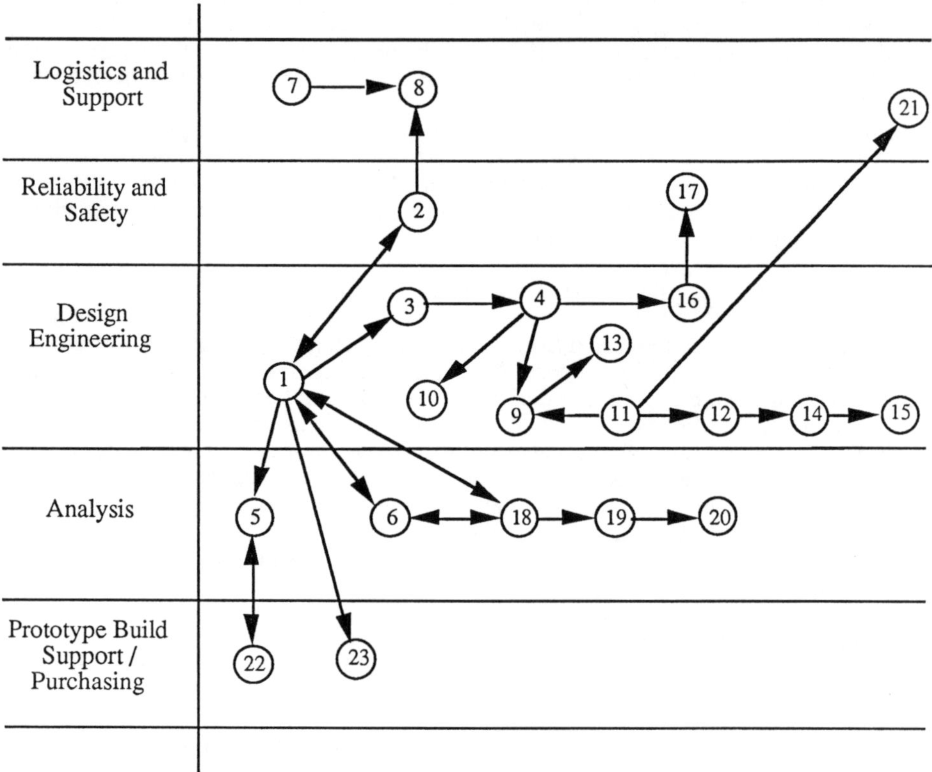

Figure 2. The design activity network indicating various activities and corresponding departments playing major role in performing these activities

Table 1. The hierarchical structure of the design organization

	Responsibility	**Resource Allocation**	**Operation Mode**
Design Organization	Estimate due dates for various designs Determine global schedule, due dates for milestones in the design process	Determine manpower and skills required for the design of the product	Static
Individual Design Stage	Check capacity, work in process, decide which activities to perform first according to the set priorities	Identify resource requirement of design activities waiting for service	Dynamic
Individual Design Activity	Check feasibility of due dates for the information required	Allocate available personnel to design activities according to skills requirement and preferences	Dynamic
Design Personnel	Perform design activities with utmost efficiency and try to maintain the schedule of design activities		Dynamic

48

The horizon and period of decision making forms a hierarchy that closely follows the organizational structure used in the design organization. The hierarchical structure of the design organization is shown in Table 1. In the proposed approach, the system operates in a static decision making mode at the product / design project level and in a dynamic decision making mode to translate the goals of the top level schedule into the actions at the lower level. Proper communication between the design organization level and the lower levels will ensure that the global schedule is realistic. At each level of the design organization the responsibilities need to be well defined. The identification of the resource requirement is done at all levels.

Activities in each design phase can be identified by the department playing major role in performing those activities. These departments can also be indicated on the design activity network, as shown in Figure 2.

3.2. DUE DATES FOR INFORMATION REQUESTS IN A PULL SYSTEM

The design activities are performed in order to fulfill the information requests. The due dates for these information requests can be interpreted as the latest finish time of the corresponding design activities. The latest finish time of all the activities in a design phase are derived by the demanding activities. For example consider activities 1, 2, ..., 9 in Figure 3.

The information requests specify what information is needed and by when. The design activities are performed by various departments in the design organization. When new information is received the waiting requests are modified accordingly. For example, consider the network of activities in Figure 3. Suppose that activity 8 is waiting for input from activity 5, activity 6 and activity 7. Suppose that the activity 5 is completed before activity 6 or activity 7 begin. Then the time at which the information is received by activity 8 from activity 5 may change the information request made by activity 8 at activity 6 and activity 7.

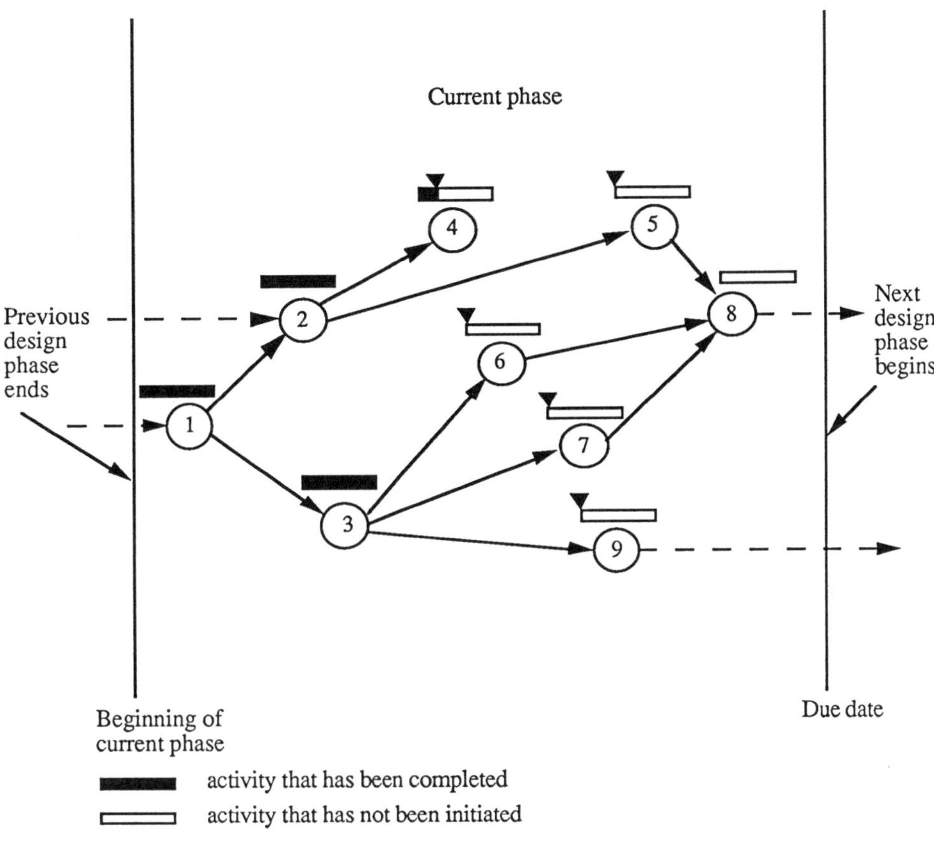

Figure 3. Due dates in the design activity network

In case of design activities it is difficult to determine the duration of activities (d_i) in advance. The estimates of the activity duration can be expressed using the notation of PERT (Badiru 1988). The time estimate of each design activity is based on three values. The following notation is used:

s_i	start time of activity i
f_i	finish time of activity i
$LS(i)$	latest start time of activity i
$LF(i)$	latest finish time of activity i
d_i	estimated duration of activity i
l_i	optimistic time estimate
u_i	pessimistic time estimate
m_i	most likely time estimate

Let the duration d_i of activity i may assume any value between lower and upper limits (l_i and u_i).

$$0 \leq l_i \leq d_i \leq u_i$$
$$s_i + l_i \leq f_i \leq s_i + u_i$$

The most likely estimate need not coincide with the midpoint, ($l_i + u_i / 2$), and may occur to its left or to its right. Because of these properties beta distribution is used to estimate the duration of activities in project networks (Moder and Phillips 1970). The same principle is used to find the mean estimate for the duration (d_i) of activity i .

$$d_i = \frac{l_i + u_i + 4m_i}{6}$$

In practice values of the parameters l_i, u_i and m_i are not available, but what is available are the subjective estimates of these three parameters. These estimates are sample values that are subject to error of unknown magnitude. In addition, the estimates of duration of individual activities combine in a complex manner to yield estimate of the variance of the duration of the design phase. Errors in the duration of the design activities introduce largely undetermined errors in the duration of the design phase.

3.2.1. *Simple Method to Determine the Latest Finish Time of Design Activities.* The start time of any design phase can be assumed to be zero. The due dates for phases (milestones) in the design process are assigned by the global schedule.

The method presented in this section follows the analysis by Dodin (1984) and the heuristic method developed by Ord (1991). Here the simple approximation developed by Ord (1991) is implemented in backward direction to determine the latest finish time of the design activities.

The duration of design activity is denoted by a random variable T. The range of this non negative random variable is partitioned into k subsets with boundary points

$$0 \leq t_0 \leq t_1 \leq ... \leq t_k < \infty$$

such that,

$$P(t_r \leq T \leq t_{r+1}) = \mu_r$$

The discrete random variable T_j is used to approximate the activity time distribution.

$$P(T_j = \mu_{jr}) = p_{jr} \qquad r = 1, \ 2, \ ..., \ k$$

The procedure of considering two convolutions at a time rapidly increases the number of partitioned subsets. This problem was resolved by Ord (1991) by collapsing the distribution. The values of latest start time and latest finish time can be estimated by using this approximation scheme and working backwards in the design activity network. Since in the methodology presented in this paper dynamic scheduling policies are used and hence the schedules of design activities are closely monitored, the approximation considered above is good enough for the purpose of providing fair estimate of the latest finish time of the design activities.

Example

Consider the example design activity network of a design phase shown in Figure 3. The design activities 8 and 9 are the final activities in the design phase. Suppose that the due date for the design phase is 20 and hence the latest finish time of activities is also 20. Suppose that the activity 8 has duration 2, 3 or 5 with probability 1/3. Then the latest start time of activity 8, i.e. the random variable $(LF(8) - d_8)$ may take up the values 18, 17 or 15 with probability 1/3. These values also give the distribution of the latest finish time of activity 5 and activity 7.

Suppose that the activity 5 has duration 1, 3, 4 or 5 with probability 1/4 and activity 7 has duration 2, 3, or 4 with probability 1/3.

In order to determine the latest start time of activity 5, all possible combinations of the latest finish time distribution and the distribution of duration of activity 5 are considered. The computed boundary points are sorted in ascending order. The latest start time of activity 5, $(LF(5) - d_5)$ may take 12 equally likely values:

$$(10, 11, 12, 12, 13, 13, 14, 14, 14, 15, 16, 17)$$

Here the method developed by Ord (1991) of merging the sets of k is used. The obtained values reduce down to:

$$(11, 12.67, 14, 16)$$

Similarly the latest start time of activity 7, $(LF(7) - d_7)$ may take values:

$$(11, 12, 13, 13, 14, 14, 15, 16)$$

which reduce down to (12, 13.67, 15.33). Hence the approximate values,

$$E(LS(5)) = 13.42 \quad \text{and} \quad E(LS(7)) = 13.67$$

can be obtained.
In this way, one can work backwards in the design activity network to obtain the approximate values of the latest start time and duration of design activities.

3.3. SCHEDULING OF DESIGN ACTIVITIES WITH MULTIPLE RESOURCE CONSTRAINTS

The problems of scarce resources are categorized in one of the following two classes (see Elmaghraby 1977):
1. Each resource is available in fixed amounts and one is interested in evaluating the impact of such scarcity on performance of the project.
2. Unlimited resources are available at some price and the problem is of determining optimal amounts to be acquired to achieve the due date.

The parallel heuristic approach (Moder 1975) seems to be particularly useful for scheduling design activities. In this approach, one time interval and all the activities that can be scheduled in that interval are considered. The activities are ranked in some

52

manner and scheduled according to resource availability. The activities are delayed otherwise.

The PERT and CPM methods assume that the relationship between the level of resources and duration of activities is known a priori for all activities. With certain modifications, the problem of scheduling of design activities can be viewed as a multiple-resource constrained project scheduling problem.

The major difference between the project scheduling problem and design activity scheduling problem is that in project scheduling, the decisions are taken in a static mode. In the scheduling problem of design activities, duration of activities are not known. There is usually large product to product variation in the design activity duration. Since each of the new design problems is unique in its nature, it is not possible to predict activity duration in advance.

The traditional project scheduling problems seek to minimize objective functions such as project make span, weighted tardiness of jobs, and so on, subject to the constraints on flow and resource utilization. The primary difference and analogies between project scheduling problem and the problem of scheduling design activities are:

1) The objective is to find a schedule that maximizes the number of activities with earliest due date simultaneously, subject to resource constraints.

2) Some of the resource constraints are analogous to those in traditional job shop scheduling. Many shared resources are encountered. A shared resource with availability m is analogous to m identical parallel machines.

3) Temporal constraints that result from design activity's effects being a precondition to a later design activity do not have analogs in traditional project scheduling.

By solving the problem of scheduling design activities, this paper implies finding a subset of activities at every activity completion, which can be started concurrently, such that the weighted lateness of design activities is minimized.

3.3.1. *Search Process* . A number of partial schedules, that represent the decisions for some subset of activities are created. The partial schedules are feasible, satisfying both precedence and resource constraints. The partial schedules are considered at time instants which correspond with the completion of one or more project activities. A partial schedule PS_j at time instant j consists of the set of activities that have been scheduled at j.

The following notation is used in the algorithm for solving the problem considered in this paper:

$(R_1, ..., R_n)$	vector indicating availability (status) of resources at time t
$(r_1, ..., r_n)_i$	vector of resource requirements for activity i
D	due date for the current design phase
w_i	weight of design activity i
C_i	estimated completion time of activity i
s_i	start time of activity i
A_j	set of activities which are in progress at time j
F_j	set of design activities which are completed at time j
U_j	set of design activities for which scheduling decisions are not yet been made at time j
S_j	set of activities whose preceding design activities have been completed

At time instant j, all activities in the design phase can be grouped into three sets, A_j, F_j, and U_j. These three sets are mutually disjoint. All the required information to begin the activity is available from the preceding activities. The activities in this set can start at time j if the resource constraints are not violated.

In the dynamic scenario, the problem of scheduling design activities under multiple resource constraints is reduced to the problem of scheduling N independent activities.

In order to resolve the resource conflicts, it is necessary to decide which combinations of activities are to be delayed. At each time instant j, subsets s^k, $k = 1$, ..., m of set S_j are considered. A set s^k is the subset of S_j and it represents the combination of design activities from set S_j such that addition of one more activity in s^k from set S_j would result in resource conflict. The selection of one of such subsets, represents the decision of delaying the combination of design activities in S_j and which are not in s^k. From the definition of s^k, it is clear that all subsets s^k are the minimal delaying alternatives. Demeulemeester and Herroelen (1992) proved that, in order to resolve resource conflict, it is sufficient to consider only minimal delaying alternatives.

3.3.2. *Minimization of the Estimated Weighted Lateness of Design Activities.* The objective of the problem considered in this paper is to reduce the estimated weighted lateness of design activities. Now, the decision is to be made to pick up a subset s^k of schedulable activities such that the total number of late activities is minimized.

This decision will be based on the priorities (weights w_i) assigned to the schedulable design activities, the estimated design activity time (d_i) and the latest start time (LS(i)) and latest finish time (LF(i)) determined based on the duration estimates and the milestone due dates.

Assume that one of the activities is completed at time t. As a result, some resources are made available. The resource status vector, (R_1, ..., R_n) needs to be updated. Completion of an activity means that some information request is fulfilled. This fulfillment results into some more information requests and hence few more activities may become schedulable. This means that, there is also a need to update the set S_j. For each activity in set S_j, the due date and its resource requirement vector, (r_1, ..., r_n)$_i$, are known.

Consider the activity network shown in Figure 3. The activities are represented in the Gantt chart shown in Figure 4. Suppose that completion of activity 3 is occurring at t. The set of schedulable activities at time t is $S_3 = \{5, 6, 7, 9\}$. Note that, the activity 4 is already in action and it is not considered in set S_3. The latest start time and latest finish time of four activities $\{5, 6, 7, 9\}$ are shown in Figure 4.

The result of scheduling activities 6 and 7 is shown Figure 4. Scheduling of activities begins at time t. The partial schedule depends upon the resource requirements. As a result, according to the time estimates of the design activities activity 7 finishes first at C(7). Since C(4) > C(7), the first completion will most probably occur at time C(7) after t, denoted as t*. The decision of scheduling activities 6 and 7 results in delaying activity 5. Activity 5 has the latest start time before completion time of activity 7. The subset s^k of schedulable activities is selected in such a way that this lateness is minimum. The priorities of some design activities over the other activities are expressed by assigning weights to the design activities.

The main advantage of this approach is that, it is dynamic in the sense that the system takes decision at every completion of an activity and secondly the system looks into future only one step ahead based on the available time estimates. This strategy reduces the risk on the decision maker's part of assuming too much in advance. The uncertainty is taken care of by a simple strategy: Do the best what you can at the present moment with the available information.

54

Since the subset of activities schedulable at any particular time is small (even for fairly large network of activities), the algorithm suggested below is computationally attractive.

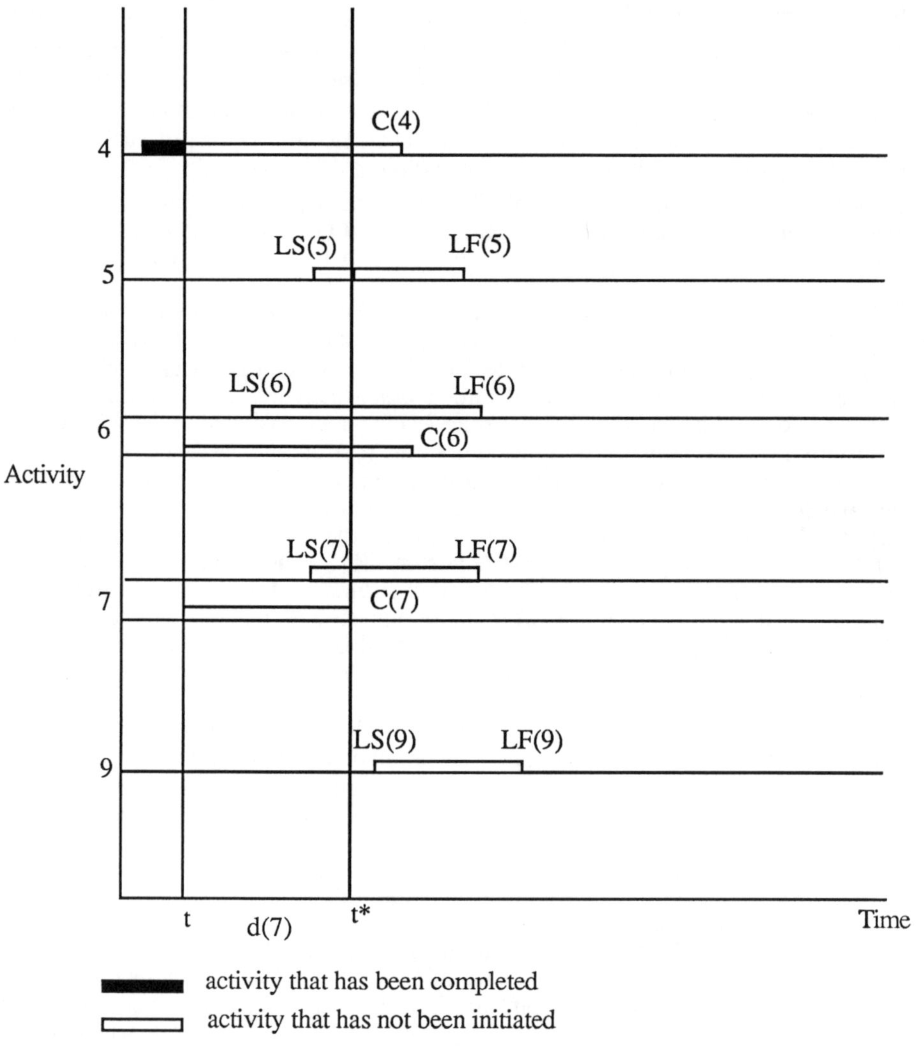

Figure 4. A Gantt chart showing latest finish time and latest start time of design activities

Algorithm

Step 0 At the completion of some activity i at time t, update the resource vector

$(R_1, ..., R_n)$ and the set S_i of all schedulable activities.

Step 1   Find all subsets s^k of S_i, such that the sum of resource requirements of all the activities in s^k does not exceed the availability of resources and addition of one more activity in set s^k will violate a resource constraint.

$$\sum_{i=1}^{m}(r_1,...,r_n)_i \le (R_1,...,Rn) < \sum_{i=1}^{m+1}(r_1,...,r_n)_i$$

Step 2   For each subset s^k, find an activity k, with the shortest estimated duration.
Step 3   Find $t^* = t + d_k$. Compute the value of objective function Z.

$$Z = \sum w_i \ \max(t^* - LS(i), 0), \text{ for all } i \notin s^k$$

Step 4   Go to Step 2 until the objective function Z for each of the subsets s^k is computed.
Step 5   Find the subset s^k with minimum Z. Schedule all activities in this subset s^k.

3.4. THE FRAMEWORK FOR SCHEDULING DESIGN ACTIVITIES

Dispatching provides the direction for selecting a right activity of the design plan. Working on the wrong activity could delay the delivery of some important outcome. Critical design stages that impact the total design time should not be allowed to be idle. Proper dispatching at the upstream design activities can ensure an uninterrupted flow of information through critical design stages.

Given a number of requests that are available for processing at a design stage, the problem is to find the sequence of requests that optimizes an objective function. The objective function could be, for example, to minimize the mean flow time or to minimize mean tardiness to satisfy delivery commitments

Scheduling of design activities at a design stage with due date constraint is a relatively simple problem. It is equivalent to scheduling n operations on a single machine. Our interest is to schedule information requests in such a way that the maximum lateness is minimized. As stated in Baker (1984), the maximum lateness can be minimized by sequencing requests such that:

$$d_{(1)} \le d_{(2)} \le ... \le d_{(i)} \le ... \le d_{(n)}$$

The schedule obtained is the Earliest Due Date (EDD) schedule. Analogously, in Step 3 of the algorithm in section 3.3.2, the design activities with earliest due date (latest finish time) are given priority along with their weights.

The algorithm discussed above is not restricted to using the earliest due date dispatching rule, rather it describes a general framework for scheduling of design and analysis activities. The algorithm makes decisions at any moment in time, based on the available information. The main emphasis of the framework is on pulling the design information and scheduling as many activities as possible based on the precedence and resource constraints.

4. Numerical Example

In this section, an example illustrating application of the algorithm presented in section 3.4 to solve scheduling problem, is presented. The design process has been divided in different phases A, B, C, and so on. Suppose that, the ongoing design phase is D and the set of schedulable activities consists of 5 activities (see Figure 5). The design process under consideration consumes 4 types of resources.

The four types of resources are as follows:
1. Design engineers
2. Analysts
3. Testing engineers
4. Computing facilities

The status of these four resources is indicated by a status vector. Suppose that the resource status vector at time t is (5   5   3   0.9), where the first three types of resources are indicated by the number of available personnel and the fourth resource is indicated by the fraction available (90%).

Table 2. Activities with their resource requirements, priorities, and estimated duration

Activity	Resource Requirements	Weights	Estimated Duration	Latest Finish Time	Latest Start Time
5	(2 1 1 0.15)	10	2	11	9
7	(2 3 1 0.2)	20	7	13	6
8	(3 3 3 0.5)	10	3	14	11
10	(4 5 3 0.25)	30	6	19	13
11	(1 1 2 0.4)	10	3	16	13

The LF time of activities can be calculated by using the activity duration estimate, due date of the design phase D and the above relationships. The LF time of activities is indexed by setting the phase D beginning time to zero. At time t, the set of schedulable activities consists of activities, {5, 7, 8, 10, 11}. The LF, LS and weights of the schedulable design activities along with the resource requirements for individual activity are shown in Table 2.

The main intention is to schedule as many activities concurrently, as possible. The different subsets s^k of set S are determined by finding different combinations of activities for which the resource constraint is not violated. Then the value of the objective function (weighted lateness) for each of these subsets is determined.

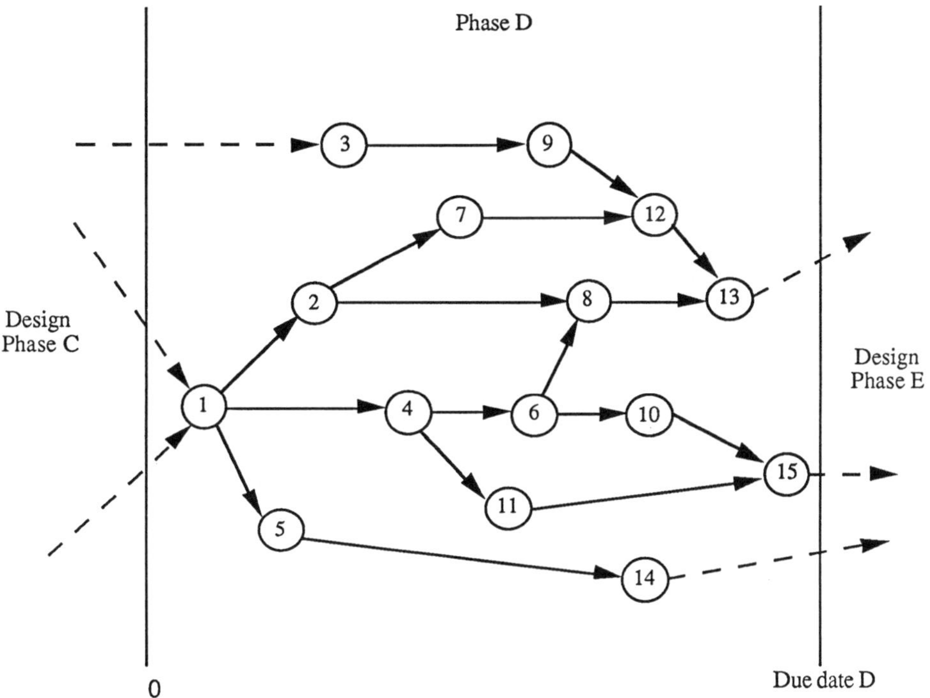

Figure 5. Phase D in the design activity network

Step 0
S = {5, 7, 8, 10, 11}
Resource status vector at time t = (5   5   3   0.9)
Step 1
s^1 = {5, 7, 11}, where the vector of resources utilized for s^1 = (3   5   2   0.75)
s^2 = {10}, where the vector of resources utilized for s^2 = (4   5   3   0.25)
s^3 = {5, 8}, where the vector of resources utilized for s^3 = (5   4   4   0.65)

Iteration 1
Step 2
t* = t + min{d(5), d(7), d(11)} = 5 + 2 = 7
Step 3
Z = 10 max{(7 - 11), 0} + 30 max{(7-17), 0} = 0
Step 4
Since all subsets are not checked, perform next iteration.

Iteration 2
Step 2
$t^* = t + 6 = 11$
Step 3
$Z = 10 \max\{(11\text{-}9), 0\} + 20 \max\{(11\text{-}7), 0\} + 10 \max\{(11\text{-}11), 0\} + 10 \max\{(11\text{-}13), 0\}$
$\quad = 100$
Step 4
Since all subsets are not checked, perform next iteration.

Table 3. The subsets of the set of schedulable activities and their weighted lateness

s^k	Total Resource Utilized	Z
$s^1 = \{5, 7, 11\}$	(5 5 4 0.75)	0
$s^2 = \{10\}$	(4 5 3 0.25)	100
$s^3 = \{5, 8\}$	(5 4 4 0.65)	20

Iteration 3
Step 2
$t^* = t + \min\{d(5), d(8)\} = 7$
Step 3
$Z = 20 \max\{(7 \text{ - } 6), \ 0\} + 30 \max\{(7 \text{ - } 13), 0\} + 10 \max\{(7 \text{ - } 13), 0\} = 20$

Step 4
All possible subsets s^k and the corresponding value of the objective function are tabulated in Table 3.

Step 5
The subset of schedulable activities with minimum weighted lateness is chosen. The subset of activities with minimum lateness is found to be $\{5, 7, 11\}$. Hence, the activities 5, 7 and 11 are scheduled simultaneously.

5.   Conclusion

In this paper a model for implementing a pull system for design activities was presented. The problem of scheduling design activities is a multiple resource constrained scheduling problem. The due dates of individual design phases are treated as milestones and by using the estimates of duration of design activities the latest finish time and latest start time of activities are determined. Since the design activities are unpredictable in their nature, a dynamic scheduling approach was selected. The developed algorithm introduces a general framework for scheduling design activities by using different dispatching rules. The *a priori* knowledge of the better estimates of duration of design activities can improve the quality of the solution generated by the algorithm.

References

Baker, K.R., 1984, "Sequencing Rules and Due-date Assignments in a Job Shop", *Management Science*, Vol. 30. No. 9, pp. 1093-1104.

Badie, C., Bel, G., Bensana, E., and Verfaillie, G., 1990, "Operations Research and Artificial Intelligence Cooperation to Solve Scheduling Problems: The OPAL and OSCAR Systems", *First International Conference on Expert Planning Systems*, June 1990, Brighton, UK.

Badiru, A, 1988, *Project management in manufacturing and high technology operations*, Wiley, New York.

Baker, K.R., 1974, *Introduction to Sequencing and Scheduling*, Wiley, New York.

Bel, G., Bensana, E., and Dubois, D., 1988, "OPAL: A Multi-Knowledge Based System for Industrial Job-Shop Scheduling", *International Journal of Production Research*, Vol. 26, No. 5.

Bell, C.E., and Park K., 1990, "Solving Resource-Constrained Project Scheduling Problems by A* Search", *Naval Research Logistics*, Vol. 37, pp. 61-84.

Christofides, N., Alvarez-Valdes, R., and Tamarit J.M., 1987, "Project Scheduling with Resource Constraints: A Branch and Bound Approach", *European Journal of Operations Research*, Vol. 29, pp. 262-273.

Davis, E.W., and Patterson J.H., 1975, "A Comparison of Heuristic and Optimum Solutions in Resource-Constrained Project Scheduling", *Management Science*, Vol. 21, No. 8, pp. 944-955.

Demeulemeester, E., and Herroelen, W., 1992, "A Branch-and-Bound Procedure for the Multiple Resource-Constrained Project Scheduling Problem", *Management Science*, Vol. 38, No. 12, pp. 1803-1818.

Dodin, B., 1985, "Approximating the Probability Distribution Functions in Stochastic Networks", *Computers in Operations Research,* Vol. 12, pp. 769-781.

Elmaghraby, S., 1977, *Activity Networks*, Wiley, New York.

Gorenstein, S., 1972, "An Algorithm for Project Sequencing with Resource Constraints", *Operations Research*, Vol. 20, No. 4, pp. 835-850.

Graham, R.L., Lawler, E.L., Lenstra, J.K., and Rinnoy Kan, A.H.G., 1979, "Optimization and Approximation in Deterministic Sequencing and Scheduling: A Survey", *Annals of Discrete Mathematics*, Vol. 5, pp. 287-326.

Graves, S.C., 1981, "A Review of Production Scheduling", *Operations Research*, Vol. 29, No. 4, pp. 646-675.

Hernandez, A., 1989, *Just-in-Time Manufacturing: A Practical Approach*, Prentice Hall, Englewood Cliffs, New Jersey, pp. 1-71.

Karp, R., 1972, "Reducibility among Combinatorial Problems, in Complexity of Computer Computations", R.E. Miller and J.W. Thatcher (Ed.), Plenum Press, New York, pp. 85-103.

Kerzner, H., 1989, *Project Management : A Systems Approach to Planning, Scheduling and Controlling*, Van Nostrand Reinhold, New York, pp. 575-653.

Kurtulus, I., and Davis, E.W., 1982, "Multi-Project Scheduling: Categorization of Heuristic Rule Performance", *Management Science*, Vol. 28, No. 2, pp. 161-172.

Kurtulus, I., and Narula, S.C., 1985, "Multi-Project Scheduling: Analysis of Project Performance", *IIE Transactions*, Vol. 17, No. 1, pp. 58-66.

Kusiak, A., 1990, *Intelligent Manufacturing Systems*, Prentice Hall, Englewood Cliffs, New Jersey.

Kusiak, A. (1991) "Concurrent Engineering: Design of Assemblies for Schedulability," C.T. Leondes (Ed.), *Control and Dynamic Systems*, Vol. 47, Academic Press, San Diego, CA, pp. 173-219.
Mallick, D.N., and Kouvelis P., 1992, "Management of Product Design: A Strategic Approach", A. Kusiak (Ed.), *Intelligent Design and Manufacturing*, Wiley, New York, pp. 157-178.
Moder, J., and Phillips, C., 1970, *Project Management with CPM and PERT*, 2nd ed., Van Nostrand Reinhold, New York.
Monden, Y., 1983, *Toyota Production System*, Industrial Engineering and Management Press, Atlanta, pp. 13-28.
Ord, J., 1991, "A Simple Approximation to the Completion Time Distribution for a PERT Network", *Journal of Operational Research Society*, Vol. 42, No. 11, pp. 1011-1017.
Patterson, J.H., 1984, "A Comparison of Exact Approaches for Solving the Multiple Constrained Resource Project Scheduling Problem", *Management Science*, Vol. 30, No. 7, pp. 854-867.
Pinedo, M., 1983, "Stochastic Scheduling with Release Dates and Due Dates", *Operations Research*, Vol. 31, No. 3, pp. 559-572.
Schrage, L., 1970, "Solving Resource-Constrained Network Problems by Implicit Enumeration - Non Preemptive Case", *Operations Research*, Vol. 18, pp. 263-278.
Shin, K.G., and Zheng, Q., 1991, "Scheduling Job Operations in an Automatic Assembly Line", *IEEE Transactions on Robotics and Automation*, Vol. 7, No. 3, pp. 333-341.
Stinson, J.P., Davis, E.W., and Khumawala, B.M., 1978, "Multiple Resource-Constrained Scheduling Using Branch and Bound", *AIIE Transactions*, Vol. 10, No. 3, pp. 252-259.

A MODEL-BASED METHODOLOGY FOR MANAGEMENT OF CONCURRENT SIMULTANEOUS ENGINEERING

H.-J. Bullinger and F. Wagner
Fraunhofer-Institut für Arbeitswirtschaft und Organisation (IAO)
Nobelstr. 12
W - 7000 Stuttgart 80
Germany

ABSTRACT

A methodology for management of concurrent simultaneous engineering and other business processes based on dynamic and hierachical models is presented together with its application within the industrial research project AMIES.

1. Introduction

Industrial enterprises of tomorow will have to face harder conditions of market and competitors. Any potential for improvement has to be found and exploited. For advantages of costs and times in the domain of product development companies try to use modern methodologies like Concurrent Simultaneous Engineering (CSE). The CSE methodology seems to be one of the basic concepts for an increase in efficiency and effectiveness of product development. CSE as a concept for the design of development processes also influenced other business processes, all hierarchies of the enterprises and leads to changes in their management and organization. The CSE targets of

* parallelization

* integration

* standardization

of the main processes leads towards process oriented structures of organization. The sucessful decoupling of the classical triangle of time, costs and quality seems to be possible with the extension of the CSE to a holistic process orientation of the complete enterprise /1/. Lean production and management are additional positive aspects of this extented CSE philosophy.

S. Y. Nof (ed.), *Information and Collaboration Models of Integration*, 61–67.
© 1994 *Kluwer Academic Publishers. Printed in the Netherlands.*

A methodology based on models of business processes for the management of product development and other related business activities will be presented in this paper. The application of this methodology in the international R&D project AMIES for supplier integration in automotive industry is shown as a typical example.

2. Challenge

The main challenge in nowadays product development projects are the three main factors time, costs, quality, their interaction and the measures to reduce their coupling (Fig. 1).

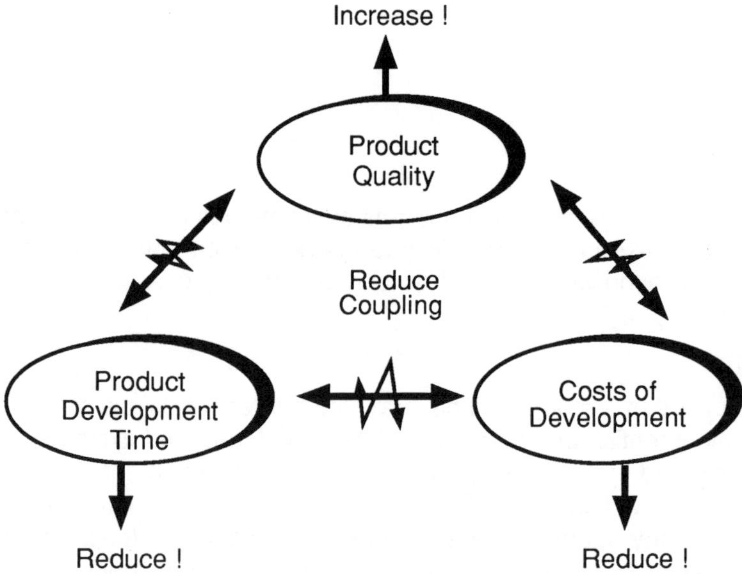

Figure 1: Triangle of time, costs and quality

One of these measures can be a transparent design of all the business processes and their costs inside the main development process. But how can this goal be achieved?
There is a list of questions:

- How to get a clear idea of the current process with all its complexity?

- How to document, analyse and evaluate it?

- How to evaluate alternatives, different scenarios and perform a necessary "what-if analysis"?

- Where and how to change the organization and technical architecture to the new development process?
- And how to support this task with a user-friendly software tool?

With the experience from various CSE projects /2/ and the background of modern simulation methods and systems /3/ a methodology was developed to solve these problems.

3. Methodology

The fundamental idea of the methodology is simple: *"Create a dynamic modell of the business processes and use it for improvements."* This development of the business process models must be supported by an easy understandable method and a comfortable computer-aided tool. In analogy to well-known production process models and their simulation /4/ these ideas were applied to engineering, product development and other business processes /5,6/. The steps of the methodology (Fig. 2) are similar:

1. Analysis, Modelling and Structuring of the Processes.
   With organigrams and available process or quality manual a rough analysis is made. A moderated interview with interactive "on-line" modelling the analysis team together creates the first model. Afterwards this current state model must be hierarchically structured and documented.

2. Grafical Animation and Validation.
   During a review together with the interview partners and supported by a grafical animation the modell is validated.

3. Simulation and Evaluation.
   The simulation and evaluation of the model shows quantitative figures like load of resources, delay and runthrough times and the performance of the investigated process.

4. Determination of Potentials for Optimization
   With the results from the simulation and mathematical evaluation potentials for improvements in time and costs can be detemined.

5. Development and Evaluation of Alternative Scenarios
   The possible improvements are used to build models of new and alternative scenarios and processes. From an evaluation of this alternatives the best one can be selected as the new target process.

6. Specification of the new Process Architecture
   Using this new process model the architecture of the development or other business process can be specified. Organizational, personal and technical measures and a functional description of new information technology (IT) systems are the results.

7. Organizational and Technical Implementation

When the measures and the new IT systems are implemented the new process model can be validated again. The model can be used for teaching and training and as an input to CASE tools.

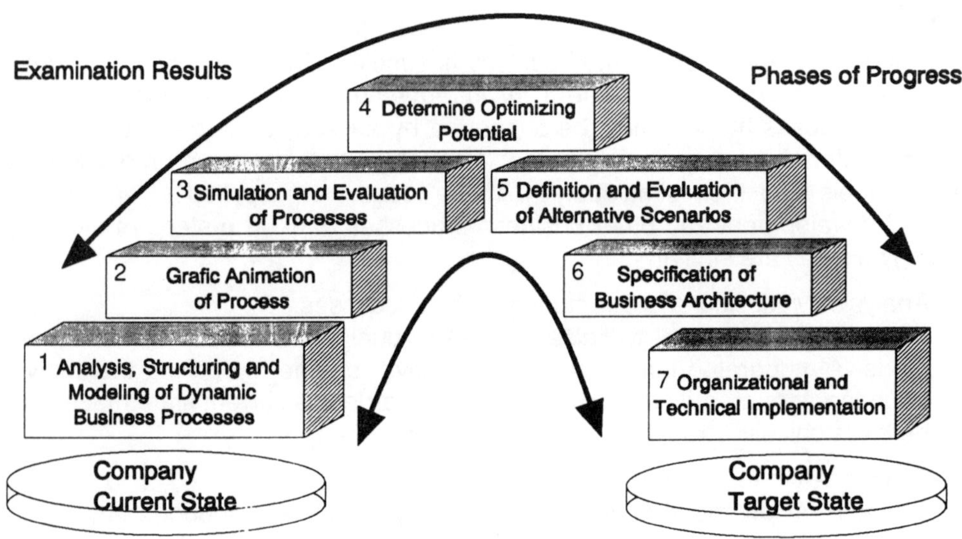

Figure 2: Steps of the Method

The methodology is currently using the Petri Net tool PACE /7/ as software platform. This interactive modelling and simulation tool is based on extented, timed and colored Petri Nets /8/ and embedded in the comfortable Smalltalk-80 environment. With this kind of Petri Nets a simple, flexible and very powerful formalism for the description of discrete, distributed and parallel systems is used. Since decades this kind of Petri Nets models are used for the decription and evaluation of hard- and software system. For better acceptance the graphical standard element of circles, squares and tokens can be substituted by user-defined individual icons. Fig. 3 shows a hardcopy of a typical application with some non-standard icons.

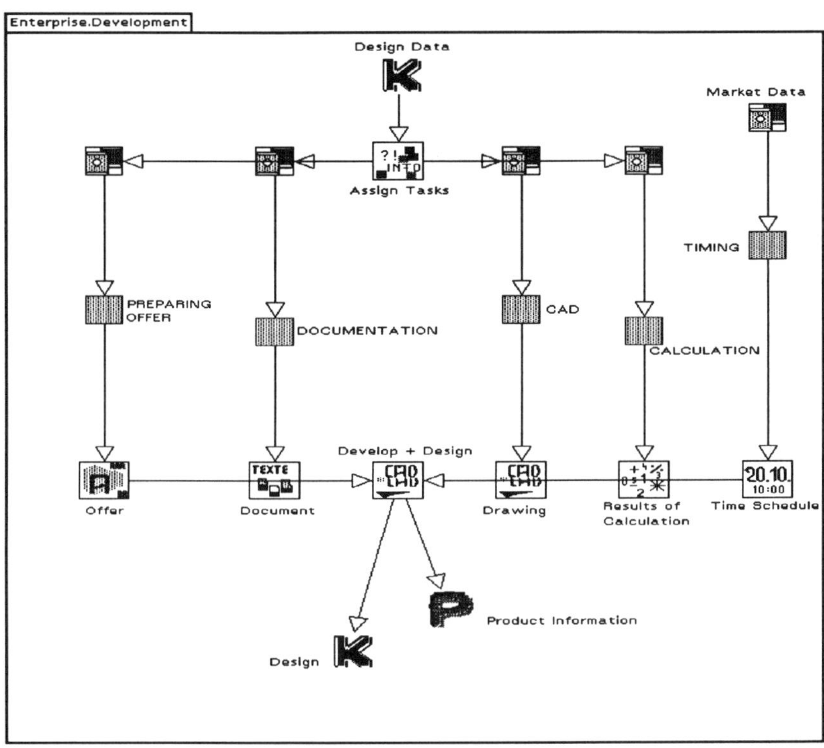

Figure 3: Hardcopy of the top level of a development process

4. AMIES Project

In the current 90's the European automotive manufacturers challenge is to reduce the overall development time of vehicles from five years to at least the japanese ratio of only three years. At the same time the estimated costs for the development of a new car from nowadays about 920 MECU have to be decreased together with an improvement in quality.

The Commission of the European Community (CEC) is supporting the project AMIES - **A**utomotive **M**ulti-company **I**ntegrated **E**ngineering **S**ystem - inside the Brite-EuRam 2 program. This AMIES project has the main objective to develop a methodology, guidelines and tools to support in the development of automotive parts in an heterogenous and complex multi-company context. The joint application of CSE and Total Quality Management and Information Management was named Integrated Engineering (IE) and will focus on the management level of the development process. The expected result of an Integrated Engineering System (IES) consisting of new methods, new kinds of organization, guidelines and computer-aided tools will be applied by three industrial partners. So the main goal of the AMIES project is to develop this IES for saving costs and time with increasing quality in car development. The consortium in this 30 man year project consists of six partners from three European countries:

- SEAT as a spanish car company and main coordinator of the project,

- INERGA Plasticos as a spanish plastic part producer,

- Constructiones Mechanica MARES as a spanish mould maker,

- SEMA Group as an European software company with their dependancies in France and Spain,

- LABEIN as a spanish and

- IAO as a german research institute.

An average passenger car consists of around 400 plastic parts with the instrument panel being one major bottleneck in the overall car development process. The specification and application of the IES will be based on the development process of a concrete instrument panel. For example, the current state analysis is following the development project of the instrument panel from the new SEAT IBIZA which will be released in May 1993.

The application of the model-based methodology in the AMIES project starts with first step of the project: analysis of the current situation. The development process of the instrument panel is modelled and analysed trough the three involved companies. The result is a business process modell of the development inside this companies including the interfaces and cooperation among them. This currently (April 1993) performed task is focused on aspects like project management, decision making and information logistics. Under the aspect of "Time to Market" special emphasis is given to the times the different activities and subprocesses consume.

5. Summary

In analogy to models used in production management a model-based methodology for the management of CSE, development and other business processes was presented. This methodology using dynamic and hierachical Petri Net models was developed during various industrial CSE projects and is now sucessfully applied in an international multi-company development project.

6. Acknowledgements

We want to thank Dr. J. Warschat who gave the first ideas for this methodology and H. Mutzke who has developed it further. We thank all partners of the AMIES project and also the CEC for supporting it under contract Nr. BRE2 CT92-0194. Special thanks to Dr. R. E. Schöplin from GELAG who is supporting us with the PACE software.

7. Bibliography

/1/   Bullinger, H.-J.: Triaden-Management - Produktentwicklung im Blick-winkel veränderter Strukturen. In: Bullinger, H.-J. (Ed.): 4. F&E-Forum Offensivstrategien für die Produktentwicklung. gfmt-Gesellschaft für Management und Techologie-Verlags KG. München. 1992.

/2/   Warschat, J.: Just in Time - Product Management. In: Proceedings of the International Conference on Computer Integrated Manufacturing (ICCIM). Singapore. 1991.

/3/   Zeigler, B. P.: Multifacetted Modelling and Discrete Event Simulation. Academic Press. London o.a. 1984.

/4/   Salvendy, G. (Ed.): Handbook of Industrial Engineering (2nd ed.). John Wiley & Sons. New York a. o. 1992.

/5/   Mutzke, H.: Beurteilung einer Methode sowie eines Werkzeuges zur Modellierung und Analyse von Geschäftsprozessen. Diploma Thesis. Fachhochschule Pforzheim. 1993.

/6/   Wagner, F. and Fischer U.: Methoden und Werkzeuge zur Analyse und Auslegung objektorientierter Systeme. In: Bullinger, H.-J. (Ed.): IAO-Forum Objektorientierte Informationssysteme II. Springer. Berlin a. o. 1992.

/7/   PACE User's Manual. GPP Gesellschaft für Prozeßrechnerprogram-mierung mbH. Oberhaching bei München. 1990.

/8/   Brauer, W., Reisig, W. and Rozenberg, G. : Petri Nets: Central Models and their Properties. Advances in Petri Nets 1986, Part I. Springer. Berlin a.o. 1987.

II. Integration in Computer Integrated Manufacturing/Enterprise (CIM/E)

LIFE-CYCLE SUPPORT OF A NEW GENERATION OF OPEN AND HIGHLY CONFIGURABLE MANUFACTURING CONTROL SYSTEMS

R H Weston, M Leech, P Clements and A Hodgson
Department of Manufacturing Engineering
Loughborough University of Technology
Loughborough, Leicestershire
LE11 3TU, England

ABSTRACT. The need for a new generation of highly configurable, extendable and integratable manufacturing control systems is identified and described. Current generation software products have a number of shortcomings which severely constrain solutions and prevent them from meeting certain of the requirements identified. Based on a description of those constraints the paper explains the rationalle for a new integration methodology which can support key aspects of the life-cycle of both new and old generations of manufacturing control system. Also included is a brief description of a case study use of the methodology.

1. The Need for a New Generation of Manufacturing Control Systems

Contemporary manufacturers are increasingly forced to seek a market edge by delivering better products than their competitors or by supplying them more quickly or cheaply. Obtaining such an edge inevitably implies the need to resolve conflicting requirements. For example, it is typically necessary to minimise inventory levels and achieve improved resource utilisation whilst remaining highly responsive to consumer demands. To further complicate matters often, even within a single manufacturing organisation, there will be a need to supply a spectrum of product types each subject to its own market requirement. Thus it may be highly advantageous to manufacture products within batches (of discrete parts) across the full manufacturing continuum illustrated by Figure 1, which can be considered to range from 'Project Manufacturing' to 'Repetitive Manufacturing' [Dinitz 90]. Clearly, the environment required for each class of manufacture (be it 'engineer to order', 'make to order', 'assemble to order' or 'make to stock') will require its own distinct set of operating characteristics.

This paper considers ways in which recent advances in the understanding of manufacturing methods can be implemented and supported, through the use of a new generation of computational tools which can underpin the variety of requirements facing contemporary manufacturers. The need is highlighted for a new generation of manufacturing control systems (MCS) which, for a given manufacturing system, can establish and control a set of operating conditions suited to the type (or types) of manufacture involved. As the title MCS would imply, focus will be on the manufacturing cycle (ie, making products which have previously been designed), but means of facilitating the integrated operation of various forms of MCS with business, product design and logistic systems will also be considered.

S. Y. Nof (ed.), Information and Collaboration Models of Integration, 71–88.
© 1994 *Kluwer Academic Publishers. Printed in the Netherlands.*

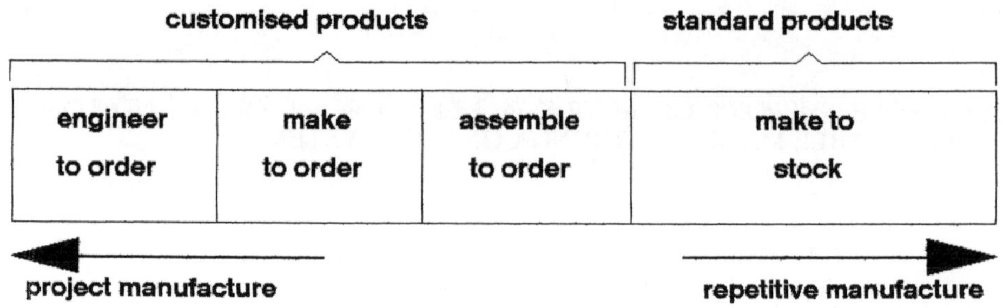

Figure 1. The Discrete-Parts Manufacturing Continuum (after Dinitz 90)

Significant activity world-wide has advanced a variety of manufacturing methodologies [Waterlow and Hodgson 92] each with their relative strengths. For example, that of materials requirements planning (MRP) lies in mid to long term planning, whereas Just In Time (JIT) and Optimal Technology (OPT) focus on the execution of short term plans. Often the highly complex nature of requirements in even a single manufacturing organisation cannot closely be met by adopting just one of these methods, leading optimally to the need to adopt some combination of them. Any implementation of chosen manufacturing methodologies must also recognise the need to positively enable change, arising from changing market forces, environment factors and technological advances (which themselves can give rise to shifts in the balance between manual and automated operations). Therefore there is a need for more flexible and configurable forms of MCS which can be easily tailored to individual sets of manufacturing requirements. There is also a growing need for more integratable forms of MCS, in the sense that they can more effectively realise business requirements by interoperating synergistically with other enterprise components, which themselves are also likely to be a mixture of computer and people systems. For example, next generation forms of MCS must interoperate across any de facto compartmentalisation of business, engineering and production activities. In addition, with the need for more timely, accurate and integrated decision-making there emerges a secondary set of requirements related to matters of 'concurrency' and 'information explosion'. A need for concurrency arises out of a requirement to be more responsive, which itself leads to the need to decompose what conventionally are homogeneous, centralised MCS processes into a number of distributed ones, which need to operate in a co-ordinated manner.

2.   Limitations Of Contemporary Manufacturing Control Systems

The builders of contemporary forms of MCS can utilise a raft of stand-alone software packages, each designed to address some focussed aspect of the MCS problem domain. Almost invariably, each software product will maintain its own computer information system (i.e a database or data files) and rely heavily on operator interaction via terminals to facilitate decision making and the input and output of production information.

Computer aided production management (CAPM) software will very frequently form the hub of current generation MCS, thereby providing decision-support capabilities which enable the planning of manufacturing operations. Contemporary proprietary CAPM software will typically comprise one of a wide range of MRP II software packages, or possibly some form of finite capacity scheduling package, where the package chosen maintains its own integral information model. This model will, in some way, describe shop-floor production, e.g. in terms of operations, bills-of-material (BOMs), stock levels and works-order processing. Appropriate choice of a CAPM package will depend upon the class (or classes) of manufacture involved and the degree with which its underlying information model reflects the realities of the particular set of shop floor operations it is required to mimic.

Other common current generation MCS proprietary software can take the form of computer systems to facilitate: computer aided process planning (CAPP), shop floor data collection (SFDC), cell control, financial planning and cost estimating, time and attendance recording and reporting, etc. Each of these types of software package can be used in conjunction with CAPM software to form a more functionally comprehensive MCS. In this way it is possible to achieve better (more informed) and quicker (more responsive) decision making. However, in practice, major problems will arise when attempting to integrate the operation of contemporary MCS software building blocks. Indeed, invariably the building blocks will have been designed in a way which (i) mitigates against achieving any high degree of interoperability between any chosen set of software components, and (ii) will lead to very high MCS system design and build costs, both in terms of the manpower resources and long installation times involved.

Many of these problems stem from a distinct lack of common purpose, and hence homogeneity, relating potential MCS software building blocks. Each block will have been produced to reflect the designer's view of a particular aspect of the MCS problem space, which is highly complex and unfortunately in its entirety is not yet well understood. The heterogeneity of the building blocks will arise in a number of respects which include differences in the computer hardware used, the underlying operating system, the software language used to write the software, the human interface systems provided, the computer interface capabilities provided (e.g. network and messaging protocols supported), the mechanisms used for storing and accessing information (e.g. file-stores relational databases and knowledge bases), the models used to represent the functional behaviour of a plant and the models used to represent information flows and their relationships.

Over the last decade significant progress has been made towards improving the 'hardware portability' of MCS software building blocks, where software vendors have sought to adopt the use of de jure and de facto computer network, operating system, database, fourth generation language and graphical user interface standards. Essentially this trend is enabling the functionality implemented, by any particular piece of application software, to be separated from (i) the computer hardware on which it is run and (ii) the data on which it operates. The result is that certain problems associated with installing, using and changing individual MCS software building blocks can be alleviated. In seeking much greater enhancements in the interoperability between chosen MCS software building blocks, such trends also provide a basis on which to build. However, as yet, available solutions do not meet certain of the core sources of heterogeneity (previously listed) which arise out of a lack of understanding of collective MCS needs and hence an absence of common function and information models which can describe generally applicable forms of MCS.

3.　　MCS Software Interoperation:　Contemporary Practice

Conventionally, interoperation between MCS software building blocks is achieved on a pair-wise basis. This involves the establishment of a one-to-one interface between each interoperating pair of MCS components. As depicted by Figure 2, in general interoperation will involve the need to connect (i.e. communicate electronically between) remotely located computer systems, to enable MCS components to functionally interact (according to an agreed set of rules) and to enable the MCS components to share information of common interest. Using this approach two MCS components (such as a CAPP system and a MRP II package) could interoperate to some degree by establishing a 'bespoke interface' between them. This would typically involve adding additional computer code to one or both components and providing a special 'driver' for either or both components, essentially thereby connecting the two software packages so that they can interact and share information. This pair-wise approach to systems integration results in the incorporation of knowledge (concerning the need to interoperate) into individual application software and its associated drivers. This will include knowledge of other application software, data sources, data access mechanisms, communication protocol, communication channels, data formats, and data structures [Lim 92 and Kaul et al 89].

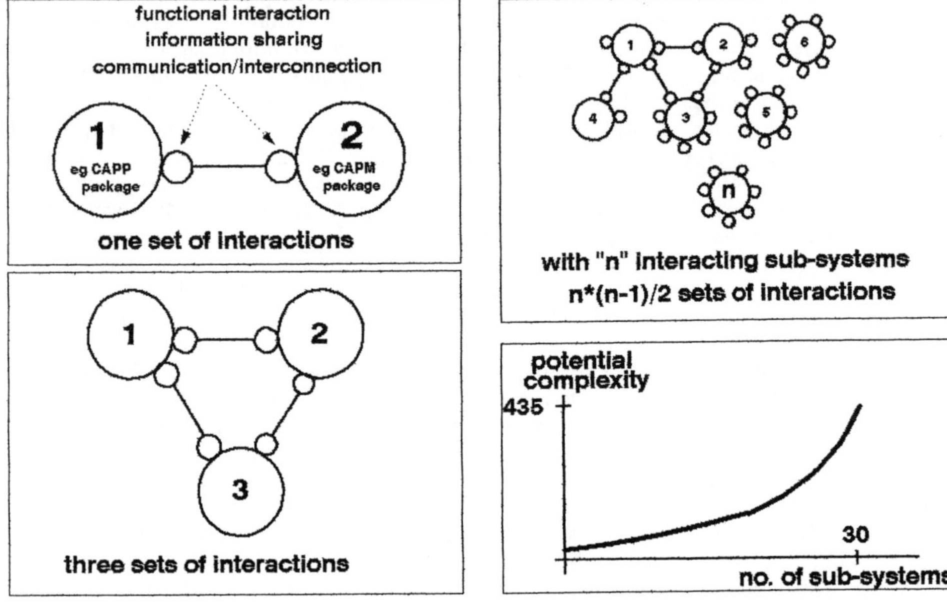

Figure 2. 'Pairwise' Integration - leading to an exponential growth in complexity

This inbuilt interaction knowledge can be the cause of very significant integration complexity problems and will inevitably result in solutions which can be classified as 'hard integration', inasmuch that the cost of subsequent modification and change may be so great as to render the solution obsolete as soon as requirements change significantly. Also, using pair-wise integration approaches, this situation is greatly compounded when systems are implemented which include a significant number of interoperating MCS components, where the complexity may well grow non-linearly in proportion to the number of interfaces (i.e. N(N-1)/2). This indicates a square law relationship between CIM system scope (in terms of the number, N, of MCS components) and potential integration complexity. In assessing the importance of this observation the reader should not assume that N will typically be a small number. As mentioned previously the trend towards breaking down MCS function blocks into a number of smaller concurrently operating units will be fuelled by both a technology push (via increases in available computer power) and market pull (the need for better and faster forms of MCS). For example a large number of decision-support tools could be required to structure and co-ordinate the work of people, operating in a number of roles and located in various parts of a manufacturing enterprise.

The costs and difficulties associated with the installation, maintenance and upgrade of software in a modern manufacturing enterprise are now sufficient to represent a major limitation on a company's capability to improve its systems. Current approaches to software interoperability and information sharing, based on the use of crude bespoke pair-wise interfaces compound such difficulties. Those approaches are also time-consuming and unreliable in operation, often requiring the down-loading of data in batches and, after processing, uploading of the resultant data. The approach can also result in multiple, inconsistent versions of data. In addition, the replacement of one item of software may result in a requirement to recreate all bespoke interfaces.

4.   A Methodology for Creating Next Generation Forms of MCS

This section reports on a new approach to creating MCS which has emerged out of a major, UK ACME/SERC funded, research project carried out by the Systems Integration (SI) Research Group at Loughborough University of Technology (LUT). The research had derived a methodology which can be used to guide and support key processes associated with the design, build and operational life-cycle phases of an MCS. The SI group has also produced a family of 'integration enablers', which underpin the methodology by automating what they perceive to be the most important of the build and change processes. The methodology so derived has already been used to produce wide-scope, highly functional forms of MCS which are more readily modifiable, extendible and capable of incorporating emerging manufacturing methods and IT standards than predecessor contemporary MCS solutions.

4.1   BACKGROUND RATIONALE TO THE MCS METHODOLOGY

The MCS methodology has been designed to meet the following requirements:

(a)   A capability to integrate contemporary or 'as is' MCS software systems (a pre-requisite of any industrially acceptable approach), i.e. existing investment in an installed base of legacy software components needs to be protected. This presents problems relating to:
- heterogeneity of the computer hardware operating systems, communication protocols, information representations and storage mechanisms;
- data fragmentation across a variety of data repositories;
- lack of knowledge of the underlying implicit functional and information models.

(b)   A capability to integrate the operation of a new generation of more open (or 'to be') MCS components which may be decomposed into more atomic functional units and to consider their:
- functional properties;
- information and interaction requirements;
- level of granularity.

(c)   Means of formally structuring and supporting design, build and change processes associated with MCS, in a manner which supports (a) and (b) and enables:
- improvements in system synergy in a manner which can more effectively match business requirements;
- significant reductions in engineering effort and leadtimes.

(d)   Means of separating out and decomposing MCS integration problems from various perspectives into more readily soluble parts. At the macro level there was considered to be a need to:
- target the operation of integrated MCS on an 'open systems' integrating infrastructure.
- separate the functional and information architectures during much of the life cycle support;
- break down the life cycle support into three phases - design, build and runtime, with change straddling all three.

4.2   OVERVIEW OF THE METHODOLOGY

The essential elements of the methodology are depicted in Figure 3. It incorporates five distinct but inter-related stages, each of which is described below:

(1)   Classification and specification of MCS: Based on the use of a set of generic reference models and user-defined requirements. Outputs consist of a particular MCS functional model and information model.

(2)   Generation of a functional architecture: A set of tools is invoked to assist in the creation of the functional architecture which is consistent with the output of (1).

(3)   Generation of an information architecture: The information structure is generated in a form which is consistent with the information model generated by (1).

(4)   Use of an integrating infrastructure: This facilitates MCS configuration in a flexible data-driven manner, where the integrating infrastructure is used to separate out specifics (i.e. realities associated with implementation details) of a particular MCS from more abstract (and hence generalised) models of MCS used during design and build processes. Thus the integrating infrastructure is charged with resolving differences in the physical system relating to heterogeneity, distribution and data fragmentation. Indeed the use of the integrating infrastructure is key to the methodology. As the integrating infrastructure assumes responsibility for maintaining knowledge of integration details (such as networks used, the hardware and operating systems on which an

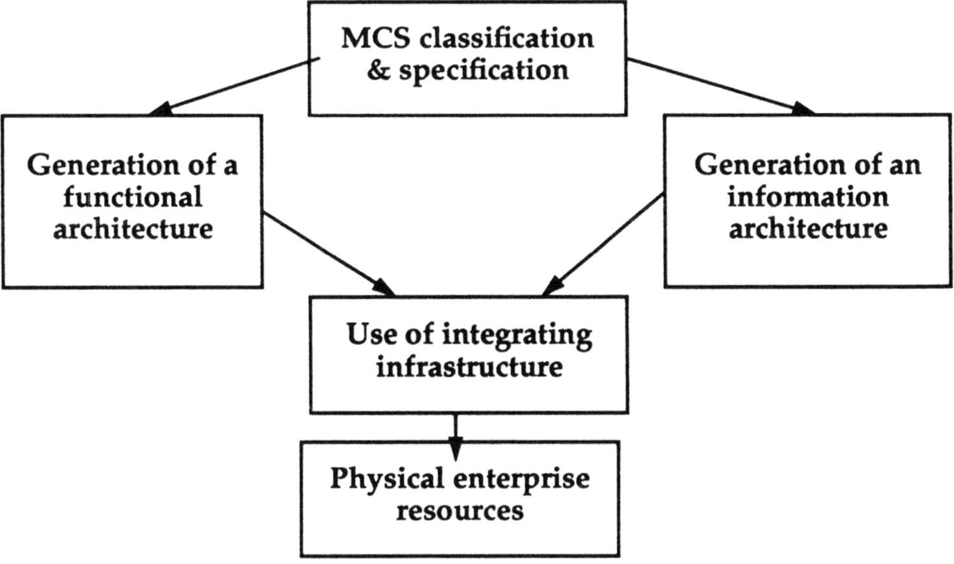

Figure 3. Overview of the MCS Methodology

MCS component is run, the location of an information fragment, etc, etc) the MCS components themselves need only have knowledge of how to use the integrating infrastructure (i.e. NOT OF EACH OTHER). Essentially this leads to a linear relationship between system scope (in terms of the number N of MCS components) and complexity, as depicted by Figure 4.

(5)   Physical enterprise resources: An essential element of this methodology is the ability to flexibly map 'as is' and 'to be' software applications onto system resources, i.e. databases and computer hardware.

4.3   IMPLEMENTATION OF THE METHODOLOGY

The creation of integration enablers to underpin the use of the methodology is an ongoing process. Significant progress has already been made with implementation work, focussed to-date on prime areas of need, identified in conjunction with UK MCS software vendors and MCS users who collaborate with the SI Group. The prime areas of advance so far have led to new ways of:

· dealing with legacy systems
· enabling build and runtime processes, particularly with regard to information issues

Significant effort has, nevertheless, been directed towards each component part of the methodology.

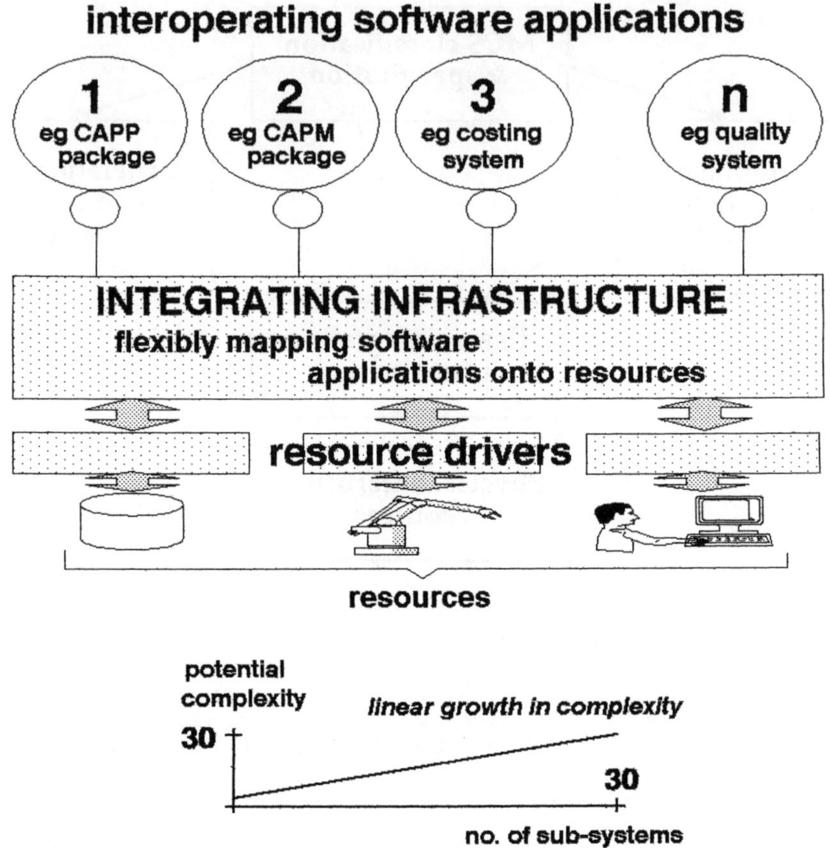

Figure 4. Use of an Integrating Infrastructure - copying with complexity

4.3.1 *Classification and Specification of MCS Requirements.* A number of alternative approaches to MCS classification and specification have been studied.

One set of approaches has investigated means of specifying specific purpose (or particular) MCS models, this based on general purpose structured design methods (such as Yourdon) [Hodgson et al 91] and manufacturing oriented modelling methods (such as IDEF and CIM-OSA). The CIM-OSA work has been based on the use of an extended CIM-OSA CASE tool produced by the SI group [Aguiar and Weston 92]. The use of this tool has proved to be very encouraging particularly in regard to the creation of MCS behaviour models. These attach temporal characteristics to more static MCS function models via the use of a Petri-Net simulation tool.

In addition, complementary approaches have aimed to identify and characterise different classes of MCS, where an aim has been to draw on previous ACME CAPM research fundings to identify a set of generic MCS models which provide guidelines during MCS design processes. Indeed a free-standing, limited-capability prototype has been written. A selection of relevant company parameters is elicited from the user and, based on these, outline information structures are produced, compatible with the EXPRESS information modelling language. At present, this system has only a small number of `generic sector models'; however, the intention is to demonstrate the methodology, not to produce a substitute for the knowledge of experienced researchers and users in the MCS area.

4.3.2 *Generation of a Functional Architecture.* The functional architecture work has been conducted for two primary purposes: (1) to assess the feasibility of producing reference functional architectures for MCSs (which could guide high level design processes) and (2) to consider which set of modelling methods and tools could support such an activity. This work has led to the definition of MCS reference functional models with varying degrees of genericity and completeness. Subsequently functional modelling has been achieved using the extended CIM-OSA CASE tool.

It has become clear that it is also necessary to attach behavioural descriptions to the static function models, thereby assigning temporal characteristics (or dynamic properties) in terms of the synchronization and sequencing of functional blocks of MCSs (software application code). Prototype solutions have been built using Petri Net modelling tools; this has led to the modelling of SMT production lines and an ability to simulate shop and cell control functions before building such systems [Aguiar and Weston 92].

In parallel, the requirements relating to functional modelling have been identified by conducting bottom-up implementation studies. Here, two main threads of activity have identified (a) the nature of functional building blocks (and their 'interface' requirements) and (b) a set of 'functional services' which, at runtime, draw on the services of the integrating infrastructure (CIM-BIOSYS) to execute (i.e. sequence and synchronise) software applications in a data-driven manner [Singh and Weston 93]. The main feature of (a) is the generic and modular construction of the building blocks, which allows software re-useability; whilst (b) provides a stable interface via which to integrate the functional building blocks.

4.3.3 *Generation of an Information Architecture.* A number of integration enablers have been produced to support the creation of an information architecture as illustrated by Figure 5. The enablers are as follows:

(a)   EXPRESS-to-SQL Compiler: This allows the system integrator to define conceptually the global information model that is to be supported by one or more MCS components. The compiler has recently been enhanced to include new features defined by EXPRESS standards committees [Clements 92], in particular the ability to support multi-schema EXPRESS models has been implemented. This work has attracted much interest, and the software has been released to a number of UK Universities.

(b)   STEP Parser [Clements 92]: This has been created to enable population of physical databases with real data in the format specified by the EXPRESS model.

(c)   View Provider: This is a key area of innovation and provides the application writer with an ability to request data contained within the database without knowledge of how the data is actually stored within the database. This facility, used in conjunction with the CIM-BIOSYS integrating infrastructure, provides a

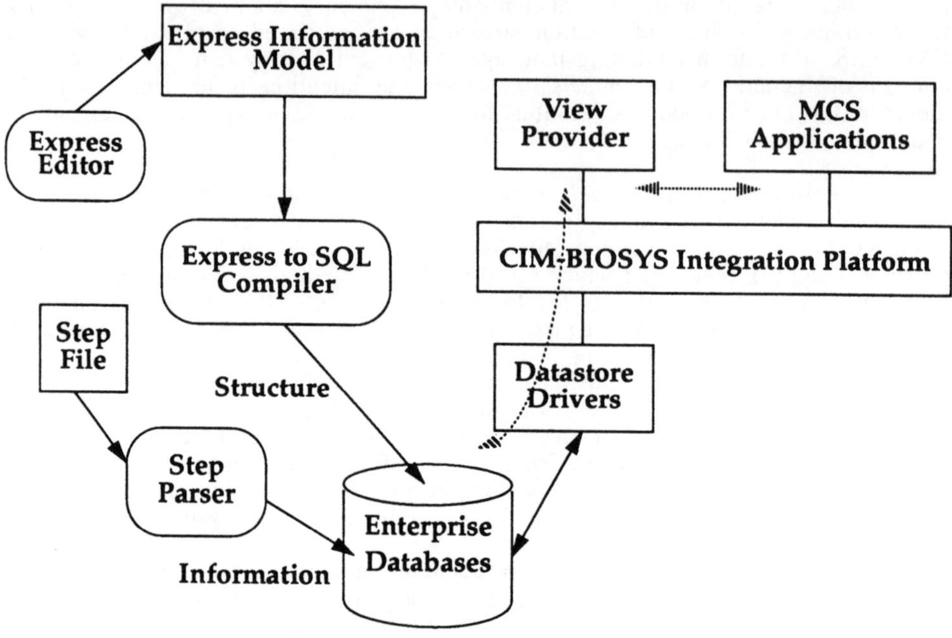

Figure 5. Information Architecture Toolset

very powerful means of resolving differences between the various heterogeneous and distributed information sources found in contemporary forms of MCS. Currently the view provider works over distributed homogeneous databases.

(d)   EXPRESS Editor: This is a tool to allow the dynamic editing of the global information model as specified by the EXPRESS model. This tool is necessary as the CIM-BIOSYS integration infrastructure needs to support changes in the information model during the lifetime of the system. This tool is still at an early stage of specification and implementation and its final version will be a graphical based editor running on OSF/Motif [Clements 92].

4.3.4 *Use of an Integrating Infrastructure.* The MCS methodology is targeted on the CIM-BIOSYS (CIM - Building Integrated Open Systems) integrating infrastructure which has been produced and successively enhanced by the SI group since 1988. Essentially this project has been a user of CIM-BIOSYS rather than a developer of it. However, the process of building demonstrator MCS systems on CIM-BIOSYS has identified the need for:

(a)   a set of complimentary integration enablers, which underpin the MCS function and information architectures

(b)   more formally structured application interfaces to CIM-BIOSYS

(c)   enhancements of the CIM-BIOSYS configuration and operator interface services.

4.3.5 *Manufacturing Enterprise Resources.*   Common manufacturing enterprise resources include sources of information (typically stored in databases and filestores of various kinds), which can be accessed by manufacturing software. In order to achieve access via an integration platform, a consistent set of services is required. The research of the SI Group has looked at this problem in two complimentary ways, as described below:

(a)   *Generic datastore drivers:* As it needs to function as a general purpose integration tool, the CIM-BIOSYS integration infrastructure has no built-in databases, hence a method of accessing external databases is required. Generic datastore drivers allow the platform to access information in proprietary databases in a consistent way. During the grant period, drivers have been created for three very commonly used proprietary databases, all of which provide a SQL [Clements 92] user interface; the databases are: Ingres, Progress, and Oracle. The user interfaces to these databases differ in various ways, but the driver deals with these problems and hides them from the platform. An application on the platform uses a simple set of information services to access information, and has no need to know which database, or even which driver, is required. For example, if a database is changed from say Ingres to Oracle all the applications will run without any modification, provided the new database holds the required information, the format of which irrelevant.

(b)   *Alien application shells*: These CIM-BIOSYS applications provide another way of controlling alien (or non-CIM-BIOSYS conformant) applications which often run above databases. One example of such an alien application is Manufacturing Control Code (MCC), which is a proprietary production scheduling package which runs over the Oracle database. Its information can be accessed via datastore drivers, but a means of controlling its functionality is required. A shell has been written to control Manufacturing Control Code, so that many of the important functions such as scheduling etc. can be performed and controlled via the CIM-BIOSYS platform. Other shells have been written to run 4GL applications on the Progress database, and also to a limited extent to control the proprietary MFG/Pro production planning and control package written in Progress 4GL.

The drivers and the shells created have proved valuable when building MCS demonstration systems. However, as with many experimental systems, there have been some problems and limitations; these have been primarily due to implementation difficulties associated with peculiarities of custom-built closed software, rather than inherent weaknesses in the MCS methodology.

5.   Case Study Use of the Methodology

The underlying concepts of the MCS methodology were derived through gaining practical experience of designing, building and changing a number of forms of MCS. Most of these systems were created as university-based proof-of-concept demonstrators, this also providing means of evaluating the use of successive versions of integration enablers.

One MCS demonstrator system recently produced is illustrated in Figures 6A and 6B. Essentially the creation of the demonstrator was achieved in three largely distinct phases, as follows. In phase 1 concentration was on a study of the problems of flexibly linking typical 'as is' MCS components, whereas in phase 2 a number of more open and easily integratable ('to be') MCS components were included. In phase 3 the functionality of the complete MCS was significantly extended to illustrate the power of the methodology in providing steps towards much wider-scope MCS.

5.1   DESCRIPTION OF FUNCTIONALITY

In phase 1 concentration was on establishing a degree of interoperation between a proprietary Finite Capacity Scheduling package (known as Manufacturing Control Code or MCC) and a cell controller, with common information stored in an Oracle relational database package [Welz and Singh and Weston 93]. This first phase system facilitated the distributed planning and control of manufacturing operations with scheduling, dispatching and monitoring achieved with different time horizons and spheres of influence.

In phase 2, 'to be' MCS components were included within the demonstrator system, the design of which was fully conformant with the methodology. In particular, CIM-BIOSYS conformant software applications were created to enable order entry, scheduling and cell control in a highly interoperable manner and where system reconfiguration and change could be easily achieved [Leech 92].

The phase 3 enhancement in system functionality centred on the inclusion of 'to be' cost modelling software which can be used to produce comprehensive costing information related to product manufacture using either form of cell controller (i.e. it can be used to establish costing data for either phase 1 or phase 2 sub-systems) [Shaharoun et al 92].

The detailed layout of the functional elements included in each phase are depicted by Figure 6B. The demonstration runs on a number of Sun workstations running the CIM-BIOSYS integration platform over UNIX. Three proprietary databases are used, namely Oracle, Ingres and Progress, each of which is controlled via a separate CIM-BIOSYS datastore resource driver.

5.2   DESCRIPTION OF THE MCS SOFTWARE APPLICATIONS

In phase 1, a generic software shell was written to encapsulate the functionality included within the MCC package and to provide an enhanced human interface to the system, so that it could be partially integrated. In addition a resource driver was created for the Oracle database to provide a programmable interface via the use of SQL access mechanisms. MCC can then provide the cell controller, via relatively open access to information from its Oracle database. This includes work orders, process routes and stock levels etc. In this scheme the cell emulates production processes and updates the database appropriately. This first phase system offers a basis for experimentation and alteration, providing opportunities to investigate methods of achieving hierarchical and distributed scheduling in a pseudo factory environment.

During phase 2, fully conformant applications were written to facilitate the entering of orders, scheduling these orders and to emulate the operation of a cell involved in various production processes. All file and database accesses are via the CIM-BIOSYS integration platform and strictly follow the SI Group methodology. The open software

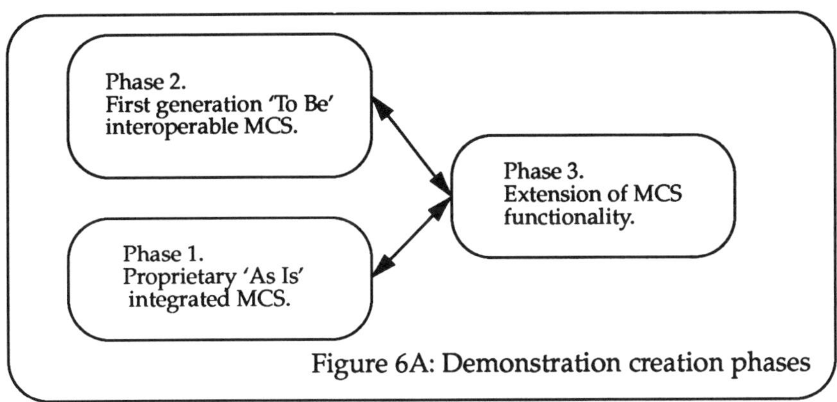

Figure 6A: Demonstration creation phases

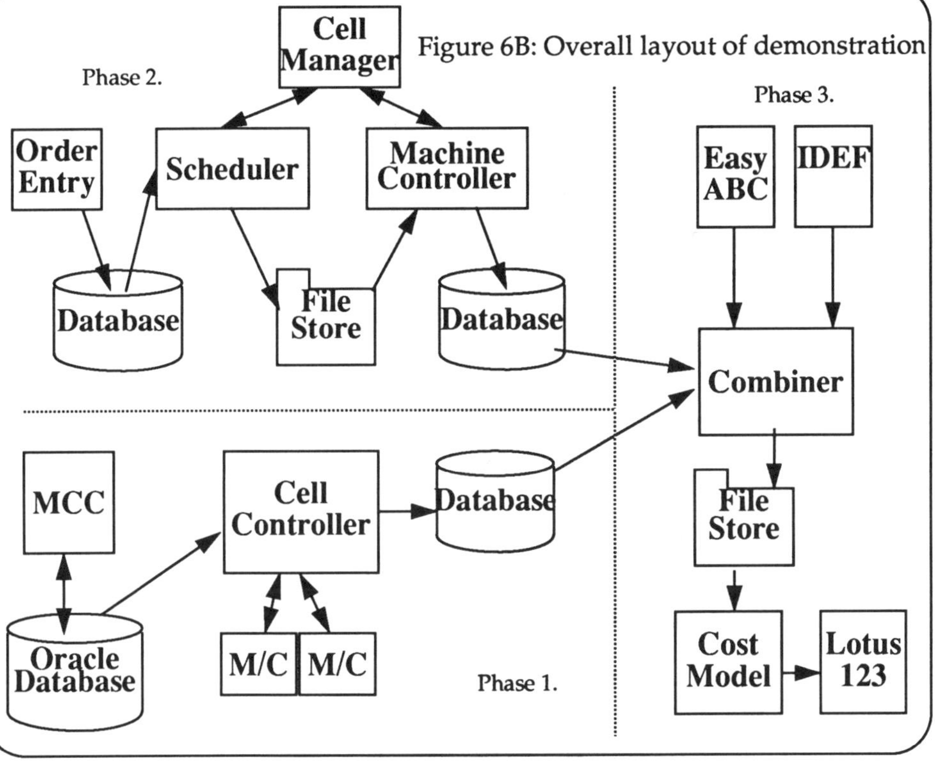

Figure 6B: Overall layout of demonstration

applications were coded up manually having defined their integration requirements using IDEF modelling tools.

Phase 3 has focussed on ABC cost modelling as a means of demonstrating the way in which the MCS methodology can support system extension in a graceful and flexible manner. Inputs can be from various sources as shown in Figure 6B. Information produced from the other two phases is used to provide planned and actual product costings using an overall cost model of a factory. Output can be transferred to Lotus 123 spreadsheet software for further processing and analysis which enables the production of graphs and pie charts.

5.3   IDENTIFICATION OF NEEDS

The results of designing, building and then running the demonstration system have led to the identification of a number of general requirements and needs that are essential to building flexibly integrated MCSs, some of which are listed below.

- An integration platform which has the ability to positively support change in a manufacturing system is essential.
- Phase one indicates that information models of proprietary systems such as MCC are required to support interoperation and to enable a move towards more open forms of MCS.
- The demonstration has highlighted the need for behaviour control mechanisms which allow greater flexibility when controlling application interaction [Singh and Weston 93].

6.   The MCS Methodology - Its Main Strengths and Weaknesses

6.1   MAIN STRENGTHS COMPARED TO CURRENT INDUSTRIAL PRACTICE

Compared with conventional industrial practice, LUT's MCS methodology has been shown to facilitate major improvements in regard to:

(a)   Enabling greater synergy between MCS components: this through more timely and improved information sharing and by flexibly structuring (the synchronisation and sequencing of) MCS operations.

(b)   Providing mechanisms for placing the operations of MCS systems within a strategic framework.

(c)   Allowing the granularity (i.e. effective operational domain) of individual MCS components to be significantly reduced, by enabling the decomposition and distribution of MCS functionality; this is a vitally important step towards higher functionality and more open MCS.

(d)   Significantly reducing the time and engineering effort involved in building and incrementally extending the functionality and components of integrated MCSs.

(e)   Providing mechanisms for enabling change in a variety of forms.

6.2   MAIN STRENGTHS COMPARED WITH CURRENT RESEARCH PRACTICE

Compared with other methodologies currently advanced by the international research community, the MCS methodology defined here has the following features and advantages. Figure 7 is included to clarify the points made.

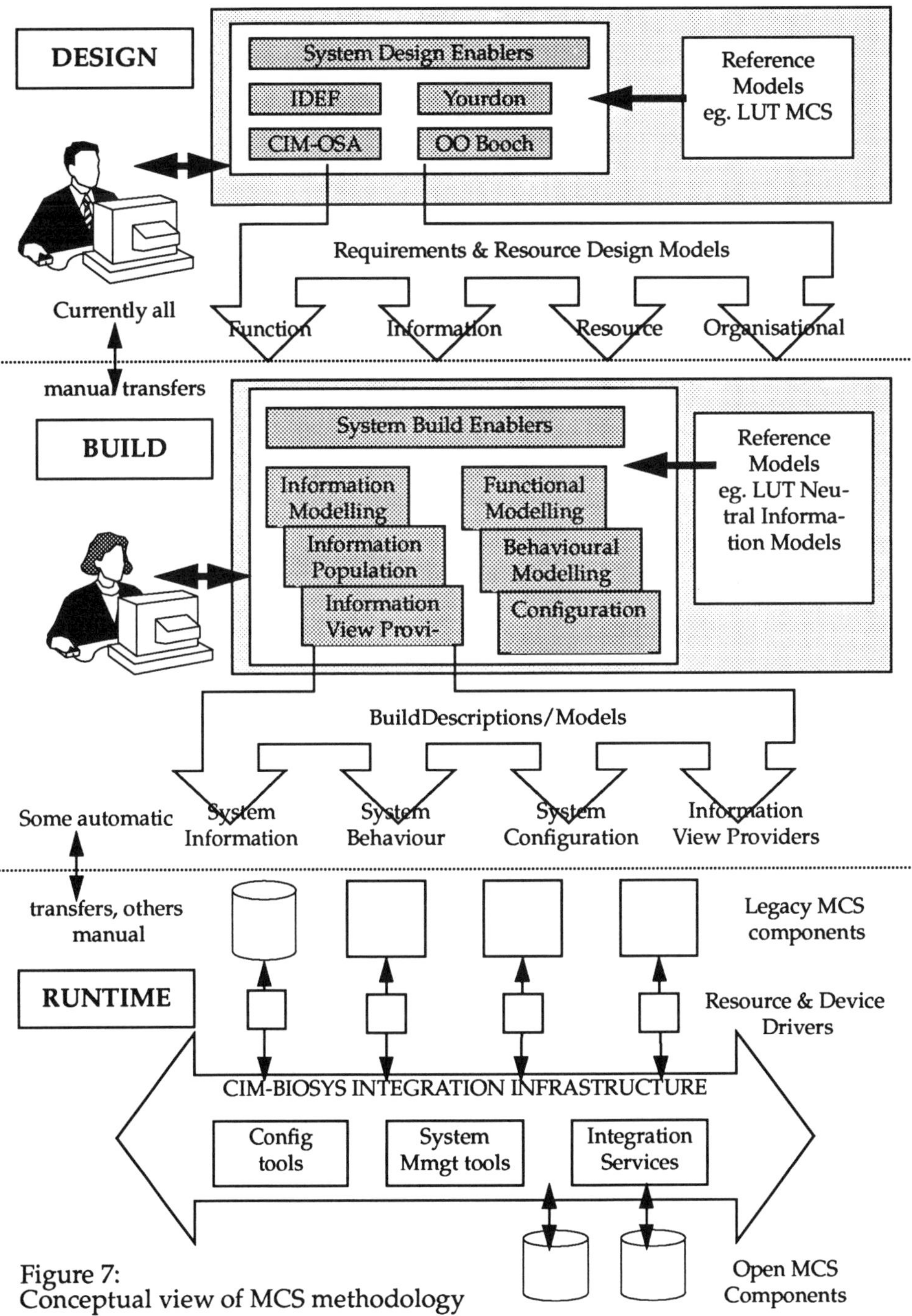

Figure 7:
Conceptual view of MCS methodology

(a)   The MCS methodology unifies and extends a number of frameworks, architectures, methodologies and standards and, as a result, has a wider life-cycle coverage than any single methodology known to the SI Group.

(b)   The MCS methodology encompasses means of bringing together 'as is' (legacy) MCS components with 'to be' (next generation) open MCS software. It is therefore a pragmatic approach which can be used now to enable MCS integration at 'brownfield sites', whilst offering a migration path towards next generation open software systems.

(c)   The MCS methodology does not over-prescriptively define:
·   particular structural relationships between MCS components,
·   the use of particular MCS components,
·   the use of particular MCS standards.

However, it does 'enable' the use of emerging standards and standard frameworks. Thus, potentially the MCS methodology is widely applicable, being capable of supporting various manufacturing control scenarios and facilitating implementation in a wide range of companies. The methodology provides a life-cycle framework into which MCS components can be plugged via fundamental (and hence general purpose) integration mechanisms. The methodology also allows users to define and adopt more prescriptive structural solutions, effectively via the use of reference models and the creation of appropriate system build tools. For example one member of the group (located at NIST) has investigated means of combining the NIST MSI architecture with CIM-BIOSYS integration enablers [Gilders 92]. SI Group members believe that their approach, of enabling standards and unifying their use rather than attempting to force their adoption, is the only way of achieving widely applicable practical solutions which do not over-constrain technological advance.

6.3   CURRENT WEAKNESSES AND CONSTRAINTS

Before considering current weaknesses and constraints, it is important to draw a distinction between the LUT MCS methodology and the current implementation of that methodology.

The MCS Methodology: Although the SI Group is confident that the adoption of the methodology can provide an important advance on current practice, it is much nearer 'completeness', in terms of formal definition, in the arenas of system build and runtime than in system design. This is partly due to the need to focus project activity so far, but it is also due to (i) the complexity of the MCS design problem, (ii) current limitations of publically available manufacturing modelling technology, and (iii) and the belief that more industrial benefit can be gained by focusing on MCS implementation and change processes. There is much more work to be done in the MCS design arena to facilitate the creation of 'executable models' and in formally describing the interfaces between MCS design and build processes. The work so far has in broad terms identified the nature of the design/build interface and it is anticipated that this will lead to more formal MCS descriptions over the next few years. A further important area in which research is required relates to the definition and adoption of various forms of MCS reference models which would ultimately allow MCS vendors to create more open

MCS components. Here, for example, derivatives of MCS structural definitions (such as the NIST MSI model) and proof-of-concept LUT MCS neutral information models could prove to be an excellent starting point.

The current implementation of the MCS methodology: Although it has already been possible to implement sufficient of the methodology to prove the feasibility of its underlying concepts, it has been necessary to constrain to some extent the resultant MCS integration enablers in terms of their scope, their portability and the various international, de jure and de facto standards they support. For example, it has proved necessary to target much of the software (constituting the MCS integration enablers) to run on Sun UNIX workstations and to restrict the range of database technology supported to that most commonly used industrially. This will place some immediate exploitation constraints on the methodology (e.g. in the form of alpha and beta site release of enablers), particularly for example to SMEs seeking pc-based solutions. However, the SI Group is confident that the work done so far can lead naturally to many new and repackaged existing enablers, which collectively support MCS design, build and runtime. As an example, stripped-down versions of the CIM-BIOSYS integrating infrastructure and associated enablers have already been successfully produced to run on Motorola hardware and OS9 operating systems and are being used to support the control of high speed printing machinery. In addition, ongoing LUT research is investigating the role within the MCS methodology of other emerging standards such as MANDATE [Clements 92].

Much of the evaluation work aimed at testing the suitability, completeness and overheads associated with the MCS methodology and its implementation, has focused around the build and test of MCS. However, a series of investigations has sought to test more formally the 'overhead' associated with the integration enablers. Major attention has been paid to the assessment of performance relating to runtime issues, but other studies have focused on the usability of the integration enablers.

The runtime performance testing has proved to be very encouraging. For practical forms of MCS, the overhead associated with the methodology (in terms of processing and access times) represents a small fraction of the overheads due to the execution of the application systems themselves.

As yet it is not practical to implement the whole methodology within a CASE tool. However, by the end of quarter 3 of 1994, key aspects of it will have been implemented using the Ipsys meta-CASE tool.

7.   Acknowledgements

The authors wish to fully acknowledge the support and encouragement of the ACME Directorate of SERC. They also fully acknowledge vitally important inputs to the conceptual thinking and application study work from Awalludin Shaharoun, Valdew Singh, Frank Welz, Marcos Aguiar, Binglu Zhang, Ian Coutts, Shaun Murgatroyd, Jack Gascoigne and John Edwards.

8.   References

Aguiar, M. W. and Weston, R. H. 1993. CIM-OSA and stochastic time petri nets for behavioural modelling and model handling in CIM systems design and building. To be published in *Proc. I. Mech. E.*, Part B, J. Eng. Manuf. in Aug 1993.

Clements, P. 1992. The Application of EXPRESS modelling and tools within an integration platform, *2nd EXPRESS users group*, Dallas, Oct. 1992.

Dinitz, M. 1990. Configure-To-Order: an industry challenge, *Industrial Engineer*, July 1990, pp 21-22.

Gilders, P. 1992. Application of a European Infrastructure to the NIST MSI Demonstrator Software, *Conf. on Flexible Automation and Management '92*, Virginia Tech., USA.

Hodgson, A., Ryan, A., Leech, M. J., Clements, P., Weston, R. H. 1991. Design and Implementation of Information-Driven MCS, *Int.Conf. on CIM*, Singapore, Sept 1991.

Kaul, M., Drosten, K., Newbold, E. J. 1989. View System: Integrating Heterogeneous Information Bases by Object-oriented views. *Proc. IEEE Int. Conf. on Data Engineering*. 1989.

Leech, M. 1992. Functional Decomposition, *LUT SI Group Internal Publication IR19*, March 1993.

Lim, B. S. 1992. CIMIDES-A Computer Integrated Manufacturing Information and Data Exchange System, *IJCIM*, Vol. 5, No. 4 & 5, pp 240-254.

Shaharoun, A. M., Hodgson, A. and Weston, R. H., 1992, Cost Modelling in Advance Manufacturing Systems, *Proc. of ICMA*, Hong Kong, 1992.

Singh, V. and Weston, R. H. 1993. New generation of 'open' manufacturing control systems for 'seamless' integration in CIM, *Proc. of ICCIM 92 Conf.*, Singapore, Sept 1993.

Waterlow, G. and Hodgson, A. 1992. Editors Special Feature on Computer Aided Production Management. *Special Issue of of Computer & Control Engineering Journal*. Vol. 3, No. 2. March 1992.

Welz, F. 1993. Software Interoperability with MCSs, Graduate Dissertation, LUT Press, March 1993.

A THEORETICAL MODEL TO PRESERVE FLEXIBILITY IN FLEXIBLE MANUFACTURING SYSTEMS

P. VALCKENAERS and H. VAN BRUSSEL
Katholieke Universiteit Leuven
Mechanical Engineering Department
Celestijnenlaan 300B
B-3001 Leuven
Belgium

ABSTRACT. The manuscript presents a formal model of flexibility for solutions of problems in general. Next, it elaborates a state transition model of the real world and discusses the relationship with the previous model. Three design principles follow from the above. Finally, examples in the domain of flexible manufacturing illustrate the theory.

1. Introduction

The importance of flexibility in manufacturing systems is growing continuously. Manufacturing organisations increasingly face unpredictable demands for customised and complex products without defects that they have to produce in small quantities, under time-based competition, and at low cost. Moreover, technological innovations and developments constantly create new opportunities in this competitive environment.

In sharp contrast to its importance, flexibility is still an ill-defined concept, both in general and within manufacturing systems. This failure to understand the precise nature of flexibility makes it difficult to identify flexibility, to assess or quantify its value within the manufacturing environment, and to preserve useful flexibility without 'overdoing' it. This paper describes a formal theory on flexibility that starts to remedy this situation.

2. Problem Statement

2.1. LIMITATIONS OF TOP-DOWN DEVELOPMENT

Automation and integration efforts typically follow the principles of traditional top-down and hierarchical development [AMRF]. This approach failed to scale up in the past, in spite of numerous attempts and considerable effort. The *watchmakers' parable,* from

S. Y. Nof (ed.), Information and Collaboration Models of Integration, 89–104.
© 1994 *Kluwer Academic Publishers. Printed in the Netherlands.*

90

the book entitled "The Sciences of the Artificial" by H. A. Simon, illustrates why this is not so astonishing. This parable goes as follows:

> *There once were two watchmakers, named Hora and Tempus, who made very fine watches. The phones in their workshops rang frequently; new customers were constantly calling them. However, Hora prospered while Tempus became poorer and poorer. In the end, Tempus lost his shop. What was the reason behind this?*
>
> *The watches consisted of about 1000 parts each. The watches that Tempus made were designed such that, when he had to put down a partly assembled watch (e.g. to answer the phone), it immediately fell into pieces and had to be reassembled from the basic elements.*
>
> *Hora had designed his watches so that he could put together subassemblies of about ten components each. Ten of these subassemblies could be put together to make a larger subassembly. Finally, ten of the larger subassemblies constituted the whole watch. Each subassembly could be put down without falling apart.*

This simple story reveals that the ability to develop large complex systems depends on two elements in fundamental and universal ways.

The first element is the environment. How frequently and how seriously will the environment disturb the developer? Typical 'disturbances' in the CIM/E domain are:

- Technological evolution (and revolution).
- Introductions of new products.
- Strong and unpredictable fluctuations of the demand for members within the current product range.

The contemporary manufacturing environment increasingly provides these disturbances.

The second element is the ability to develop stable subsolutions (subassemblies, subsystems) that survive these disturbances. When the environment disturbs development activities more frequently, the need for stable subsolutions increases. For the developers of large and complex systems, this means that they need modules of appropriate size that permit the rapid development and adaptation of the overall system to the environment. Such systems evolve through the gradual replacement and addition of modules that survive themselves the changes in the environment (when seen individually) much longer than the large system itself (without evolution).

In a traditional top-down and hierarchical approach, developers design and optimise subsystems for a context that is defined because of the top-down approach. This results in subsolutions that are unable to survive disturbances from the environment significantly better than the overall system. The parable above reveals that this makes the emergence of large systems in demanding environments impossible. Our experience confirms this: automated systems are either small (e.g. islands of automation) or enjoy a stable environment (e.g. mass production systems). Systems that were developed and optimised with a particular environment in mind have proven to be highly unstable 'subassemblies.'

A short analysis shows that top-down development is hitting against the proverbial brick wall. Consider a system with n components. Assume that the probability of surviving the next time slot for these components is p, where $p < 1$, and that the survival of the components is mostly unrelated. In this case, the probability that the system survives the next time slot is p^n. In other words, the probability of survival for the overall system decreases exponentially with the number of components. Very rapidly, the expected lifetime of the system will drop below the level were it is able to repay its development and maintenance costs when the number of components increases.

The parable also reveals that hierarchy comes naturally in large and complex systems because it drastically increases the probability of their emergence and survival. However, this does not mean that it is necessary to postulate a hierarchy when starting to develop an artificial complex system. To the contrary, it means that a hierarchy will emerge that reflects hard constraints in the environment; it reflects the presence and absence of stable and suitable subsolutions. Postulating a hierarchy up front may cause problems instead of solving problems. Developers must look for possible natural hierarchies.

2.2. STABLE CIM/E SUBASSEMBLIES

In the domain of CIM/E, there exist subsystems that are sufficiently large but unsuitable for integration. Likewise, there exist components that are suitable for integration but too small for the development of large complex systems in demanding environments. Typical representatives of the former category are the 'islands of automation' whereas mass production systems are typically built with representatives of the latter category such as motors, pallets, conveyors, … Apparently, designers fail to preserve the suitability for integration (or flexibility) during the development of members of the former category.

In addition, there is a strong synergy between the stability of automated systems and their suitability for integration. Automation requires repetitive use to be viable in an economic context. The survival of automated systems depends on their ability to serve a large number of users, which allows them to profit from economies of scale and to rapidly go up their learning curve. In other words, emergence and survival of CIM/E subsystems depend on their ability to fit into a large number of manufacturing systems, at different places within a manufacturing system, and for a sufficiently long time during which neighbouring components will be replaced. This goal proved to be unattainable until today.

The next paragraph elaborates a theory that reveals how this suitability for integration is lost during the design of Hora's subassemblies. Three design principles follow from this theory. It describes how to develop systems that cope with uncertainty about the future context(s) in which they will have to function adequately.

3. Design theory

A radically different approach is required for the construction of large, modular, multi-hierarchical, distributed CIM/E systems that are capable of evolution: bottom-up development. However, it is insufficient to develop subsystems that provide the required functionality. It is necessary to develop subsystems that will cope with integration demands that are, to a certain extent, unknown during their design and development.

3.1. FLEXIBILITY OF SOLUTIONS

3.1.1. Definitions and One Axiom. This paragraph formally defines the flexibility of solutions for problems in general. In CIM/E, this flexibility reflects the ability of solutions to accommodate integration requirements, which emerge after their development.

1. The symbol ϑ denotes the state space of the universe under consideration.
2. The symbol Ω denotes the set of all constraints on elements of ϑ.
3. The following axiom on Ω and ϑ is postulated :

   $\forall\, x \in \vartheta, \forall\, y \in \Omega : \sigma(y,x) \in \{\text{ true, false }\}$ where $\sigma(y,x)$ denotes a predicate that is true if and only if state x satisfies the constraint y.
4. By definition, a problem P is a subset of Ω.
5. $\forall\, x \in \vartheta, \forall\, P \subset \Omega : \sigma(P,x) \Leftrightarrow (\forall\, y \in P, \sigma(y,x))$.

   $\sigma(P,x)$ is pronounced as x *satisfies* P.
6. By definition, a solution S to a problem P is a subset of Ω for which every member of ϑ satisfying S also satisfies P. In other words, solutions reject all states that are rejected by the problem but, they may reject additional states that are permitted by the problem.
7. By definition, an environment $\mathcal{E}$ is a set of subsets of Ω. Any solution S to a problem P with an environment $\mathcal{E}$ will be confronted with a single member E of this environment. However, only $\mathcal{E}$ is known when the solution S must be developed. An environment defines the range of possible situations that solutions might have to face; it models uncertainty. When there exists a state x, member of ϑ, satisfying both S and E, the solution S is said to *survive the confrontation* with member E of the environment $\mathcal{E}$.
8. The flexibility of a solution S relative to an environment $\mathcal{E}$ is defined by the *survival trace* of the solution S over environment $\mathcal{E}$, where this survival trace is the subset of environment $\mathcal{E}$ containing the elements that S survives.
9. By definition, a solution S is *more flexible* than a solution T relative to environment $\mathcal{E}$ if and only if the survival trace of T is a subset of the survival trace of S. Note that this definition acknowledges that flexibility inherently implies a partial order, not a full order. Quantifying flexibility requires a mapping from survival traces to economic values; this implicitly imposes a full order upon the solutions in which the inherent

partial order ought to be embedded. This mapping indicates how well the traces cover each possible future.

3.1.2. *Discussion and Two Design Principles.*

Discussion and Two Design Principles. Developers create solutions through a sequence of design decisions. Whenever they take a design decision, they reduce the number of states of the universe that satisfy the next solution in this sequence. When a decision eliminates all remaining states that satisfy both the previous solution and a member of the environment, the flexibility of the solution decreases. Therefore, design decisions may cause integration problems afterward. This leads to two design principles for the bottom-up development of flexible solutions that are suitable for integration:

P1. *Design decisions are guilty until proven innocent.* It is necessary to prepare the decisions but commitment should be avoided or postponed when possible. It is dangerous to commit to arbitrary choices that simply satisfy the current requirements since, in bottom-up development, unpredictable integration requirements will emerge later and the arbitrary decisions probably will be inconsistent with these emerging requirements. In this context, designers must also be aware of arbitrary design decisions that sneak in their solutions together with more justifiable ones.

P2. *Look for design decisions that maintain flexibility.* Every design decision that brings a developer closer to a final solution without reducing flexibility significantly, is a safe decision. Commitment to such decisions is unlikely to cause integration problems afterward. Moreover, it reduces the arbitrariness of subsequent design decisions. Therefore, developers must take the possible safe decisions in the beginning of their design sequence. An important source for these safe decisions is the environment itself. Constraints that are present in most of its elements correspond to safe design decisions.

During bottom-up development, the designers of Hora's subassemblies have to create solutions that offer functionality at minimal conditions. Their results should resemble laws of nature in the micro-cosmos defined by the environment $\mathcal{E}$. Laws of nature represent something unavoidable whereas manmade systems typically have a more arbitrary character. Alternative, designers can see each design decision as a reduction of this micro-cosmos in which the conditions of their solution are unavoidable.

3.2. WORLD MODEL

It is not trivial and often impossible to rigidly apply the above principles to real-world situations. A model of our universe reveals the underlying causes.

3.2.1. *The Real World.*

The Real World. This paragraph lists a number of properties of our universe that affect our (in)ability to preserve flexibility. The model depicts the world as a (not finite) state-transition machine with the following characteristics:

1. The universe always is in exactly one state.
2. Every state has precisely one time co-ordinate.
3. Every transition corresponds to an increase of this time co-ordinate.

94

4. Every transition has a probability coefficient indicating how likely it is to occur given that the universe is in its initial state. This coefficient represents the laws of nature and instinctive behaviour. Typically, there are several transitions possible in any given state of the universe.

5. Problem solvers (human individuals) influence the probability coefficients through their conscious actions. These actions do not influence the progressing of the time co-ordinate.

   Typically, problems are formulated in function of this time co-ordinate. Therefore, the progress of the time co-ordinate can be an uncontrollable cause of failure. In contrast, design decisions are controllable potential causes of failure.

6. Every action requires effort. Problem solvers can only spend a limited amount of effort in function of time. These limitations make it impossible to solve problems in an infinitesimal period of time. It necessitates a sequence of actions that spans a finite period of time.

   Note that the effect of actions is sequence-dependent (not-commutative). Also note that observation of the state of the universe requires actions. This means that problem solvers have to face significant amounts of uncertainty, which is another uncontrollable possible cause for failure.

   Finally, the universe contains resources that amplify the effect of the problem solvers' actions. There are natural resources and manmade resources or artefacts. Today, manufacturing relies heavily on both types.

3.2.2. Discussion and Another Design Principle. The above model reveals two facts. The first explains the difficulty to apply the above design principles. The second clarifies the difference between taking a design decision and commitment to a design decision. These facts lead to a third design principle.

Our universe always is in one particular state. This implies that flexibility inherently is a property of intermediate solutions relative to some final solution(s) as well as relative to an environment. Whenever a problem solver comes close to the actual solving of a problem or task, he is unable to preserve flexibility. Therefore, problem solvers should not expect that they can maintain flexibility in the final phases of their activities. Fortunately, the uncertainty also will decrease.

Flexible intermediate solutions have to be embedded in the concreteness of the current state of the universe. In other words, flexible intermediate solutions need representations within the current state of the universe. The discovery, construction, and support of these representations is the essence of developing flexible systems. This is not trivial and it is the second reason why the application of the first two design principles can be difficult.

Flexibility will increase the complexity of a representation. Usually, this requires some additional effort. However, the price of flexibility varies greatly within the manufacturing world. There is brainware, software, firmware, hardware, and ironware. The cost of creating and, most importantly, reproducing complex representations of flexible solutions varies orders of magnitude when we cross the boundaries between these domains.

Also note that flexible intermediate solutions mostly serve for multiplc final solutions. For instance, flexible manufacturing systems serve to satisfy streams of production requests. This re-use-ability can compensate for the increased complexity of the flexible intermediate solutions. In these situations, the evolution toward a final solution must not destroy the representation of the intermediate solution(s). In other words, there is a need for persistent representations of flexible intermediate solutions.

The second important fact is that problems are solved through non-commutative sequences of actions. Whenever a problem solver takes a design decision, he/she can easily change his/her mind immediately afterward without difficulty. More serious problems occur when he/she wants to do this later. Typically, most of the effort that was spend after the decision will be lost. The decision has built up inertia and the problem solver is committed to his earlier decision. This leads to a third design principle:

P3. *Avoid to build up inertia for any unsafe design decisions.* This principle means that earlier design decisions that reduce the flexibility of intermediate solutions are not an acceptable justification to allow later design decisions to imply the same loss of flexibility (since it is already lost). In other words, every design decision must justify the possible reductions of flexibility, which it may induce, on its own. Of course, if safeguarding flexibility is too costly, design decisions should be consistent and reduce flexibility in the same way whenever possible.

This third principle must be seen in the perspective of the large differences in the reproduction cost of flexible, and therefore complex, subsystems in the different technological domains that were mentioned earlier (brainware, software, firmware, hardware, and ironware). It typically is extremely expensive to safeguard flexibility in ironware. On the other hand, it often is trivial to develop software that is suitable for integration with an entire family of such ironware where a data file defines the actual configuration. It is tempting to simplify the software development when the ironware configuration is known at design time. However, the small savings will prove very costly when, afterward, a design flaw is discovered in the ironware or when the ironware needs upgrading. Moreover, it will prevent re-use of the software for other systems, which implies that the development will not benefit from economics of scale and go up its learning curve at a slow pace. Finally, note that this edge of software over ironware, when it comes to the reproduction of complex subsystems, is typical for situations where the proposed concepts are applicable.

3.3. REMARK ON THE RESULTING SUBSYSTEMS

Stable subsolutions, Hora's subassemblies, in CIM/E typically are flexible systems that are suitable for integration. These are inherently intermediate solutions, not final solutions. To the average end user, they are less attractive because they come only halfway; they fail to offer the final solution. Therefore, flexible systems often will have to be autonomous systems as well. Their autonomy enables these systems to postpone design decisions until the environment provides sufficient information to take the proper decisions. In other words, the analysis above confirms human intuition, which already relates flexibility and adaptivity with autonomy.

4. Examples

In CIM/E subsystems, information that is embedded within the software implies a model of parts of the manufacturing enterprise. The following paragraphs highlight how, in existing practice, this software (implicitly) uses models which correspond to unnecessary loss of flexibility. More flexible alternatives are given. An elaboration of the above theory to assess and understand the flexibility of manufacturing systems as a whole can be found in [KOBE].

4.1. TRAJECTORY GENERATION

This first example illustrates how potentially harmful constraints can sneak into a design while hiding behind another decision that has a solid justification. It concerns trajectory generation within off-line programmable robot systems. The example restricts itself to robot tasks without sensor feedback, which constitutes over 90 % of all robot applications in industry today.

In a robot system, the trajectory generation subsystem generates data from a robot path specification. The control subsystem uses this data to make the manipulator execute the movement that is specified. Typical path specifications, supported by existing robot systems, are:

- Joint-interpolated motion, in which every robot joint moves from the current position to the target position at a speed proportional to the distance it must cover.
- Linear Cartesian-interpolated motion, in which the robot end-effector moves along a straight line from its current position to the target position.
- Circular Cartesian-interpolated motion, in which the robot end-effector moves along a circular path. This is only supported by advanced robot systems.

The data from the trajectory generation subsystem typically consists of a sequence of robot joint positions where the manipulator control system requires a target position for every joint at a fixed frequency of, for instance, 100 Hz (10 millisecond intervals).

In principle, it is possible to perform trajectory generation off-line (on beforehand). This has the important advantage that the required computations must not be performed under real-time constraints and that the computations must not be repeated every time the robot executes the motion. Unfortunately, trajectory generation produces a large amount of data that will often exceed the memory capacity in a typical robot system.

Historically, the above difficulty is used to justify the decision to perform trajectory generation exclusively on-line. A more flexible design would support off-line trajectory generation in combination with asymmetric compression and decompression of the data (i.e. the compression requires much more computation than the decompression which must happen in real-time). The compression technique can rely on the physical limitations of robot manipulator such as acceleration and deceleration limits, position sensor resolution, and on the required accuracy for the robot task. The flexible design has the following advantages:

1. More complex trajectory specifications become possible since trajectory generation and compression happen off-line. Decompression does not depend on the complexity of the trajectory.
2. A higher frequency for the transmission of data to the control subsystem becomes feasible (e.g. 1000 Hz). This permits the execution of trajectories that vary more rapidly over time and it simplifies the task of the control subsystem.

Today, many advanced robot applications simply are uneconomical or even technically impossible because of this decision to perform trajectory generation exclusively on-line. Especially the integration of robot systems with CAD-based robot programming and/or process planning systems suffers from this lack of flexibility. In addition, there exist other obstacles, beyond the scope of this text, that hinder this integration [VALCK].

4.2. DEVICE DRIVERS FOR MANUFACTURING EQUIPMENT

This second example compares the traditional model of programmable equipment, which is embedded in software for manufacturing systems, with a more flexible alternative, which is commonly used in administrative/office type environments.

4.2.1. *The Traditional Model.* In manufacturing systems, most programmable equipment supports its own programming language. Numerically controlled machines have NC programming languages, robots have off-line programming languages, and programmable logic controllers support so-called ladder logic. Their users employ programming systems, which accompany the equipment, to generate the required programs. In CIM systems, the production management and control subsystem, which supervises these devices, typically uses the following commands or functions:

1. Downloading of a program from a database or a file system to a device.
2. Uploading of a program from a device to a database or a file system.
3. Deleting of a local copy of a program in a device to free some storage space.
4. Ordering a device to make a downloaded program *ready-for-execution*. This typically consists of loading the program from secondary storage in the device into its main memory.
5. Making a device execute a program, which should be ready-for-execution.
6. Handling notifications from the devices which mostly indicate errors or the completion of an earlier command. For instance, devices may notify the supervising system when they finish executing their program.
7. Executing queries, such as getting a list of downloaded programs or the machine status.

Implicitly, this organisation implies the following modus operandi. First, the users develop the necessary programs. These users will adapt this set of programs when required. Second, the supervising system, probably assisted by humans, ensures that the proper programs are downloaded to the proper devices. Finally, the production management and control system makes the devices execute the proper programs to accomplish the required production. When devices have not enough capacity to hold all the required programs, this supervising system will upload/delete and download programs. Here, some optimisation may be necessary. This organisation originates from a situation where the device had to operate in a stand-alone mode.

4.2.2. *The Flexible Alternative.* In the administrative/office environment, computer systems also incorporate a collection of devices: printers, plotters, displays, keyboards, mice, tape streamers, hard disks, and speakers. They are known as input/output (I/O) devices. The organisation in this environment has a different view of devices. It sees them as objects that can execute a set of commands which make the functionality of a device available on the computer platform. There is significantly less duplication of the computer functionality in the devices.

The organisation in the administrative/office environment has a different historical background. It originates from the early days of the computer era. In those days only one program at the time was present on a computer. This program had to perform all functions including those that are nowadays performed by the operating system. The software routines that performed input/output functions were especially important. Each new I/O device had its own characteristics, requiring very careful programming. The solution to this problem was to write, once, a set of subroutines that knew how to drive that device and simply have everyone use these subroutines. Such a subroutine set is called a device driver. On modern computer platforms, the device drivers are integrated into the operating system.

In the administrative/office environment, the above organisation still applies. Application software uses device drivers to issue commands to the devices. This organisation typically functions as follows. The operating system initialises the devices at start-up and arbitrates amongst the users of the devices. The current user is a computer process that issues commands to the device.

The important difference with the organisation in CIM systems is the granularity of the commands. Consider for instance NC machining. In the administrative/office organisation, there would be a process that consumes data, which specifies a production recipe, and that issues commands corresponding to a single line in an NC program. Note that this organisation existed in the early days of NC programming (direct numerical control) but was abandoned because it required a computer that was too expensive at the time.

4.2.3. *Discussion.* The two different views on programmable equipment have significant consequences in two domains: software development and process planning.

For software development, the latter organisation signifies that considerably less software must be developed on device-specific computer platforms in device-specific programming languages. Software developers profit from the quality, availability, familiarity, and competitive prices of mainstream computers and software environments. Software development is very sensitive to these issues [DiNitto, Rushinek]. Economies of scale and a rapid ascent of the learning curve are extremely important for computer systems, operating systems, and programming tools. Personnel training aspects cannot be ignored either. The organisation based on device drivers minimises the fragmentation which is characteristic for the present manufacturing environment.

In the terms of the theory on flexibility, the device driver solution is farther from a final solution but it avoids the arbitrary design decisions that are unavoidably taken when it would become a complete programming system. The device driver organisation is suitable for integration, even with programming systems that do not yet exist (C++ version 7, Ada 200X, Eifel 4.0) and with operating systems that are still under development (e.g. based on CHORUS or MACH). Programmable manufacturing equipment typically is capable of stand-alone operation, but unsuitable for integration with today's and tomorrow's computer platforms and software development systems.

Another problem arises in the domain of process sequence planning and program generation. The traditional manufacturing organisation violates the first design principle since it unnecessarily forces early decisions. This causes integration problems, which reduce the agility of the manufacturing system. Consider the following situation. A developer has to generate a program for NC milling machines to manufacture a workpiece that has 20 complex cavities (pockets) where the tolerances amongst these cavities are very loose. In principle, the workpiece can be made, provided that roughing operations precede finishing operations for each cavity, in any sequence of the following actions:

- Every rough cutting operation.
- Every finishing operation for an entire cavity.

This represents more than 10^{36} possible sequences from which the NC programmer or process planner must select one sequence.

In addition, the workpiece may (but must not) temporally leave the NC machine after each of these operations. This means that the programmer must select one possible partition amongst all possible partitions of the selected sequence. To make things worse, there may be several options to produce some of these cavities (e.g. cutting a cavity in one pass with a dedicated tool or in two passes with ordinary tools). In other words, process sequence planners and NC programmers face an enormous number of options from which they must select a single one.

The real problem is that, when the NC program must be generated, the information to select an adequate sequence and partition is unavailable. Hence, the constant increase in computer power will not solve this problem. The main constraint which the sequence selection must take into account is the availability of manufacturing resources. A workpiece should leave the NC machine when a rush order arrives, which is by definition unpredictable. Likewise, the operation sequence should adapt itself to the availability of the tools. This means that multiple orders for the same type of workpiece often should have different NC programs, and that NC program optimality depends on the composition of the current order book.

With the device drivers, it is possible to develop a production control system that uses non-deterministic process plans, and that selects suitable sequences/partitions when the information to make good choices is available. The next paragraph discusses a prototype control system that does exactly this for robotic assembly. To apply this approach to NC machining systems, it is necessary to have NC controllers that can download, prepare, execute, and delete small programs (which execute a single operation) concurrently such that there is no idling or an unrequested motion.

4.3. FACCS

This third example discusses the prototype production control system FACCS, which was developed at the mechanical department of the K.U.Leuven [VALCK].

4.3.1. Genericity. The prototype FACCS is generic with respect to the underlying system. In other words, the design is suitable for integration with any member of an entire family of production systems. FACCS can control any production system that fits the following model:

- The underlying production system consists of a set of loosely coupled stations connected by a transport system that handles workpieces, tools, and components or consumables.
- The transport system supports random and independent routing. This means that each mobile object can follow an arbitrary route through the system regardless of the routes followed by the others. The transport system may be biased toward the nominal or expected routings (more capacity and faster), but it must support adequate capacity and acceptable speed for any possible routing request from the control system. The control system issues transportation requests to the transport system, which notifies the control system when it completes a request.
- Each station has a repertoire of indivisible operations. There are no restrictions on the presence or absence of operations in the repertoire of the stations. The control system FACCS launches these operations but is unable interrupt when the station performs them. The stations must not accomplish a given goal at any cost. When a problem occurs, the station signals the corresponding result code. In case of a testing operation, these result codes are inherently unpredictable.
- The agility of the overall production system increases when the granularity of the indivisible operations decreases. However, there is a lower limit to this granularity. The intermediate products or workpieces must be transportable and storable at the end of every indivisible operation.
- There is no requirement regarding the type and degree of automation in the stations. There can be a mix of hard automation, flexible automation, mechanisation, manual production. However, the control system must be able to issue requests for the execution of indivisible operations and must receive reliable feedback on the result.

The control system genericity allows to spread its development cost across multiple production systems. What is more important, it enables the control system to survive the evolution of the underlying production system. The design of FACCS reflects the first principle of the theory above. The decisions that imply a specific configuration for the underlying system are delayed until this configuration becomes known. The design also reflects the third principle. The target for the development of FACCS was a specific flexible robotic assembly system (see figure 1). The resulting control system only relies on properties of this specific target when it reduces the complexity of the design significantly. Notice how the model of the underlying system imposes properties on the production system as a whole, but relies nowhere on the presence of a specific production process (assembly) or type of automation in the stations.

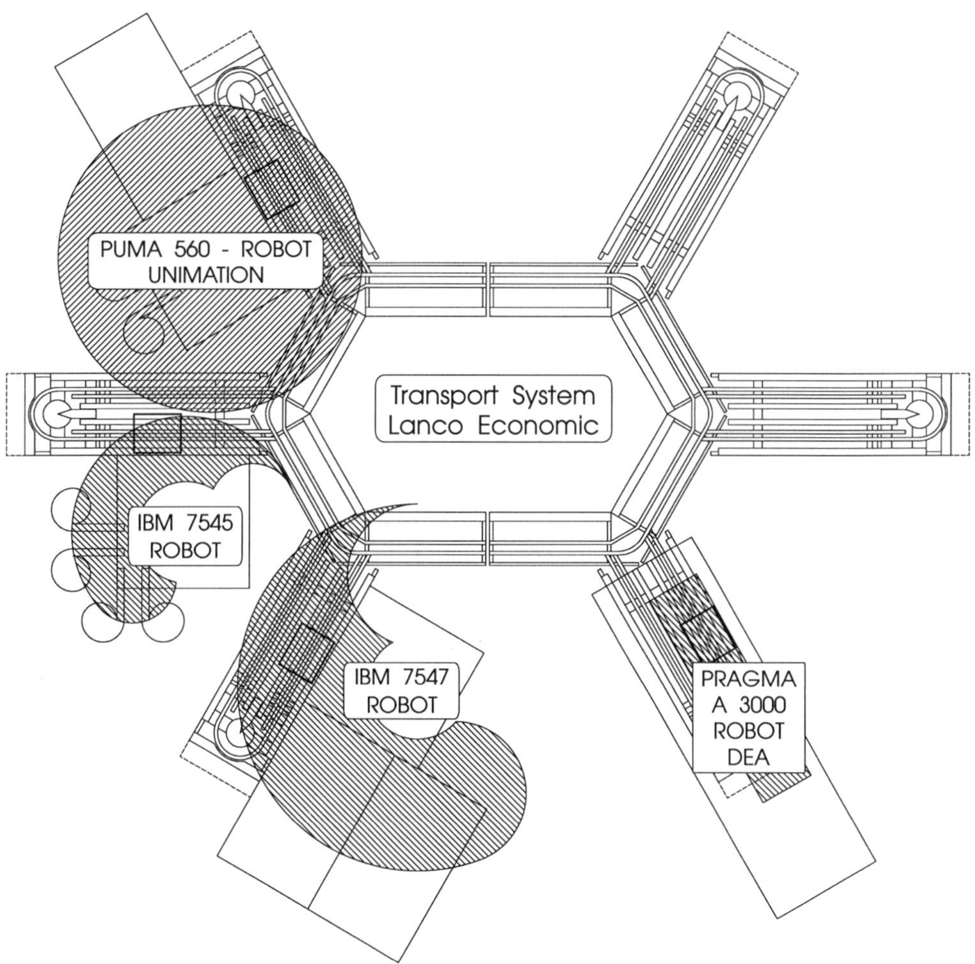

Figure 1: The Target Flexible Assembly System.

However, the flexibility of the design translates itself in distance from a final solution. Users must implement the model, which is described above, and connect it to FACCS. This was done for the specific target assembly system.

4.3.2. *Product Range and Production Load.* The decisions related to the product range and the production load are even further delayed. FACCS allows its users to define new products at run-time. Likewise, orders for any product, which is known to FACCS, can enter the system all the time.

FACCS supports mixed-model production; each product carrier can contain a different product at any given instant. This is a result of applying the third design principle; although most underlying production systems require some change-overs to make all the products in their product range, it is not necessary to increase the inertia of these limitations by introducing a similar constraint in the control system. A change-over, seen through the eyes of the control system, is equivalent to indivisible operations that become respectively unavailable/available at the stations that are involved. An important advantage of this flexible design is that, while a production batch is leaving the production system, the next batch can enter the system with the change-overs as a moving 'boundary' between them.

The control system FACCS supports product definitions that may specify any finite combination of the available indivisible operations. The control system supports non-deterministic product definitions that permit to express the following properties of the corresponding process sequence plan:

1. Precedence relations amongst the indivisible operations.
2. Mutually exclusive choices amongst the indivisible operations.
3. Mutually exclusive choices amongst sequences of indivisible operations.
4. The effect of the operations' results on the actually allowable sequences. For instance, the result code from a testing operation will signal whether rework is necessary. Notice that in this situation the proper sequence of operations can only be determined at run-time.

Details of the product definitions can be found in [VALCK]. These product definitions also indicate which mobile resources (product carriers, tools) are needed for the execution of the indivisible operations.

This support for non-deterministic process plans by FACCS enables process planners to specify as many possible sequences as they desire. They no longer face a combinatorial explosion of options from which they must select a single sequence/partition before the required information is available.

4.3.3. *Autonomy: a Sample Scenario.* FACCS utilises the freedom of the non-deterministic product definitions to handle unpredictable events autonomously. The example below illustrates this. Note that it only demonstrates the capability to handle one particular type of disturbance (malfunctioning equipment). Similar results are obtained when handling rush orders [VALCK]. Larsen reports on simulations that demonstrate the usefulness of non-deterministic product definitions for production scheduling [CIRP].

The sample scenario consists of printed circuit board assembly during an unmanned shift. The assembly system must place 35 components on each board. In the sample scenario, stations 1 and 2 are equipped to place all 35 components. During the unmanned shift, the system has to assemble a number of printed circuit boards. Figure 2 shows the production of 12 printed circuit boards during the entire unmanned shift to keep the details in the picture discernible. In reality, the assembly of a single board requires a few minutes, while a shift lasts several hours; a production of 300 boards is closer to reality.

Figure 2 shows GANTT-charts that corresponds to a feeder jam for the third component (in the deterministic process plan) at station 1. The feeder jam happens in the beginning of the unmanned shift. According to the non-deterministic process plan, the assembly system can place the remaining 34 components in spite of the feeder jam.

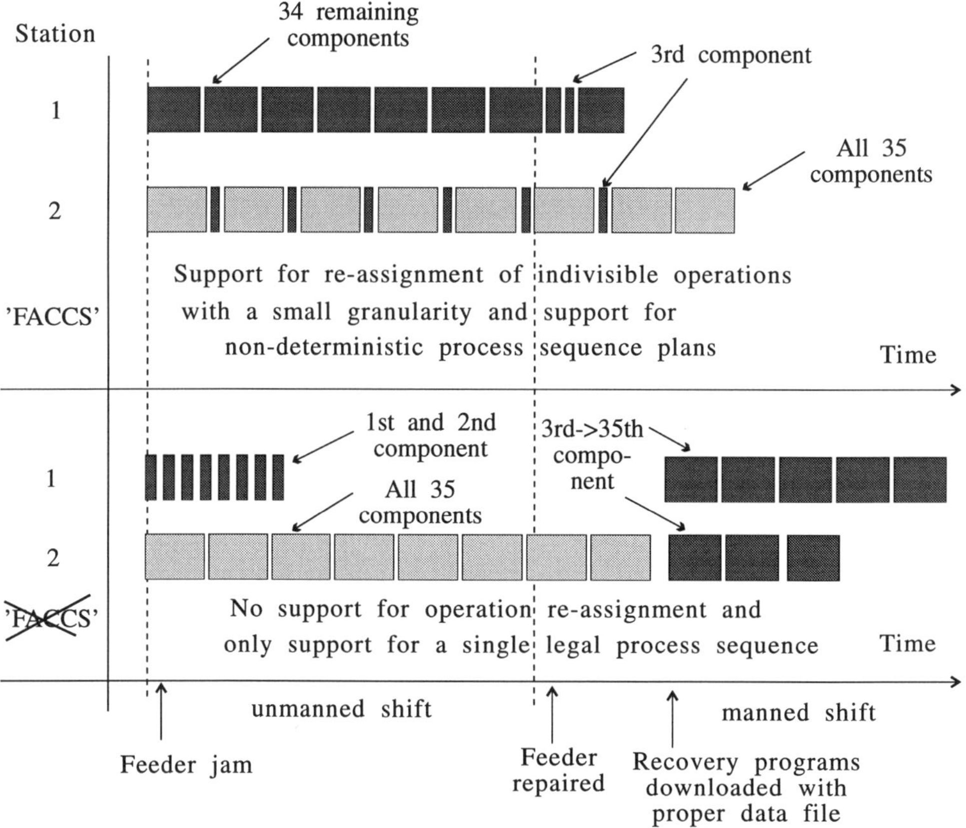

Figure 2: The Sample Scenario.

The control system FACCS autonomously re-assigns the operation that becomes unavailable at one station to the remaining alternative. It does this at the granularity of the indivisible operations and uses the non-determinism in the product definitions to deviate as little as possible from the nominal production plan. With the control system FACCS, the feeder jam has little impact on productivity. The system produces 11 boards during the unmanned shift. With a conventional control organisation, only 6 boards are complete when personnel arrives in the morning to fix the feeder. It is not difficult to make the above scenario even less favourable for the traditional control system. For instance, assume a second feeder jam, this time on station 2.

5.   Summary

This manuscript presents a theory on flexibility. Three design principles follow from this theory:
1. Commitment to design decisions is harmful until proven innocent.
2. Always look for design decisions that maintain flexibility first.
3. Loss of flexibility caused by earlier design decisions is no justification for later decisions to do the same.

The theory also reveals that flexibility merely implies a partial order on solutions, that flexibility inherently is a property of intermediate solutions, and that flexible intermediate solutions need a representation in the concrete state of our universe. The latter two properties reveal a close relationship between autonomous behaviour and flexibility.

Three examples illustrate the theory. The first shows how harmful decisions sneak in a design. The second compares two views of programmable equipment in manufacturing systems. The third discusses the control system FACCS that was developed according to the above design principles.

Ongoing research in this domain focuses on holonic systems. These are distributed systems with autonomous subsystems that behave in function of global goals [KOEST].

6.   References

[AMRF]   A. Jones and C. McLean, *A proposed hierarchical control model for automated manufacturing systems*, Journal of Manuf. Sys., Vol. 5, No. 1, 1986.

[CIRP]   N. Larsen, *Effects from use of alternative routings*, 24th CIRP SEMINAR on Manufacturing Systems, June 11-12, 1992, Kopenhagen, Denmark.

[DiNitto]   S. DiNitto, Jr., *The future directions in programming languages*, Proceedings of the Intern. Conf. on computer languages, IEEE CS, 1988.

[KOBE]   P. Valckenaers and H. Van Brussel, *Formal Models of Flexibility for Manufacturing Systems*, Proceedings of the RMMS conference, Kobe, 1992.

[KOEST]   Arthur Koestler, *The GHOST in te MACHINE*, Arkana Books, 1989.

[Rushinek]   Avi Rushinek and Sara Rushinek, *What makes users happy?*, Communications of the ACM, July 1986, Vol. 29, No. 7, pp. 594-598.

[SIMON]   Herbert A. Simon, *The Sciences of the Artificial*, 6th ed., MIT Press Cambridge (Mass.), 1990, ISBN 0-26269073-X

[VALCK]   Paul Valckenaers, *Flexibility for Integrated Production Automation*, PhD. Thesis, 1993, ISBN 90-73802-11-3.

DESIGNING COLLABORATIVE SYSTEMS TO SUPPORT REACTIVE PROBLEM-SOLVING IN MANUFACTURING

ANANT BALAKRISHNAN
Sloan School of Management
M. I. T.
Cambridge, MA 02139
email: anant@mit.edu

RAVI KALAKOTA
ANDREW B. WHINSTON
University of Texas at Austin
Austin, TX 78712-1175
email : abw@emx.cc.utexas.edu

PENG SI OW
IBM, Austin
Austin, Texas

ABSTRACT. This paper develops and illustrates a methodology for building systems to support manufacturing coordination and collaboration. The paper begins with a case study to understand the information requirements for reactive problem-solving in manufacturing. We focus on the communication mechanisms, decision-making processes, and implementation issues for problems that arise due to materials shortage in electronics assembly. We then describe a conceptual framework for collaboration support, emphasizing parsimony and generalizability, that incorporates the problem-solving requirements that we identified through the case study. Finally, we describe a prototype system, implemented on a client-server architecture, that incorporates many of the features needed to support reactive problem-solving.

1. Introduction

To maintain their leadership in an increasingly competitive marketplace, manufacturing and service organizations alike are emphasizing customer satisfaction–providing good value and quick response to meet customers' changing needs. In the manufacturing arena, achieving and maintaining world-class standards has forced companies to move beyond cost control, inventory reduction, and quality improvement campaigns to agile manufacturing–providing the right product or service quickly at low cost, with high quality, and in small lot sizes–through just-in-time manufacturing, greater supply chain integration, drastically reduced time-to-market, assimilation of total quality management, emphasis on teamwork and organizational learning, adoption of activity-based costing, organization redesign and business process reengineering. The task is especially challenging in industries such as electronics fabrication and assembly that are characterized by rapid technological changes, and dramatically shrinking product lifecycles. The increasing interdependence and integration of all functions along the innovation and supply chain–design, materials, production, distribution, suppliers– requires more information sharing and coordination within and across organizational boundaries in order to respond swiftly, but cost effectively, to the changing environment especially since managers can no longer rely on traditional mechanisms such as inventory and time buffers to absorb uncertainties and shocks. Information technology, broadly defined to include hardware and software of all kinds, is seen as the key enabler to facilitate the necessary organizational, cultural and process changes to accomplish the ambitious organizational transformation and quantum leap goals. The information system must provide relevant and accurate information rapidly to the right people so that they can make informed decisions collaboratively; it must eliminate the overhead

105

S. Y. Nof (ed.), Information and Collaboration Models of Integration, 105–133.
© 1994 *Kluwer Academic Publishers. Printed in the Netherlands.*

associated with accessing, exploring and analyzing data for routine as well as unstructured scenarios, facilitate collaboration, and provide support mechanisms for continuous improvement and organizational learning. Recent advances in IT, such as client/server computing, distributed database servers and the proliferation of Local (LAN) and Wide Area Networks (WAN) provide the necessary computing and communication infrastructure to support the collaboration and coordination needs of the emerging tightly-coupled organizations.

This paper attempts to understand and address one component of the computer-communication system needed to support coordination, namely, supporting the process of reactive problem-solving in procurement and manufacturing operations. Reactive problem-solving refers to mechanisms for monitoring and correcting production and delivery processes to accommodate unanticipated changes in the environment. Quick response and speedy resolution of problems is critical for tightly-coupled systems such as just-in-time (JIT) or continuous flow manufacturing systems that operate with minimal or no safety stocks and safety times. A disruption in any segment of the chain propogates to the rest of the system, and so problems must be resolved quickly in order to maintain acceptable asset utilization levels. Indeed, one of the cornerstones of the JIT philosophy is the culture of continuous improvement, driven by a process of deliberately reducing the in-process inventory levels to expose problems which in turn motivates and focusses team efforts for operations improvement. An effective process to identify and resolve problems is, therefore, an essential prerequisite for such a system. We emphasize that although our main focus is reactive problem-solving, data gathered from this process forms the basis for long-term improvement, and so one of the functions of the coordination system includes providing the linkage between short-term problem solving and long-term improvement activities. We might also note that, in the absence of a well-designed and established resolution process, reactive problems often tend to motivate managerial actions that are solution-centered, restrict innovation, and limit the number of alternatives that are considered due to intense time and performance pressures.

The literature on organizational problem-solving and decision making focusses mainly on describing normative processes (Nutt [1984]); however, little attention has been directed towards issues of how organizational subunits collaboratively solve unstructured reactive problems in an uncertain and dynamic environment, and how information technology can best support this class of problems. By "unstructured problems" we mean contingencies that are sporadic or have not been previously encountered in quite the same form, and for which no predetermined and explicit set of ordered responses exist. According to Mintzberg *et al.* [1976], managers deal with unstructured problems by factoring them into familiar components, and use their experience to provide problem-solving shortcuts to aid in formulating collaborative decisions. A collaborative decision, in this context, is defined as a specific commitment of resources to an action by an agent(s) involved, and a decision process is a set of actions taken by agents that begin with problem identification and end with a specific schedule of action.

Developing a coordination support system for reactive problem-solving first requires understanding current operations, characterizing the operational problems and their information requirements, examining the current problem resolution processes, and identifying opportunities for computer-based support. Accordingly, we began this study by working with a large manufacturing facility that assembles electronic and mechanical components on printed circuit boards (using mainly surface mount technology). To limit the scope of the study, we decided to focus on materials (i.e., components) shortage problems. In our experience with many companies in the electronics assembly industry, materials shortage is a pervasive problem, leading to frequent disruptions in production schedules, unanticipated setups on the line, and slippage in delivery dates. Materials shortage is also a "multifunctional problem", requiring inputs and action by suppliers, material managers, production supervisors, quality control personnel, sales representatives, and customers. Therefore, this problem provides a rich domain to observe, conceptualize, and validate systems to support manufacturing coordination. As part of our study, we interviewed managers and observed several production meetings at the facility. This experience proved invaluable, enabling us to classify the types of problems that managers face, and develop a conceptual problem-solving framework. Section 2 presents our observations in the plant from the point of view of collaboration and coordination for problem-solving. Based on our experience at the electronics assembly facility and a review of the literature on individual and organizational decision-making, we developed a framework for reactive problem-solving that clarifies the different decision support needs and system roles during different phases of the process. This framework is consistent with our view of manufacturing problem-solving as a dialectic discourse between multiple participants who must work together to provide inputs, evaluate alternatives, and execute action plans. Section 3 reviews the relevant prior literature on the motivation and modes of coordination, task interdependence, and the effect of the types of uncertainties encountered in reactive problem-solving. In Section 4, we present our conceptual framework of reactive problem-solving. We decompose the process into four phases–problem identification, diagnosis, solution formulation, and implementation. For each phase, we briefly describe related concepts covered in the literature on organizational decision-making, identify the information and coordination needs, and discuss how to provide the necessary computer support for dialectic discourse. Section 5 outlines the coordination support prototype that we designed and implemented. The design of this prototype uses theoretical constructs from computer supported collaborative work (CSCW) and distributed computing. Section 6 discusses promising directions for future work, including field studies and experiments needed to validate the prototype.

2. Case Study: Materials Shortage in Electronics Assembly

This section provides a brief overview of the manufacturing operations at the electronics assembly plant we observed, and discusses the nature of day-to-day operational problems with particular emphasis on materials shortage, and the current mechanisms to address these problems. The discussion should help us understand who are the various agents concerned with identifying, resolving, and implementing solutions to materials shortage contingencies, what are their respective motivations, interactions, and information needs, and what is the current and desired problem solving process. We can then address issues of how to best support this process.

2.1 OVERVIEW OF OPERATIONS

The plant, which we call Electronic Card Assembly and Testing (ECAT), has a staff of 2,000, working three shifts, and assembles approximately 6,000 electronic circuit boards daily for a variety of machines, primarily PC and UNIX workstation cards. The ECAT division is an intermediate stage in a vertically integrated electronics company. The plant receives bare printed wiring boards and components from various internal (i.e., other divisions of the company, not necessarily colocated) and external suppliers, and it ships assembled boards to other system assembly divisions as well as direct to customers. The boards use predominantly surface mount technology (SMT); the core of the assembly process consists of solder place application, component placement, soldering and testing operations in sequence. ECAT assembles a diverse set of printed circuit boards. It operates several SMT assembly lines operating in parallel. Each line, called Pull Production Lines or PPLs, contains the necessary equipment to fully assemble, test, and pack printed circuit boards. To minimize setups and maximize equipment utilization, each PPL assembles a limited (predetermined) set of similar products that share many common components and require similar processing steps. However, products might be dynamically reassigned from one PPL to another in order to accommodate contingencies such as the breakdown of a line or high priority for shipping certain products. The PPLs attempt to operate as continuous flow manufacturing (CFM) operations, with a strong emphasis on short-cycle manufacturing.

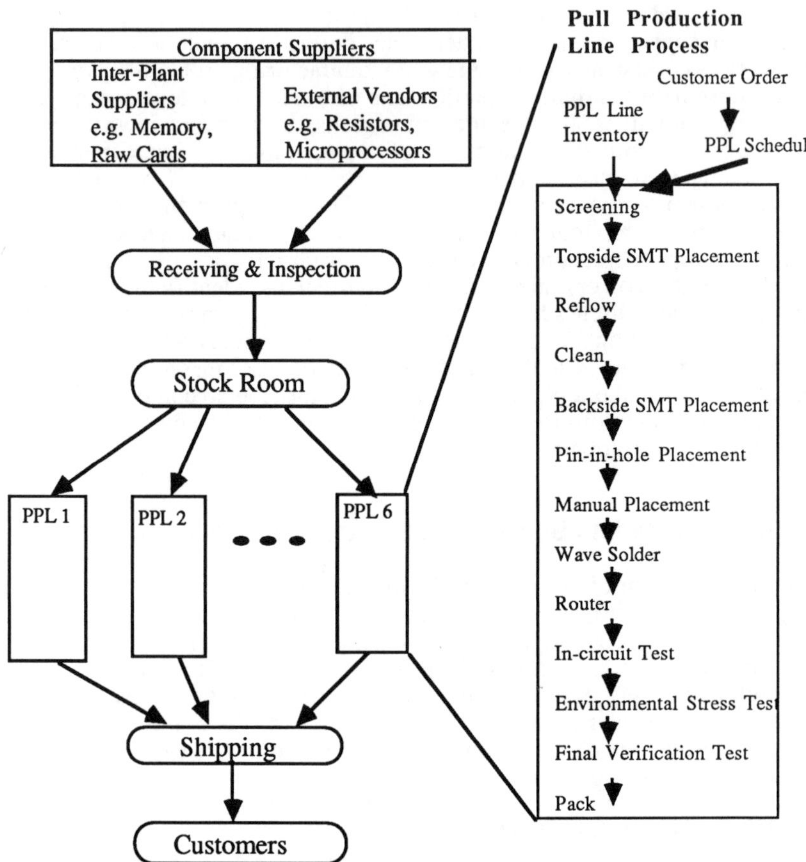

Figure 1 : Overview of the Manufacturing Process used in the Case Study

Each PPL has a manager who is responsible for the quality, throughput, flow time, cost, and delivery performance of that line. To exploit economies of scale and coordinate the different lines, support functions such as purchasing (materials), production planning and control (floor control), and quality are "centralized" for the entire plant. The various functional managers–materials, PPLs, floor control, accounting, quality, distribution–report to the plant manager. Figure 1 shows the process flow and interrelationships between the various functions in the plant (for convenience and clarity, the figure shows only one PPL). Floor control is responsible for scheduling the lines, setting production targets, and providing training to direct employees. Distribution handles the receipt of components, delivery of components to the lines, as well as shipping finished products to customers. The materials organization is responsible for procuring boards, components and supplies, and ensuring adequate supply to meet the PPLs needs. Component inventory represents the largest class of assets in the plant; hence, unit cost and inventory levels are important performance metrics for materials management. Managing component inventories effectively is critical to the business due to significant economies of scale, obsolescence, and long procurement lead times for certain parts.

2.2 PLANNING AND PROBLEM-RESOLUTION MECHANISMS

ECAT employs an elaborate planning process to develop and monitor long, medium and short-term plans in consultation with other divisions of the company. Every month, the Scheduling department generates a build plan, with inputs and approvals from the materials, production and sales organizations. This plan is fed into the MRP system which then creates parts order recommendations sufficient to cover the period until next regeneration–normally four or five weeks (see figure 2). Any event that causes an anamoly, disruption or alteration to the plan, e.g., schedule changes, inventory variance, scrap, unplanned usage, inventory record error, and so on, must be accommodated through reactive scheduling; we refer to such contingencies as reactive problems.

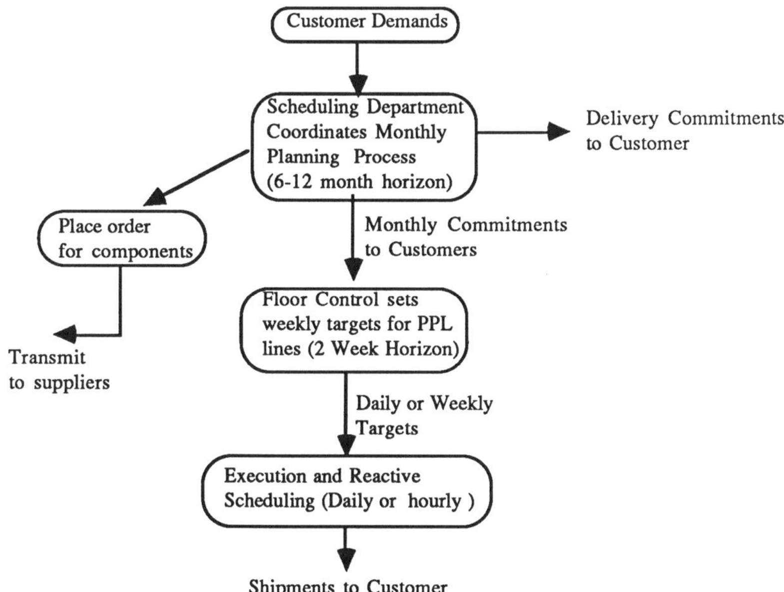

Figure 2 : Manufacturing Planning, Scheduling and Execution Process

Although some of these day-to-day problems can be avoided (e.g., by improving material control), many contingencies are inevitable due to the dynamic marketplace and the close interdependence between vendors, ECAT and its customers. In general, reactive problems arise due to internal or external events. Examples of internal events include machine breakdown, poor process yeild, absenteeism, design changes, and poor planning practices. External events refer to sudden shifts in the marketplace, vendor quality and delivery problems, and so on. We can also view reactive problems as unanticipated disruptions of resource availabilities; this view leads us to classify problems according to the affected resources, namely, labor (e.g., absenteeism), equipment (e.g., malfunction or breakdown), and materials (e.g., quality or availability).

Figure 3 shows some common operational problems and contingencies–the root causes, their chain reactions or consequences, and the impact on performance metrics–in electronics assembly. Although the problems displayed in Figure 3 are not exhaustive, this figure captures a large percentage of day-to-day production problems at ECAT and many other electronics assembly facilities. We will return to Figure 3 when we specifically address material shortage problems in Section 2.3.

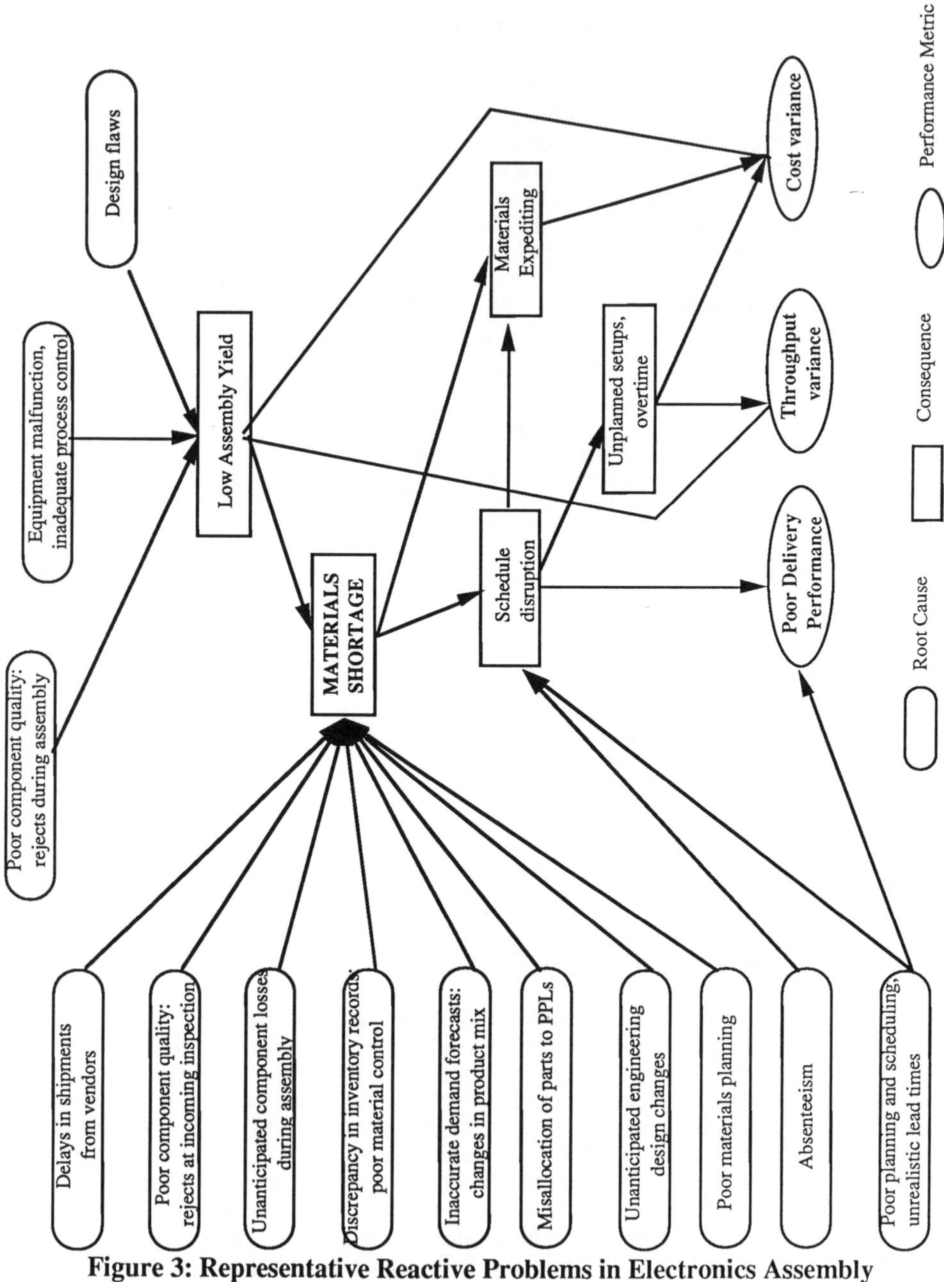

Figure 3: Representative Reactive Problems in Electronics Assembly

Continuous Flow Manufacturing permits ECAT to be responsive to customer needs with minimal inventories. However, this mode also makes the PPLs very susceptible to perturbations in demand, material shortages, yield variations, machine breakdowns and other stochastic events. Consequently, production is monitored very closely in formal, scheduled meetings (during shift changeovers, daily in the morning and afternoon, and weekly) where day-to-day production problems and weekly performance are discussed. In addition, the floor control group is responsible for monitoring performance continuously (to eliminate delays in reporting problems and taking corrective action), and the plant has several expediters to address urgent materials and production needs.

At the start of each shift, every PPL spends 30 minutes reviewing component inventory levels, work-in-process, progress with respect to weekly production targets, and problems encountered during the last shift. At the 9:00 a.m. production status review meeting, the PPL manager raises problems that require assistance from manufacturing support groups, such as materials shortages, information system failures, and process degradation. Attendees at this meeting typically include all the PPL managers, the floor control supervisor(s), and representatives from each of the support groups–materials, production control, engineering, distribution, quality, planning and shipping. Because of the close relationship between ECAT and its sister plant that supplies the (bare) printed wire board (PWB), a representative from that supplier is also present to report on PWB production and delivery schedules. The topics discussed at this meeting are:

(i) New or ongoing problems that surfaced during the shift changeover meetings. An appropriate support group might be assigned the responsibility to resolve each problem. In turn, the support group might request the formation of a cross-functional team with representatives from other groups to further investigate this problem.

(ii) Teams responsible for resolving prior problems provide a status report on their work. This report often includes a description of the solution so that the various support groups have an opportunity to review the new proposal for any unforeseen problems. For example, an engineering solution to reduce stress test time in response to a bottleneck at this operation may be challenged by the quality group because of potential impact on the quality of the finished product.

(iii) The group reviews the status of resources and processes to identify potential problems, and communicate changes in production targets. The agenda typically includes (see Appendix A for more detail):
  (a) a report on critical parts;
  (b) quality problems reported by customers;
  (c) process yields;
  (d) shipping discrepancies; and,
  (e) printed wiring board fabrication and delivery schedule

In general, no one records the minutes of the 9:00 am meeting, but copies of transparencies for each presentation are distributed to all attendees. Also, soft and hard copies of reports are also mailed to distribution lists that extend beyond the meeting participants. Depending on the urgency of problems raised during the meeting, short-term or stopgap solutions might be proposed while the team works on a more permanent, comprehensive solution to the original problem.

ECAT also holds a daily 12:45 p.m. meeting that lasts about an hour. Participants at this meeting includes the plant manager and other senior managers. The purpose of this meeting is to review long-term objectives, and acquaint senior management with the more severe short term production problems. In the latter case, senior managers are responsible for provide guidance to the operating managers by prioritizing the problems that are critical to the business. Information about impact on customers, the plant's reputation, and possible future contracts are communicated to the operational groups. New processes and new product introductions are also discussed. The overall

112

performance measurements of the plant are reviewed regularly at this meeting. These include financial measurements, such as revenues, expenses and return on assets as well as non-financial indicators such as production volumes, headcount and serviceability.

The use of monitors, expediters, meetings, and ad hoc teams for coordination and problem solving is quite commonplace in industry. Our purpose in attending the meetings at ECAT was to gain a better understanding of the purpose, characteristics, pitfalls and decision support opportunities for these generic coordination mechanisms. Meetings serve a valuable purpose in breaking down the "functional silos" and to establish camaraderie and a common purpose by bringing together managers from different functions that can act upon or are impacted by various contingencies. Meetings also serve as a means to communicate and share information, and a forum to provide input and generate alternatives. However, the size of the group, its composition, and the frequency of meetings must be carefully planned to ensure that they facilitate dialog and discussion, are taken seriously by managers, and do not degenerate into rituals. Meetings are expensive in terms of managerial time. So, we must consider ways to supplement meetings, for instance, by using information systems for routine communication, information sharing, agenda-setting and preparation. Our second observation deals with the types of issues covered during meetings. In general, we can classify the agenda items into two broad categories–routine proceedings and extraordinary (or unstructured) items. Routine activities include, for instance, reviewing daily production performance and the critical components list, although this review might spawn some extraordinary items. Again, to improve the efficiency of the meeting, we might consider providing appropriate computer-communication support (e.g., distribute the performance report and critical component list prior to the meeting so that only extraordinary items need to be addressed). Third, daily meetings tend to focus on the many current and pressing problems of the day. Consequently, short-term solutions that alleviate the symptoms temporarily might get institutionalized (because the original stimulus is temporarily out of view, and recent problems get higher priority) without pursuing longer term solutions to completion. Without formal mechanisms and timetables to review pending issues and track the frequency and impact of various problems, this process can potentially create a vicious circle, resulting in much wasted time and effort. For instance, unresolved problems will resurface after some time, and also cause other problems that are, in turn, addressed through additional fixes and so on. Therefore, information systems can play an important role to improve coordination effectiveness by recording decisions, tracking progress on various projects, providing data on previous problem instances, and so on.

2.3 COLLABORATION IN MATERIALS SHORTAGE PROBLEMS

Let us now examine a particular class of problems which we refer to as "materials shortage". By materials shortage we mean that a needed component is not available on time to commence assembly of a product as scheduled. Note that, in surface mount assembly, the production of a particular board can begin only if all the components needed for that board are available. Consequently, the shortage of even a single part can disrupt production, force lines to shut down, delay shipments, and increase unit costs (due to unanticipated setups, overtime for assembly workers, and expediting costs). As Figure 3 suggests, materials shortage is a frequent occurence, requiring considerable managerial attention and time. The problem is acute at ECAT because it uses a vast number of components (over 7000 components) to assemble over 120 board types. Each board requires 100 to 300 components, and the plant consumes approximately 4 to 5 million components per day. The direct costs plus opportunity costs for each PPL can exceed tens of thousands of dollars per hour; hence, ensuring an uninterrupted supply of the correct sets of components at the right time is critical.

As Figure 3 illustrates, materials shortage has many root causes–delayed deliveries from vendors, poor quality of components, low assembly yield, inaccurate demand forecasts, changing customer requirements, engineering change orders, errors in the MRP and materials control system, poor planning, and so on. The consequences of materials shortage are also far-reaching, from disrupting production schedules, causing unnecessary setups, and incurring overtime to delaying customer shipments. Ultimately, these consequences adversely impact all the performance metrics of the plant–delivery performance, througput, yield, and cost. We wish to highlight some important features of materials shortage problems:

(i) Almost all the functions in the plant can potentially cause materials shortage: from vendors who supply poor quality parts late, materials planners who do not plan adequate safety stocks, engineering organizations that issue frequent design changes, production personnel who are unable to control the processes and ensure consistent yield, marketing people who cannot provide good forecasts, to customers who change their requirements at short notice.

(ii) Resolving materials shortage problems requires cooperation by many different functions: vendors might agree to expedite shipments, material planners might arrange for emergency shipments from other facilities, PPLs might reschedule their production, and the sales organization might negotiate with customers to delay their shipments.

(iii) Reactive problem-solving addresses the immediate need, i.e., procure the needed component as quickly as possible, and postpone the product requirement as much as possible. On the other hand, the long-term solution must address the root cause. Consequently, the players involved in developing and implementing the reactive solution might not necessarily be responsible for or participate in the long-term improvement process.

(iv) Although "materials shortage" is a common phenomenon, it is not a routine or repetitive problem since each occurence of materials shortage might pertain to a different component and have a different root cause.

Thus, materials shortage requires effective coordination and collaboration among all the functions in the plant, and addressing this problem quickly is critical because it can have a considerable adverse impact on plant performance (especially in tightly-coupled systems such as the CFM environment). We wish to address the following questions: What is the nature of the collaborative process needed to resolve materials shortage problems? What are the necessary information and material flows? What is the nature of technology required to support the collaborative requirements? At ECAT, the collaborative process takes place primarily in the production status meeting; Appendix A contains the minutes of one such meeting.

Let us examine the mechanisms and processes at ECAT to address materials shortage or "critical parts" problems. Because this problem is quite common, ECAT has specially designated an individual as the *critical parts coordinator* (CPC). Early each morning, the CPC uses a computer system (called the Logistics System) to identify components with remaining inventory less than or equal to 5 days (at predicted usage rates), and distributes the critical parts report (via electronic mail) to about 25 production control (PC) analysts. Each PC analyst deals with a particular subset of components, working directly with the supplying divisions or the corporate procurement center (for parts supplied by outside vendors). The PC analyst is responsible for ensuring that the supply pipeline has sufficient orders for his/her parts to cover production needs, while monitoring and controlling inventory costs. When they receive the critical parts list, the PC analysts check the status of orders, and inform the CPC regarding anticipated problems with the supply stream (and the reasons for these problems). Typically, they will confer with the suppliers to ensure that orders that are due in the immediate future will arrive as scheduled, expedite orders if necessary, or simply reconfirm that the inventory quantities recorded in the plant's materials logistics systems are correct. For the most critical components, the CPC often calls the respective PC analysts to ensure that he has adequate information by 9:00 a.m. to report on the current status and acquaint managers with potential problems during the production status meeting. At the meeting, manufacturing managers might challenge the inventory quantities specified in the inventory report, particularly the quantity of materials on the manufacturing floor itself. In these cases, the CPC requests (after themeeting) a production control person to take a physical count of the inventory, or search for the missing quantities.

When the follow-up actions for a critical part suggest an imminent shortage, Floor Control initiates and coordinates actions on many fronts to react to this contingency. The Floor Control organization has a team of 5 members assigned to each PPL. When a materials contingency arises, the team that is responsible for the affected PPL is immediately notified, as is the senior materials planner responsible for that component. The materials planner will work with the PC analysts handling the problematic components to find extraordinary ways to alleviate the supply problem. For instance, one alternative might be to look for other sources, e.g. at sister plants, other suppliers and distributors. Another would be to look for substitute components that are acceptable to board designers and engineers. Meanwhile, a Floor Control team member prepares contingency schedules for the affected PPL in case the materials organization is unable to procure the component or a suitable substitute on time. Rescheduling the PPL involves assessing the impact of delaying the production and delivery of the affected product on the customer, and deciding which other product must be assembled in its place (given the current setup of the PPL, the due dates of different orders, etc.). Floor Control needs to keep close tabs on the progress of the materials organization's efforts to determine if changing the production schedule is inevitable. Conversely, the Floor Control person might discover that the schedule disruption has no effect on the customer, e.g. if the customer currently has abundant stock of this card anyway, in which case the materials organization must be informed so that they can refocus their efforts and avoid disproportionate expense to expedite the component.

If the cost to procure an emergency shipment of a critical component exceeds a certain threshold, senior management needs to be involved to decide if the additional expense is worthwhile and affordable. If materials shortage cannot be avoided, the planning/floor control group must contact the customer to re-schedule the order, and also work with the PPL to implement a change in the production schedules. Materials shortage problems are further complicated because several parts might simultaneously become critical, and certain parts are common to multiple products. Suppose the plant has insufficient supply on two or more components needed for the same product, and suppose these components are assigned to different PC analysts (for instance, if they are supplied by different vendors, and one is a mechanical component,while the other is a passive or active electronic device). What if one of these components is available (say, from a distributor) at a higher price? Should the plant purchase this component to tide over the current emergency? In theory, procuring this component at higher cost is worthwhile only if the other component can also be expedited; otherwise, the board assembly must be postponed anyway. When each PC analyst performs purchase negotiations independently, the plant might incur additional cost of expediting components without actually increasing throughput or delivering the product on time. Another complication arises when a critical component is common to several different products (that are assembled, say, on different PPLs) that are concurrently scheduled, and current supply is inadequate to meet the immediate needs of all the PPLs. In this case, Floor Control must also decide how to allocate the available inventory among PPLs, and which PPL(s) must be rescheduled. This process involves extensive consultation, interactions and negotiations with several PPL managers, customers, etc.

The materials organization classifies components into two categories: key and non-key parts. By definition, a non-key part has minimal dollar value; the plant does not maintain rigorous inventory records for these parts, and often the plant negotiates a "Contract Line Delivery" (CLD) with the supplier, i.e., the supplier is responsible for maintaining a specific inventory level. Materials management devotes considerably more attention to monitoring inventory levels–at the stock room and the PPLs–for the key parts. Figure 4 shows the rough sequence of information flows and events for shortage of a key part. Although this figure represents the interactions as a sequential process, resolving materials shortage problems is very much a concurrent, and collaborative effort with simultaneous exploration of many different alternatives.

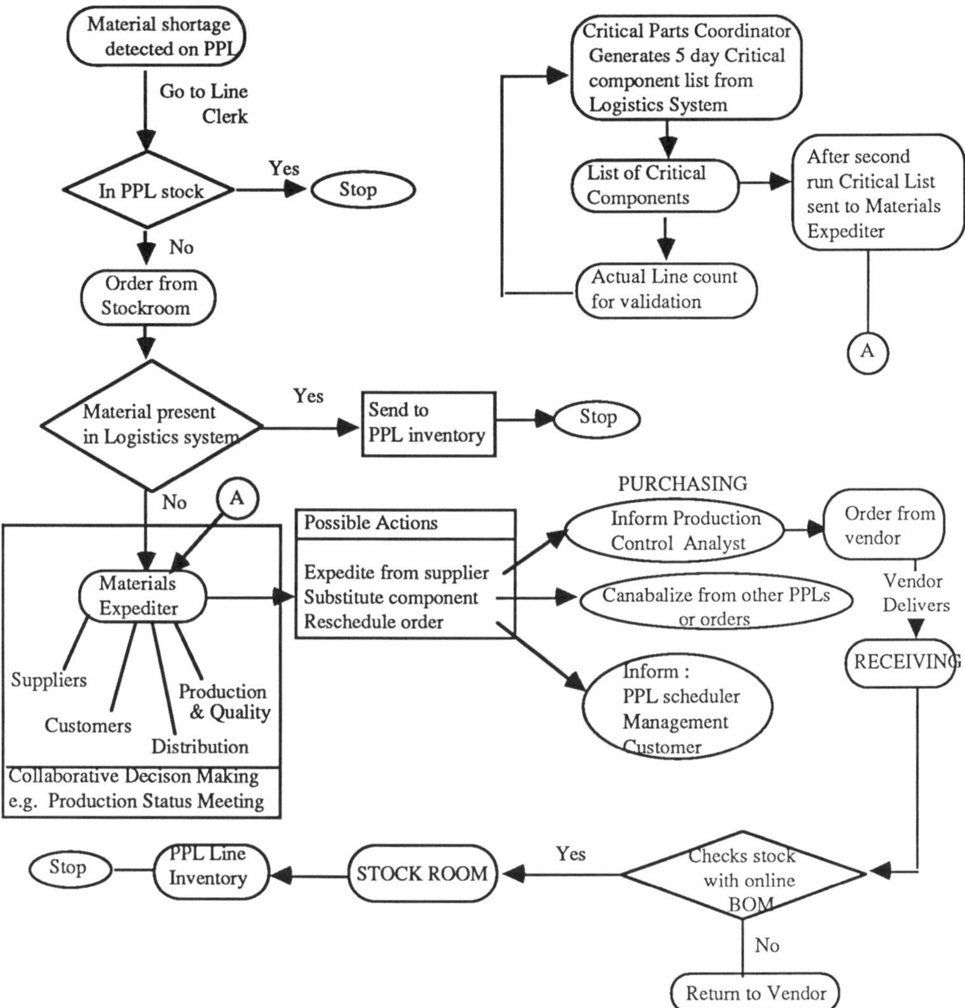

Figure 4: Materials Shortage Workflow Diagram

Our previous discussion focussed on the case when the CPC identifies potential materials shortage (prior to its actual occurrence) by monitoring the 5-day critical parts list. If the inventory records are inaccurate, then materials shortage might manifest itself much later, when the assembly worker gets ready to setup the PPL for a new product but discovers that one or more components needed for the product are not available for loading on the placement machines. In this case, the worker contacts the PPL line clerk to obtain the necessary part. If the part is not available in PPL inventory, the clerk queries the logistics system to find out whether the the plant's stock room has this part. If found, the clerk places an order through the logistics system, and Distribution delivers the part to the production line. Otherwise, the clerk escalates the problem to a higher level by informing the materials expediter or planner about the part shortage. The materials expediter is a senior manager in the materials management, whose sole responsibility is to solve these crises in conjunction with other groups. The materials expediter faces three possible courses of action: (i) inform the appropropriate PC analyst to expedite delivery of that part, (ii) transfer parts from other PPL lines that might have this component in their inventory, or get the part from sister plants, and/or (iii) postpone the assembly of this particular board to a later date, and inform the management, customer and PPL scheduler. Deciding which action to pursue requires coordination with different teams, functions and organizations. The decision process is collaborative; the materials expediter serves primarily as a mediator and facilitator in this process.

To summarize, this section has examined a common reactive problem–materials shortage–in the electronics assembly context. By observing actual operations, we have generated insights about the causes, mechanisms, solutions, and coordination needs to resolve this class of problems. An extensive multi-product, multi-line facility such as ECAT faces additional reactive problems due to "shortage" of other resources (e.g., labor or machines) and changes in demand. For instance, when a PPL goes down due to, say, breakdown of a placement machine, the management must decide how to reallocate production to the remaining PPLs, and also simultaneously take actions to repair the machine. This contingency requires interactions between different PPL managers, process engineers and maintenance crews, sales, customers, distribution, and scheduling. Similar multi-function collaboration is needed when customers change their orders, or engineering releases a design change. In general, the planning and reactive problem-solving process must be geared towards dynamically matching supply and demand to achieve the following objectives (Bang, Decker, Gisler and Haugen [1990]): (i) scheduling final production to current customer demands so that final product is delivered JIT to the customer, (ii) managing customer orders to be consistent with the supply schedule, (iii) implementing inventory buffers in the supply pipeline to respond to global sourcing and changing customer and market requirements, and (iv) using scheduled delivery pulls and forecasts to communicate delivery and start requirements to suppliers based on customer order requirements. In turn, the scheduling function must account for constraints imposed by machine capacity, operation sequence, lot size, waiting time, and workforce availability. Next, we examine the literature on coordination and collaborative problem-solving to better understand how to support the complex interactions and decision-making processes in tightly-coupled environments such as ECAT.

3. Coordination for Contingencies

Coordination refers to the means for integrating or linking together different organizational subunits whose decisions and actions are interdependent in order to achieve common goals (Van de Ven, Delbecq and Koenig [1976]). As we noted in the introduction, achieving agile manufacturing increases the interdependence of functions. The actions of one subunit not only defines the available options of other subunits but also impacts their performance, and ultimately affects the effectiveness of the organization (McCann and Galbraith [1981]). Coordination is necessary to ensure goal congruence and induce complementary, rather than contradictory, actions. Understanding the interdependence in an organization is, therefore, an essential first step in developing coordination support systems.

Coordination arrangements to integrate "boundary spanning" workflows (compared to, say, intrafunctional coordination) are especially important and challenging. Often the different subunits have differing perceptions of interdependence, and disagree over strategies because: (1) asymmetries exist in their status and access to information, (2) different subunits have varying performance and reward criteria, (3) some subunits are pursuing long-term goals and proceed systematically (e.g., by identifying and addressing root causes) whereas others have a short-term orientattion, (4) task ambiguities stimulate unilateral actions to reduce ambiguities through informal sources, and (5) difference in training and knowledge bases create semantic barriers and perceptual diversity (McCann and Galbraith [1981]). To overcome these difficulties, a large amount of ad hoc or informal coordination actually takes place outside structured arrangements, for instance, to resolve short-term production problems. Ad hoc coordination mechanisms include face-to-face communication, group discussions, bargaining and negotiation, and incentive mechanisms. The critical question is how to design information technology to support ad hoc coordination structures.

3.1 MODES OF COORDINATION

Researchers have proposed several classification schemes for coordination mechanisms (March and Simon [1958]; Thompson [1967]; Argote [1982]). Three general task coordination methods are: programmed coordination, non-programmed coordination, and coordination by feedback. Applications of programmed versus nonprogrammed coordination differ primarily in terms of the extent to which activities and problems can be specified in advance. In *programmed coordination*, the activities are dictated by organizational goals and objectives, and plans are specified in advance by the organization. Thus, programmed coordination is characterized by mechanisms such as pre-established plans, schedules, forecasts, formalized rules, policies and procedures, and standardized information and communication systems. The common element of these mechanisms is that a codified blueprint of action is specified in advance; once implemented their use requires minimal communication between subunits responsible for performing the individual tasks. In *non-programmed coordination*, on the other hand, activities are not specified in advance but usually occur as a result of some emergency. For instance, this type of coordination might apply to the materials shortage problems we described in Section 2. Coordination by *feedback* combines both the programmed and non-programmed modes; it is defined as the mutual adjustments made to an original plan based upon new information. Organizations often use two operational modes to develop plans and make mutual adjustments: a personal mode and a group mode. In the personal mode, individuals make mutual task adjustments through vertical or horizontal communication, whereas in the group mode, a collection of individuals use scheduled or unscheduled meetings to make mutual operational adjustments. The type of computer support best suited for non-programmed and feedback coordination in complex and dynamic environments depend on whether the interactions and responses to different contingencies are relatively well-structured and share common patterns. If the coordination mechanisms needed to resolve the problem are fairly well-understood, then we can encapsulate them in software to ensure quick and efficient handling. But if the interactions and sequence of actions are complex, sequence and state-dependent, and vary widely from one scenario to the next, then collaborative technology might be the most appropriate computer support.

3.2 EFFECT OF UNCERTAINTY

Since we are primarily interested in coordination for contingencies, we must understand the nature and causes of the uncertainty that causes these contingencies, and how manufacturing organizations cope with uncertainty. Uncertainty appears in two forms:

(i) lack of information necessary to make the decision, and (ii) lack of information about the outcome of the decision or the perceived risk associated with the decision. Lawrence and Lorsch [1967] showed that firms simultaneously employ multiple modes of coordination, such as hierarchical communications, rules, plans, and spontaneous contacts to deal with uncertainty. Duncan [1972] showed that departments have multiple decision-making structures, some of which are used for routine situations, such as hierarchical authority and rules, and some for exceptional situations, such as ambiguous roles and participation; organizations facing a high degree of uncertainty still use hierarchical structure but to a lesser extent. Van de Ven *et al.* [1976] examined the effect of task uncertainty, task interdependence, and size on the coordination modes, and found that task uncertainty has the strongest influence on the use of rules, plans, vertical and hierarchical communications, horizontal communications, unscheduled and scheduled group meetings. Increasing work flow interdependence (team arrangements) results in lesser use of the hierarchy, and also reduces the importance of plans as a means to coordinate activities within organizational units. The effects of task uncertainty and interdependence on the effectiveness of different modes of coordination across boundaries are not clear especially when the task itself is actually a composite entity that has multiple activities as antecedents. Argote [1982] examined the effect of input uncertainty on programmed and non-programmed coordination in emergency hospital units, and found that programmed means of coordination are most often used in units experiencing low uncertainty while non-programmed coordination is most appropriate for units experiencing high uncertainty.

Although the literature presents sufficient evidence to show that the level of uncertainty characterizing an organization's operation impacts the choice of means of coordination, there is little evidence to show that the actual methods used to coordinate work groups are designed with uncertainty as a critical design factor (Argote [1982]). One explanation for this phenomenon is that designing the coordination mechanisms is a complex activity that is affected by many variables, and focusing on a single dimension such as uncertainty is insufficient to capture its nuances (Gresov [1989]). Also, these mechanisms evolve in response to a diverse set of requirements, and their design is affected by multiple contingencies often involving tradeoffs that hinder a fit to all of them (Gresov [1989]; Gerwin [1981]).

Another form of uncertainty in manufacturing pertains to the definition of the problem itself, and a lack of understanding of the problem's root causes. Problem diagnosis is especially difficult in environments with many different explanatory and dependent variables. Problem diagnostic decision trees provide a useful way to classify problems into a known universe of objects, and the tree metaphor can capture meaningful relationships between the object's class and value of its attributes (Quinlan [1993]). To construct these diagnostic trees, both inductive and deductive reasoning techniques have been used. Currently, diagnostic trees are limited in scope because of the lack of methods to deal adequately with a constantly changing environment, and lack of information about key attribute values (although machine learning algorithms such as, ID3, C4 and C5 (Quinlan [1993]) are being used to address this problem, this area is still in its infancy). The problem of diagnosis is compounded by the lack of adequate process understanding and reliable information about the manufacturing processes. Information technology can facilitate diagnosis by eliminating some of these obstacles.

4. Conceptual Framework for Reactive Problem-Solving

This section describes a conceptual framework for the reactive problem-solving process, based upon the literature on individual decision-making (primarily, Newell and Simon [1972] and Dewey [1910]), organizational decision making (Mintzberg *et al.* [1976],

Witte [1972]), and our case study of electronics assembly operations. We decompose the process into four phases–problem identification, diagnosis, solution, and implementation– that each have distinct information and coordination needs. We describe how to support the diagnosis and solution phases using a dialectic model.

Dewey's [1910] framework of reflective thought, which has served as the foundation for several subsequent theories, consists of five phases: (1) suggestion, where the mind leaps to possible solution; (2) intellectualization of the felt difficulty into a question; (3) development of hypotheses; (4) reasoning or elaboration of theses; and (4) testing of hypotheses (Mintzberg *et al.* [1976]). Simon [1965] postulated the trichotomy of intelligence-design-choice. Witte [1972] examined 233 organizational decision-making processes associated with data processing equipment and found no evidence for the sequence of actions suggested by previous frameworks. He found that the decision-making process consists of a plurality of subdecisions, and communication activities dominate this process. Mintzberg *et al.* [1976], building on Witte and Simon's work, described seven routines to conceptualize the strategic decision-making process observed in a field study of 25 decisions. The framework used in this paper is consistent with Mintzberg *et al.* model, and is also related to plan-do-check-act cycles and other multi-step methods discussed in the total quality management literature.

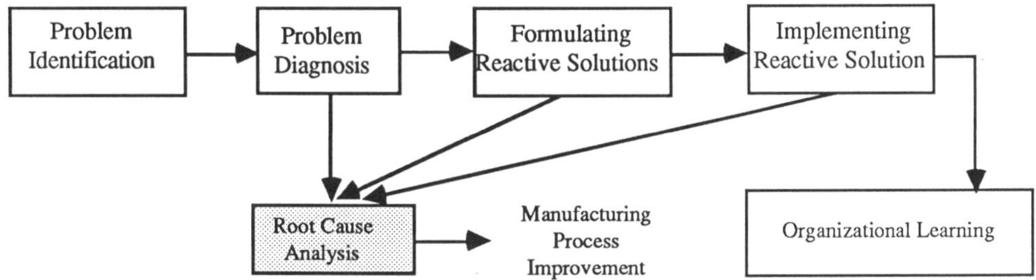

Figure 5 : Model of Reactive Problem-Solving in Manufacturing

Figure 5 presents our view of the reactive problem-solving process. The process begins by observing symptoms and recognizing that a problem exists and must be addressed. The specific problem must then be identified and defined. Following problem identification, the causes underlying the problem must be determined using aids such as fishbone diagrams. The nature and depth of problem diagnosis might differ for reactive problem-solving versus long-term improvement. The third phase consists of developing a strategy to resolve the immediate problem, and formulating the necessary action plans. This phase requires iteration if the problem is multi-functional, the organization faces many different alternatives, and different functions must participate to evaluate proper tradeoffs and agree on a plan that meets the organization's goals. The coordination requirements–information sharing, communication, collaboration, and synchronization of decisions and actions–are most intense during this phase. Finally, the implementation phase consists of executing the reactive action plan, and reporting the results. The diagnosis, solution, and implementation phases generate inputs to root cause analysis and long-term improvement; the experience with collaborating and implementing the reactive solution also provides the opportunity to learn and improve upon the problem-solving process itself.

Observe that the process that we have described applies to reactive problem-solving (after a problem has occurred) but not necessarily to contingency planning (before problems occur). Organizations that operate in dynamic and highly uncertain environments often draw up contingency plans in anticipation of various problems. These plans effectively correspond to programmed coordination; they specify the course of action for different organizational units under each scenario. In the manufacturing setting, contingency plans might correspond to establishing inventory and time buffers between different stages of the manufacturing process so that problems in one stage do not affect upstream or downstream stages. As we mentioned in the introduction, competitive pressures are forcing companies to eliminate such safety buffers (to ensure quick response and flexibility); consequently, manufacturing managers must increasingly rely on quick reactive problem-solving capabilities to respond to contingencies. Next, we consider each of the four problem-solving phases of Figure 5 to understand the coordination needs.

4.1 PROBLEM IDENTIFICATION PHASE

Problems surface when managers detect a threat or observe a variance in their performance metrics (i.e., deviation in a performance metric is the typical symptom). Anticipating and identifying problems quickly raises issues of choosing the right metrics and leading indicators that can predict problems, developing the ability to recognizing the core problem from a multiplicity of symptoms, and communicating problems quickly to the appropriate level once they are detected. For instance, ECAT uses the 5-day critical list as a leading indicator to detect potential material shortage problems. Due to deficiencies in the material control system, some impending materials shortage problems might go unnoticed, in which case the problem is detected only when the assembly worker starts setting up the machines. The problem must then be communicated swiftly to the materials manager, PPL manager, and floor control supervisor so that they can initiate corrective actions. once a problem is detected, is it communicated swiftly to the appropriate level.

In general, problems can be classified into two broad categories: "tame problems" and "wicked problems." (Rittel and Webber [1973]) Tame problems have: (i) well constrained goals; (ii) can be formulated with precision; (iii) are recurring and not unique; (iv) have algorithmic solutions. Wicked problems on the other hand, have: (i) complex and vague goals; (ii) can only be stated in an approximate way, and the definition of the problem changes with time, resources and context; (iii) are unique and the solution is not transferable to another situation; and (iv) algorithmic solutions are not possible. We are concerned mainly with so called "wicked", unanticipated problems that involve discovering and interpreting signals from the manufacturing environment to accurately define the problem. These problems have no one correct answer but a set of possible solutions each of which can be justified by taking various supporting positions. Problem identification is thus a form of interpretation, as individuals are constantly trying to understand and process the information that is available to determine if corrective action is needed and to identify the root cause. Research suggests that many organizations are informal and unsystematic in their interpretation of the environment (Daft and Weick [1984]). In a crisis, the organization might respond actively, but the lack of interpretation systems adversely affects the quality of problem identification and diagnosis. Preventive problem-solving, which involves identifying problems before they escalate into crisis, also requires sound interpretation systems.

4.2 PROBLEM DIAGNOSIS PHASE

Problem diagnosis entails understanding, through appropriate data gathering and analysis, the factors that precipated the problem and the cause-effect relationships. Drucker [1971] argues that the emphasis on careful problem diagnosis is one factor that contributes to the success of Japanese manufacturing enterprises. Adams [1986] describes the pitfalls of the "hit-and-run" approach to problem solving:

".. the natural response to a problem seems to be to try and get rid of it by finding an answer- often taking the first answer that occurs and pursuing it because of one's reluctance to spend the time and mental effort needed to conjure up a richer storehouse of alternatives from which to choose. This hit-and-run approach to problem solving begets all sorts of oddities- and often a chain of solution-causing-problem-requiring-solution, ad infinitum. In engineering, one finds the "Rube Goldberg" solution, in which the problem is solved by an inelegant and complicated collection of partial solutions."

In the reactive setting, problem diagnosis must often rely on incomplete and inaccurate information, and so the support system must permit managers to make informed decisions with limited information; it must also overcome managers' reluctance to use formal diagnosis under intense time and performance pressures.

We can distinguish problem diagnosis activities along two dimensions: spatial and temporal. In the spatial dimension, problem diagnosis consists of two parts: local diagnosis and global diagnosis. Local diagnosis has a limited locus, usually intra-functional. For example, a process engineer might resolve a process problem with the help of specialized technicians, or PPL scheduler and the PPL inventory line clerk might resolve a material shortage problem without escalating it. Global diagnosis, on the other hand spans departmental boundaries. If the final decision affects an entire process such as materials flow from the from the suppliers to the PPL line inventory, global diagnosis is done by a cross-functional team. In the temporal dimension, we distinguish between short-term and long-term (or root cause) diagnosis. Short-term diagnosis usually focusses on factors and alternative that are relevant to a reactive solution to tide over the immediate contingency (however, even reactive problem solving might require some cursory root cause identification). Long-term diagnosis, on the other hand, is a detailed, systematic effort to identify the root causes of a problem, and correct or eliminate these root causes through design, process or operations improvement. As Figure 3 illustrates, problems such as materials shortage can have many different root causes. Indeed, multiple causes and solutions are not uncommon, and deciding which cause to address requires dialog and discussion among various functional managers. For instance, a "bridge" defect on a printed circuit board (when a solder bump spans two adjacent solder pads) might be due to either poor board design or improper soldering. Different managers might have different opinions regarding he true root cause of the problem and its resolution. We therefore need a generally accepted process for deciding the root cause. Forcing the managers to diagnose the root cause each time a problem occurs will make them cognizant of persistent problems, leading them to champion and develop long-term improvements.

In the root cause phase, a dialectic structure is necessary to facilitate and structure the exploration of the problem space. A dialectic model is most appropriate because in a regular communication approach the process and structure of problem analysis never gets captured or documented in an explicit representation (Hashim [1991]) for future use and organizational learning. One such dialectic model, developed to deal with the category of wicked problems, is Issue Based Information Systems (IBIS) (Rittel and Weber [1973], Hamalainen *et al.* [1992]). We have extended the vocabulary of the IBIS method to support root cause analysis. The model is parsimonious and explicitly accounts for all the activities performed in the local and global diagnostic phases such as identifying the problem, scoping the investigation, developing hypotheses, and corroborating them. Our variant of issue nets (Rittel [1980]) consists of the following elements: **HYPOTHESIS** : A set of one or more interrelated CLAIMS that are proposed as a solution to the problem. A **CLAIM** is a part of a HYPOTHESIS that deals with a specific subproblem or symptom. A **CAUSE** is proposed as a explanation of a CLAIM in a hypothesis. An **ARGUMENT** questions or supports a proposed cause. A **REFERENCE** supports a hypothesis, proposed cause or argument. **DATA** (may be statistical) forms a significant part of root cause analysis and can be used to support a hypothesis, proposed cause or argument.

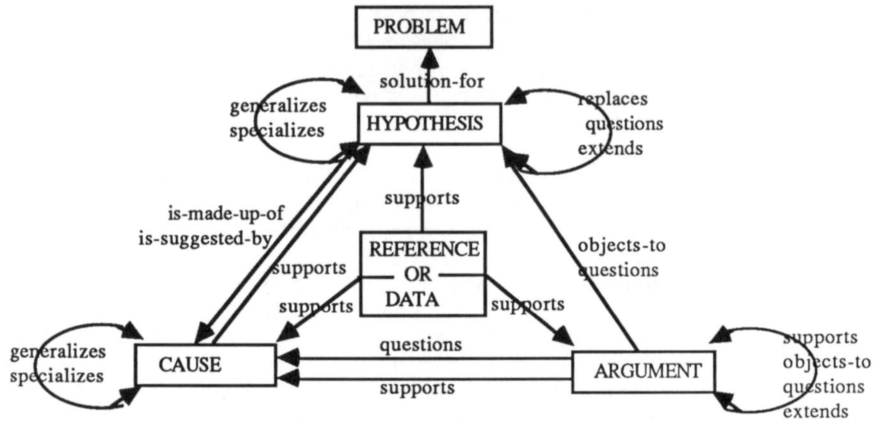

Figure 6: Vocabulary of Root Cause Discussion

Figure 6 shows the relationships between the nodes representing the elements of system. The dialectic approach also enables the creation of an active, longitudinal problem list with discussions about the problems, their causes and solutions. Since all the information is recorded, this system permits managers to even learn about problems that are currently beyond their domain of expertise; it contains historical information on trends, diagnosis, and solutions, and provides the ability to segment and view the database along several dimensions such as, date, time, frequency and problem category.

4.3 REACTIVE SOLUTION FORMULATION AND EVALUATION PHASE

Evaluating potential solutions to the problem and developing an action plan requires generating many different alternatives, assessing the effectiveness of each alternative, and exploring the actions needed to implement the strategy. The process of deciding the best course of action must balance the need to "optimize" the solution, while meeting stringent deadlines and resource constraints. The solution phase uses two basic techniques: search and design. Search is evoked to find past solutions to the problem whereas design refers to developing solutions to a novel situation or modifying past solutions to fit current needs. Deciding which solution to adopt requires information from different functions. For instance, to reschedule a board (i.e., delay its assembly) due to materials shortage, we need information about backorders and customers' willingness to wait (from sales), anticipated delivery dates from suppliers (from materials management), flexibility of production schedules (from floor control) and so on. Similarly, to expedite a shipment from a vendor, we need information on how much the vendor can ship, when will it arrive, and how much it will cost.

According to Winograd [1987], conversations for clarification, possibilities, and orientation form the fabric of cooperative work. Thus, in this phase we orchestrate a dialog, using the dialectic approach, between the players involved in making a decision. For instance, depending on the problem and its causes, managers in sales, production scheduling, purchasing, and other line managers need to participate in generating and evaluating alternative reactive solutions. Therefore, based on the problem and its hypothesized cause, the conversation must have specific dialog extensions to IBIS that caters exclusively to this phase of problem solving (see Figure 7). This model accounts for all the activities performed in the formulation and evaluation phase by explicitly representing tasks, commitments, and roles that make up the flow of work in the solution. The derivative of the issue nets that we use consists of the following elements: **SCHEDULE**: A set of one or more interrelated ACTIONS that are taken to solve the problem. These **ACTIONS** are limited by a set of **CONSTRAINTS**. For example, tradeoffs between inventory, material handling costs and the potential consequences of lost production or delayed deliveries must be carefully considered in ACTION generation. In response to a contingency, managers must evaluate a variety of ACTION choices, but recognize CONSTRAINTS on resource availability and time. ACTIONS are specific coordinated tasks o be performed by various departments. ACTIONS impact and are affected by the performance metrics (both explicit and implicit) of different subunits. If metrics are consistent across subunits, then all subunits can act coherently in formulating the solution, but if they are contradictory the solution planning phase might raise conflicts and intensify problems of coordinating ACTIONs. The problem-solving team must anticipate the **CONSEQUENCEs** of each ACTION to determine if the proposed actions adversely affect performance metrics and create conflicts of interest. A **POSITION** as in IBIS, is a general response to a proposed ACTION. An **ARGUMENT** questions or supports a proposed ACTION or POSITION taken by a discussant. Agreement on specific ACTIONs results in a **COMMIT** by each department to execute the respective actions in proper sequence. These COMMIT's are necessary to ensure accountability and tracking in the workflow.

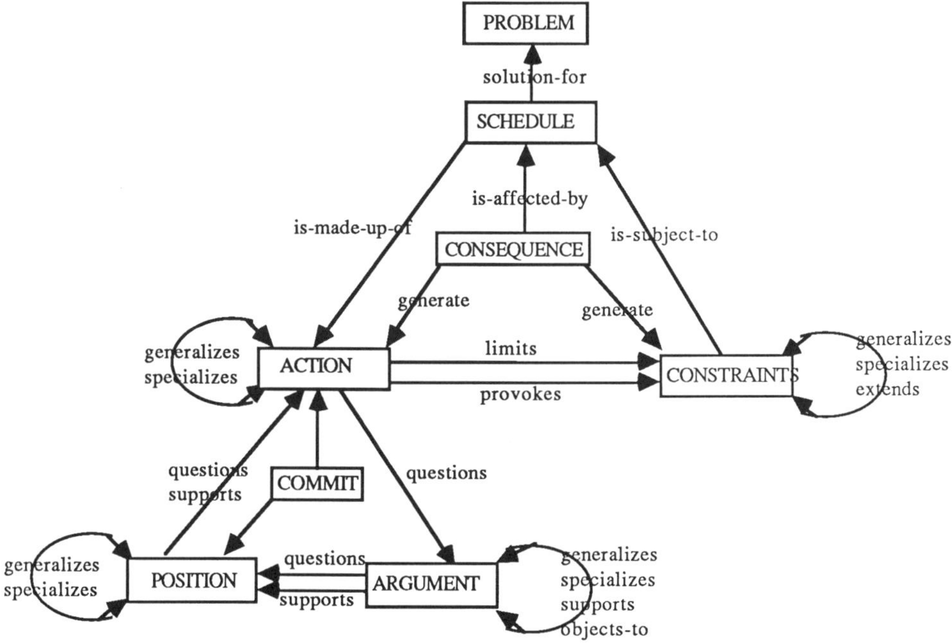

Figure 7: Vocabulary of Reactive Solution Formulation

In addition to these primitives–actions, constraints, consequences, positions, arguments, and commits–the dialog is governed by two kinds of rules: usage rules and regulating rules. *Usage rules* define the application of these primitives. Regulating rules

124

specify: (a) conditions under which a dialog can be initiated; (2) the nature of contributions to the dialog; (3) the terms for ending the dialog (Hamalainen *et al.* [1992]). Every COMMIT action has certain pre-and post-conditions. Pre-conditions are preceding actions that must be completed and resources that must be available (i.e., in the terminology of programming languages, the parameters of thefunction) before this COMMIT can proceed. Post-conditions represent the system state that other agents in the process assume to hold when this particular COMMIT is honored.

4.4 REACTIVE SOLUTION IMPLEMENTATION PHASE

As the supply chain becomes more closely integrated, reactive solutions require monitoring and action by various agents along this chain, and so coordination plays an important role in the effective implementation of solutions. For instance, production scheduling requires accessing process routing and timing information, keeping track of component inventories and vendors' performance, and providing monitoring capabilities of run-rates, yield rates, and work-in-process. Integration of existing systems would enable managers and planners to track transactions spawned by tasks. The ability to create applications triggers that are activated on certain conditions or actions allow for the distributed automation of task events. Automatic forwarding of transactions from one queue to another facilitates more flexible scheduling and control. Although workflow computing is in its adolescence, it is evolving quite rapidly with the agent-task model, making automated coordination practical. Once the reactive plan goes into effect, the effects of the actions must be assessed for correctness. If the solution has satisfactory outcomes, then it must be standardized and stored for future reference when similar problems recur. The implementation phase consists of the following activities : (i) notify agents about actions that need completion; (2) provide agents with specific information and tools to complete actions; (3) remind agents with alerts, triggers and follow-ups to synchronize workflow; (4) provide instantaneous access and visibility of the status of the workflow; and, (5) provide feedback on the success and failure of various actions in the schedule so that the organization can learn.

4.5 LEARNING FROM THE REACTIVE PROBLEM-SOLVING EXPERIENCE

Experience from problem-solving is central to the creation of knowledge and organizational learning. According to March [1991], organizations accumulate knowledge over time from their members, storing it in their norms, procedures, rules and forms. Accelerating this learning process has prime importance to cope with the increasing rate of change in technology, markets, and the operating environment. Senge [1990] identified two types of learning: generative learning, which pertains to creativity and innovation, and adaptive learning, which focusses on coping with changes in business environments. Argyris and Schon [1978] have proposed two types of organizational learning called: (i) single-loop learning, in which organizations adjust their behavior relative to fixed goals, norms, and assumptions, and (ii) deutero or double-loop learning, in which, goals, norms and assumptions, as well as organizational behavior, are open to change. Generative and deutero learning are more closely related to long-term improvements where detailed and systematic efforts are made to identify and correct root causes of problems, while adaptive and single-loop learning are necessary for reactive problem-solving. In principle, the frequency, severity, and persistence of day-to-day operational problems must serve as inputs to prioritize long-term improvement projects, and the experience with reactive solutions provides data for root cause analysis and long-term solution planning. It is useful to understand the difference between organizational learning and organizational memory. Learning is a process which affects the cognition and preferences of managers, and in turn influences future actions. Managers learn by understanding and reflecting on current processes and improvement projects. Understanding involves both inferring new knowledge, and discarding obsolete and misleading knowledge. Organization memory, on the other hand, is the capability to organize, store and manage information that is generated in an organization. IT has traditionally supported organizational memory through the use of databases but has not adequately addressed the problem of organizational learning.

5.   Prototype Implementation

In this section, we describe the design and implementation of a prototype that demostrates in one way in which the requirements described in the foregoing sections have been satisfied. The prototype design is based on Lotus Notes and Lotus Notes API's (Application Programmable Interface -- A set of C language routines that allow users and value-added resellers (VAR) to build extensions to the base system) implemented in Microsoft C, which provide an asynchronous, integrated application environment based on the client/server paradigm with facilities to construct free-form (words, numbers, and pictures) document databases, user interfaces, mechanisms for storing structured documents, connectivity through networking and communication services, electronic mail and database replication. Notes is example of a class of software programs called Groupware and more broadly Computer-Supported Collaborative Work (CSCW). While research in CSCW tends to focus on the individual work group, a broader organizational perspective needs to be taken in the case of reactive problem solving, as evident from our case study, in order to support coordination and task interdependencies between work groups from different functional areas. To support this broader view we saw the need to embedd the framework in workflow technology. Workflow technology is that aspect of groupware concerned with keeping track of tasks, events, work, or projects, along their various stages toward completion. This involves more than just entering project information in a database but actually coordinating the activities. This involves basic project-management tasks, such as tracking project beginning and ending dates, and also utilizing the computer network's capability to send and receive messages to ensure that the projects arrived at the next location. It also provides capabilities to perform calendering and agenda-setting functions (for regular meetings), and distribute advance information to participants through electronic mail.

Client-Server Overview : As self-managing teams become increasingly prevalent in manufacturing, it becomes important that the knowledge of every member or client station be collectively leveraged by others in the team and across teams to solve problems quickly and in some cases, concurrently. A cornerstone of such an architecture is the client/server relationship, an arrangement whereby one program - on the client - obtains the services of another program - on the server - by means of a well-defined interface. Although client and server processes can use the same hardware/software platform, usually the client's functionality resides on the workstations and takes advantage of the graphical user interface capabilities provided by these workstations, with the server functionality residing on larger, more powerful systems designed to support the resource sharing required by this environment. The client platforms are networked to the server platforms through the local area network (either Ethernet or Token Ring and in our case, Wireless Ethernet WaveLan), and multiple servers in the network can be arranged hierarchically or in a peer-to-peer structure, extending the client/server architecture throughout the enterprise. In our prototype we have a combination of a client portion that interacts with the user, and a server process that acts as a software "engine" managing shared resources. The client initiates the exchange by establishing a "conversation" through interprocess communication, for example named pipes; with the server so that it can send transactions to the server (A transaction is a request for data or services). The server responds to these transactions by providing data or services to the client.

Information Architecture. The enterprise wide information architecture in Figure 8, consists of several subsystems to support the diverse capabilities required for reactive problem-solving. The enterprise level, production planning and logistics system contains all the information about production, schedules and part information. The Notes server (OS/2-based) is linked to the enterprise system through a gateway or a bridge. The Notes server provides the collaborative infrastructure to support group interaction between clients (Windows-based) in the problem solving process. To support the information needs, all the parts information is stored in an external database and imported into Notes to be displayed in a form. The user can query the database by either inputing a query in a Notes form or selecting a predefined canned query. These queries would then be translated into API calls to the database server which would process the query and transfer the appropriate records back to the user. The advantage of this implementation is the clear separation of the interface, storage and the query processing capabilities. Notes is restricted in its ability to present a view of the database in a stuctured format that is easy to manuever around in.

One of the advantages of the information architecture is the ability to identify and prevent problems before they escalate into crisis. For example, the Notes server automatically dials up the enterprise server containing the production schedules and parts information at prespecified intervals and execute SQL queries that finds parts which are below the critical threshold thereby automatically triggering a materials shortage workflow. This is a proactive approach and aims at identifying problems before they cause a crisis. The Notes server would then dispatch electronic mail to the agent who "owns" the part and all the other people involved in that particular workflow. To help resolve the crisis, the server process would include historical information about this problem and action taken to resolve it in the electronic mail message.

In case of the materials shortage workflow outlined in section 2.2, the problem can be identified either by the PPL line clerk or by the critical list coordinator. The PPL line clerk selects "PRODUCTION PROBLEM" from the menu of problem types and fills up the template form that is generated before posting the component shortage problem in the "MATERIALS" discussion-base (See Figure 7). If the problem is well known or has occured previously then the workflow managers spend little time in the diagnosis phase and move expeditiously to the schedule formulation phase. On the other hand, if the problem is more complex involving part interdependencies then more participants get involved in the preliminary diagnosis phase by selecting the "RESPONSE" menu item from the list of problem types and respond to the original problem thereby stimulating a discussion to unravel the interdependencies. In the prototype, this is accomplished by the use of the response form template to highlight the dependencies before the schedule formulation phase. If the discussion facilitator feels that more root cause diagnosis is required he moves that particular problem with its corresponding discussion to the "ROOT CAUSE" discussion-base.

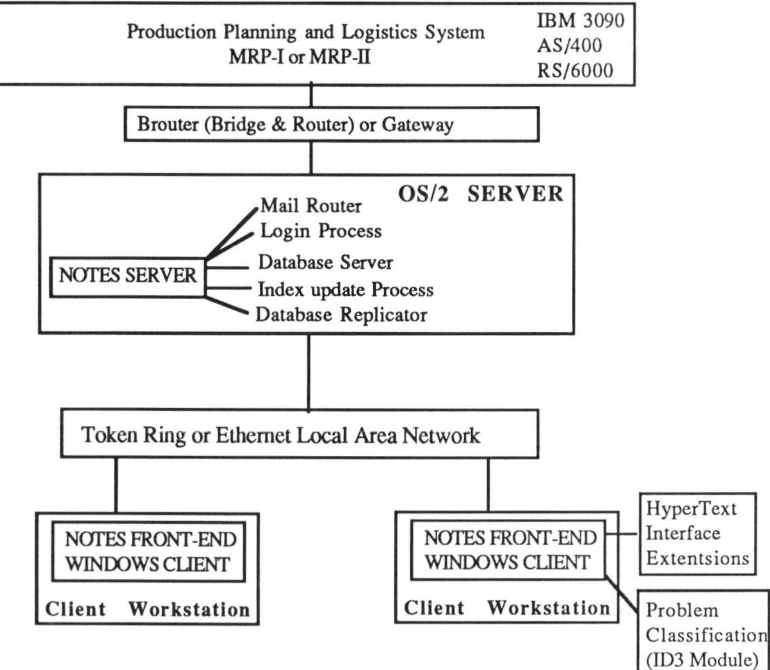

Figure 8. Overview of Prototype Architecture

In the root cause phase of framework, the various agents involved in the process use the language specified earlier to unravel and understand the cause of the problem. The implementation of the root cause phase has been done in Microsoft C/C++ using the Foundation Classes object library. The root cause analysis uses the hypertext format of knowledge reprsentation for efficient navigation in the discussion-base. The hypertext interface process on the client machine augments the limited facilities of the Notes client process by using database API's to interface into the document databases stored on the Notes server to retrieve artifacts which are the intermediate and final products of a discussion and display them. The building and manipulation of hypertext networks, which are the byproducts of the argumentative discussion on the root causes and scheduling phases, is enhanced by the graphical tool. This hypertext interface tool is similar in functionality to MCC's rIBIS (real time IBIS) and GROVE (GRoup Outline Viewing Editor) (Conklin and Begeman [1988]). The client interface provides support for two main grouware concepts: (1) group window: This stands for the a set of processes that allow the dynamic configuration and synchronication of views of the different members of the workgroup. This enables each member to see immediately what other members in the group are posting or observing in their workspaces. This is accomplished by the index update process which continously monitors and updates users screens; (2) views: This is a partial representation of the shared environment in different formats enabling the user to structure the discussion database according to his current needs. The prototype, however, does not support another common group concept of concurrent editing as this was not required in the case of reactive problem-solving.

In the scheduling phase of the framework, agents first try to understand the actions needed to solve the problem, tasks that make up the actions, interdependencies of tasks and the constraints that limits the extent of action. This is done using a hypertext interface similar to the one outlined above but using a language specifically tailored for this phase of problem-solving. The COMMIT action is a critical action in this phase as it assigns responsibility for agent actions for later accountability. In the implementation phase, agents perform tasks within the ACTIONs specified in the SCHEDULE. The scheduling and actual implementation phase are closely interlinked as the schedule may have to be changed when there is a disruption in the workflow of ACTION chain caused by some agent not keeping his COMMIT as pledged in the SCHEDULE. This missed COMMIT may trigger actions which suspend subsidiary workflows which are dependent on the completion of that particular ACTION. This would cause a recursive adjustment in the time for completion of the primary SCHEDULE to compensate for the disruption. In later versions, workflow facilitators can use a graphical workflow mapping tool to visualize workflow with its internal structure, plus the links and interdependencies between the various tasks and the status of the workflow with different representations for the various stages of COMMIT completions. Organizational learning phase is currently done by human agents who evaluate and understand patterns in the discussion-bases and then revamp the solution database to reflect the experience gained.

5.1 LIMITATIONS AND EXTENSIONS

A requirement that has not been implemented currently, is transparent linkages from the Notes environment to production planning/control systems that exist on other machines to retrieve information stored in their databases and prevent redundant manual keyboard entry (due to the immense amount of data being processed, it would be impractical to even think about this) leading to increased data accuracy and timeliness. This integration is conceptually simple but technically difficult to implement due to lack of protocol compatibility between the system architectures. This is important as the lack of availability of current data or lack of data integrity manifests itself in poor decisions resulting in material shortages and excess inventory. As industry moves closer to realizing the automated factory, most manufacturing organizations are using input and output devices such as relays, photoelectric scanners, bar code readers, proximity scanners, and timers and counters interacting with computers and databases to obtain upto date information. The ability to retrieve this information from external enterprise databases would extend the capabilities of the Notes collaborative environment and improve decision-making.

In the information architecture, Figure 8, at the client end we indicated the presence of the problem classification module that organizes the known materials shortage problems, their root causes, the discussions, and reactive actions in the form of a diagnostic decision tree (Quinlan [1993]). Each node in the tree represents a problem. The children of each node represent the subproblems that may be causes of the above parent problem. Each cause may itself recursively have subproblems and subcauses under it thereby framing the discussion as an inductive task. The basis is a universe of objects that are described in terms of a collection of problem attributes. Each attribute measures some important feature of an object and will be limited here to taking a (usually small) set of discrete, mutually exclusive values. For example, if the object is material shortage and the classification task involved excess (unanticipated) final demand for products using this part, the attributes might be demand forecasting, safety stocks, lead times. In order to classify an object, we start at the root of the tree, evaluate the test and take the branch appropriate to the outcome. The process continues until a leaf is encountered, at which time the object is asserted to belong to the class named by the leaf. If the attributes are adequate, it is always possible to construct a decision tree that correctly classifies each object in the training set, and usually there are many such correct decision trees. The essence of induction is to move beyond the training set and have the ability to classify objects not only from the training set but other unknown object as well.

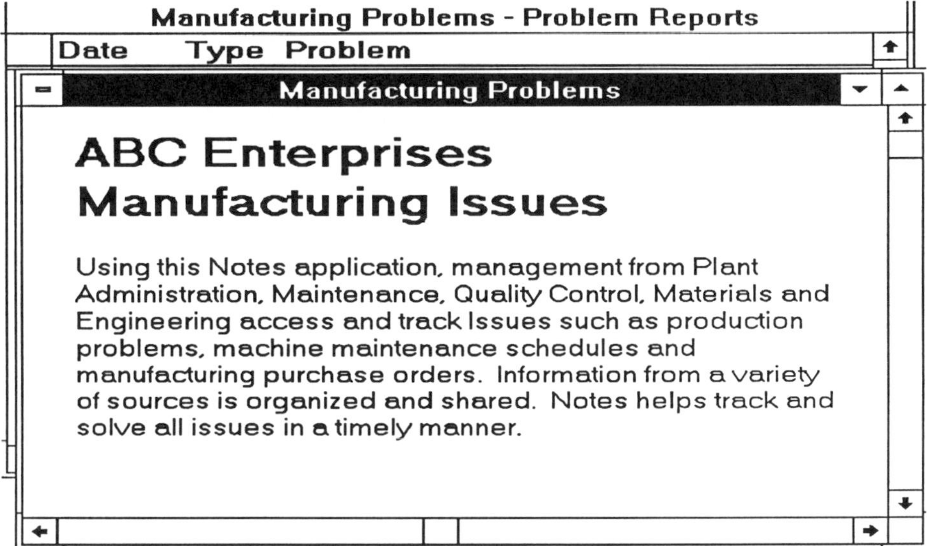

Figure 9.1: Overview of Collaborative Discussion Database in Lotus Notes

Figure 9.2: A listing of the various manufacturing departments involved in problem-solving

Date:	05/18/93	**Subject:**	Expedite shipment
From:	Bill Curry	**Dept:**	Purchasing
Part Number:	Microprocessor-386		

I suggest that we try to expedite the shipment from our sister plant in Japan.
They are also using the same component and since they are not using Just-in-
Time, their inventory levels may be higher than ours. Call Yoshi Nagasaki and
check if they have surplus supply. He was very helpful the last time I called
him.

Figure 9.3: An example of a response to a request to expedite shipment

Summary

How does one link the realm of production planning and control with the dynamic nature of real-time control and problem-solving? As a solution, we have outlined the need for an information architecture that can support at one level the collaborative dialectic needs of dialogue oriented real-time work meshed closely with the capabilites of MRP systems at another level. The manufacturing control system needs to be flexible, modular, on-line, and interactive and self-adjusting, in that it constantly replans and reallocates existing inventories and other resources to meet changing requirements. Manufacturing inventories are, therefore, minimized relative to the management-imposed master schedule, lot-sizing policies, and the constraining factors of manufacturing lead times. They attempt to balance the conflicting demands of maximum customer service and minimum inventory investment by linking company operations together with closed-loop information management, data collection, and feedback functions. Hence, it would be invidious to disentangle the collaborative requirements of real-time problem-solving from the planning/control functionality. Some of the problems that we envision in a discussion environment are : (1) information overload - information filters have to constructed which prevent unnecessary information from the reaching managers. The challenge is to provide information filters which can be dynamically configured for the needs of each manager instead of being statically built into the system; (2) rapid increases in the size of the problem databases - production problems occur at a steady rate and this could result in a tremendous information explosion making the management of databases difficult both from a storage and access perspective. Problems which on the surface appear to be new but are actually instances of old problems need to be categorized and managed carefully as otherwise the capabilities of the system are not being optimally used; (3) effectiveness of past decisions - the success and failures encountered in the coordination of the solution needs to be monitored and recorded for posterity.

As manufacturing organizations strive to achieve quality standards such as Six Sigma Quality, they must become adept at solving manufacturing problems and translate this expertise into long-term improvements. With the increasing adoption of JIT, manufacturing is being approached with a systems view that includes not only the production activities but also processes that linked to the shop floor. Also Mechatronics systems, a composite word that denotes the union of mechanical, computer and electronic engineering, is increasingly being used in the design of sophisticated products. Our goal is to use case studies in manufacturing as a vehicle for generating broader generalizations about the problems of coordination inherent in multiorganizational systems and the need for supporting informal mechanisms therein. The modal tendency of contemporary discussions of coordination problems has been to assess the interests of the agents involved and then evaluate the attractiveness to each agent of various alternate solutions in light of the costs and benefits involved as a precursor to predicting behavior. Here, we focus reactive problem-solving process and mechanisms through which communications take place, solutions are sought and implemented. The main concern is on the actual observed behavior and processes of coordination rather than on the motivation and interests of agents. This paper has been motivated by the excessive amount of time and cost involved in resolving reactive and recurring problems associated with manufacturing coordination and scheduling,. These problems lure managers into interventions that focused on obvious symptoms not underlying root causes of the problem. These interventions often produce short-term benefit but long-term malaise, and foster the need for more symptomatic interventions.

References

Adams, J.J., 1986. Conceptual Blockbusting: A guide to better ideas, 3rd ed. Reading MA: Addison-Wesley.

Argote, L., 1982. "Input Uncertainty and Organizational Coordination in Hospital Emergency Units", Adminstrative Science Quarterly, Vol. 27, pp. 420-434.

Argyris, C. and D.Schon, 1978. Organizational Learning: A Theory-in-Action Perspective. Reading, MA : Addison-Wesley.

Bang, D., J. Decker, R. Gisler and D.Haugen, 1990."Synchronizing Manufacturing and Materials Flow", Proceedings of 9th International Electronics Manufacturing Technology Symposium, pp. 252-255, IEEE.

Conklin J. and M.L.Begeman, 1988. "The right tool for the job: Even the system design process falls within the realm of Hypertext", Byte, pp. 255-268.

Daft, R. and K.Weick, 1984. "Towards a Model of Organizations as Interpretation Systems", Academy of Management Review, Vol.9, No. 2, pp. 284-295.

Dewey, J., 1910. How we think. Boston: D.C. Heath.

Drucker, P., 1971. "What we can learn from Japanese Management", Harvard Business Review, Vol. March-April, pp. 110-122.

Duncan, R., 1972. "Characteristics of Organizational Environments and Perceived Environmental Uncertainty", Adminstrative Science Quarterly, Vol. 17, pp. 313-327.

Gerwin, D., 1981. "Relationships between Structure and Technology", In Handbook of Organizational Design, Nystrom and Starbuck (eds.) Vol. 2, pp. 3-38.

Gresov, C., 1989. "Exploring Fit and Misfit with Mulitple Contingencies", Adminstrative Science Quarterly, Vol. 34, pp. 431-453.

Hamalainen, H., C.Holsapple, Y.Suh and A.B.Whinston, 1992. "Structured Discourse for Scientific Collaboration: A Framework for Scientific Collaboration based on Structured Discourse Analysis", Journal of Organizational Computing, Vol. 1 (1), pp. 1-26.

Hashim, S., 1991. "What: An Argumentative Groupware Approach for Organizing and Documenting Research Activities", Journal of Organizational Computing, Vol. 1 (3), pp. 275-302.

Lawrence, P. and J. Lorsch, 1967. "Differentiation and Integration in Complex Organizations", Adminstrative Science Quarterly, Vol. 12, pp. 1-47.

March, J.G. and H.A.Simon, 1958, Organizations. New York: Wiley.

March, J.G., 1991. "Explorations and Exploitations in Organizational Learning", Organization Science, Vol. 2, No. 1, pp. 71-87.

McCann, J. and J. Galbraith, 1981. "Interdepartmental Relations", In Handbook of Organizational Design, Nystrom and Starbuck (eds.), Vol. 2, pp. 60-84.

Mintzberg, H., Duru Raisinghani and A. Theoret, 1977. "The Structure of "Unstructured" Decision Processes", <u>Adminstrative Science Quarterly</u> Vol. 21, pp. 246-275.

Newell, A. and H. Simon, 1972. <u>Human Problem Solving</u>. Englewood Cliffs, N.J. Prentice-Hall.

Nutt, P., 1984. "Types of Organizational Decision Processes", <u>Adminstrative Science Quarterly</u>, Vol. 29, pp. 414-450.

Quinlan, J. R., 1993. <u>C4.5 : programs for machine learning</u>. San Mateo, California

Rittel, H. and M.Webber, 1973. "Dilemmas in the General Theory of Planning", <u>Policy Sciences</u>, Vol. 4, pp. 155-169.

Senge, P.M. 1990. "The Leader's New Work: Building Learning Organizations", <u>Sloan Management Review</u>, Fall 1990, pp. 7-22.

Thompson, J. D., 1967, <u>Organizations in Action</u>. New York: McGraw-Hill.

Van de Ven, A., A. Delbecq and R. Koenig, 1976. "Determinants of Coordination Modes within Organizations", <u>American Sociological Review</u>, Vol. 41, pp. 322-338.

Winograd, T., 1987. "A language/action perspective for the design of cooperative work", <u>Human-Computer Interaction</u>, Vol. 3(1), pp. 3-30.

Witte, E., 1972. "Field research on complex decision-making processes-The phase theorem", <u>International Studies in Management and Organization</u>, pp. 156-182.

Appendix A
Minutes of a Production Status Meeting

The scene is a production status meeting at the manufacturing plant of a multi-national company assembling electronic components into circuit boards. The participants, at this meeting, are middle-level managers' and their representatives. The functional areas represented were production, materials, quality, shipping, and raw card vendors. This meeting is the second of a series of three official production status meetings held in a day. The first meeting is held at 7.00 am, the second at 9.00 am and the third at 12.45 PM. The duration of the meeting is, typically, an hour. At precisely 9.00 am, the meeting was brought to order by the facilitator and the floor was yielded to the production manager. Each meeting typically involves 5 to 10 people (not many are able to attend all the discussions).

The critical problem, affecting production, was the shortage of component "X" in the inventory. The short-term solution adopted was to use component "Y" instead. A short discussion ensued about the root cause of the problem. The materials manager pointed out that this is a recurrent problem and it occurs, at least 15 times a month and that there is no *historical* data available to know which components are the culprits and to analyze for patterns. One of the comments made by the production manager was, "we have a hole in the process and have no idea what is causing the problem? Engineering change (EC) process or the status bill." The materials managers suggested that mandatory EC be included in the status bill to ensure adequate safety (bonepile) stock in future ordering. Another suggestion made was that before the ""commit to build" was made, stock of components must be available of the entire bill on hand. To ensure that this suggestion would solve the problem, one manager pointed out, that they should have some criteria for estimating the number of "fused" cards in the assembly process and that it should be incorporated into the process description. The final consensus was that there was an underprojection on the status bill and acceptance of the fact that it was a recurrent issue.

In an analysis of this part of the meeting, the following were apparent:
• production managers displayed a lack of understanding of the material ordering workflow, and lack of planning for the shortage
• lack of coordination was evident between the production and materials managers about critical components before the shortage actually arises
• because of intense time pressure the root causes were not adequately explored.

The quality manager had nothing to report and the floor was yielded to the shipping manager. The shipping manager reported that a shipment was stopped due to detection of faulty EPROMs (erasable programmable read only memory) chips. The floor was then taken by the Kanban manager from the raw card vendor. This vendor makes the specific components that are used in making the final assembly. The first problem was raised when certain projected figures about lines "tiger" and "cougar" were disputed by the materials managers. They pointed out that a decision was taken at 4.45 PM on the previous day about prioritizing only "tiger" and not "cougar" with other Kanban managers. The Kanban manager pointed out that she did not receive any information about the decision. Vendor quality was also discussed briefly.

In an analysis of this part of the meeting, the following were apparent:
• Lack of quick and effective information transfer of decisions taken between the materials management and the Kanban management.
• Information on specific products and their status on the production line should be available to each interested manager.
• People leaving early from a meeting have no way of knowing the decisions that were taken on relevant problems after they left the meeting. This creates a time delay in the effective transmission of information concerning the decisions.

The floor was then taken over by the materials manager who passed out a sheet containing the five day critical component list. The mangers most interested were the PPL managers. Components which had zero or one day coverage were of the most concern and expediting efforts were discussed. Details were discussed about transferring stock for a critical component from surplus list in Japan to the local plant but were informed that another sister plant had already placed a hold order on them. One production manager complained that he had run out of a part (10,000 required) which was not on the critical list. He was informed by the materials manager that a line count on the previous day had indicated 5,000 outstanding indicating that the production people apparently cannot locate the part.

In an analysis of this part of the meeting, the following were apparent:
• No root cause discussion took place to analyze materials shortage.
• Different levels of management were interested in different levels of details about the various components.

The meeting was finally terminated with the production status goals and exit outlooks for the entire week and target goals for the day.

Meta-Models for Integrating Production Management Functions in Heterogeneous Industrial Systems

Agostino Villa   Paolo Brandimarte   Mario Calderini

Dipartimento di Sistemi di Produzione ed Economia dell'Azienda
Politecnico di Torino
Corso Duca degli Abruzzi 24
10129 Torino Italy

Abstract

Several heterogenous processes can be found in industrial practice, such as plants involving partly continuous, partly discrete manufacturing and companies composed by departments managed by different production planning philosophies. An efficient management of heterogeneity calls for decentralization. This paper addresses a distributed approach to job-shop scheduling in discrete manufacturing, by analysing conditions for a robust integration of a set of autonomous multiple agents for the shop-floor control in a time-varying manufacturing environment. Instead of machines processing jobs according to a plan established by a global controller, a population of intelligent local controllers should cooperatively operate in order to achieve both individual goals (the best possible utilization of machining and transportation resources) and the overall system target (minimum waiting time).

1   Introduction

An emerging problem concerning production management is the development of more and more sophisticated architectures for planning and coordinating production flows and processing capacity in industrial environments whose structure (e.g., either layout or production management strategy) could be modified over time. The need for process modifications, either 'pushed' by the product market evolution or 'pulled' by the new technologies market proposals, suggests to investigate

S. Y. Nof (ed.), Information and Collaboration Models of Integration, 135–145.

the feasibility of incremental innovation programs. The final aim is to apply gradual modifications to the plant components in such a way to have a continuously improving 'refreshing' of the manufacturing process [8,20].

The need for a gradual innovation of the process requires:

- to adopt a modular organization of the plant, such that local modifications will not induce great influence on the remaining part of the plant;

- to account for plant components of different nature, such as processes involving partly continuous partly discrete manufacturing (paper and food industry, for instance), and processes managed by different production management procedures (e.g. some by MRP tools and the others by JIT logic).

This second feature of 'easy-to-be-innovated' industrial plants refers to heterogeneous processes. When dealing with heterogeneous processes, the idea is to develop and apply modular approaches to production management, in order to let local controllers assure an efficient management of each plant component, whilst a sufficient information pattern could assure a robust integration among local strategies.

The paper approaches the problem of optimizing the integration of a set of autonomous multiple agents for the shop-floor control in a highly uncertain and changing manufacturing environment. The goal is to develop a distributed scheduling and control system for discrete manufacturing, consistently different from traditional approaches. Instead of machines processing jobs according to plans established by a global controller, a population of intelligent local controllers operates cooperatively to achieve both individual goals and the overall system target. Thus a higher degree of robustness, simplicity, maintainability and mainly of modularity will result.

Centralized scheduling and control approaches have supported a great deal of successful work [9]. However, they do not account for the uncertainty and the complexity of a real-world manufacturing environment, mainly in a situation characterized by fast variations in both the product market and the new technologies' market. The present recession in the industrial environment reveals two 'sources of weakness' of the Western manufacturing companies: long time-to-market and poor quality standards. Awareness of these sources of low competitiveness has recently spurred great interest towards two new key-words: 'lean production' and 'total quality'. 'Lean production' means organization of the manufacturing plant so that the necessary resources (machines, items, tools, fixtures, personnel, time) could be intensively utilized for processing the required loads at the right time. Moreover, targets of 'lean production' are zero inventories, reduced flow times, short time-to-delivery. In summary, 'lean production' calls for a strong cooperation of clearly motivated agents in 'pulling' production [19]. 'Total quality' means utilization of the shop-floor resources in an efficient Production Activity Control (PAC) system in such a way that both the processed and the produced items can be carefully monitored, malfunctionings can be immediately detected and, in case of frequent failures, qualified resources can be easily embedded in the plant, in a short-transient progressive substitution of the old inefficient plant components.

The distributed control system approach recognizes the fact that a very complex manufacturing plant, whose structure can change over time either due to failures or to innovation programmes, is beyond direct control. Only in presence of a network of distributed controllers, with a strong cooperation among local control agents, effective control strategies can be envisaged. This requirement has suggested researches in different complementary fields, such as the ones of cooperative systems [1], heterarchical structures [2], object-oriented programming concepts [3], real-time negotiation of resource assignments [4] and opportunistic scheduling [5]. The cooperative system approach suggests the organization of many interacting subsystems, which can have proper independent goals and operation modes, but in such a way that cooperation could assure a collectively determined behaviour of the whole system. Its application needs a clear statement of the interactions among local controllers and of the conditionings among them, in order to avoid conflicts (the effect of which is dramatic for the cooperative system operations). The heterarchical structure provides a potential network of interactions among local controllers, through a reduced set of higher-layer coordinators interacting together according to specified rules. Also for heterarchical structures, the interactions among sets of local controllers have to be carefully handled. Object-oriented programming is a representation environment which supports the description of subsystems and local controllers but does not supply any theoretical tool for the controller design. Real-time negotiation appears to be the approach to distributed coordinated controls specifically oriented to an effective management of the interactions among local controllers. Moreover, it seems to be the theoretical approach to distributed control system design which should be coupled with the object-oriented approach to the effective organization of both individual local controllers ('individual objects') and of their interrelations ('communications among objects').

The main objective of this paper is to outline an expandable modular shop-floor control system suitable for assuring an effective real-time coordination of a set of local controllers, also denoted agents (each one associated to a specific manufacturing resource, either 'resident' such as a machine, or 'movable' such as an AGV).

Negotiation offers a reasonable way for local agents to use their own local knowledge in achieving reasonable performance and, by using knowledge shared with contiguous agents, in maintaining system robustness, modularity and stability. The other fundamental contribution comes from the ability to decompose the overall scheduling problem into a set of sub-problems. Suitable mathematical models of the interactions among sub-problems must be derived, with the aim of assigning each local agent the simplest possible optimization sub-problem, accounting for the overall goal of the global system in a mathematical form allowing a suitable approximation. Based on this model of the interactions among agents, an opportunistic approach for coordinating agents must be designed, in such a way that contiguous agents find a suitable compromise through a negotiation of respective aims and targets.

2 Logical Description of the Decentralized Approach

The goal of the paper is to propose a network of many cooperating control agents, each one associated to a resource in a manufacturing plant, and such to manage the resource in order to achieve the best possible performance, while assuring a good performance of the overall system. The network of local controllers has to be designed in such a way to assure the easiest possible embedding of a new plant component, either resident (machine) or movable (AGV), such to innovate the existing plant with lowest possible cost. This feature will define the self-organizing nature of the control network to be designed, and will specify the modularity of the manufacturing plant to be managed/innovated.

The new design concept is the following: since manufacturing systems are distributed both in space and in organization, the decision-making in production control lends particularly well to the distributed problem solving approach.

Distributed problem solving, a field in which AI-based results have to be joined to LSS theory results [20], deals with the interactions of groups of intelligent agents attempting to cooperate in order to solve a given control problem. A distributed problem solver has been formally defined as a network of loosely coupled, asynchronous, semi-autonomous, intelligent problem solving agents, cooperating to fulfil the global mission of the overall system to be controlled [14]. 'Asynchronous' means that the agents are designed to operate concurrently. 'Cooperative' means that, because no individual agent is able to solve the overall control problem for the entire system by itself, the agents have to work together as a team and to exchange knowledge about tasks, goals, results, and constraints, in order to contribute in the overall problem solution. 'Distributed problem solving' requires that the global control task can be decomposed into sub-tasks, among which the number of mutual interactions is not too large: this fact makes possible to find individual solutions fairly independently. A cooperation mechanism is necessary to combine the global solution from the local solutions of individual sub-problems, and to coordinate the distribution of sub-tasks to the agents. Several authors (among whom, Boboam [15], Descotes-Genon, Ladet et al. [16]) note that the problem of computing the optimal decentralized control strategy in a manufacturing system usually becomes untractable, and suggest to construct the system model directly within a framework of decision-making agents. Then, interactions variables, which describe the influence among sub-systems and then the related agents, are defined. Since each agent has a local proper model of the sub-system which it must control, and since the agent uses the local model for estimating the future consequences of its own control decisions, suitable bounds on the overall system performance have to be derived, in order to assure the stability of the whole set of controllers. As communications improve the agent's ability to reduce its uncertainty about the impact of its own control decisions on the behaviour of both the local sub-system and the other ones, the bounds on the overall system performance are expected to improve as communication increases. That means: an efficient communication assures an effective solution of local control sub-problems and a good coordination among local control decisions.

To authors' opinion, a good integration among local agents can be searched for through application of the contract net approach [6]. A contract net consists of a suitable negotation process among local agents, in such a form to distribute parts of a global control problem (i.e., control sub-problems) among the different agents, and to handle the agents' interactions through a protocol of 'requests', 'bids', and 'orders'.

First approaches using contract nets assumed to apply, for each finished product, an MRP II system which determines when to begin production and which tasks need to be performed. As soon as a task has been completed, the agents negotiate among themselves to determine who is the candidate for the subsequent task [17,18].

In recent developments of the contract net approach due to Solberg [4], a job comes to the system with its own process requirements: process plan, priority and objective (i.e., performance measures to be optimized for obtaining an efficient processing of the considered part, such as lead-time, cost, quality, etc.). These individual part specifications have to be supplied by some off-line procedures including CAPP, RCCP and part route planning. Each functional unit or job is considered as a system entity. Each entity is connected to a control unit. A control unit is realized by a computer which communicates with the other control units through a communication network. When a job enters the plant, it tries to fulfil the processing requirements to achieve the best possible set of weighted objectives (time, cost, quality) by negotiating its processing date and time with the 'resource agents' (related to resident and movable equipments). A resource agent acts as a vendor of its processing capability, and tries to sell its services to maximize its profit. Each offered service will give rise to a 'price' for the job: price should serve as an invisible hand to guide the negotiation towards improvement of the system performance, so that the control/scheduling system is expected to function based on an equilibrium of prices and objectives.

A similar approach has been adopted by Moriwaki et al. [10,11,12,13]. The logical components of the manufacturing systems are again divided into 'equipments' and 'workpieces'. The manufacturing equipments and the workpieces have autonomous decision-making functions, to determine their own suitable processes in the manufacturing system on the basis of their respective status. The manufacturing equipments have the function of determining their operation plans. The design and modification of the manufacturing process plans are carried out by the individual workpieces. The production schedules are obtained through a negotiation, structured in therms of an AI-based decisional tree, between the manufacturing equipments and the workpieces.

As it can be easily understood by accounting for a real manufacturing plant with thousands of items, and also recognized by analyzing the simple examples reported in the above mentioned papers, the main drawback of this approach including each workpiece as a part of the system to which an individual autonomous agent is associated, is the extremely large dimension of the network of agents, and the very high volume of information flowing through the network itself.

In order to overcome this significant drawback, the contract net approach proposed in this paper is based on the following concepts:

1. The network of agents is composed by two different types of agents, namely master agents associated to resident resources (the machining units), and slave agents associated to the movable resources (AGVs).

2. The real-time negotiation is driven by the occurrence of events, namely either initiation of a job at some machine or completion of a job. Every time an event occurs, a negotiation must be fired among the master agent which the event has occurred on and the neighbouring agents, in order to select the most convenient request for a part transportation.

3. A negotiation consists of the solution of a local optimization problem, searching for the best performance of a master agent, when the performance index is weighted according to approximated information about the current status of the overall system. In case of the event 'initiation of a job', the negotiation must support the selection of the job to be processed. In case of the event 'completion of a job', the negotiation must support the selection of the downstream machine to complete the job.

4. Every time a negotiation has to be made (through solution of the related local optimization problem), availability of slave agents have to be analyzed. This analysis can ask for the solution of a lower-level sub-problem for optimizing the selection of the movable resource and related slave agent, required for transportation needs. This last sub-problem has to be handled by the agent associated to the movable resource involved in the negoatiation, as well as by all agents controlling movable resources currently staying in the neighbourhood of the machine under direct control.

These main concepts allow to envisage a relevant innovation with respect to existing approaches, also based on contract net: in fact, in the case here proposed, the potentiality of the contract net is associated to the specific structure of a manufacturing plant, thus specializing the agents included and, as important gain, significantly simplifying the net. In practical implementation, agents are related to machines and also to transportation equipments. Agents assigned to machines could perform local planning of the process, real-time machine control, forecasting of production orders, and accounting. Agents associated with movable equipments could manage demands for transport and storage, product procurements and local order entry.

The negotiation protocol can be briefly outlined as follows. When a part has been completed by a machine, and it is ready for being processed by another machine, each downstream master agent receives a 'request for bid', i.e. a request to verify if it can process the part. An agent accepting the request for bid has to solve its own local optimization sub-problem, to estimate the time required for processing the part, based on the present state of the upstream storage, and accounting for the part priority through some penalty function (a bid is a cost functional parametrized by time of delivery and volume of production). Once solved, the agent sends a requests for transportation to the closest AGVs. Each slave agent accepting a request has to verify its availability, and send its expected transportation time to the master agent. Based on both information contents, each agent can present its own cost to the master agent which has originated the negotiation, and the best solution will be chosen. Once selected the downstream machine, the original master agent sends it a 'purchase order' which specifies desired time of delivery and volume of production. This forces the agent associated to the selected machine to reserve appropriate capacity, and the same happens for the selected AGV. Each time the required capacity is no more available, a 'cancel message' must be sent, and the negotiation process must be started again.

3 Concepts for a Mathematical Approach to Modular Scheduling

The design of the above outlined network of autonomous agents for dynamic scheduling requires to state and analyze the scheduling problem in formal terms. The main lines of the mathematical approach can be summarized as follows:

1. to define the general problem of job-shop scheduling for a plant in which not only local schedules have to be generated, but also transportation alternatives have to be handled;

2. to analyze a suitable approximation of the stated general model, such to allow its decomposition into local sub-problems, each one associated to a resident resource (a machine) and accounting for potential constraints induced by current full utilization of movable resources (AGVs);

3. to analyze which conditions on the individual solutions of the local scheduling sub-problems guarantee that the global solution of the overall scheduling problem presents a desirable performance;

4. to analyze which conditions the decentralized modular solutions of the local sub-problems have to satisfy so that heterogeneous sub-systems could be efficiently managed by using different production management philosophies (for istance, different scheduling rules).

Given the above points, it appears that a fundamental issue for a successful implementation of a decentralized scheduling architecture is the use of approximate mathematical models in order to avoid myopic behavior.

In fact, an opportunistic real time scheduler may exploit relatively local information in order to optimize system performance: for example, when downstream buffers are full, jobs with long processing times should be selected in order to avoid blocking, whereas short jobs should be selected when downstream buffers are empty in order to avoid starvation. However, a purely distributed scheme may not be able to guarantee overall system performance. On the other hand, a purely centralized scheme is not able to cope with complex issues such as tooling, transportation, sequence dependent setups etc. The right compromise should be met.

Further reasons to consider a certain degree of centralization are:

- the need to have an overall state information of the work in process in order to release materials to the shop floor;

- the need to predict the behavior of the system in order to set realistic due date when negotiating with the customer; to this aim, rough cut models are used in [21] in order to assign due dates.

If we think in dynamic programming terms, what we need in order to obtain an effective yet modular architecture is an approximation of the cost-to-go. Consider for example the problem of minimizing total waiting time for a set of jobs in a multi-stage system; the overall waiting time of each job i is the sum of the waiting times on each stage p:

$$W_i = \sum_p w_{ip}.$$

However, simply minimizing the waiting time on each stage (e.g. by the SPT rule) does not assure an overall good performance. The need to anticipate interactions is clearly exacerbated when complicating issues (trasportation, etc.) are taken into account.

One possible approach to cope with these problems has been proposed in [22], based on a continuous approximation of the material flow, which allow to formulate an aggregate scheduling problem, whose output is a reference trajectory to be fed to real time schedulers. The solution of the aggregate scheduling problem does not constrain shop floor control, but it yields a way to compute non myopic priorities: in other words, the reference trajectory plays the role of a surrogate cost-to-go.

Continuous flow approximations have been often adopted in the context of repetitive manufacturing. In particular, in [7], this allows to obtain stability conditions ensuring suitable work in process levels.

The problem with continuous flow approximations is that they are not suitable to cope with small batch production. In this case single-stage decomposition approaches can be pursued: such approaches have proved very successful when used as heuristics for static scheduling problems [23]. The extension of such approaches is needed in the following directions:

- real time reactive scheduling;

- investigation of stability conditions constraining the behavior in terms of material release to the shop floor, due date negotiation, work in process levels and flow times.

4 Perspectives for Future Works

The proposed concepts for a mathematical formulation of the modular scheduler simply allows to outline the main features which the resulting network of local controllers should present.

1. The production management structure at the shop-floor has to be composed by the combination of a higher-layer CAPP procedure and the network of agents, thus resulting in a relevant simplification of the existing and proposed architectures (no intermediate management neither coordination is necessary).

2. The network of local agents should act as a 'pull mechanism', since each master agent (machine) negotiate with others to receive any needed part/component. This process propagates backward through the plant to the machines which can have the required supplies in their buffers. This fact makes the network a special implementation of a Just-In-Time manufacturing system.

3. Finally the proposed decentralized scheduling should perform a dynamic allocation, responding to changing business conditions, because capacity is reserved dynamically when new products are requested.

In order to allow a complete exploitation of all features, the following issues will be analyzed in more detail in the future:

- analysis of the efficiency of the scheduling activity;
- analysis of the robustness of the distributed control strategies, despite degradation of some component either resident or movable;
- estimation of the transient behaviour, fired by either changes of the production demands or by relevant modifications of the manufacturing plant.

References

[1] Nof S.Y., and Papastavrou J. D.:*Decision integration fundamentals in distributed manufacturing topologies*, IIE Transactions, 24 (1992) 27-42.

[2] Hatvany J.: *Intelligence and cooperation in heterarchical manufacturing systems*, Proc. 16th CIRP Int. Seminar on Manufacturing Systems, Tokyo, (1984).

[3] Joannis R., and Krieger M.: *Object-oriented approach to the specification of manufacturing systems*, Computer Integrated Manufacturing Systems, 5 (1992) 133-143.

[4] Yuh-Jiu Lin G., and Solberg J.J.: *Integrated shop floor control using autonomous agents*, IIE Transactions, 24 (1992) 57- 71.

[5] Newman P.A., and Kempf K.G.: *Opportunistic scheduling for roboting machine tending*, Conf. Artif. Intell. Appl., Miami, FL (1985).

[6] Baker A.D.: *Complete manufacturing control using Contract Net: a simulation study*, 1988 Int. Conf. on Computer Integrated Manufacturing, Rensselaer Pol. Inst., Troy, USA (1988).

[7] Perkins J.R., and Kumar P.R.: *Stable, distributed, real-time scheduling of flexible manufacturing-assembly-disassembly systems*, IEEE Transactions on Automatic Control, 34 (1989) 139-148.

[8] Villa A., Rossetto S., Lucertini M., and Telmon D.: *Methodological approach to planning and justifying technological innovation in manufacturing*, Computer Integrated Manufacturing Systems, 4 (1991) 221-228.

[9] Brandimarte P., and Villa A. (Eds.): *Optimization Models and Concepts for Production Management*, to be published, Gordon & Breach Science Publishers.

[10] Moriwaki T., Sugimura N., and Wirjomartono S.H.: *Production scheduling in autonomous distributed manufacturing systems*, PED, Vol. 56, Quality Assurance Through Integration of Manufacturing Processes and Systems, ASME 1992, 159-170.

[11] Moriwaki T., Sugimura N., Martawirya Y.Y., and Wirjomatrono S.H.: *Object-oriented modeling of autonomous distributed manufacturing systems and its application to real-time scheduling*, Proc. of the ICOOMS'92 (1992) 207-212.

[12] Okino N.: *A prototyping of bionic manufacturing system*, Proc. of the ICOOMS'92 (1992) 297-302.

[13] Ueda K.: *An approach to bionic manufacturing systems based on DNA-type information*, Proc. of the ICOOMS'92 (1992) 303-308.

[14] Decker K.S.: *Distributed problem-solving techniques: a survey*, IEEE Transactions on Systems, Man and Cybernetics, 17 (1987).

[15] Roboam M.: *Intelligent networking architecture: AI in integrating the enterprise*, Computer Integrated Manufacturing Systems, 4 (1991) 3-17.

[16] Devapriya D.S, Descotes-Genon B., and Ladet P.: *Petri net based node structures for distributed problem solving in FMS control*, Computer Integrated Manufacturing Systems, 5 (1992) 229-238.

[17] Shaw M.: *A distributed scheduling method for Computer Integrated Manufacturing: the use of Local Area Networks in cellular systems*, Int. Journal Production Research, 25 (1987) 1285-1303.

[18] Parunak H.V.D.: *Manufacturing experience with the contract net*, in: *Distributed Artificial Intelligence*, M. Huhns (Ed.), Pitman, London (1987).

[19] Villa A., and Watanabe T.: *Production management: beyond the dychotomy between 'push' and 'pull'*, Computer Integrated Manufacturing Systems, 6 (1993) 53-63.

[20] Villa A.: *Hybrid Control Systems in Manufacturing*, Gordon & Breach Science Publ., London, U.K., 1991.

[21] H. Luss, and M.B. Rosenwein: *A due date assignment algorithms for multi-product manufacturing facilities*, European Jou. of Operational Research, Vol. 65 (1993), pp. 187-198.

[22] P. Brandimarte, W. Ukovich, and A. Villa:*Factory Level Aggregate Scheduling: a Basis for a Hierarchical Approach*, Proc. 1992 IEEE Conf. on CIM, RPI, Troy NY, pp. 413-422.

[23] J. Adams, E. Balas, and D. Zawack:*The Shifting Bottleneck Procedure for Job Shop Scheduling*, Management Science, vol. 34, No. 3 (1988), pp. 391-401.

III. Methods for Planning and Evaluating the Integration and Collaboration

ENTERPRISE INTEGRATION: A TOOL'S PERSPECTIVE

J. POLITO
Sandia Nat'l Labs
Albuquerque, NM 87185
USA

A. JONES
NIST
Gaithersburg, MD 20899
USA

H. GRANT
NSF
Washington, DC 20550
USA

ABSTRACT.   The advent of sophisticated automation equipment and computer hardware and software is changing the way manufacturing is carried out. To compete in the global marketplace, manufacturing companies must integrate these new technologies into their factories. In addition, they must integrate the planning, control, and data management methodologies needed to make effective use of these technologies. This paper provides an overview of recent approaches to achieving this enterprise integration. It then describes, using simulation as a particular example, a new tool's perspective of enterprise integration.

1. Introduction

A revolution has taken place in manufacturing plants over the past two decades. This revolution has been fueled by the introduction of new computer and production technologies on the factory floor and in the front office. From FAX machines to computer-controlled, high-precision machine tools and from Computer Aided Design systems to Automated Data Collection systems, these technologies promised to increase quality, productivity and profits. Utilizing these new technologies, many companies have succeeded in substantially upgrading/automating individual production and office functions.   They have been less successful when it comes to integrating all of these functions together within a single enterprise. The integration of multiple enterprises has progressed at an even slower pace.

In (Jones et al 1989) the authors expressed a commonly held view at the time, that the foundation for achieving this integration was a system architecture. This system architecture consists of four separate but related architectures: business management, production management, data management, and communications management. Business management includes all of those functions normally associated with running the business aspects of the company. Production management includes all of the functions related to the design, fabrication, and inspection of parts. Data management includes all functions related to the delivery of accurate and timely information to the production management processes. Communication management includes those functions required for the reliable transmission of messages between computer programs.

Each of these architectures has been the subject of intense world-wide research, at universities, individual companies, research institutes, and government laboratories. Numerous papers have been published which report the results of these efforts, but the fact is that only very low levels of integration have been achieved in real manufacturing companies. This paper provides some insight into 1) issues related to the design and implementation of each of these individual architectures and 2) reasons for the inability of enterprises to integrate these individual solutions into an overall system architecture.   It

The U.S. Government right to retain a non-exclusive, royalty-free licence in and to any copyright is acknowledged.

S. Y. Nof (ed.), Information and Collaboration Models of Integration, 149–167.
© 1994 *Kluwer Academic Publishers. Printed in the Netherlands.*

then proposes a new focus for enterprise integration - a concentration on integrating key manufacturing software tools. This paper does this by way of a specific example, simulation.

The two most renown public efforts in this area have been undertaken by the National Institute of Standards and Technology (NIST) and The European Strategic Programme for Research and Development (ESPRIT). The NIST effort is centered in its Automated Manufacturing Research Facility (AMRF). The ESPRIT effort is centered in Project 688 - Computer Integrated Manufacturing Open System Architecture (CIM-OSA). This paper will touch on some of the work done at NIST. Those interested in finding out more about CIM-OSA are referred to two excellent sources: "The Proceedings of the First International Conference on Enterprise Integration Modeling (Petrie 1992) and a book titled "Open System Architecture for CIM" (ESPRIT 1989).

2. Issues in business and production management

Taken together, business and production management includes all functions needed to run a manufacturing enterprise. Many companies have invested a large amount of time and money to develop their own proprietary architectures. While these are typically separated in the real world, most government and university architecture-related research efforts have lump them together into a single factory control architecture. In the following sections we summarize the state of the art in this effort. Much of what follows is excerpted from (Jones et al 1989).

2.1 CURRENT APPROACHES TO DESIGNING FACTORY CONTROL ARCHITECTURES

Hierarchical organizational structures have been used to provide the coordination necessary to manage production activities in traditional human-based factories. The number of levels and the responsibilities of the individuals at each level can vary dramatically from one company to another. In many small companies, all decisions are made at the top, and subordinates simply implement various decisions at their own level. In most larger companies, people at every level are expected to make certain decisions, based on input from a superior, and exert the control necessary to have subordinates execute their decisions.

Early attempts to design and implement factory architectures for modern manufacturing enterprises have used the same hierarchical approach (Jones and Whitt 1985). Their designs are based on three principles (Albus, Barbera and Nagel 1981). First, levels are used to reduce the size and complexity of the problem and to limit the scope of responsibility and authority. Each "level" consists of a combination of people and computers which decompose commands from a supervisor at the next higher level into procedures to be executed by other entities at that same level and subcommands to be issued to one or more subordinate levels. Second, decision making and control always resides at the lowest possible level. That is where the most complete, up-to-date, and deterministic information is available. Third, planning horizons decrease as you go down the hierarchy. At the higher levels, the horizon can be months or years. At the lowest level, the horizon can be less than a second.

There are, however, three attributes that distinguish one hierarchy from one another. First, there is the number of levels and the assignment of functions to levels. In most designs, this "decomposition" depends on both the complexity of a given function and the actual physical configuration of the system. The Advanced Factory Management System (Liu 1985), which is based on the approach of grouping similar machines close together, has four levels. The Automated Manufacturing Research Facility model (Jones and

McLean 1986) is based on a group technology approach to shop floor layout and has five levels. The Factory Automation Model (Ottawa 1986), which attempts to accommodate both, has six. The second characteristic involves the identification and direction of control paths. Control relationships can be assigned once and remain static or they can be assigned dynamically as the situation dictates. The question of direction is independent of assignment. Most hierarchical architectures allow only vertical control flow. This means that each control module can have only one supervisor who issues commands. This supervisor always resides at the next higher level.

The last distinguishing characteristic is the method of handling data. In early hierarchical architectures data handling was viewed as one of the control functions at each level and is included in the existing control hierarchy. This means that the data needed to carry out a given command is either part of the command itself or tightly coupled to the control path. The consequence of this was that the only exchange of information, regardless of the content was between a supervisor and its' subordinates. Modern communication technologies have obviated the need for this restriction. In fact, there is a totally separate data management architecture which serves the control hierarchy. In this case, control modules must access the data needed to execute a command from the data management system.

During the last several years, many researchers have turned their attention to the heterarchical control strategies first advocated in (Hatvany 1985, Duffie and Piper 1986). In this approach, there are no supervisors per se. All entities are treated as co-operating equals in an ongoing process to plan, schedule, and control the manufacturing system. Decision are reached through a complicated series of broadcasts, bids, and negotiations.

2.2 WHERE ARE WE NOW

Although numerous designs (Jones and Whitt 1985, Jones 1990) have been proposed, major problems have arisen in the implementation of each design in real manufacturing enterprises. There have been and continues to be problems developing automated decision-making methodologies which adequately duplicate the way human beings conduct the negotiations needed to make complex decisions. Even if we could automate these decisions, we do not know how to integrate the results into the distributed computing environments that result from either a hierarchical or heterarchical architecture. Another major implementation problem arises because factory control architectures are typically imbedded as part of an existing factory. No suitable migration strategies have been developed which will allow the required modular implementation. In addition, there is no way to determine whether a given design is appropriate for a particular application before, during, or after initial implementation. No performance metrics or software tools are available which are designed to test a given factory architecture. We agree with (Nof 1988) that a combination of quantitative/qualitative performance measures and analysis tools must be developed which can be used to compare different designs and select the "best".

3. Issues in data management

The main purpose of a data management system in a manufacturing enterprise is to support functions in the business and production management architectures with timely access to all required data. There are, however, many characteristics of a manufacturing enterprise which makes this a difficult task. We now present a brief discussion of three of these characteristics.

The computer and data systems used in modern manufacturing enterprises are purchased from a variety of vendors over a long period of time. This implies that data is likely to be physically distributed across a network of local heterogeneous computer systems. These local systems will have a wide range of capabilities and restrictions for access, storage, and sharing. In addition, many of these systems were developed in-house or purchased from vendors that no longer support those particular products. These so-called legacy systems cause enormous integration problems. In many cases, the integration of these systems is performed manually. That is, the output from one system is manually re-entered in to another, because electronic integration is impossible. Finally, manufacturing data comes in all shapes and sizes ranging from gigabytes of CAD data down to bits of sensor data.

A second major characteristic is the fact that control computers for shop-floor equipment make real-time decisions. If the data is not present when it is needed, erroneous decisions or no decisions may be made, resulting in processing delays and reduced plant throughput. To complicate matters, some of that data may be shared by several users with different "real-time" access requirements. But, it is important to realize that data delivery, like material delivery, takes time and must be included in the planning of each job. This means that data is quickly becoming a critical resource which must be scheduled. Poor "data scheduling" will lead to delays, bottlenecks, and idle equipment. Therefore, the scheduling decisions made by the data manager have a direct impact on the scheduling decisions made by the production scheduler. This implies the need for coordination between the data scheduling function in the data management architecture and its counterpart in the production management architecture.

The third major characteristic is that many functions use parts of the same data package. This means that providing a uniform structure for these data packages will be a key to successful integration of these functions. Two of the most important of these data packages are product data and process plans. Electronic product data includes all of the raw data needed to design, fabricate, test, inspect, and maintain a product during its entire life cycle. A draft international standard (STEP) now exist for this product data (ISO10303 1992) . STEP is critical to external relationships with both customers and suppliers and the internal integration of many production management functions. Externally, it allows for the reliable, and unambiguous transfer of product information. That transfer can take place between customer and producer, or between two or more facilities involved in the manufacture of a single complex product. Internally, it is the main input to the process planning function which provides the integration of many downstream production functions. Major changes will be required in existing process planning systems to fulfill this role. All of the data (resources, timings, routings and alternatives) required to transport, fabricate, inspect, and ship a part must be included in the total process plan for that part. The total plan should be decomposed into subplans having the same generic structure. One or more subplans will be used by each production function to make decisions about each "job" it executes. Efforts are currently underway in ISO TC 184 SC 4 WG 3 to develop a standard process plan model.

3.1 WHAT CONSTITUTES AN ARCHITECTURE

A data management architecture must address three major concerns: data modeling, database design, and data administration.

Developing a "conceptual data model" of all the information involved in the enterprise is critical to the success of any integrated data management system. Because the amount of information is so large, a "divide-and-conquer" approach to performing the analysis must be taken. Experts on individual business and production functions will perform the analysis and develop a conceptual information model for each functional area. Then the resulting "component" models must be integrated into a single "enterprise model". The enterprise model is the conceptual representation of the global information base. To be successful, this integration must be based on the identification of common real-world objects and concepts, rather than trying to identify the "common data". A "conceptual model", therefore, must represent the relationships between information units as they apply to the real world, rather than the structured and limited relationships between these units as they are stored in a data system. Several powerful modelling techniques now available allow representation of the real-world objects themselves, as well as the information units which describe and distinguish them (Nijsen 1989, Peckman 1988, and Ullman 1988).

Once the global enterprise model has been constructed, we have the problem of mapping this model onto live databases. We must now choose systems, organizations and representations for the information units in the model. This process is called database design, and it is a difficult process to automate, particularly for manufacturing applications. It must result in databases which are consistent with the model and tuned to the timing and access requirements of the functions that use them. Since much of that data must be shared, two problems result: partitioning and replication. Partitioning means that some production functions must simultaneously access information stored in two or more databases. Replication means that some data must be stored simultaneously in two or more different databases, and maintained consistently. The available options for the placement of databases in the CIM computer system complex, and for the selection of specific data management systems to support them, are dictated to a large extent by the architecture of the "global" data administration system.

There are three control architectures for data administration systems which have been used with varying degrees of success in various business applications: centralized data and control, distributed data and control, and hybrid systems. The totally centralized approach is the traditional design, the simplest, and the most workable. There are currently available high-speed, internally redundant, fault-tolerant, integrated centralized systems. But even if such systems can keep pace with the growing demands and time constraints of automated production systems, the centralized architecture is not workable from the point-of-view of subsystem autonomy. The canonical architecture for the totally distributed approach consists of local data management systems which process locally originated and locally satisfiable requests and negotiate with each other to process all other requests. In this case, difficult problems of concurrency control, distributed transaction sequencing and deadlock avoidance occur and must be resolved by committee. While there has been considerable research in these areas, satisfactory solutions have not been found. The hybrid architecture attempts to combine the best features of both centralized and distributed architectures. Subsystem autonomy and high throughput are achieved by allowing local data systems to process locally originated operations on local data. Operations which transcend the scope of a local system are sent to a centralized "global query processor" for distribution to the appropriate sites. The global query processor acts as a central arbiter for resolving the characteristic problems of distributed transactions and for handling configuration changes in the data administration system itself. There are a number of ways of implementing such an architecture, but they are all characterized by standardized interfaces between the local data systems and the global query processor.

3.2 WHERE ARE WE NOW

There are many new commercial data systems available which are of this hybrid variety. But, in general, they have not yet resolved the control problems associated with distributed transaction management to the extent necessary to provide robust support for manufacturing applications (Thomas 1988). Unless the local system is aware of potentially global consequences of local changes, and can propagate those correctly, the integrity of the global information bases is always in doubt.

In the last decade, several object oriented database management systems have been commercialized (Morris et al 1992). These new systems provide many enhanced features like more advanced data modeling techniques and a more sophisticated transaction management. But, they have not yet reached the level of maturity seen in relational database systems.

Several standards are emerging which will have an impact on the development of data systems for manufacturing enterprises. These standards fall into two categories: those designed to make access and integration easier, and those for data representation. Some examples of standards falling into the first category include Structured Query Language (ISO/IEC 9075 1992), Remote Data Access (ISO/IEC 9579 1991), and Information Resource Dictionary System (ANSI -IRDS 1988). Some examples of standards falling into the second category include DFX (Autodesk 1992), Initial Graphics Exchange Specification (IGES5.1 1991) and Standard for the Exchange of Product Model Data (ISO10303, 1992).

In summary, while there have been numerous advances in data management and numerous standards adopted, there are still no commercially-available, distributed data systems that allow arbitrary data repositories. This means that we are still a long way from a commercially-available solution to the legacy problem for manufacturing companies.   However, there have been prototypes developed in the research lab. One of the earliest and, perhaps best known, is the Integrated Manufacturing Data Administration System (IMDAS) developed at NIST (Libes and Barkmeyer 1988).   IMDAS implements a hybrid four-level hierarchical planning and control architecture for information which provides a common interface to user programs and a common interface to the underlying data repositories. The interfaces transparency from the actual data repositories. That is, the actual location of data and the work done to satisfy a user request is completely transparent to the user. Conceptually, the user simply sees a single logical database managed by the IMDAS. This single database can contain several different physical data repositories ranging from commercial databases and file systems to "home-grown", application-specific data systems. The major unresolved problem with IMDAS and all other similar types of distributed systems is distributed transaction management (Thomas 1988).

4. Issues in communications

Communications can be divided into two classes: those WITHIN computer systems, and those ACROSS computer systems. The first type is often referred to as "interprocess communication" while the second is often called "network communication". Interprocess communication is dependent on features of the operating system. Many systems provide no such facility at all, or provide only for communication between a "parent" process and "child" subprocesses which are created by the parent. On such systems the coordination of multiple business, production, and data management activities is extremely difficult. On the other hand, properly implemented "network communication" software should provide for the case in which the selected correspondent process is resident on the same computer system as the process originating the connection. That is, the proper solution for the future is to make local interprocess communication a special case handled by the network software.

The accepted paradigm for network communication is the Open Systems Interconnection Reference Model (OSI) which separates the concerns of communication into 7 layers (Day and Zimmerman 1983). Traditionally, "network communication" has meant concentration on the lower four layers and "exposure" of the Transport layer to production management programs. The important aspect of this model is that it formalizes and separates the logical process-to-process link (in layers 5-7) from the physical network service considerations (in layers 1-4). By exposing only the Application layer service, we insulate them from the networking concerns. Consequently, we believe that it is meaningful and proper to build an Application layer interface which is common to ALL program-to-program communications, both "local" and "networked".

4.1 CIM network architecture

A great deal of flexibility is created by implementing the OSI model. On one hand, a single physical medium can multiplex many separate process-to-process communications. On the other hand, a given process-to-process connection can use several separate physical connections with relays between them. Ideally, all stations on the network implement common OSI protocol suites in the intermediate layers (3-5) and some globally common protocols for moving data sets in layers 6 and 7. In addition, other standard application layer protocols will be shared among systems performing related functions. The choices of protocol suites in the Physical and DataLink layers and the connectivity of individual stations will vary. They will depend on the physical arrangement and capabilities of the individual stations, and their functional assignments and performance requirements. There may be one physical network, or many. All of these separate physical networks, however, must be linked together by "bridges" that implement the proper Network layer protocols. This results in a SINGLE LOGICAL NETWORK on which any given production management or data management process can connect to other processors regardless of location. We note, that because this architecture is layered, multiple subnetworks become transparent.

4.2 Where are we now

There are now many standard protocol suites for the DataLink and Physical layers, and there will soon be more. They all provide frame delivery and integrity checking; some provide for reliability and recovery, others defer those considerations to the transport layer. The real distinguishing characteristics among these standards are the signalling technologies and the sharing algorithms. Loosely speaking, the signalling technology determines the raw transmission speed, the relative immunity to electronic noise, and the cost. The sharing algorithm determines the nature of network service seen by the station. There are generally three choices:

a) connection to one other station or one other station at a
time, with fixed dedicated bandwidth (point-to-point, time- and
frequency-division);

b) connection to multiple stations simultaneously, with
variable bandwidth with fixed lower and upper bounds depending
on the number of stations connected to the medium (token bus and ring);

c) connection to multiple station simultaneously, with variable
bandwidth from zero to the bandwidth of the medium depending on
the traffic generated by all stations connected to the medium (CSMA/CD).

In general, engineering and administrative activities, which have infrequent and variable communications requirements, can tolerate and use the type (c) services more effectively. The production control activities, which have frequent and regular messaging requirements, however, prefer type (a) or (b) services.

There are also several "standard" protocol suites for the intermediate layers as well. But in this area, the differences are historical rather than functional. It is clear that the existing intermediate layer protocols will be THE standard in the near future. In the upper layers, standards are still evolving. Here the only problem will be to determine the suite of protocols necessary to a given production management or data management function.

The Manufacturing Automation Protocols (MAP) concept of one physical bus connecting all factory-floor stations may be appropriate for some manufacturing facilities. It is not, however, general enough to meet all communications requirements of the factories of tomorrow. However, the "enterprise networking" concept, connecting MAP control networks with Technical Office Protocols (TOP) engineering networks, demonstrates that a more generalized manufacturing network architecture is, in fact, currently practical. We believe that this will lead customers to demand, and vendors to produce, products consistent with that architecture (MAP and TOP 1988).

It is likely that emerging physical networking technologies will, in time, make the physical layer standards selected by MAP/TOP obsolete. This will lead to the addition or substitution of subnetworks with new physical and datalink protocols to current manufacturing networks. Nevertheless, the transparent, multiple, subnetwork architecture can result in little or no impact on networks already in place and on process-to-process communication. At the same time, adherence to at least the layering, but preferably also the intermediate layer protocol suites, in various types of "gateway" machines, provides for the transparent interconnection of subnetworks based on proprietary, or nonstandard protocol suites in the lower layers.

5. Remarks

The key benefit of enterprise integration lies in the potential for consistently delivering the right information to the right people at the right time, regardless of the physical location of the information or those who need it. To achieve this benefit requires the solution to all of the problems described in the sections on information and communications management. But, for this benefit to have maximum impact, this concept of "just in time" information delivery must be coupled with the ability to perform "just in time" information processing and analysis. This requires the solution to all of the problems described in the section on business and production management. Remember, that these problems are divided into two classes: choosing the right architectures, and advancing the various technologies needed to implement those architectures.

The fact is that much more progress has been made on individual technologies than on the integration of those technologies. This is particularly true for business and production management. New and better software tools appear in the marketplace everyday. But, the integration of tools has been hampered because there are no standard architectures for vendors to build to. Consequently, each individual manufacturing enterprise must pay large sums of money to develop its own architectures to use as the basis for integration. And, unless something is done to decouple integration from architectures, enterprise integration will continue to be a costly, if not elusive, undertaking.

We believe that significant progress can be made in enterprise integration, in the short term, by shifting the focus away from this top-down approach. The time has come for to begin a bottom-up "integrated tools" approach. In other words, we should begin to answer the question "What does it take to integrate the software modeling/analysis tools needed to perform key business and production functions?" Examples of these key functions include manufacturing engineering, production planning and control, systems design, and cost estimating. Each of these has three important characteristics: each can be decomposed into several well defined subfunctions; modeling/analysis tools exist to perform these subfunctions; and these tools cannot be integrated together electronically today. How do we address the third of these characteristics?

In the remainder of this paper, we concentrate on one of the most important of these modeling/analysis tools - simulation. (Much of what is covered in the remainder of this paper applies to all such tools.) In section 6, we discuss 1) why simulation is important as an enabling technology for enterprise integration, 2) why simulation is not currently fulfilling that role, and 3) what new technological advances are needed before simulation can fulfill that role in the future. In section 7, we turn our attention to other barriers to the successful integration of simulation technology into the enterprise.

6. Simulation as an Enabling Technology for Enterprise Integration

6.1 WHO NEEDS SIMULATION MODELS?

There are three principle groups within the enterprise who must have access to models of products and production: 1) Product and Process Design Teams, 2) Production System Designers, and 3) Production Managers. Each of these groups will use these models in different ways and for different purposes, but they all share a need for easy access and inexpensive analysis.

6.1.1 *Product and Process Design Teams*. Product and process design teams are responsible for developing new products and their associated manufacturing processes. They need models that evaluate performance of the product versus product specifications, and, to an increasing degree, they require models that assess manufacturability and cost. The simulations performed by these teams often represent the dynamic behavior of the product or of a production process; they often represent the physical and chemical interactions of the production processes, and frequently the models are highly detailed. Such models are needed at three levels of detail: system, subsystem/assembly, and unit process.

System level models relate design characteristics to performance. For example, a systems level model of an automobile will relate "crashworthiness" to structural design characteristics. Such a simulation might be a complex, computationally intensive finite element model. Another example would be a model of radio frequency emissions from a computer's motherboard. These simulations are used to negotiate product requirements and cost trade-offs with customers, to perform virtual prototyping, and to reduce the number and variety of actual prototypes required.

Subsystem/assembly models focus on the cost, performance, and manufacturability implications of individual design decisions. These models are used to evaluate the ability of existing manufacturing processes to produce the product at acceptable quality and yield and to allocate design space, tolerance, and error budgets to avoid costly, technology-limiting configurations. These simulations may be statistical models that use process capability knowledge, heuristic/knowledge-based models of manufacturability, or they may involve computationally intensive process modelling of the interactions between product features and process characteristics. Examples include 1) tolerance analysis of total indicated runout for multi-axis mechanical assemblies and 2) prediction of functional yield for printed wiring boards from combined knowledge of statistical process capability and cross-trace electromagnetic interference.

Unit process models optimize the parameters of production processes to produce the required product characteristics, to prevent damage to the product from the process, and to reduce cost or environmental impact. Examples include finite element analysis of the solidification of encapsulants and glass-to-metal seals to reduce residual stresses and cracking, optimization of electro-plating control strategies to reduce environmentally damaging heavy metal wastes, and computer-aided casting design to produce good cast parts the first time.

6.1.2 *Production Systems Designers* . Production System Designers are responsible for developing production capability to manufacture products. They need simulations and models to plan capacity, develop control algorithms, and insure quality and flexibility. The simulations performed by this group generally represent the flow of materials to and from processing machines and the operations the machines perform. The critical issues that are assessed include throughput, location of bottlenecks, impact of machine unreliability, and cost. In addition, the flexibility of the production facility is often a key design consideration. It is rare for a production cell, shop, or plant to manufacture only a single product. The ability of the facility to handle a mix of products is often a key design concern. In fact, today, plants are being designed to economically produce a large mix of products each of which is made in small lots.

Production systems inherently include a large number of static and dynamic policy opportunities. Static, or one-time, decisions are usually the responsibility of the designer and include the answer to such questions as: which is better,

     1) a single, high capacity machine with low operating costs but which completely interrupts production when it is off line, or

     2) several smaller, lower capacity machines with greater total operating costs.

The performance of production systems hinges critically on these policy decisions, and intuition is frequently a poor guide to making optimal selections because of the complexity and stochastic nature of the interactions among subsystems. Furthermore, the performance objectives are highly correlated - sometimes positively and sometimes negatively. Minimizing cost and maximizing flexibility are two such negatively correlated objectives, because flexible, highly reliable equipment is expensive. As a result, analysis tools such as simulation are critical to capturing the trade-offs and helping the decision-maker select the best among the many alternatives available when configuring and running a production system.

6.1.3 *Production Managers* . Dynamic decisions, which include real-time operating policies and control algorithms for routing of automated guided vehicles, scheduling and sequencing procedures for production, queuing disciplines for work in process, tool management policies, and inventory control are made by production managers. They are the people who are responsible for day-to-day operation of production facilities. They can use simulations and models to perform master scheduling, production scheduling and sequencing, control of production equipment, factory reconfiguration, and troubleshooting. They are primarily concerned with the day-to-day dynamic policy decisions related to real time scheduling and sequencing of production to satisfy immediate delivery requirements. This is a challenging job because the shop floor environment is never static. Production machines become unavailable, processes go out of control, discrepant source materials must be accommodated, materials do not arrive on time, due dates change, scrap rates fluctuate, and the product mix changes wildly. Under these conditions, sequencing of production and the allocation of resources within the plant become challenging tasks. The level of detail that the Production Manager must deal with is usually significantly greater than the Production System Designer. Furthermore, the time horizon is usually much shorter. Production Management focuses on policy horizons of a single day or a fraction of a day whereas Production System Designers typically develop optimal designs for weeks or months of operation.

6.2 COMMON OPERATIONAL NEEDS FOR MODELS AND SIMULATIONS

Each of the three groups of simulation users has different application needs, and they use a variety of modeling methodologies. Nevertheless, in creating, using, and maintaining their models, these groups perform many similar activities, and they must all deal with the shortcomings of today's modeling technologies. These shortcomings arise because of several factors: inadequate descriptive methodology, poor interoperability among models, little or no information integration, clumsy user interfaces, and few intelligent decision-making aids. All of these factors contribute to the difficulty involved in the entire process from model building to output analysis process. Listed below are problem areas that exist with current simulation methods and enterprise technology.

Modelers should not have to translate system definition data that exists in one form into a special descriptive language that is unique to the simulation program. Simulation programs should work directly with the system definition information available to the user and modeler. A typical case is that of the product designers for whom this information is frequently a CAD representation. Examples of this type of technology are automated meshing programs for finite element analysis that operate directly on 3-D CAD models of the parts to be analyzed. Little is being done to address the problem on a broad scale, however. For example, in the case of factory modeling, shop floor layouts must still be translated into simulation software- specific descriptive formats, and data entry is still mostly manual (albeit interactive).

Users should not have to reenter information that is or should be in computer readable form. Most factories have information systems which contain or could contain process plans, equipment characteristics, order information, lead time data, shop floor status, inventory status, and resource availability. In the case of performance and process simulations, data such as material characteristics, electrical properties, component data, and other performance information is readily available in parts data bases and other sources which could be electronically accessible. This low level (if any) of integration between enterprise information bases and simulation continues to be a major reason for the high cost of developing, validating, using and maintaining models.

Users should not have to be experts in the design of statistical experiments and the subsequent analysis and interpretation of results from those experiments in order to safely and effectively use simulations. The design of experiments to yield valid results and the interpretation of outputs are largely procedural activities. Intelligent processing modules could relieve much of the burden of performing these chores, and they could bring simulation analysis within the reach of many potential users who are not experts in these arcane topics.

Users should not have to manually assure the accuracy and currency of all of the data and configuration information embedded in a model each time it is used. Models should be reusable without extensive checking and reentry of data which is readily available electronically. This issue is particularly acute in the case of Production Management where shop floor data and machine status may change daily.

Expert modelers should not be required to perform maintenance on the model each time a slightly different scenario is to be analyzed or a different performance parameter is to be examined. With today's simulation methods and architectures, it is difficult to separate the model from the analysis. It is also difficult to expand or contract the modeling viewpoint in order to change the focus of the analysis. Models are too frequently constructed (and chartered!) with only a single use in mind. This leads to an inseparable relationship between representation and data collection elements. In other words, the level of detail in the model is tailored specifically and narrowly to yield the performance data in question. To some, this may appear to be an admirable and even necessary economy to satisfy the customer's immediate needs at minimum cost. But, simulations, like most other

160

software, are seldom used only once. Moreover, subsequent uses almost always require enhancement to the original mode. Today, experts are often needed (sometimes at high cost) to build models that a user or manager would consider to be minor extensions of an existing model. For example, a production model of a work center that contains several cooperating work cells cannot be modified easily to focus on only one of the cells. Nor can it be focused on a group of machines within a single cell. The reason is that today's modeling methodologies are not conducive to building different levels of aggregation within a single model.   In other words, the monolithic architecture of today's simulation methods does not facilitate multiple views without reconfiguration of the model by experts.

6.3 A TECHNOLOGY RESEARCH AGENDA FOR THE NEXT GENERATION SIMULATION TOOLS

If simulation modeling/analysis methodologies were structured properly, then simulation could be as pervasive tomorrow as MRPII and CAD are today. Both MRPII and CAD are complex software systems that perform mathematical operations which are completely hidden from the users. Few users understand the details of the functions performed by these programs. Yet they are able to use MRPII and CAD effectively and routinely in their daily work. There are three main reasons. First, these applications have automated the labor intensive chores associated with manipulating data. Second, they have hidden the theoretical aspects of their functions from the users. Third, and perhaps most important, they can be utilized in a manner which is completely within the domain expertise of the users.   Modeling and simulation must take a similar approach if it is to be an institutionalized and operationally useful tool. Several capabilities need to be developed before that will happen.

6.3.1 *Interfacing.*   Simulations will need to interface seamlessly and automatically with other enterprise information systems. Manual gathering, analysis, formatting, and entry of data for simulations is an enormous barrier to the use of simulation. It is so burdensome that even users who have a nonrecurring need for analysis may decline to build the model because the data collection is so expensive. In the real time environment of a production plant, simulation as a management tool is out of the question unless direct links to shop floor status information are imported automatically. .

6.3.2 *Simpler Construction and Maintenance.* Descriptive methods are needed which simplify the construction and maintenance of most simulation models. Once configured, the information that is contained in and manipulated by MRPII and CAD systems is maintained by users. A similar condition must be achieved for simulation. This implies that the system definition language must be contained in the domain expertise of the user. Today's simulation definition languages were developed to facilitate programming; indeed, these languages are equivalent to programming languages. Even those systems that use pictorial, interactive model building schemes have their roots in making the translation to a working program easy. They do not focus on capturing a dynamic representation of the user's design intent.   Today's system definition languages fall into two categories. The first group tends to be static and descriptive in nature like NIAM (Nijsen and Halpen 1989) and IDEF$_0$ (Mayer 1979). These languages show relationships among functions, but capture none of the rich dynamics of interaction. The second class includes most of the simulation "languages." These tend to be programming language oriented and are structured around and constrained by artifacts of the underlying simulation engine. The contrast between the world views of Systems Dynamics (PughIII 1973) and GPSS (Schriber 1973) illustrates this clearly. The way in which a system is described by these two languages is determined by the computational scheme that will be used in the simulation. A fresh approach is needed which combines the best features of both categories. This new representation technique must have three important characteristics:

1) a tailorable external view that is simple enough to be learned
by users of simulation who work in a diverse set of domains,

2) a hidden abstraction layer which maps the representation elements
of the external view to a finite set of enterprise concepts, and

3) a dynamic modeling layer which can link operational models
of the enterprise concepts into a coherent simulation of the user's intent.

6.3.3 *Distributed simulation architectures* . Distributed simulation architectures are needed which will enable multiple views and which will allow local construction and maintenance of models. A distributed architecture means that submodels interact through a well defined protocol with other submodels to achieve a combined simulation. Several benefits are achieved with this approach. First, the smaller submodels can be understood, built, and maintained at less expense and risk than a monolithic model. Second, the owner of a subsystem, such as a work cell, can be the owner of its submodel. Distributed ownership will mitigate against monolithic, single purpose models. The submodels will be designed to represent actual operations (at some defined level of detail) rather than be tailored to provide a specific performance parameter. Finally, a "plug and play" capability is conceptually possible in which the view of a simulation will change depending upon which submodels are included. It will be possible to focus a model on several machines within a cell as easily as performing an analysis that involves multiple cells.

6.3.4 *Statistical Design and Analysis*. Functions are needed for data reduction, experimental design, and output interpretation which are architecturally separate from the models. Users must be shielded as much as possible from the theoretical issues associated with the quality and meaning of the data that is input to and generated by the simulation. This separation also mitigates against embedding artifacts within the model to address specific data collection needs and therefore limiting the model's scope.

6.3.5 *Verification and Validation*. In a distributed environment, capability will be required to automatically verify the integrity of the simulation model. With distributed ownership of models, an automatic agent will be needed to verify that the correct versions of models are present, that the data sets are current, that the models selected to be a part of the simulation are coherent and compatible. In addition, real time manufacturing management applications will require that the latency of information is monitored and updated as required to create a coherent data set.

7. Simulation as Part of an Integrated Enterprise

We have argued that enterprise integration provides a common environment for data, information, and decision making. This common environment makes it possible to realize the goal of having all aspects of the manufacturing corporation managed consistently and effectively. We argued further that simulation tools have the potential to play an important role in the decision-making part of this goal. In the preceding section, we discussed several technology-related barriers to simulation realizing this potential. In this section, we will describe several other issues which must be resolved before simulation can be really integrated into the enterprise. These issues are quite complex, involving considerations of technology, culture, and corporate structure. In order for simulation to be effective in the integration of enterprise functions, it must be used at all levels in the enterprise. We begin with a description of the enterprise integration process, with special emphasis on those steps where simulation can make the process more effective. We will then describe a basic taxonomy for research in simulation and identify several important research topics to support the use of simulation in enterprise integration. Much of this can be generalized to include all tools that support enterprise integration

7.1 THE ENTERPRISE INTEGRATION PROCESS

The use of simulation in enterprise integration is based on the premise that simulation models of all aspects of the corporation can be constructed. These simulation models can provide the means to capture decision processes, data and knowledge about how an enterprise is designed to operate. The models can also be used to increase knowledge and understanding about an enterprise. They provide a vehicle for understanding and managing the continuous change and improvement of manufacturing operations. The need for this process is described in detail in (Hayes et al 1988). Pritsker (Pritsker 1991) also addresses the need to provide simulation-based modeling tools to help in managing the "capacity" of a manufacturing enterprise.

To understand how simulation can be used to enhance enterprise integration, we must have a firm grasp of the enterprise integration process. The following summarizes the major steps in enterprise integration, which occur continuously throughout the life of an enterprise, and discusses the role of simulation in each of these steps.

7.1.1 *Define the Enterprise Structure.* The first step in Enterprise Integration is to completely define the components of the enterprise and the relationship of each component to the others. This is a dynamic, or changing, representation but should reflect the current status of the organization. It would include a specification of the objectives of each component, their information requirements, and the flow of information from one component to another. Typical components in enterprises include executive management, research and development, product and process design, production or factory floor control, finance, and marketing.

Simulation models of these components and their relationships can be constructed. These models can be detailed or general, depending upon the specific analysis objectives. These models should at some level of detail, span enterprise components. For example, the models might include data flow representations between marketing and the factory floor, to investigate information system requirements to support the policies to be implemented. Other models might be focused in great detail, to represent, say, the scheduling and control of the manufacturing facility.

Along with the identification of the role of simulation in each of these components of the enterprise, the characteristics of the model users should also be specified. The model users should be involved intimately in the model development process and will flesh out the representation of the structure of the Enterprise.

7.1.2 *Identify Current and Future Enterprise Characteristics.* Enterprises change and evolve in a variety of ways. The strategic plan of a corporation may change in response to certain market conditions or opportunities. The enterprise may change in response to internal forces or characteristics. Key personnel can leave to explore other interests, new and dramatically different equipment can be purchased, or the location of the company can be changed. In planning for a successful, integrated enterprise, it is important to have the capability to represent all possible feasible "states" and to analyze their impact. The state of the enterprise includes not only the resources and information in the enterprise, but also the management procedures and decision processes which comprise it. These can all be effectively represented and analyzed using simulation models.

7.1.3 Develop the Roadmap for Enterprise Change. To reach a desired future state, an enterprise must often evolve through a variety of intermediate states before reaching it. A migration plan, or roadmap, is typically developed to support this evolution. The roadmap describes how each component will change over time. For example, if a new product line is to be introduced, manufacturing must have both capacity and capability to support it. This may happen over several months, perhaps including out sourcing as a interim measure until manufacturing develops the capabilities to support it. Other examples might include the development of new marketing capabilities for existing products, or changes to the information system of the company to allow important decision making information to be available.

Simulation can be used iteratively to represent the various interim states of the enterprise, and predict their performance. This might require several sets of models, each reflecting a step in the evolution of the enterprise. These models would need to interact as well, to communicate objectives and performance outputs.

7.1.4 Manage the Enterprise. As enterprises operate, changing or remaining static, they must be managed. This requires management tools to schedule and control the enterprise. This is needed on a variety of levels, from the Board Room to the Factory Floor. The models which were built to describe, and perhaps design, the enterprise can also be used to manage it. To be effective, these models must be linked into sources of data and information about the enterprise in real time.

One of the best examples of the use of simulation for enterprise management is factory floor scheduling. Tools have been developed which are effective in bringing simulation to the factory floor for use in day to day production management (Grant 1988) But other parts of the enterprise could benefit from this technology as well. Finance might use information flow and personnel models to schedule various accounting system updates. Marketing might use simulation tools to manage the schedule for the introduction of a new product in conditions of reduced resources.

7.2 A TAXONOMY FOR SIMULATION RESEARCH IN ENTERPRISE INTEGRATION

Simulation research in support of enterprise integration is broad and involves many parts of the enterprise. The development of a taxonomy for simulation research will be helpful in organizing the research and describing the relationships of various components of the research and where they should be focused. This section provides a basic taxonomy of research areas in which simulation research should be expanded to support enterprise integration.

7.2.1 Enterprise Modeling. Simulation tools in this area of the taxonomy are concerned with broad enterprise management issues and are typically, but not always, less detailed and more subjective in nature. They address problems concerned with strategic planning, and broad enterprise design issues. They also include issues of technology management, and management of the corporate culture. These high level modeling tools would tend to drive models developed in other sectors of the taxonomy.

7.2.2 Process and Product Design. Simulation modeling in this area is concerned with the development and introduction of new products as well as the development and refinement of production processes. A typical example of an application in this area would be the evaluation of the impact of a new product on existing production facilities. Other topics include production resource planning, process/product design integration, and detailed process simulation. Design alternatives can also be evaluated, regarding their impact on production.

7.2.3 *Factory Floor/Production.* The simulation and modeling tools in this sector of the taxonomy are concerned with managing and controlling the factory floor. Scheduling and dispatching is the most important example. Hierarchical simulation is also important, where varying levels of detail are included in the models, from factory wide representations, to detailed cell models. These hierarchical models would communicate between layers to develop local schedules, or reschedule broader areas as the performance of the enterprise demands.

7.2.4 *Information Infrastructure.* This sector of the taxonomy is concerned with the underlying information systems which are needed to support the various models described above. Real time databases would be needed to store up to the minute information on the factory status as it evolves. These databases may be distributed, logically or physically, as the characteristics of the enterprise demands. Network and hardware technology would also be needed to integrate the various models in the enterprise. Both the database and the network technology should be extended specifically for simulation applications.

7.2.5 *Simulation and Modeling Support.* The final sector of the taxonomy is concerned with the development of support tools to make the application of simulation technology easier and more cost effective. It would include tools for statistical analysis of simulation output, visualization of output data in new and varying forms. It would also include research in special purpose simulation languages, which perhaps could be developed by the application engineer, for his specific enterprise. Finally, user interface technology is needed to support the user's interaction with the large variety of models present in any enterprise.

7.3 RESEARCH PRESCRIPTION FOR SIMULATION IN ENTERPRISE INTEGRATION

The taxonomy described above broadly describes the major area of research required for the effective application of simulation in enterprise integration. The following is a list of specific research topics which should be addressed.

1. Research in the development of tools which support the integration of models with daily management tools and technology to better support the daily use of models.

2. Research in the interfacing of models to information systems so that this is easy and flexible for users to incorporate and modify.

3. Develop tools and technology which will support the integration of simulation with other management tools.

4. Develop output analysis tools which are more easily applied and imbedded in simulation models so that users can generate statistical analyses with confidence.

5. Research simulation tools and methodologies to build enterprise wide evaluation models to support strategic planning functions.

6. Research the development of tools which can integrate multiple views into enterprises including financial/accounting, sales, marketing, management, information, and manufacturing.

7. Research in the development of simulation and modeling tools which can effectively and automatically model change in dynamic systems, at various levels of complexity and detail.

8. Summary and Conclusions

In this paper we have concentrated on enterprise integration. We looked at two different approaches: a top-down architecture-based approach, and a bottom-up integrated-tools approach. Most of the research into enterprise integration has focused on the former. We reviewed the basic strategies and outlined some of the outstanding problems associated with designing and implementing these architectures. We concluded that enterprise integration can be achieved this way but, because there are no standard architectures, it tends to be company specific, thus, very costly. We then argued that a new integrated-tools approach should be undertaken. We used simulation as an example of one such tool. We described both the technological and the integration issues which must be resolved before simulation can become integrated into the manufacturing enterprise of the future.

9. References

ALBUS, J., BARBERA, A., AND NAGEL, N., 1981, "Theory and Practice of Hierarchical Control", **Proceedings of 23rd IEEE Computer Society International Conference,** 1981.

AMERICAN NATIONAL STANDARDS INSTITUTE, "Information Resource Dictionary System", **Document ANSI X3.138,** New York, NY, 1988.

AUTODESK, INC., **DXF: AutoCAD Release 12 Reference Manual,** 1992.

BARKMEYER, E.,"Some Interactions of Information and Control in Integrated Automation Systems", **Advanced Information Technologies for Industrial Materials Flow,** Springer-Verlag, New York, 1989.

DUFFIE, N., and PIPER, R., "Non-hierarchical Control of a Flexible Manufacturing System", **Proceedings of the International Conference on Intelligent Manufacturing Systems,** Budapest, 1986

DAY, J. and ZIMMERMAN, H., "The OSI Reference Model", **Proceedings of IEEE,** Vol. 71, No. 12, 102-107, 1988.

ESPRIT Consortium AMICE, (ed.), **Open Systems Architecture for CIM,** Springer-Verlag, Berlin, 1989.

GRANT, H., "Simulation in Designing and Scheduling Manufacturing Systems", **Design and Analysis of Integrated Manufacturing Systems**, National Academy Press, 1988.

HATVANY, J., "Intelligence and Co-operation in Heterarchical Manufacturing Systems", **Robotics and Computer Integrated Manufacturing**, Vol. 2, No. 2., 101-104, 1985.

HAYES, R., WHEELWRIGHT, S., and CLARK, K., "Dynamic Manufacturing: Creating the Learning Organization", **New York: The Free Press**, 1988.

IGES/PDES ORGANIZATION, **Initial Graphics Exchange Specification, Version 5.1**, National Computer Graphics Association, Fairfax, VA, 1991.

INTERNATIONAL STANDARDS ORGANIZATION, "ISO 10303 Industrial Automation and Systems and Integration - Product Data Representation and Exchange - Overview and Fundamental Principles", **Draft International Standard ISO TC184/SC4**, 1992.

INTERNATIONAL STANDARDS ORGANIZATION, "ISO/IEC9075 Database Language SQL", 1992.

INTERNATIONAL STANDARDS ORGANIZATION, "ISO/IEC 9579 Open Systems Interconnection - Remote Data Access (RDA) Part 1", **Global Engineering Documents**, Irvine, CA, 1991.

JONES, A., BARKMEYER, E., and DAVIS, W., "Issues in the Design and Implementation of a Systems Architecture for Computer Integrated Manufacturing", **International Journal of Computer Integrated Manufacturing**, Vol. 2, No. 2, 65-76, 1989.

JONES, A., and MCLEAN, C., " A Proposed Hierarchical Control Model for Automated Manufacturing Systems", **Journal of Manufacturing Systems**, Vol. 5, No. 1, 15-25, 1986.

JONES, A. and WHITT, N., (ed.), **Proceedings on Factory Standards Model Conference**, National Bureau of Standards, Gaithersburg, Maryland, USA, 1985.

LIBES, D. and BARKMEYER, E., "The Integrated Manufacturing Data Administration System (IMDAS)", **International Journal of Computer Integrated Manufacturing**, Vol. 1, No.1, 44-49, 1988.

LIU, J.,"The CAM-I Advanced Factory Automation System", **Proceedings on Factory Standards Model Conference**, National Bureau of Standards, Gaithersburg, Maryland, USA, 1985.

MAP and TOP Version 3.0 Specifications, Society of Manufacturing Engineers, Detroit, MI, USA, 1988.

MAYER, R., "Unified SDM: The ICAM Approach to systems & Software Development Proceedings", **IEEE Computer Society International Computer Software Applications Conference,** 331-336, November 1979.

MORRIS, K., DABROWSKI, C., and FONG, L., "Databnase Management Systems in Engineering", **NISTIR 4987,** Gaithersburg, MD, 1992.

NIJSEN, G., and HALPIN, T., **Conceptual Schema and Relational Database Design:A Fact Oriented Approach**, Prentice Hall, Englewood Cliffs, NJ, 1989.

NOF, S. and MOODIE, C., (eds), "Recommendations for Future Research Directions in the Automation of Materials Flow", **Advanced Information Technologies for Industrial Materials Flow**, Springer-Verlag, New York, 1988.

PECKMAN, J., and MARYANSKI, F., "Semantic Data Models", **ACM Computing Surveys**, Vol. 20, N0. 3, 153-189, 1988.

PETRIE, C., (ed), **Proceedings of the First International Conference on Enterprise Integration Modeling**, MIT Press, Cambridge, Massachusetts, 1992.

PRITSKER, A., "Manufacturing Capacity Management Through Modeling and Simulation", **National Academy of Engineering Forum on Foundations for World Class Manufacturing Systems**, 1991.

PUGH III, A., **Dynamo II Users Manual**, MIT Press, Cambridge, MA, 1973.

REFERENCE MODELS FOR MANUFACTURING STANDARDS, Internal Report N51, ISO 184/SC 5/WG 1, Ottawa , 1986.
SCHRIBER, T., **Simulation Usng GPSS**, John Wiley and Sons, New York, 1974.

THOMAS, G. et al., 1988, "Heterogeneous Distributed Data Systems for Production Use", ACM Computing Surveys Special Issue on Distributed Data Systems, **Association for Computing Machinery**, New York.

ULLMAN, J., "**Principles of Database and Knowledgebase Systems, Vol. 1**", Computer Science Press, Rockville, MD, 1988.

ARCHITECTURE CONSISTENCY FOR CIMOSA IMPLEMENTATION

K. Kosanke,
ESPRIT Consortium AMICE,
Stockholmer Str. 7
D-71034 Böblingen/Germany

ABSTRACT. The paper provides a short introduction to enterprise integration and to CIMOSA (CIM Open Systems Architecture) focussing on the architecture validation and management of its evolution. With application of CIMOSA spreading in industry and research feedback on experience has to be consolidated' and changes have to be introduced in a controlled way. The AMICE project has introduced a change procedure applied to all requests for change and used to consolidate and approve necessary modifications on the architectural specifcations.

1. Introduction

1.1. THE NEED FOR ENTERPRISE INTEGRATION

Today's world markets on consumer and investment goods as well as on know-how and capital are very competitive and unpredictable in terms of trends on customer demands, technology developments and financial and social conditions. The frequency of change is increasing rather than decreasing and enterprises will come under more and more competitive pressure on product line innovations and cost. Product life time as well as overall product volume are decreasing fast as customer demand more and more product diversities. Cost competition gets increasingly severe and for many enterprises it will become a real challenge to survive and grow in such competitive environments.

To endure and flourish in these global markets of the future, enterprises have to be able to manage the effect of external change in their internal operation, as well as adapting in real time all its external co-operation's in the industrial networks of vendors and customers in a global economy[1]. For real time 'management of change' internal and external environments have to be viewed concurrently and impact of change has to be evaluated for all aspects of the enterprise operation.

Skilful management of change is constraint by the availability of the right information, to the right people, in the right form, at the right time and in the right place. Information which today exists, but is neither accessible, nor in a compatible form for additional processing and decision support. Therefore, information on information availability, easy access and compatible representation of the required information are

---

[1]   CALS (Computer Aided Acquisition and Logistic Support) is a current example of this requirement

S. Y. Nof (ed.), Information and Collaboration Models of Integration, 169–177.

prerequisites for 'management of change'. Prerequisites which can only be fulfilled through a clear understanding of the enterprise internal and external operations (an up-to-date enterprise model) and a compatible information technology support for information access, transformation and processing (an integrating IT infrastructure). Standardised information representation are especially needed for external co-operation support and for representation of information needed in those transactions and in their control and monitoring. But such common representation of enterprise models will significantly improve the efficiency of internal operation as well.

1.2.   THE SOLUTION

Both, compatible models of enterprises supported by an integrating IT infrastructure[2] will be major enabling technologies which together will fulfil the requirements of 'management of change' which are:

-       support and ease inter- and intra-enterprise information interchange and enable advanced use of information technology,
-       enable information exchange in a consistent and understandable manner,
-       foster active co-operation between industrial enterprises and
-       provide decision support for detecting and managing change.

The ESPRIT project AMICE has defined and developed an architecture for definition, specification and implementation of Computer Integrated Manufacturing systems - CIMOSA (CIM Open System Architecture) - which will support 'management of change'. Such an architecture will become the cornerstone of an industrial infrastructure for industry wide use of information and information technology. Business value for the industrial user will be a significantly improved support for decision making processes at all levels of management and operation, resulting in real-time 'management of change'. More efficient use of resources and better asset management will improve enterprise productivity and open up new markets with increased opportunities. For the IT industry business values will be greatly enlarged markets for their new product lines due to the ease of access and use of information inside and across enterprises. Again this will provide significant growth opportunities for all vendors involved.

CIMOSA has been developed under ESPRIT of the Commission of the European Communities with the aim on enterprise integration and specific focus on the manufacturing industry. The ESPRIT Consortium AMICE developed the CIMOSA concepts under previous ESPRIT contracts (ESPRIT I, II and III) during the years 1985-1992[3]. During the ESPRIT II phase three additional consortia with a total membership of 20 CIM users, vendors, implementers and research organisations have launched CIMOSA application projects covering a wide range of industrial applications in discrete manufacturing and process industries and adding to the validation of the architecture and its technical specifications.

The ESPRIT Consortium AMICE itself consists of a large number of European organisations and currently has 15 active members:

> *Aerospatiale (F), British Aerospace (UK), Bull (F), CAP Gemini SESA (B), FIAT (I), GRAI (F), Hewlett Packard (F), IBM (D), INRIA (F), ITALSIEL (I), ITEM (CH), NLR (NL), Siemens (D), University of Valladolid (E), WZL (D).*

The project work on CIMOSA has been conceived as pre-normative work from the start of the project. Beside the development work on architectural framework, enterprise

---

[2]   The success of the automobile as the major means of transportation was due to establishment of a sufficient infrastructure (roads and network of gas stations) and their accompanying traffic rules, regulations and standardisation's (location of steering wheel, gas and accelerator panel, etc.)

[3]   ESPRIT Projects 688, 2422 and 5288

modelling and integrating infrastructure, external promotion in standardisation bodies as well as in the public domain has always been handled as high priority items. All validation work of the AMICE project has been carried out on examples from the main areas of manufacturing enterprise operations - product development, production planning and production.

The project has achieved significant recognition and acceptance for this framework for enterprise modelling in both standardisation and in the public domain. The basic contents of the Modelling Framework (generic building blocks) have been developed for the requirements and design modelling level. Other parts of the framework have been populated as well and work on building block types has been started to provide more details of the user language for enterprise modelling.

The AMICE project has published its results on CIMOSA in two books by Springer[1][3] and numerous publications and presentations by project members. The CIMOSA specifications have already been made available to co-operating projects and standardisation for quite a while and the latest version has recently been published by the ESPRIT Consortium AMICE as the CIMOSA Technical Baseline[2].

Results from project work in standardisation have been a European Pre-Norm[4] for the Modelling Framework which has become the base for the international standardisation in ISO TC 184/TC5/WG1 and a recent submission to CEN/CENELEC for an Integrating Infrastructure framework supporting model based enterprise operation control and monitoring. In addition, the ESPRIT Consortium AMICE provides inputs to standardisation work (CEN/CENELEC) further detailing the modelling framework and thereby facilitating derivation of standard compliant product specifications. International recognition has been achieved through participation in numerous professional events and especially through the EC/US collaboration on enterprise integration. Results from the latter have been made publicly available as well[5].

With the CIMOSA implementation work spreading to more and more companies and with focus not only on modelling applications, but on modelling tools as well, the need for consistent evolution of the architecture specification becomes more and more apparent. Therefore, the AMICE project has established a special Workpackage responsible for the consistent evolution of the CIMOSA specifications. A formal change procedure has been introduced to manage change requests on the CIMOSA technical specifications to be evaluated, formally accepted and the CIMOSA Technical Baseline updated accordingly. Consolidation of change requests from the AMICE external projects is enabled by the CIM Europe supported Special Interest Group on Co-operation for CIMOSA Validation and Promotion (SIG-CCVP). The four ESPRIT projects AMICE, CIMPRES, CODE and VOICE are represented in the SIG-CCVP, with the objectives to resolve issues encountered during the application of CIMOSA and plan joint promotion activities to increase awareness and acceptance of CIMOSA.

2.   CIMOSA Content

CIMOSA provides both, an reference architecture that supports enterprise model engineering and an integrating infrastructure aimed on model execution for model based operation control and monitoring.

172

The architectural Framework of CIMOSA is shown in Fig 1. It consists of a Reference Architecture which supports the engineering of particular enterprise architecture's. The CIMOSA Reference Architecture provides a collection of generic constructs and partial models which allow to structure and design particular enterprise models. These constructs permit model engineering supporting all phases of enterprise systems and operation life-cycle (from Requirements Definition through Design Specification to Implementation Description - see Fig 2). Starting from enterprise objectives and constraints the enterprise model

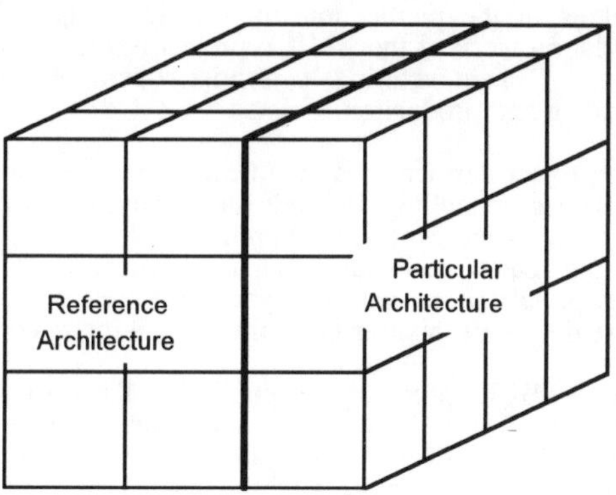

Fig 1: CIMOSA Architectural Framework

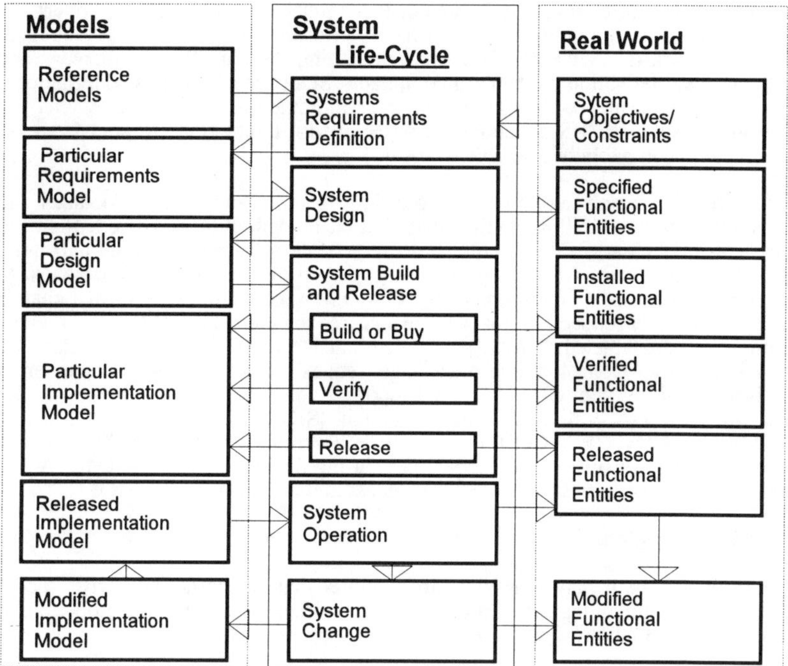

Fig 2: CIMOSA Model and System Life-Cycle

will be derived specifying required processes and resources (Functional Entities) and taking into account the existing ones. A formal release process will authorise the operational use of the verified implementation description model. The enterprise modelling process follows the architectural framework (Fig 1) and allows the user to optimise the model working with specific views of the model contents supporting his specific modelling needs. Four different views have been identified (Function, Information, Resource, Organisation), however, additional ones may be defined if required. An overview of the generic modelling constructs is provided in Fig 3.

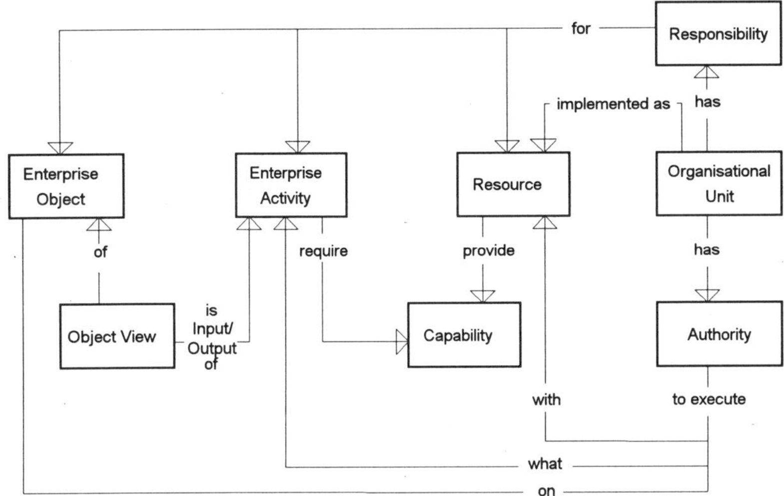

Fig 3: CIMOSA Modelling Constructs

CIMOSA provides for evolutionary modelling of the enterprise through the concept of enterprise domains and corresponding processes. Based on process modelling, enterprise operations are modelled as sets of co-operating processes exchanging results and triggering each other through events (see Fig 4). Particular enterprise domains and their connection to other domains may be modelled and these models may later be integrated into larger networks of co-operating domains respectively their particular domain processes. Processes may be detailed in terms of functionality to the level of control required for the particular enterprise domain.

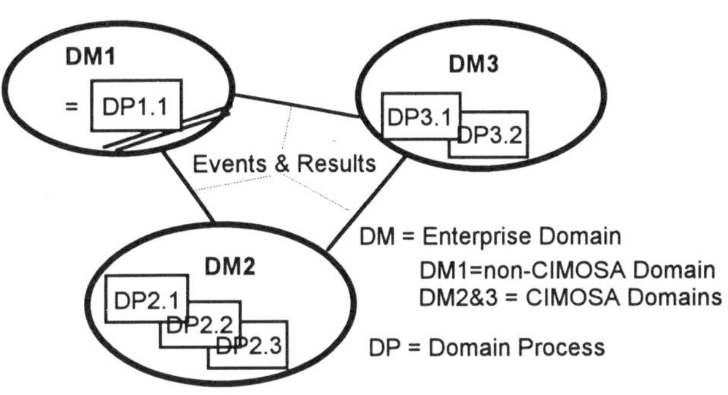

Fig 4: Co-operating Enterprise Domains

Functionality itself is defined through its different inputs and outputs (Function, Control and Resources). The model captures not only the enterprise functionality but its dynamics - the flow of control - as well. Models may include non-CIMOSA domains (DM1 in Fig 4) providing thereby a graceful migration to a fully model based enterprise operation.

174

The CIMOSA Integrating Infrastructure consists of a set of generic services aimed at supporting model execution for real time model based operation control and monitoring. Fig 5 shows the five major service entities which make up the Integrating Infrastructure. The released implementation model (lower right in the CIMOSA architectural framework) is executed through the Business Entity with the support of the Information, Presentation and Common Entities. The latter provide access to the information needed and the resources used employing the IT facilities provided by the enterprises IT resources. The Management Entity (not yet available) will accommodate overall system services like IIS configuration, performance, security management, etc. The Integrating Infrastructure makes use of existing and emerging IT standards (OSI, MAP, ODP, etc.) as much as possible. It is not the intention of this work on CIMOSA to develop competing standards but to complement and enhance established and emerging standards.

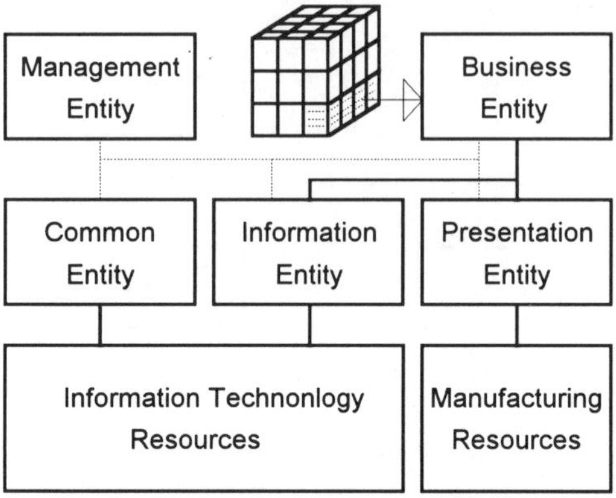

Fig 5: CIMOSA Integrating Infrastructure

CIMOSA is described in a very recent publication by Springer[1] and its current set of specification is available from the ESPRIT Consortium AMICE[2].

3.   CIMOSA Validation by AMICE

Validation of project results has been one of the main tasks throughout the development process. Scenarios and case studies from the industrial partners have used to evaluate the applicability and usefulness of the architecture. Case studies have covered different industries (aerospace, automotive, electronics), different parts of the product life cycle (product development, pre-production, parts production and assembly). Validation results have always been used to improve the architecture specifications and several versions of the AMICE Formal Reference Base have been published internal to the project.

Work on enterprise modelling has been amended by basic developments on computerised tools. The aim was to define and verify basic requirements for computerised modelling support rather than to develop a tool itself. The work carried out on the CAEE tool has let to a limited prototype demonstrating the use of requirements modelling level constructs for model engineering. This work is continued by one of the Consortium partners, but outside the AMICE project.

The following validation efforts have been carried out over the years by the AMICE project.

1987   Illustrative Example on Shop Floor Operation Modelling
1988   Aerospace Industry Case Study on Shop Floor Operation
        Automotive Industry Case Study on Pre-Production Planning
        Electronic Industry Case Study on Product Development (Printed Circuit Board Design)
        Electronic Industry Case Study on Part Production (Printed Circuit Board Manufacturing)
1989   Early Prototypes of IIS Services (demonstrated at ESPRIT Conference 1989)

1990   Aerospace Industry Case Study on Shop Floor Operation (enhancements of previous study)
Automotive Industry Case Study on NC-Part Program Management

1992   Prototype of Machine Dialogue Service in co-operation with a partner of the VOICE project

1991   Participation in Case Studies of ESPRIT Projects CIMPRES, CODE and VOICE (see co-operating projects)
Participation in VOICE efforts on validation of the Integrating Infrastructure.
continuous: application studies in member companies with feedback to the project.

Current project work is focused on implementation of CIMOSA pilot applications for both model engineering and model based operation control and monitoring. Results from this work together with feedback from validation efforts from outside the AMICE project will be used to further enhance the CIMOSA specifications. A new version of the Technical Baseline will be published at the end of this project (early 1994).

In addition the current work is aimed at demonstrating business benefits through CIMOSA model engineering. Comparison with current best praxis in the enterprise modelled will identify these benefits in measurable terms for both improved business operation (improved visibility of operation content, impact of change, others) and better model engineering (engineering time, flexibility for re-engineering, adaptability to changes in scope of business evaluation, others).

4.   CIMOSA Consistency Management

The ESPRIT Consortium AMICE has documented the specifications of CIMOSA in its Formal Reference Base (FRB). This document contains not only the description of the CIMOSA concepts, its modelling constructs and the Integrating Infrastructure service entities, but specific specifications in the form of definitions of terms, template formats and formal descriptions as well. Fig 6 shows an example of a template specification of a CIMOSA generic building block (Enterprise Activity) indicating the type of information gathered in CIMOSA based enterprise models. The technical specifications of CIMOSA have been made available to co-operating projects and standardisation bodies for quite a while and have recently be made available to the public at large[2].

Numerous implementation and exploitation efforts have been started on enterprise model engineering, model based operation control and monitoring and modelling tool developments. Such efforts are carried out not only in the AMICE project itself and its member organisations, but in co-operating ESPRIT projects (CIMPRES; CODE, VOICE) and ESPRIT external initiatives as well. These efforts identify needs for modifications and enhancements of the CIMOSA technical specifications. These requests for changes are than evaluated, consolidated, approved, formally accepted, implemented in the common set of technical specifications and communicated to interested parties as well as to the public. The latter has to be done by new releases of the CIMOSA Technical Baseline.

To provide a consistent 'management of change' for the CIMOSA technical specifications the AMICE project has introduced a formal change procedure (documented in the project FRB). The principles of this change procedure have been adopted by the Special Interest Group on Co-operation for CIMOSA Validation and Promotion and will be used for all incoming change request internal and external to the AMICE project. Fig 7 shows the change procedure contents. Forthcoming change requests should not only indicate the problem encountered in the implementation work, but should propose a potential solution as well. Evaluation of the change request will be by members of the AMICE Workpackage on architecture consistency with consultancy of specific experts from the relevant work areas. The resulting proposal will be presented to an AMICE Technical Co-ordination Team (TCT) consisting of AMICE Workpackage leaders, project managers from co-operating ESPRIT projects and project experts invited according to competence needed.

<table>
<tr><td colspan="2">B11-2200. Enterprise Activity</td></tr>
<tr><td>Version: 4.0</td><td>Author: WP-D</td></tr>
<tr><td>Status: Author-Released</td><td>Last Update: 92.08.25</td></tr>
</table>

ENTERPRISE ACTIVITY

Type:	[relevant category - select from list <name>]
Identifier:	[EA-<Unique Identifier>]
Name:	[name of Enterprise Activity instance in the form: <adjective> <noun> <adjective>: qualifying the Enterprise Activity instance <noun>: related to the scope of the Enterprise Activity]
Design Authority:	[name of a person and department with the authority to design/maintain this particular instance]
A. Functional Description	
OBJECTIVES:	[non-empty list of objectives which must be fulfilled by the Enterprise Activity instance in the form: <identifier of Objective/Constraint> / <name of Objective/Constraint>]
CONSTRAINTS:	['NIL' or list of constraints applicable to this Enterprise Activity instance in the form: <identifier of Objective/Constraint> / <name of Objective/Constraint>]
DECLARATIVE RULES:	['NIL' or list of Declarative Rule instances applicable to this Enterprise Activity instance in the form of: <identifier of Declarative Rule> / <name of Declarative Rule>]
FUNCTION DESCRIPTION:	[optional; textual description of the tasks performed by this Enterprise Activity]
INPUTS	
FUNCTION INPUT:	['NIL' or list of Object View instances (defined in the Information View) in the form of: <identifier> / <name>]
CONTROL INPUT:	['NIL' or list of Object View instances (defined in the Information View) in the form of: <identifier> / <name>]
RESOURCE INPUT:	[must be left undefined at this modelling level]
OUTPUTS	
FUNCTION OUTPUT:	['NIL' or list of Object View instances (defined in the Information View) in the form of: <identifier> / <name>]
CONTROL OUTPUT:	['NIL' or list of Event instances, occurrences of which could be generated during the course of occurrences of the Enterprise Activity instance, listed in the form of: <identifier> / <name>]
RESOURCE OUTPUT:	['NIL' or list of textual statements indicating requirements on data collection on resource usage]
ENDING STATUSES:	[non-empty list of Ending Status values and their signification in the form: <value> : <signification> where <value> is mandatory and is a 0-argument predicate and <signification> is optional and is text]
REQUIRED CAPABILITIES:	'NIL' or list of Capabilities required by occurrences of this Enterprise Activity instance]
B. Structure Description	
WHERE USED:	[non-empty list of identifiers and names of Domain Process and/or Business Process instances employing this Enterprise Activity instance in the following form: <identifier of Domain Process or Business Process> / <name of Domain Process or Business Process> Can be 'NIL' for partial instances]

Fig 6: CIMOSA Specification (Enterprise Activity Template)

The approved change are documented in the Formal Reference Base and distributed periodically either as pre-releases or a new release of the CIMOSA Technical Baseline.

The procedure has been applied internally to the AMICE project and has proven to be very efficient to manage the change requests identified in the CIMOSA validation work. It supports both the control of the change process a whole and the consolidation and approval part in particular. Number of requests as well as time elapsed in the request processing requires a formal procedure as identified above.

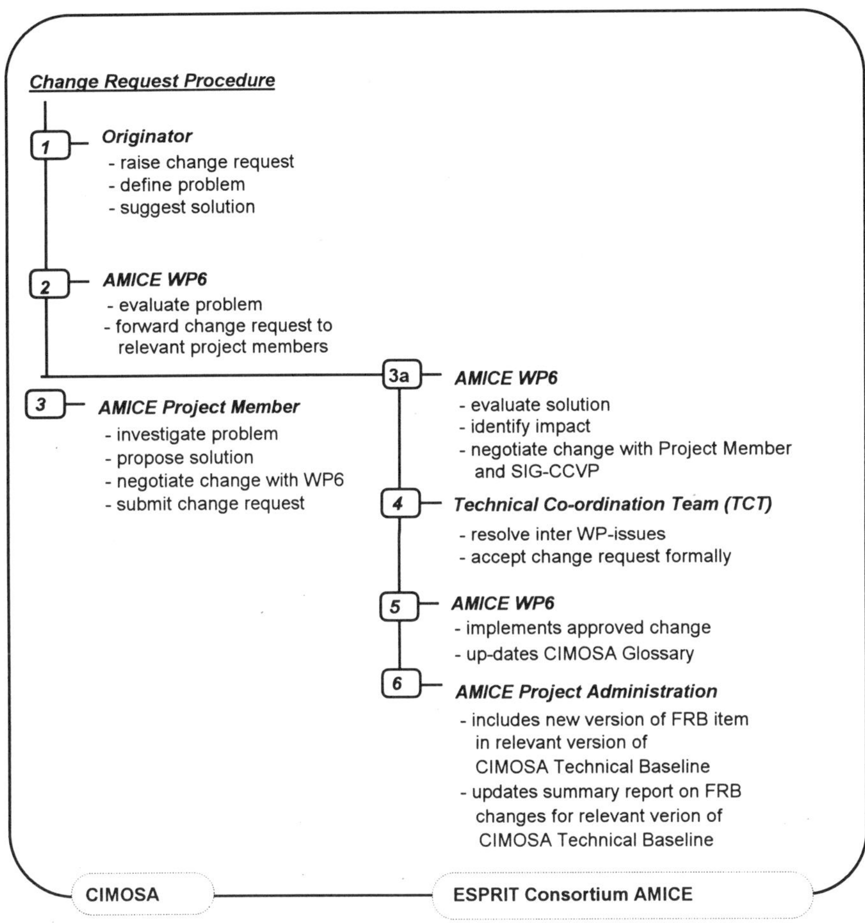

Fig 7: AMICE Change Procedure for Change Requests on CIMOSA Technical
Specifications

5. References:

[1]   ESPRIT Consortium AMICE (Eds.), CIMOSA Open Systems Architecture for CIM, Springer 1993

[2]   ESPRIT Consortium AMICE (Eds.), CIMOSA Technical Baseline, private publication 1993

[3]   ESPRIT Consortium AMICE (Eds.), Open System Architecture for CIM, Springer 1989

[4]   CEN/CENELEC ENV 40 003: Computer Integrated Manufacturing Systems Architecture Framework for Modelling

[5]   Charles Petrie (Ed.), Enterprise Integration Modelling, Proceedings of the First International Conference, MIT Press, 1992

COOPERATION REQUIREMENT PLANNING FOR MULTIPROCESSORS

Venkat N. Rajan
Industrial Engineering Department
Wichita State University
Wichita, KS 67260-0035

Shimon Y. Nof
School of Industrial Engineering
Purdue University
West Lafayette, IN 47907-1287

ABSTRACT

Although cooperation among multiple manufacturing machines seems intuitively desirable, it is necessary to find what specific cooperation can and should be accomplished. Cooperation Requirement Planning (CRP) is the process of generating a consistent and coordinated global execution plan for a set of tasks to be completed by a multi-machine system based on the task cooperation opportunities, requirements and interactions. In this sense, CRP is an essential part of integration in manufacturing. CRP is divided into two main steps: CRP-I which matches the task requirements to machine and system capabilities to generate cooperation requirements, task precedence, machine operation and system resource constraints. CRP-II uses the results of CRP-I to generate a task assignment and coordinated, consistent global execution plan. In this chapter, we describe the CRP-II methodology in which each task is considered to be a goal and the CRP-I process is viewed as generating alternate plans and associated costs. Some open research directions are discussed.

1. INTRODUCTION

Manufacturing planning has traditionally considered tasks whose processing requires the independent operation of a single machine. However, the development of robotic devices has provided the ability to handle tasks whose requirements are beyond the capabilities of a single machine[1, 2]. Additionally, limited integration of process planning

Acknowledgment: Partial support by NSF Grants DMC-8719845, Models of Cooperative Robotics , and DDM-9214143, Collaborative Integration.

S. Y. Nof (ed.), Information and Collaboration Models of Integration, 179–200.
© 1994 *Kluwer Academic Publishers. Printed in the Netherlands.*

and scheduling functions has constrained the flexibility of manufacturing systems[3].

Cooperation among machines to perform assigned tasks is considered an essential attribute of intelligent manufacturing and assembly systems. The cooperation paradigm improves the flexibility and reliability of the given cell. The traditional view of machines operating independently to perform assigned tasks limits flexibility to the range of processing capabilities of an individual machine and the overlap in these capabilities among the machines in a given cell. Such overlap provides flexibility while making routing decisions. However, by allowing cooperation, an added dimension of flexibility is realized due to the additional tasks that can be performed by cooperating machine sets. System reliability is also improved due to the overlap in machine capabilities. By integrating the process planning and scheduling functions, the process capability information of individual machines and of cooperating machine sets is readily available. Tasks assigned to failed machines can therefore be dynamically reassigned to other machines or cooperating machine sets to complete them successfully.

The Cooperation Requirement Planning (CRP) methodology presented in this research[4] is aimed at developing an integrated planning system using the cooperation paradigm. CRP is divided into two steps:

1. *Cooperation Requirement Planning I (CRP-I)*: This step of the planning process takes as input the product and cell description, and generates the *cooperation requirements matrix* (CRM) whose elements represent the performance capabilities of machine sets for the processing tasks. It also generates the machine, and cell constraints.

2. *Cooperation Requirement Planning II (CRP-II)*: The cooperation requirements matrix and the various constraints are used to determine the assignment of tasks to machine sets and to generate a consistent and coordinated global execution plan.

CRP-I generates the task cooperation requirements based on their processing requirements. A *job* is defined to consist of a set of *tasks*, and each *task* is defined to consist of a set of *actions*. In the following discussion, *machine* and *processor* will be used interchangeably. The cooperation requirements are defined in terms of cooperation modes, to specify the single and multiprocessor needs of the tasks, as[1, 2]: *mandatory task cooperation* for tasks requiring the participation of two or more processors in *sequence* or in *parallel*; *optional task cooperation* for those tasks that can be independently completed by one or more processors of *similar* or *different* types in the cell; and *concurrent task cooperation* for those tasks that can be performed optionally, but by using two or more cooperating processors, better performance is achieved.

We will use the multi-robot assembly domain to illustrate the CRP methodology. The framework for the methodology is shown in figure 1. The aim of this methodology, is to enhance the flexibility and reliability of a given multi-robot cell. Therefore, it is assumed that cell information is pre-specified and available in a database. The database consists of: geometric, operational, and physical information. Geometric information specifies the workcell layout, and robot kinematic capabilities. Operational information specifies the non-geometric processing capabilities. Physical information specifies physical attributes. In addition, the database contains information on the various resources in the workcell, robot status, and allowable cooperation sets. The input to the system is an assembly description. It also consists of geometric, operational, and physical elements: geometric information of the assembly components; a high-level operational task plan that specifies the required tasks and associated non-geometric parameters; functional assembly constraints imposed by the user, and physical information of the assembly components.

The geometric description of the assembly is used to determine the mating directions and assembly precedence constraints. The CRP-I process maps the geometric, operational, and physical characteristics of the assembly tasks to the equivalent robot and cell capabil-

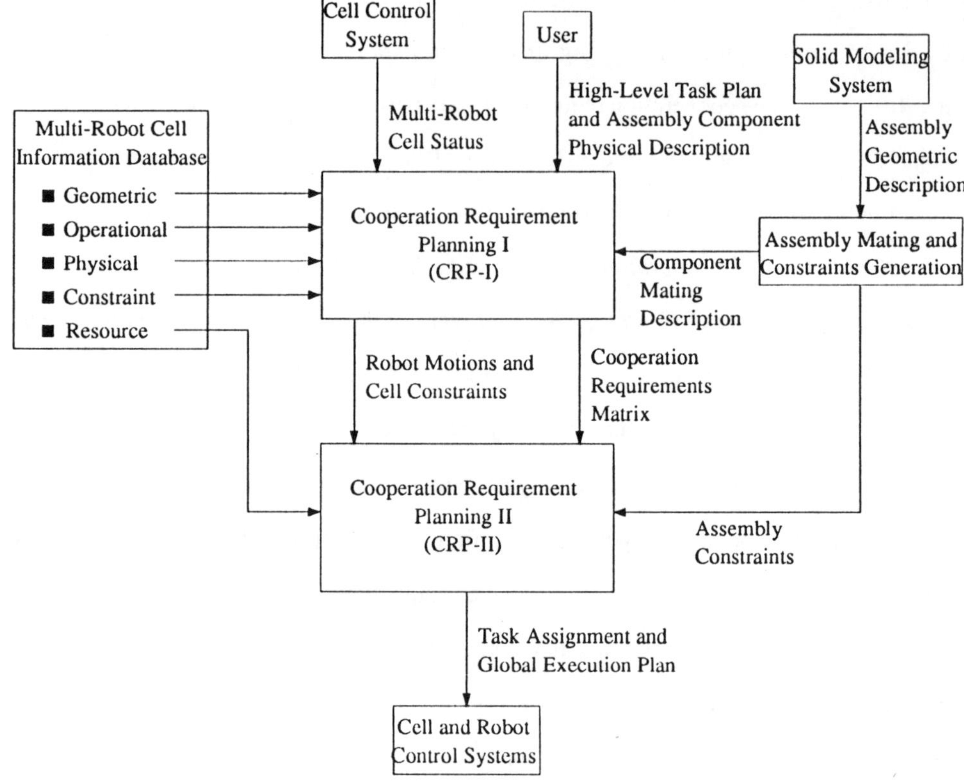

Figure 1: Cooperation requirement planning framework for multi-robot assembly.

ities to determine the *cooperation requirements matrix*. The robot and cell constraints are also generated during this process. The detailed CRP-I methodology is presented in [4, 5].

2. CRP-II: TASK ASSIGNMENT AND GLOBAL EXECUTION PLAN GENERATION

The outputs of the CRP-I process are used as inputs to the CRP-II module. The CRP-I process generates the cooperation requirements matrix which provides the best capability measures of various robot sets for each of the tasks in the assembly. Let $N(R)$ be the set of robots in the cell, and $N(T)$ the set of assembly tasks. Then, the cooperation requirements matrix can be expressed as a mapping as follows:

$$C: S \times N(T) \to \Re^+,$$

where S is the set of allowed robot sets, and $C(i,j)$ represents the capability measure of robot set i for task j. Let M be the number of elements in $N(R)$. Then, S is a subset of the power set (excluding the null set) of $N(R)$, i.e., the maximum number of elements in S will be $2^M - 1$. In general, due to cooperation limitations imposed externally, or due to physical limitations between cooperating robots, the set S is much smaller than the power set. When a robot set i cannot perform a task j, the corresponding capability measure $C(i,j) = 0$.

Various constraints and related information generated during the CRP-I process are also used as input to CRP-II. The assembly constraints specify a partial ordering on the assembly tasks. The robot constraints are specified as the set of feasible task execution directions for each assembly task. Finally, the cell constraints specified by the resource requirements for each task plan of the robot set i -task j pair, are used to ensure that they are satisfied for a given assignment and that resource interactions are resolved.

The purpose of CRP-II is to assign tasks to the robot sets in the cell based on their capabilities and generate a global execution plan that is consistent and coordinated. The interactions that exist between individual task plans arise due to assembly, robot, and cell constraints. To ensure successful completion of the assigned assembly tasks, these con-straints need to be satisfied in the final consistent plan. Also, for mandatory and concurrent parallel cooperation modes, the motions of the cooperating robots have to be coordinated, to ensure that the cooperative task is successfully completed.

The task assignment and execution process can also be viewed as a problem of plan-ning for multiple goals with interactions[6]. In this case, each of the tasks represents an individual goal. CRP-I can then be viewed as generating alternate plans for each goal. Each plan is a standard template consisting of the following actions: move to the component pick location; take the component; move to the component destination location; leave the component; move to the safe position. In CRP-II we wish to identify a plan for each task (goal), resolve interactions between plans, and ensure coordination within a plan when required. The complexity of the planning problem is determined by the strategy used. The two strategies that can be used are as follows:

1. *Simultaneous plan generation strategy*: The plan selection and constraint satisfaction can be performed simultaneously. In this case, all the available plans for all the tasks (goals) are considered, and a global plan constructed by considering alternate combinations of goal plans.

2. *Sequential plan generation strategy*: The plan selection can be initially performed. This is based solely on the cooperation requirements matrix information. For a given set of plans, one for each goal, constraint satisfaction is performed.

The first method will generate an optimal global plan as a search over the entire state-space is performed. However, the complexity of the search is very high. In the latter case, by de-coupling the plan selection and constraint satisfaction processes, the search complexity is considerably reduced. However, the constraint satisfaction of the set of selected plans does not guarantee the globally optimal solution, nor is a feasible solution guaranteed. It is possible that a given assignment cannot lead to a feasible execution plan as it violates one or more constraints. In the following sections we will present algorithms for these two strategies.

2.1 Simultaneous plan generation strategy

When plan selection and constraint satisfaction are performed simultaneously, a best-first search algorithm[6] is used to generate the optimal global plan solution. The search is performed over a state space where each state represents the partial global plan for a collection of tasks (goals). A state at the ith level represents a plan for i tasks (goals). The 0 level is the empty state. An open list is maintained for the states yet to be expanded. The states in the open list are ordered in the best-first fashion. The first state in the list is selected for expansion at the beginning of each iteration. If the state contains all the assembly tasks, then the optimal global plan has been obtained and the search stops. Otherwise, the state is expanded by merging the plans for an additional task. The set of tasks to be included is formed by selecting those available tasks that do not violate fixed assembly and robot constraints.. The successor states are formed by merging the various plans for a selected task with the partial global plan. Let *OPEN* represent the open list of states, and *COST()* the cost function. The algorithm may be stated as follows:

Algorithm: Best-first Search

1. *OPEN* = $\varnothing$.

2. Find all tasks that do not have any precedence constraints. *OPEN* is the best-first order on the set of all plans for all such tasks.

3. Select the first node from *OPEN*. Call this node *FIRST*.

4. If *FIRST* includes all assembly tasks, exit with *FIRST* as the optimal global plan.

5. Find the set of feasible tasks. Call this set *TASK*.

6. For each task t in *TASK*, for each plan P of t, merge P with *FIRST* to generate all possible successor states. Call this set *SUCCESSOR*.

7. Compute the cost for each element in *SUCCESSOR*.

8. Merge *SUCCESSOR* into *OPEN* in the best-first order.

9. Go to step 3.

2.1.1 Feasible tasks

For each state to be expanded, the set of feasible tasks needs to be identified. Let F be the set of tasks already in the state *FIRST*. The available set of tasks is then $\{N(T) - F\}$. For each task t, $t \in \{N(T) - F\}$, if adding the task to the current state will not lead to violation of fixed assembly constraints of any other task u, $u \in \{N(T) - F\}$, then t is a feasible task.

2.1.2 Plan merging

The current state is expanded by merging the plans of the various feasible tasks to generate the successor states. To merge a given plan into the current global plan, the feasibility of the plan has to be determined. A plan is feasible if its robot constraint is not violated by the global plan. In other words, if the current global plan does not allow for any location motion of the given plan to be executed, then the given plan is infeasible. For a feasible plan, the types of merging actions performed will be dependent on the interactions between the partial global plan tasks and the merging task plan. The following 5 cases arise:

1. *Action combination*: When the given plan is merged after a task performed by the same robot set, the motion to the safe position in the first task is combined with the motion to the pick location of the second task, and the resultant plan consists of a direct motion from the drop location of the preceding task to the pick location of the succeeding task.

2. *Precedence relation*: If the feasible task is constrained to succeed other tasks, it can only be merged after those tasks in the global plan. The task has to be executed after the last of the predecessor tasks have been completed. In this case, the precedence interaction is between the pick and drop motions of the two tasks. Precedence interactions arise due to fixed assembly constraints and due to usage of the same robot set

to perform the tasks. In the latter case, the action combination interaction can be used to achieve limited merging. The precedence relation establishes that the actual assembly step for the preceding component should precede the assembly step of the succeeding component. The pick up of the successor component can be performed independent of the precedence constraint, subject to other resource sharing constraints.

3. *Resource sharing*: If a task requires a resource for its successful execution, then it has to succeed a partial global plan task that requires the same resource. The precedence interaction is not limited to any single action. If spatial resources are in contention, then the particular motions that access the resources are ordered sequentially. If secondary resources are required, then the actions may be ordered sequentially or occur concurrently. If the number of required units of resources are available, then the actions can occur concurrently, otherwise the actions have to be merged sequentially.

4. *Cooperative action*: When the incoming task consists of robots acting cooperatively, the task plan is merged to ensure that the cooperating robots are both available at the same time to begin the cooperative actions. Actions that precede the cooperative actions in the task can be performed in parallel with task actions in the partial global plan.

5. *Independent action*: When the incoming task is independent of the tasks in the partial global plan, in terms of assembly constraints, resource requirements, and robot sets assigned, then a high degree of task parallelism is available. The plan is executed concurrently with the plans of the tasks in the partial global plan subject to spatial constraints.

The merging process enforces assembly and resource constraints, both secondary and spatial. It eliminates unnecessary actions caused by the same resources being used in consecutive tasks. Finally, when interactions are minimal, it increases the parallelism in task execution.

It must be noted that the above discussion dealt individually with each of the interactions. In general, combinations of the first four interactions occur. As the precedence relations between tasks need to be satisfied to generate a consistent global plan, it has to be satisfied before other interactions can be considered. After the precedence relations are satisfied, resource sharing and cooperative action are considered, followed by action combination interactions. The cooperative action interaction is similar to the resource sharing interaction because the robots involved in the cooperative action become resource constraints on the task plan.

2.1.3 Plan cost calculation

The ordering of the partial global plans in the *OPEN* list is based on the best-first order. The order is based on the costs of the plans. To ensure that the planning process leads to an optimal global plan, the cost function has to satisfy the A* requirements[7]. This implies that the cost for state i should be a lower bound on the costs for all its successors. Such a cost function in conjunction with the best-first ordering in the *OPEN* list will guarantee that the optimal global plan is found.

The cost function we will use is the total time required to execute a partial global plan. This establishes the performance objective of CRP-II to be minimum makespan. To ensure that the total time of state i is a lower bound on the total time of all its successor states, we consider the effects of the merging actions performed to generate the successor states.

A partial global plan consists of various paths due to parallelism between tasks. The cost of the global plan is the sum of the costs of actions along its critical path. The cost of a successor state is determined by where the actions of the merging plan are with respect to the critical path. Let a_{ik}, $k = 1, i(n)$, be the actions in the critical path of a partial global plan i. The cost of the plan is given by

$$COST(i) = \Sigma_{k=1,j(n)} cost(a_{ik}),$$

where the function $cost()$ returns the cost of an individual action. Let us define a path A_i in the plan which is not critical. Let this alternate path be constituted of P of the actions of the critical path, and R (>0) other actions. The cost of this alternate path is then given by

$$COST(A_i) = \Sigma_{k=1,P} cost(a_{ik}) + \Sigma_{k=1,R} cost(a_{ik})$$

Let a_{jk}, $k = 1, j(n)$, be the actions in the merging plan j. The costs of successor states due to the various merging actions can then be calculated as follows.

2.1.3.1 Action combination cost

Let us assume that action a_{jl} of the merging plan is to be merged with an action a_{im} of the partial global plan. If the action a_{im} lies along the critical path, then the cost of the successor state (represented by $i+1$) is given by

$$COST(i+1) = \Sigma_{k=1,i(n), k \neq m} cost(a_{ik}) + \Sigma_{k=1,j(n), k \neq l} cost(a_{jk}) + cost(merge(a_{im},a_{jl})),$$

where $merge(a,b)$ returns the resultant action of merging actions a and b. This can be written as

$$COST(i+1) = COST(i) - cost(a_{im}) + \Sigma_{k=1,j(n)} cost(a_{jk}) - cost(a_{jl}) + cost(merge(a_{im},a_{jl})).$$

$COST(i)$ will be a lower bound on $COST(i+1)$ if and only if

$$\Sigma_{k=1,j(n)} cost(a_{jk}) \geq cost(a_{im}) + cost(a_{jl}) - cost(merge(a_{im},a_{jl}))$$

This implies that the sum of the costs of actions in the merging plan should be greater than or equal to the savings due to the merging of the two actions. Theoretically, it is possible that this condition may not be satisfied. However, in practice, the cost of a single action is much smaller than the cost of a task. Therefore, this condition is almost always satisfied, and the cost of state i is a lower bound for the cost of all its successor states.

If the merging plan does not merge with the critical path, then the cost of the successor state is given by

$$COST(i+1) = max(COST(i), \sum_{k=1,P} cost(a_{ik}) + \sum_{k=1,R,\ k\neq m} cost(a_{ik}) + \sum_{k=1,j(n),\ k\neq l} cost(a_{jk}) + cost(merge(a_{im},a_{jl}))).$$

If the process of merging the plan into an alternate path does not make the alternate path to be critical, then the critical path for the successor state is the same as the critical path for the current state, and the lower bound is satisfied. However, if the alternate path becomes the critical path, then, by definition, its cost is greater than or equal to the cost of the critical path of the current state, and therefore again the lower bound is satisfied.

2.1.3.2 Precedence relation cost

If the task plan is merged into the critical path of the partial global plan due to the constraints, the cost of the successor state is given by

$$COST(i+1) = COST(i) + \sum_{k=1,j(n)} cost(a_{jk})$$

If the task is merged into a path other than the critical path, the cost of the successor state is

$$COST(i+1) = max(COST(i), \sum_{k=1,P} cost(a_{ik}) + \sum_{k=1,R} cost(a_{ik}) + \sum_{k=1,j(n)} cost(a_j$$

As the action costs are assumed to be non-zero, it is obvious that the cost of the current state is a lower bound on the cost of the successor states.

2.1.3.3 Resource sharing cost

The resource sharing interaction can lead to a sequential or parallel interaction between tasks based on the availability of resources. If the task is merged sequentially into the critical path, then the cost of the successor state is

$$COST(i+1) = COST(i) + \textstyle\sum_{k=1,j(n)} cost(a_{jk})$$

If the task is merged sequentially into a path other than the critical path, the cost of the successor state is

$$COST(i+1) = max(COST(i), \textstyle\sum_{k=1,P} cost(a_{ik}) + \sum_{k=1,R} cost(a_{ik}) + \sum_{k=1,j(n)} cost(a_{jk})).$$

Irrespective of the partial global plan path, if the task is merged in parallel to tasks in the partial global plan, then the cost of the successor state is either the cost for the current state or the cost of the merging plan, i.e.,

$$COST(i+1) = max\,(COST(i), \textstyle\sum_{k=1,j(n)} cost(a_{jk})).$$

In all the above cases, the cost of the current state is a lower bound on the costs of the successor states.

2.1.3.4 Cooperative action cost

When the task being merged involves cooperative actions, then the task plan can be merged such that the robot usage precedence is maintained with tasks in the partial global plan. If the last action in the critical path involves one of the robots in the cooperative action, then the task is merged into the critical path, otherwise it is merged in parallel with the critical path. Therefore, if the plan is merged with the critical path, the cost of the successor state is the sum of the costs of the current state and the merging plan, i.e.,

$$COST(i+1) = COST(i) + \textstyle\sum_{k=1,j(n)} cost(a_{jk})$$

If the task is merged sequentially into a path other than the critical path, the cost of the successor state is

$$COST(i+1) = max(COST(i), \textstyle\sum_{k=1,P} cost(a_{ik}) + \sum_{k=1,R} cost(a_{ik}) + \sum_{k=1,j(n)} cost(a_{jk})).$$

As mentioned before, this cost structure is the same as that for the resource sharing interaction, because the robots involved in the cooperative actions become resource constraints for the merging plan.

2.1.3.5 Independent action cost

When the task being merged does not interact with any of the tasks in the partial global plan, then the task plan can be merged in parallel with the tasks in the partial global plan. Therefore, the cost of the successor state is either the cost of the current state or the cost of the merging plan, i.e.,

$$COST(i+1) = max(COST(i), \sum_{k=1 \, j(n)} cost(a_{jk})).$$

As mentioned previously, in general, combinations of the first four interactions occur in which case the order of invoking the interactions specified previously is used to calculate the cost of the successor states. In all such cases, the cost of the current state is a lower bound on the costs of the successor states in practice.

2.2 Sequential plan generation strategy

When plan selection and constraint satisfaction are performed sequentially, the planning process is considerably less complex. The task assignment process generates a particular assignment based on the cooperation requirements matrix. This step does not attempt to satisfy any of the constraints. Its main objective is to optimize the given performance measure. It is possible to identify a variety of performance measures for performing task assignment. Two such performance measures that have been identified are: minimum total cost assignment and minimum robot set assignment. The first finds the optimal assignment based solely on the cooperation requirements matrix. The second method determines the minimum robot set based on the cooperation requirements matrix. The global plan generation method is similar to the simultaneous search strategy, except that only a single plan exists for each goal. The minimum cost assignment solution and modified search algorithm are discussed in [4].

The simultaneous plan generation strategy shown above does not restrict the number of robots used in performing the tasks. Therefore, even if a smaller robot set is available that can successfully perform all the tasks, that method may not necessarily generate the minimum set to be the final assignment. By restricting the number of robots participating in the task execution process, robot interactions are reduced, and robots not participating in the execution of one job can be used to perform other jobs in parallel. The disadvantage of using the minimum robot set is that it may not be the optimal solution for the task assignment. This combined with constraint satisfaction could lead to a very poor solution.

The minimum robot set problem can be formulated as an integer program, and a linear programming relaxation can be used to solve it[4]. Alternatively, for small problems, an enumeration method can be used. To generate the enumeration solution, a 0-1 matrix M is defined, whose entries $M(i,j)$ are given as

$$M(i,j) \; = \; 1 \; if \; C(i,j) > 0$$

$$= \; 0 \; otherwise,$$

$i \in S, j \in T.$

Let K denote the current set size. Initially, $K = 1$. An iterative process is then used as follows:

1. For each robot set $RSET$ of size K, compute the column-by-column disjunction of the rows of all subsets of $RSET$ in the matrix M. Let the row generated by $DISJ$.

2. Compute the conjunction of all the columns of $DISJ$. Let the result be $CONJ$.

3. If $CONJ = 1$, then the robot set is feasible, else it is infeasible.

4. If there exists any robot set $RSET$ of size K that is feasible, then the minimum robot set size is K.

5. For all robot sets that are feasible, compute the total cost as the sum of the minimum costs of each column of the rows in the cooperation requirements matrix of all the subsets.

6. The minimum cost robot set is the final solution and the iteration is terminated.

7. If no robot set of size K is feasible, then set $K = K + 1$. Go to step 1.

A recursive process can be used for step 1 by computing the disjunction for *RSET* of size K as the disjunctions of all the subsets of *RSET* of size *K-1*. This reduces the complexity of the scheme considerably. Additionally, based on the mandatory cooperation requirements information, if there exists at least one task that requires mandatory cooperation, then the above algorithm can be started by initializing K to the size of the minimum robot set that can successfully complete that task.

The assignment process provides a single task plan for each task. However, due to interactions between the tasks, and robot location directions, various global plans can be constructed using the same set of individual task plans. The generation of the global plan given a set of individual task plans is similar to the simultaneous plan generation strategy, except that in this case, only one plan is available per goal[4].

3. EXAMPLE

The CRP-I methodology has been implemented using the ROBCAD™ Open-System Environment on a Silicon Graphics 4D/80GT graphics workstation. The following analysis is for an example developed using that implementation. The workcell and assembly used for the example are shown in figure 2. The cell was modeled using the ROBCAD Workcell Tool. It consists of 3 robots: an ADEPT 1 (R1) 4-axis SCARA robot, a PUMA 550 (R2) 5-axis articulated robot, and a PUMA 560 (R3) 6-axis articulated robot. The robot models and kinematic information were obtained from the ROBCAD robot library.

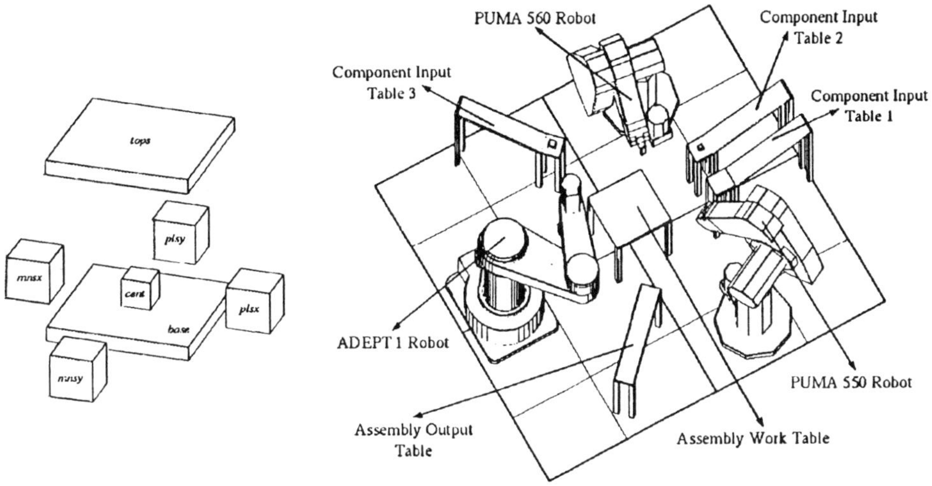

Figure 2: Example assembly and workcell used to illustrate CRP. The assembly consists of 7 components. The cell consists of 3 robots, part input tables, assembly worktable, and assembly output table.

The assembly consists of 7 components labelled *base*, *plsx*, *mnsx*, *plsy*, *mnsy*, *cent*, and *tops*. It consists of a single completely constrained component, *cent*, which is constrained by the assembly state consisting of all the other components.

The standard cell also consists of a centrally located worktable which is common to the 3 robots and is used as the location for building the assemblies. The components are input to the cell on 3 component input tables. Input tables 1 (for *base* and *tops*) and 2 (for *cent*) are common to the R2 and R3 robots while table 3 (for *plsx*, *mnsx*, *plsy*, and *mnsy*) is common to the R1 and R3 robots. When the assembly is completed, it is removed from the assembly worktable to an output table that is accessible to the R1 and R2 robots. A coarse cell space partitioning is used to describe the shared regions. The partitioning is essentially the shared workcell equipment, i.e., the assembly worktable, input tables, and output table are defined as shared regions for the robots that have these workcell elements in common as described above. The example illustrates optional cooperation among the robots to complete a given assembly.

The task plan specifies 8 tasks: 7 tasks to assemble the components, and the last task to remove the assembly from the cell. Various user-defined task precedence relations are also specified. The first task precedence relation specifies *base* as the first component to be assembled and therefore task 6 should precede all other tasks. Similarly, task 0 for assembling *tops* should succeed tasks 1 - 6, and task 7 should succeed task 0. Based on the matching of the physical, operational, and geometric requirements of the tasks to the corresponding capabilities of the robots in the cell, all tasks require optional cooperation. Due to the motion limitations of the robots, the R2 robot can perform tasks 0, 1, 6, and 7, the R3 robot can perform tasks 0 - 6, and the R1 robot can perform tasks 2, 3, 4, 5, and 7. The cooperation requirements matrix is shown in figure 3.

The cooperation requirements information and the various constraints are then used to

Cooperation Requirements Matrix:

Set	Task 0	Task 1	Task 2	Task 3	Task 4	Task 5	Task 6	Task 7
0:	0.000	0.000	5.764	5.751	5.844	5.855	0.000	5.714
1:	5.425	5.720	0.000	0.000	0.000	0.000	5.412	5.800
2:	5.551	5.297	5.452	5.551	5.615	5.519	5.550	0.000
3:	0.00	0.000	0.000	0.000	0.000	0.000	0.000	0.000
4:	0.000	0.000	0.000	0.000	0.000	0.000	0.000	0.000
5:	0.000	0.000	0.000	0.000	0.000	0.000	0.000	0.000
6:	0.000	0.000	0.000	0.000	0.000	0.000	0.000	0.000

Robot Sets:

Set 0: ADEPT 1	Set 4: ADEPT 1 & PUMA 560
Set 1: PUMA 550	Set 5: PUMA 550 & PUMA 560
Set 2: PUMA 560	Set 6: ADEPT 1 & PUMA 550 & PUMA 560
Set 3: ADEPT 1 & PUMA 550	

Figure 3:   Cooperation requirements matrix for the assembly and workcell shown in figure 2. The capability measure is execution time (in seconds).

generate the global plan. The optimal global plan is shown in figure 4 with a cost of 29.791 seconds. The minimum robot set assignment generation process is shown in figure 5. The optional cooperation requirements matrix shows that no single robot can execute all the tasks. However, any 2-member robot set is capable of completing all the tasks successfully. The minimum cost, minimum robot set solution is the R2 - R3 combination.

The flexibility in performing the assembly tasks is shown by the alternate assignments possible to successfully complete them. In a dynamic environment, depending on the currently available set of robots, the task assignment can be made based on the cooperation requirements matrix or by generating the optimal global plan. The reliability of the system is enhanced by the generation of overlapping task capabilities of the robots. For a given assignment, if one of the robots in the assigned robot set fails, the tasks assigned to the failed robot can be reassigned to ensure successful completion of the job.

The improvements in flexibility and reliability are more pronounced when one or more of the tasks that were assigned to the failed robot cannot be performed optionally by any of the remaining robots in the cell. In such a case, mandatory cooperation can be invoked to determine robot sets that can complete the task(s) in parallel or in sequence.

4. CONCLUSIONS

Cooperation requirement planning (CRP) presents a unified methodology for process planning and scheduling under the cooperation paradigm by generating task requirements, matching them to the machine set capabilities, and generating a global plan. Other issues related to CRP are:

❏ *Flexibility and Reliability*: CRP improves the flexibility of the manufacturing system by providing various processing alternatives where the task is performed by cooperating machine sets, and constraint satisfaction is performed such that the planning

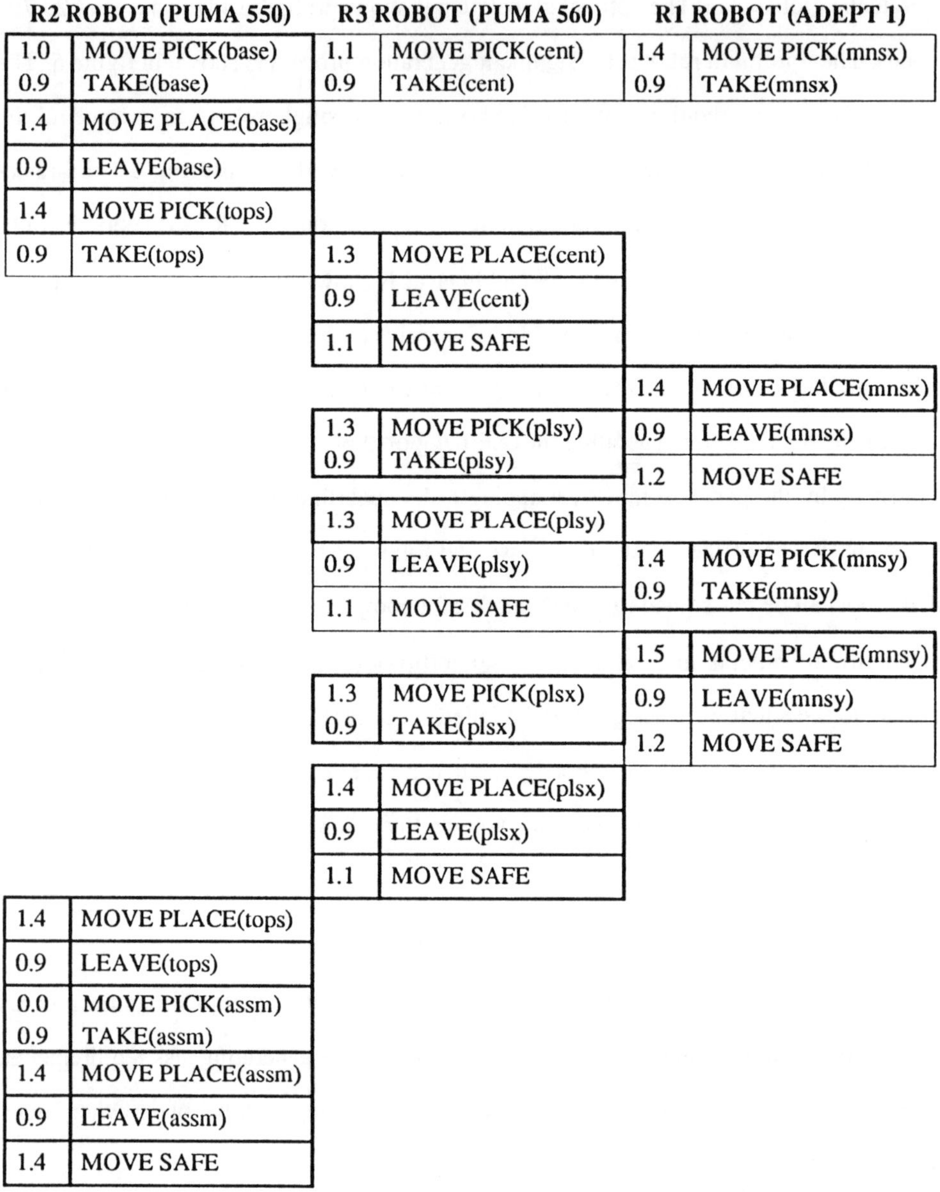

R2 ROBOT (PUMA 550)	R3 ROBOT (PUMA 560)	R1 ROBOT (ADEPT 1)
1.0 MOVE PICK(base) / 0.9 TAKE(base)	1.1 MOVE PICK(cent) / 0.9 TAKE(cent)	1.4 MOVE PICK(mnsx) / 0.9 TAKE(mnsx)
1.4 MOVE PLACE(base)		
0.9 LEAVE(base)		
1.4 MOVE PICK(tops)		
0.9 TAKE(tops)	1.3 MOVE PLACE(cent)	
	0.9 LEAVE(cent)	
	1.1 MOVE SAFE	
		1.4 MOVE PLACE(mnsx)
	1.3 MOVE PICK(plsy) / 0.9 TAKE(plsy)	0.9 LEAVE(mnsx)
		1.2 MOVE SAFE
	1.3 MOVE PLACE(plsy)	
	0.9 LEAVE(plsy)	1.4 MOVE PICK(mnsy) / 0.9 TAKE(mnsy)
	1.1 MOVE SAFE	
		1.5 MOVE PLACE(mnsy)
	1.3 MOVE PICK(plsx) / 0.9 TAKE(plsx)	0.9 LEAVE(mnsy)
		1.2 MOVE SAFE
	1.4 MOVE PLACE(plsx)	
	0.9 LEAVE(plsx)	
	1.1 MOVE SAFE	
1.4 MOVE PLACE(tops)		
0.9 LEAVE(tops)		
0.0 MOVE PICK(assm) / 0.9 TAKE(assm)		
1.4 MOVE PLACE(assm)		
0.9 LEAVE(assm)		
1.4 MOVE SAFE		

Figure 4: Optimal global plan for example. The highlighted action sequence indicates the critical path. Action times (in seconds) are indicated to the left of each action.

0-1 Matrix:

	Task 0	Task 1	Task 2	Task 3	Task 4	Task 5	Task 6	Task 7
Set 0:	0	0	1	1	1	1	0	1
Set 1:	1	1	0	0	0	0	1	1
Set 2:	1	1	1	1	1	1	1	0
Set 3:	0	0	0	0	0	0	0	0
Set 4:	0	0	0	0	0	0	0	0
Set 5:	0	0	0	0	0	0	0	0
Set 6:	0	0	0	0	0	0	0	0

1-member Sets:

Set 0: $0 \wedge 0 \wedge 1 \wedge 1 \wedge 1 \wedge 1 \wedge 0 \wedge 1 = 0$

Set 1: $1 \wedge 1 \wedge 0 \wedge 0 \wedge 0 \wedge 0 \wedge 1 \wedge 1 = 0$

Set 2: $1 \wedge 1 \wedge 1 \wedge 1 \wedge 1 \wedge 1 \wedge 1 \wedge 0 = 0$

2-member Sets:

Set 3: $(0 \vee 1 \vee 0) \wedge (0 \vee 1 \vee 0) \wedge (1 \vee 0 \vee 0) \wedge (1 \vee 0 \vee 0) \wedge (1 \vee 0 \vee 0) \wedge (1 \vee 0 \vee 0) \wedge (0 \vee 1 \vee 0) \wedge (1 \vee 1 \vee 0)$
$= 1$ (minimum cost = 45.485)

Set 4: $(0 \vee 1 \vee 0) \wedge (0 \vee 1 \vee 0) \wedge (1 \vee 1 \vee 0) \wedge (1 \vee 1 \vee 0) \wedge (1 \vee 1 \vee 0) \wedge (1 \vee 1 \vee 0) \wedge (0 \vee 1 \vee 0) \wedge (1 \vee 0 \vee 0)$
$= 1$ (minimum cost = 44.249)

Set 5: $(1 \vee 1 \vee 0) \wedge (1 \vee 1 \vee 0) \wedge (0 \vee 1 \vee 0) \wedge (0 \vee 1 \vee 0) \wedge (0 \vee 1 \vee 0) \wedge (0 \vee 1 \vee 0) \wedge (1 \vee 1 \vee 0) \wedge (1 \vee 0 \vee 0)$
$= 1$ (minimum cost = 44.071)

Minimum Robot Set Solution:

Set 5: PUMA 550 & PUMA 560 with cost of 44.071 seconds.

Figure 5: Minimum robot set solution generation using the enumeration scheme for example.

process is not over-constrained and the maximum flexibility is maintained. The reliability is improved because of reacting to machine failure by reassigning tasks to cooperating machine sets when individual machines cannot perform the tasks.

❐ *Design and Planning Interactions*: In this research, it has been assumed that the assembly and workcell designs used for generating the cooperation requirements are pre-specified. However, by modifying them, the cooperation requirements generated may be different. Therefore, the design plays an important role in the planning process which in turn can indicate design improvements.

❏ *Multiprocessor Task Scheduling*: The concept of a multiprocessor manufacturing or assembly task has been introduced. Traditionally, tasks require single processors, which is equivalent to optional cooperation. However, by generating mandatory and concurrent cooperation requirements, tasks can potentially require two or more machines to be successfully completed. As traditional scheduling literature has been limited to examining the single processor tasks, a new area of scheduling research has been introduced involving multiprocessor tasks. As this problem is NP-hard[8], heuristic methods need to be developed.

❏ *Distributed Cooperation Planning*: The CRP methodology lends itself to a variety of improvements in terms of improving its scope. The use of intelligent machines to perform a given set of tasks, with each machine having a local objective, represents an important motivation to develop distributed cooperation planning strategies[9].

5. REFERENCES

1. Nof, S. Y., and D. Hanna, Operational Characteristics of Multi-Robot Systems with Cooperation, *International Journal of Production Research.*, Vol. 27, No. 3, 1989, pp. 477-492.

2. Rajan, V. N., and S. Y. Nof, A Game-Theoretic Approach for Co-operation Control in Multi-machine Workstations, *International Journal of Computer Integrated Manufacturing*, Vol. 3, No. 1, 1990, pp. 47-59.

3. Chryssolouris, G., and S. Chan, An Integrated Approach to Process Planning and Scheduling, *Annals of the CIRP*, Vol. 34, No. 1, 1985, pp. 413-417.

4. Rajan, V. N., Cooperation Requirement Planning for Multi-Robot Assembly Cells, Ph.D. Dissertation, Purdue University, W. Lafayette, IN 47907, May 1993.

5. Nof, S. Y., and V. N. Rajan, Automatic Generation of Assembly Constraints and Co-operation Task Planning, *Accepted for Publication in the Annals of the CIRP*, Vol. 42, No. 1, 1993.

6. Yang, Q., D. S. Nau, and J. Hendler, Merging Separately Generated Plans with Restricted Interactions, *Technical Report No. UMIACS-TR-91-73*, University of Maryland Institute for Advanced Computer Studies, University of Maryland, College Park, MD, 20742, May 1991.

7. Nilsson, N. J., Principles of Artificial Intelligence, Tioga Publishing Company, Palo Alto, CA, 1980.

8. Blazewicz, J., M. Drabowski, and J. Weglarz, Scheduling Multiprocessor Tasks to Minimize Schedule Length, *IEEE Transactions on Computers*, Vol. C-35, No. 5, May 1986, pp. 389-393.

9. Rajan, V. N., and S. Y. Nof, Logic and Communication Issues in Cooperation Planning for Multi-machine Workstations, *To appear in the International Journal of Systems Automation: Research and Applications*, 1993.

BENCHMARKING AND MODELS OF INTEGRATION

MARIO LUCERTINI
Centro Volterra, Università di Roma "Tor Vergata"
via della Ricerca Scientifica, 00133-Roma (Italy)

FERNANDO NICOLÒ
Dipartimento di Meccanica e Automatica, Università di Roma III
via Segre 2, 00153-Roma (Italy)

DANIELA TELMON
TRADEOFF, consulenza e servizi tecnici per le aziende
Lungo Tevere R.Sanzio 5, 00153-Roma (Italy)

ABSTRACT. Benchmarking is an approach used for evaluating and improving the company performances, by comparing them with the best performing companies. Benchmmarking first studies the process to be improved, finds a best practice process in order to try to match two parts of the processes which have analogies, and than try to change or modify the interconnections, structures or behaviour of the part to be improved using the analogy with the best trasformation process. In the paper, we try to define benchmarking from a modelling point of view, and to outline how the related concepts can be used for evaluating the level of integration of computer integrated manufacturing systems, focusing on cost accounting subsystems.

1.   Introduction

There are several definitions of benchmarking, all based on the idea of evaluating the performance of an organized system by comparing it to exogenous entities.
The Webster dictionary defines a bench mark as: "A mark on a fixed and enduring object (as on an outcropping of rock or a concrete post set into the ground) indicating a particular elevation and used as a reference in topographical surveys and tidal observations. A benchmark is thus a point of reference from which measurements of any sort may be made."
In a business context, D.T. Kearns, executive director of Xerox Corporation, defines it as: "The continuous process of measuring products, services and practices through the comparison with its strongest competitors, with companies leaders in the field".
A definition that try to include all these different aspects can be the following:

continuing search, measurement and comparison of products, processes, services, procedures, ways to operate, best practices that other companies have developed to obtain an output and global performances, with the aim of improving the company performances.

201

S. Y. Nof (ed.), Information and Collaboration Models of Integration, 201–218.
© 1994 *Kluwer Academic Publishers. Printed in the Netherlands.*

The basic concept of benchmarking is that the methods traditionally used by management to fix their goals have often revealed uneffective, especially in rapidly changing technological and/or market scenarios. An effective way to reach best practices and best performances is to establish standards and programs on the basis of other companies' best practices. This has recently been called benchmarking.
The benchmarking kitchen recipes are then the following:

1) identify the standard or benchmark for your product and services;
2) compare your products and services to the standard or benchmark;
3) change your methods so that you may provide products and services that are better or equal to the benchmark.

The concern for performance evaluation has always existed in corporations, and has traditionally been realised on historical basis (by comparing the performance to the one of the year before) and, sometime, on competitive basis (by comparing the company to a competitor). Only in recent times some attention has been devoted to a comparison made on a functional basis, by comparing similar functions in different companies or, more generally, activities, relevant for the company performance, that are similar from a functional point of view, but concern completely different transformation processes. These activities thereby base their philosophy on the detailed knowledge of their operations, of the detailed knowledge of leading companies, competitors or not, and the study of how to incorporate the best practices, on how to reach leadership.
This philosophy is an integral part of the oriental culture, from which benchmarking, one of the main tools of quality control, takes probably its origins. Around 500 b.c. a chinese general named Sun Tzu wrote: "If you know your enemy, and you know yourself, you will not fear the result of a hundred battles". The words of Sun Tzu show the way to success in all types of businesses, and not only wars and battles. Management battles, in order to solve business problems and to survive on the marketplace, are all types of wars, that must be faught with the same weapons of the real wars: the rules of Sun Tzu.
The Japonese term "dantotzu", widely used in management handbooks, means to "make an effort to be the best of best". Hence benchmarking can be seen as the tool for accomplishing this embedding effort, as a concrete process of preventive change in procedures and operations to reach superior performances.

2.   Historical background

The origins of benchmarking can be seeked in different sectors, where these concepts have been developed indipendenly, with scarse interactions. Particular relevance have had manufacturing processes, data processing systems, accounting systems, and company practices.

Manufacturing processes
With the beginning of the XX-th century, managerial problems in production became more and more relevant, in comparison to purely technological problems. Prescriptive and planning systems were developed, generating the birth of standard costs, budget, personnel. Taylor's *scientific management* is based on a set of performance indicators, continuously mesured, compared and updated. The first manufacturing flow line, the moving assembly line,

introduced by Ford in the Highland Park plant (completely operational in 1916, after several years of gradual introduction of conveyor belts and gravity feeders), is credited to be inspired from a visit to a Chicago abbattoir plant in 1911. In this plant, the material handling system has been for a long time organized in a flow line, formed by a sequence of dedicated working cells. This layout transfer represents a classical benchmarking process. During the first world war, with the development of statistics, the first set of reliable performance measures are created. These measures introduce the concept of performance standards, emphasizing the importance of comparisons among different productive realities. More recently, the development of the total quality movement brought a more refined performance measurement system, in order to compare different situations in production and to find adeguate improvements. The growth of integration and flexible systems, in which, together with the capacity to produce, the capacity to adapt to situations varying in time is also important, makes performance evaluation and decision strategies' definition more complex. Conceptual models for evaluation and decision support become more and more complicated and require an always wider technological know how on the system to be used. Thus, synthetic conceptual models like benchmarking help to tackle with complex systems.

Data processing systems
In order to measure performance in data processing systems, from the very beginning of computer studies, evaluation tools based on benchmarking concepts were developed. In fact, although the value of a computer depends on the context in which it is used, and that context varies by application, by workload and in terms of time, nevertheless benchmarking is the basis of the computer performance evaluation process. The measurement of some main parameters of the machine (such as cycle time, response time, memory size, overall computer speed etc.) and on the use made of it (work charge, throughput, execution of predetermined programs) take place on the basis of predetermined standards and allow detailed comparisons between the efficiency of different machines. More complex is the problem of measuring effectiveness with respect to a given class of applications, that is to say the responsiveness of the processing system to the users' requirements. The performance measurement issue has been studied both for standalone computers, heterogenous systems and networks. More recently the issue of finding significant and comparable measures for machines with massive parallelism has been analysed.

Acconting systems
The first inventories financially evaluated in the modern sense, introducing an accounting and management system that takes into account their revaluation in time (at least as far as land revenues are concerned) have been thought of in some abbays towards the second half of the 3rd century. From the second half of the 5th century, the development of crafts and trades in Europe leads to the development of geographically distributed organizations, where management system is based on formal documents and comunication protocols. Methodological tools that support organizations will be developed only much later, when theorical developments of quantitative sciences (in particular statistics) and economics make available powerful and articulated conceptual models. Nonetheless, doubts on the real response to the needs of accounting systems are periodically risen.
Performance indicators actually used by corporations for budgeting refer to standards that are determined on extrapolations of the past. Without a comparative analysis of more advanced industrial practices and a consequent effort to meet those standards, the progress of productivity is gradual, evolutional, pursued only to the level seen as

acceptable for the organization. A company's productivity is reached gradually, by improving every time the worst working parts of the processes. In today's more dinamic context, where more and more attention is given to continuous improvement, there is less corrispondence between "measurement for control" and "measurement for improvement". Defining measures for improvement requires, together with new criteria and tools for performance measurement, benchmarking studies, to determine the real standards, those which quantify the best practices and the best companies.

Marketing
In marketing, tools for cross-company comparisons have always existed.Market research traditionally analyses company markets and market acceptance of products, in order to determine how customers' needs are satisfied with products and services. Competition analysis becomes of central importance in this context, studying competitors' strategies to define market activities for products and services.
In their continuous search for new arguments to improve competitiveness of their products and services, marketing managers have always looked for parameters for comparing their products to the competition's. In many countries ethical codes have been defined on what is fare and what is'nt of publishing results on comparative experiments.

Business practices
In business activities benchmarking techniques have remained at a larval state. Comparisons with other business realities were made in no systematic way; generally also the natural consequences in terms of organizational changes were not performed.
One of the first and most interesting benchmarking experience promoted by Xerox involved L.L.Bean as the benchmarking partner. L.L.Bean is a company involved in distribution and corrispondance sales of sportsware. The experiment was conducted by Xerox unit Logistics and Distribution, responsable for the management of inventories, warehouses, tools, parts and machines transportation. The area of inventory control had recently adopted a new planning system, and the transport unit was benefiting from the opportunities offered by deregulation. Improvement measures were studied for warehouses management. Xerox identified the requisition area as the most difficult of all of the operational sequence from acquisition to delivery. It was very urgent to find a new system to improve productivity in stockpiling and finished products handling operations. L.L.Bean's stockpiling system resulted to be particularly suitable   for benchmarking analysis: both companies had to develop stockpiling and distribution systems for products with different volumes, sizes and weights. This diversity precluded the use of the new ASRS system, already used by Xerox for raw materials and assembly parts. Many of the practices used by L.L.Bean were transfered to Xerox. Particularly important for the rise of productivity in Xerox' operations have been the materials stockpiling studied according to speed, to accellerate the flow and reduce the distances in routing. Also, many requisition operations have been computerized.
Other practices based on benchmarking concepts have been developed in design. Studies on competitor's products on their decomposition, to obtain information on the critical aspects of components, assembly, and their relative costs. These techniques, generally known as reverse engineering, have the advantage of not requiring agreements between companies.

In summary, the study of products, practices, markets of competing companies has always been an important aspect of management. What differentiates benchmarking from the other tools used in those environments, is that benchmmarking first studies the process to be improved and a best practice process in order to try to find two parts of the processes which have an analogous behaviour, than it try to change or modify the interconnections, structures or behaviour of the part to be improved using the analogy with the part of the best trasformation process. This can be done by studing the cause-effect relationship which produce the final results and by finding, in different environments, cause-effect chaines for which you can find an analogical correspondence and by traying to change some block or modify block interconnections of such chaines in order to improve the performances. For example, the Ford flow line and the abattoir plant are both based on assembly/disassembly processes, in which men and materials interact in similar ways: men operate standing still and materials flow. Before the benchmarking process, Xerox had a monitoring problem, solved by means of a complex sequence of men made controls. An analogous type of monitoring was solved by L.L.Bean by means of a much more simple computerized set of measures. Using the analogy, Xerox was able to change its complex monitoring system with a much more simple one, equally effective. In both cases, we modify operational constraints and we operate on intermediate variables or subchaines to produce the desired effect.

Substantially, benchmarking deals with *what* (identify analogous process parts, i.e. subchaines), *why* (identify performance indicators) and *how* (identify the new organization: interconnections, structure or behaviour of the part to be improved), practices of leader companies, having conquered leadership positions, can be transferred. This is represented in the following chart.

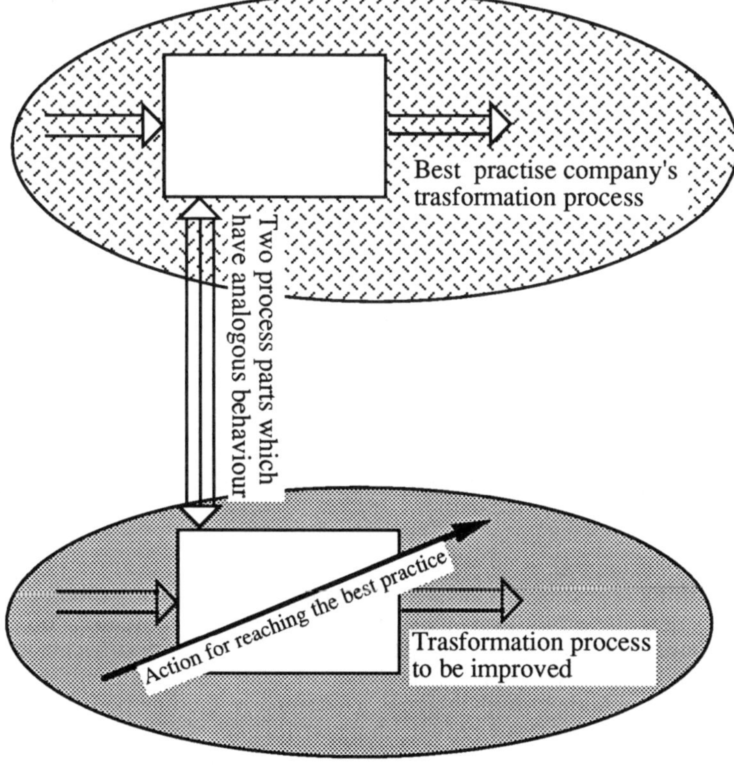

3. The benchmarking process

A benchmarking study is generally conducted according to a relatively well defined sequence of phases:

Planning
> 1. Identify benchmarking subject: the parts to be improved
> 2. Identify benchmarking partner: the best practice process
> 3. Determine data collection method/ collect data

Analysis
> 4. Determine current competitive gap and quantify it
> 5. Project future performance

Integration
> 6. Communicate findings and gain acceptance
> 7. Establish functional goals

Action
> 8. Develop action plans
> 9. Monitor implementation
> 10. Recalibrate benchmarking progress

These phases are integrated and finalized to the leaderschip achievement by a management process, where centralized and decentralized decisions and actions must be suitably coordinated.

In fact, benchmarking can be performed at all company's levels: operational unit, section, department, division, business area and cross functional business area. At each of these levels it is possible to find areas, tasks, activities, functions, processes, that can be improved. These are potential topics for benchmarking. The only requirement concerns the possibility of clearly definig the transformation process involved and associating with the process a suitable set of quantitative performance indicators.

Foe this reason, in order to simplify the analysis, generally, the benchmarking object should refer to company annual goals and improvement priorities, expressed in the strategic plan and the annual company plans.

As far as benchmarking partners are concerned, there are four tipologies of benchmarking.

1. Benchmarking on internal operations

This type of benchmarking can be developed inside a company, in multinational corporations or in companies belonging to a same holding. These benchmarking studies are the easiest to perform, if the initiative starts from top management, because informations can more easily be gathered.

2. Competitive benchmarking

The benchmarking object in this case is the product or service of a direct competitor. In this case, the most important problem is that of obtaining informations on competitor's operations; these data will always be partial because it is very difficult to obtain real informations on things such as pricing and market strategies. It is therefore necessary to use a wider notion of competition: which is the company, the function, the operation that realizes the best practices that can be of interest for us? Maybe a competitor may not even exist for a function, or a general process such as invoicing or order acquisition.

3. Direct functional benchmarking
The inquiry is in this case conducted on functional competitors or leader companies in different fields, by comparing the same global functions. This means, continuing our example in logistics, identifying companies well known for their excellence in logistics. We have already pointed out that Xerox's benchmarking partner was a company dealing with sportsware distribution.

4. Crossfunctional benchmarking
A series of basic operations, like invoicing, for instance, is common to all types of businesses and is part of many global functions. An example can be that of money counting machines originally introduced in banks, or bar codes introduced in supermarket distribution. Both of these operations are common to all businesses and several global functions (such as, for instance, logistics, to follow the materials paths).

The identification of the benchmarking object and the identification of benchmarking partners presume the existance of a conceptual model able to define performance indicators on one side, and on the other able to tranfer formats. In section 5 an outline of some basic concepts is presented.

4. **Process control: an evolving ingredient for competitive advantage and a benchmarking object**

Let us first of all see what we mean by process control, and how the modes of how this control was conducted have evolved in time. This will explain the parallel evolution of cost control and process control.
In manufacturing, a process traditionally consists of the technological tranformation, based on requirements, of raw materials in a component having predetermined physical characteristics and a relevant aspect of the quality of a process is given by the level with which manufactured products conform to requirements. A certain degree of variance, caused by men, machines, procedures and the product itself is implicit in this transformation activity. Variance control therefore has become in time the main aspect of process control and the quality of the control process can be determined on the basis of the degree by which these variances are minimized. With the progress of computer science and automation, the transformation of informations tends to become more important then the physical transformation and quality is assessed mainly on the ground of lead times, delivery times and costs (this happens also because of a growing standardization of technical requirements). We are not only talking about machines' downtime costs, but also costs of errors, rework, etc. that can be fairly easily (at least from a theorical point of view) determined in advanced control systems.
Consequently, the concept of process control evolved in time trying to incorporate manufacturing costs and lead times, bccoming therefore an important ingredient for competitive advantage.
This new, more general, concept of process control, needs a sophisticated real time information system, and a structured decision making process. All this becomes possible with the introduction in the factory of computer integrated systems (the so called CIM: computer integrated manufacturing).

This section of the paper is devoted to process control costing, in the framework of process control and cost control managed by CIM systems. The process control function is an ideal benchmarking object: a CIM system's effectiveness can partly be determined by the superiority of its cost control systems.

As R.Jaikumar points out [jai], "rapid advances in information intensive processing capability have made it possible to provide an economic basis for systematically choosing among the many different options available for controlling processes".

In traditional factories, the informations gathered on auxiliary activities are fairly scarse and the main informations surveyed for, are those on product transformation, to reduce the variable cost of production. This is so because auxiliary activities connected with the machine or with the manufacturing system are difficult to be attributed to a single product or cost center. In flexible manufacturing systems the computers, that coordinate and control the flow of products, tools, informations on the process and other resources can capture other informations on auxiliary activities. Also, many activities that traditionally were manually treated are now automated.

You can identify six phases in the history of process control. A first phase is characterized by the invention and the introduction of machines at the beginning of the 19th century. The second phase sees the introduction of dedicated machines and the interchangeability of components, in the second half of the 19th century. The third epoch is that of scientific management and work engineering in the tayloristic system. The forth, the times of the introduction of statistical process control (SPC) in the 30s. We then enter the era of numerical control and data processing, and then arrive to the sixth phase, that of intelligent systems and computer integrated manufacturing (CIM), typical of the 80s.

The first three phases are characterised by emphasis on mechanisation: the factory had to be planned for better efficiency and better control.This implied the substitution of capital for labor, and the factory gained improvement through economies of scale. In the 50s, with the beginning of computerised data processing, the human element became the core to tranform a set of single machines into a manufacturing system. In this new context, tendencies towards mechanization tended to invert, and the demand of capacities for versatility and intelligence became stronger; intelligence substituted for capital and economy of scope took the place of economy of scale.

Flexible automation (NC) is different from industrial automation, because it integrates information and material processing, separated in traditional technologies; it relies on a greater intelligence of the machine. Whileas before you could count only on the machine operator's intelligence, now flexible automation yields simultaneously: flexibility, precision, productivity, versatility, reproducibility and coordinability. As far as versatility is concerned, the machines based on microprocessor automation can produce different quantities of components, generally organized in small batches, minimizing economies of scale. The reproducibility is reached through the close integration between physical and information processing.

In a FMS/CIM environment the process is, on the contrary, under the complete control of computer programs. Product and process specifications become computational procedures that are fed in computerised programs. The specifications must anticipate and solve potential product and process problems: a well planned process does not need to be changed.

When a part manufactured using a given set of requirements is equal to every other part, the highly qualified person who writes the procedure reaches an exact reproducibility. Futhermore, the existance of procedures contributes to the transportability of product and process information: software is easily transportable, either manually or via telecomunications.

With microprocessor control, managers exert their control activity modifying procedures to introduce change. On the contrary, with CIM control becomes continuous. Although all the sistem's operations, such as part change program, decision rules for priority assignment, etc.are under a precise computerised control, dynamic contingencies are still part of the environment, and human intelligence is necessary to identify and eliminate errors, shifting from an intermittent to a continuous control.

Let us now see how cost informations fit into this evolution of process control systems.

In tayloristic systems, "the early acconting management measure....focused on conversion costs and produced summary measures such as cost per hour or per pound provided for each process and for each worker..... and involve some attribution of overhead. The goals of the systems was to identify the different costs for the intermediate and final products of the firm and to provide a benchmark to measure the efficiency of the convertion process.... and to provide incentives for workers to achieve productivity goals."

Scientific management evolved as a means to gather detailed information about the efficiency of complex processes and the people that carried them out. It aimed at finding the "one best way" to do any task. By reducing a task to a series of very small inputs, it became possible to establish "standard "rates at which material and labor should be consumed in manufacturing tasks. Standard rates provided a basis for assessing variances between actual and standard costs and for differentiating variances due to controllable conditions from those beyond management's control. This information was used to compare actual performance against perscribed performance, rather then for making judgements relative to process improvement.

At that time, it was impossible to track and to document the consumption of the different resources for every physical transformation of a material. Variance was calculated at very high levels of aggregation:information was a scarse resource and managers, intermittent observers.

5.   The decision process

Benchmarking is a tool directed to implement change, more then a tool for merely evaluating company performances. The decision making process and its link with the value of a set of performance indicators, suitably depicting the company's behaviour, is then a cornerstone of the benchmarking building.

In fact, the field measures are taken in particular points of the system, suitably related to the transformation process considered, and are in terms of quantities (flows and levels). On the other hand, performance indicators are defined on the basis of measures taken in different parts of the systems, at different times or time intervals, and adeguately elaborated. Typical performance indicators are:effectiveness,efficiency, productivity, quality of work life, innovation, profitability (or budgetability), quality.

In this framework, it is therefore important to define company goals in order to determine what and where to measure, and which are the right indicators and how they relate to measurements. To put together goals, measures and performances you need a conceptual model of the transformation process, that can be used to transform performance evaluation into improvement decisions.

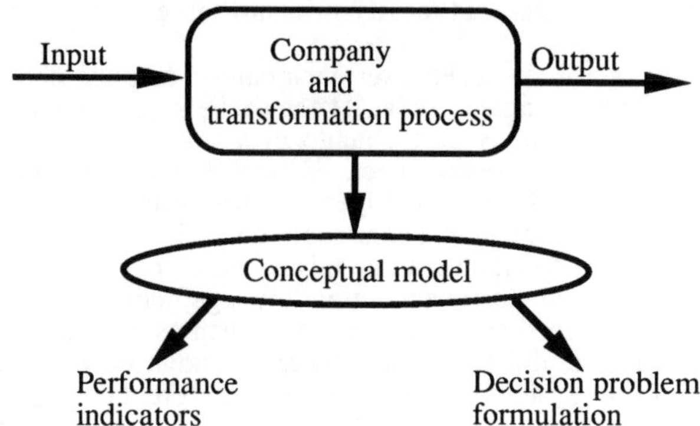

In practice, company decisions lie on different levels, and benchmarking focuses only on certain types of decision, that we may call of intermediate level.

These decisions do not concern, typically, basic company strategies, such as market selection, process selection, joint ventures, basic make or buy decisions.
In the same way, these decisions do not concern, typically, operational decisions, such as material routing and operations scheduling.

Benchmarking decisions focus on a tactical level, where you can modify organizational constraints, procedures and practices. We have had examples of this in physical material handling, distribution systems, assembly lines, production layout.

Using decision models' language, we may characterize the three levels, from operational to strategic, as follows:

Operational level
Given: environment, operational conditions, different types of technological and organizational contraints, a univocally defined objective function,...
find: the value of decision variables directly connected to the process,
such that: the performance will be optimized (throughput maximization, lead time minimization, ecc.)

Tactical level
Given: environment, structural constraints difficult to modify, a set of performance indicators,
find: operational constraints, information flows, operational procedures and the value of decision variables,
such that: to obtain good solutions.

Strategic level
Given: environment, some structural constraints, a set of interconnected decision centers, a set of basic resources and one or more strategic goals,
find: how the company is orgaized,
such that: the profitability of investments will be maximized.

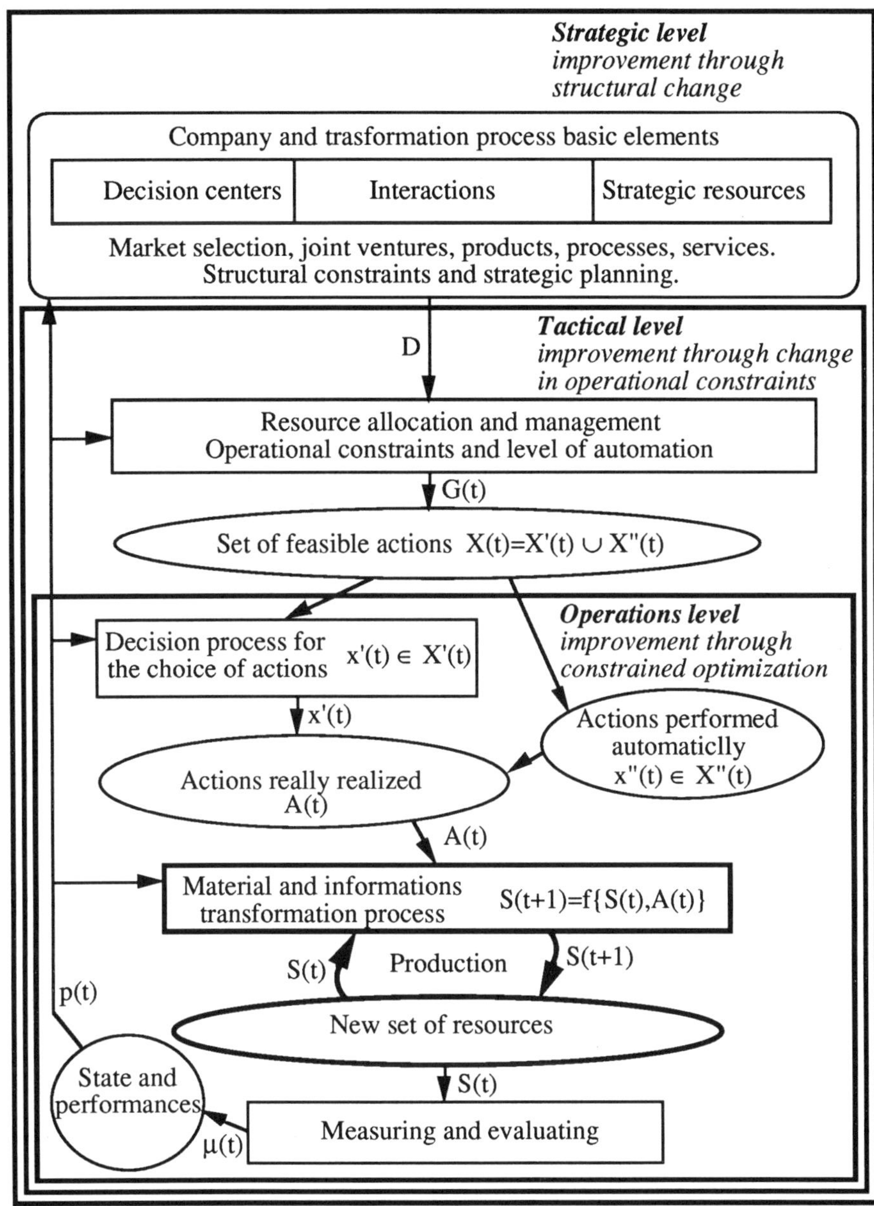

A first set of decisions concerning organizational contraints are usually the connections between input resources, activities and output resources. Notice that this is the framework for activity based costing analyses.

At this point we have to define the relations amongst activities, such as precedence constraints, concurrency, etc.
Given the resource/activity connections and the relations amongst activities, we can define the resource allocation process and plan our activities in time.

The activities can either be the output of a decision making process involving men (organized in different decision centers), machines and informations, or the output of an automatic system of decision rules, producing the actions on the ground of a set of measures of the state of the field and, if it is the case, the value of a set of parameters.

The set of decisions produced automatically and the effectiveness of the decision rules are a crucial point for the effectiveness of the whole system.

The next two charts show the links between the decision process, as outlined above, information flows and what has been said on CIM in section 4.

The tendency of modern production systems is towards an enlargement of the set od decisions taken automatically. The effect of this evolution is twofold: we need more powerfull, reliable and integrated systems to collect and manage the flow of informations; we need globally effective decentralized decision making systems.

A basic element of this evolution, in production environments, is the so called *computer integrated manufacturing* subsystem, which, generally speaking, deals with the whole management of the production informations and with the automated decisions.

The evaluation and the improvement of such CIM systems are therefore crucial points for the factory effectiveness and suitable set of performance indicators and improvement metodologies must be found.
As far as the performance indicators are concerned, the simple measures of computational power, memory size, comunication network capacity, etc., are not enough, and more powerfull indicators considering explicitely the goals of the system and the needs of the end user, must be introduced.
To move in this direction, we propose a set of cost-accounting indicators as follows.

A parameter that all CIM systems must be able to calculate is the total amount of money spent for inputs (resources) used in a given time interval (i.e.one year) to produce a given global output (formed by all products, informations, etc. produced in the time interval).
A more sophisticated set of monitoring parameters, concerns the activities required for each single output (or output type), the inputs required for each activity and the amount of money required for each input. Generally speaking, if the tracking system is good enough, the sum of all input costs plus the cost of overheads (the activities that cannot be attributed to a given output), must be equal to the total amount of money spent during the given time interval. This set of parameters correponds to the standard output of a good activity based costing system.

A set of parameters, seldom available in present systems, concerns the possibility to compute the cost of a new product, given a fairly complete specification of the product, the production process, the market forecast, the input costs, etc. This function requires a tool which simulates the behaviour of the system and its response to different external actions. Such a tool requires a conceptual quantitative model of the production system.

THE PRODUCTION SYSTEM

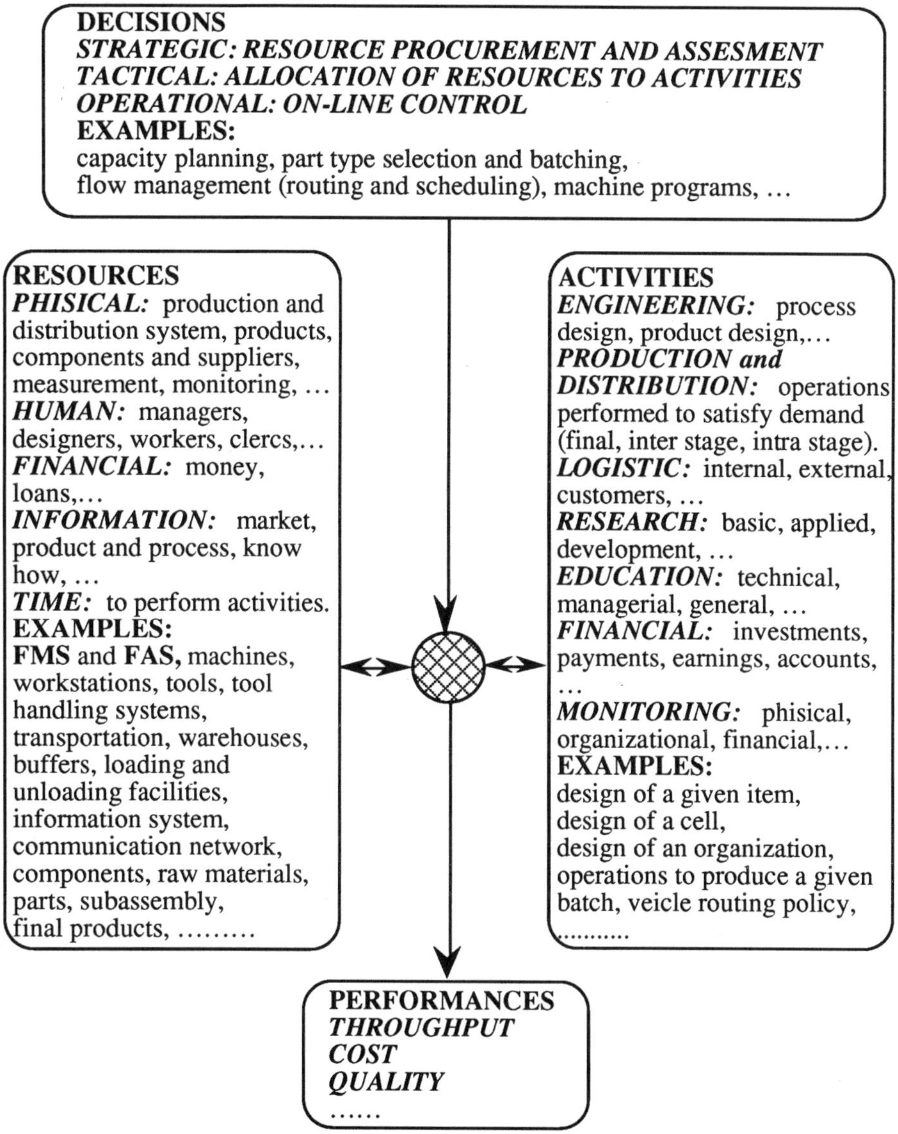

DECISIONS
STRATEGIC: RESOURCE PROCUREMENT AND ASSESMENT
TACTICAL: ALLOCATION OF RESOURCES TO ACTIVITIES
OPERATIONAL: ON-LINE CONTROL
EXAMPLES:
capacity planning, part type selection and batching,
flow management (routing and scheduling), machine programs, ...

RESOURCES
PHISICAL: production and distribution system, products, components and suppliers, measurement, monitoring, ...
HUMAN: managers, designers, workers, clercs,...
FINANCIAL: money, loans,...
INFORMATION: market, product and process, know how, ...
TIME: to perform activities.
EXAMPLES:
FMS and **FAS,** machines, workstations, tools, tool handling systems, transportation, warehouses, buffers, loading and unloading facilities, information system, communication network, components, raw materials, parts, subassembly, final products, ...

ACTIVITIES
ENGINEERING: process design, product design,...
PRODUCTION and DISTRIBUTION: operations performed to satisfy demand (final, inter stage, intra stage).
LOGISTIC: internal, external, customers, ...
RESEARCH: basic, applied, development, ...
EDUCATION: technical, managerial, general, ...
FINANCIAL: investments, payments, earnings, accounts, ...
MONITORING: phisical, organizational, financial,...
EXAMPLES:
design of a given item, design of a cell, design of an organization, operations to produce a given batch, veicle routing policy, ...

PERFORMANCES
THROUGHPUT
COST
QUALITY
....

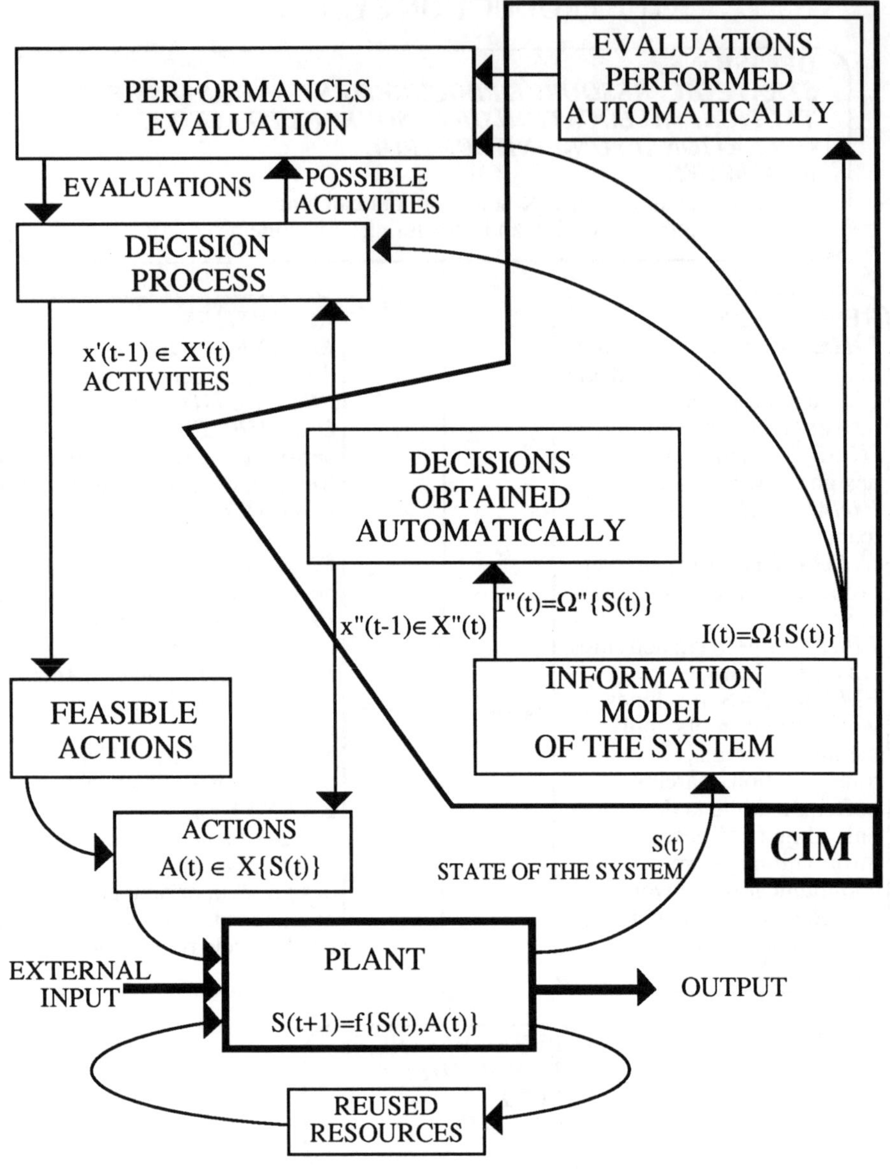

PERFORMANCES EVALUATION
EVALUATIONS PERFORMED AUTOMATICALLY
EVALUATIONS
POSSIBLE ACTIVITIES
DECISION PROCESS
x'(t-1) ∈ X'(t) ACTIVITIES
DECISIONS OBTAINED AUTOMATICALLY
x"(t-1) ∈ X"(t)
I"(t)=Ω"{S(t)}
I(t)=Ω{S(t)}
INFORMATION MODEL OF THE SYSTEM
FEASIBLE ACTIONS
ACTIONS A(t) ∈ X{S(t)}
S(t) STATE OF THE SYSTEM
CIM
EXTERNAL INPUT
PLANT S(t+1)=f{S(t),A(t)}
OUTPUT
REUSED RESOURCES

6.   Computer integrated manufacturing and activity based costing

Activity based systems aim at correcting the deficiences of the traditional costing system, introducing three main goals: assign costs to activities, or actions as we have previously called them, assign costs to cost objects, and produce not strictly financial auxiliary informations on activities.

Conventional cost system presumes that products cause costs. The correct activity based cost assumption is that the cause of cost is not the product as such, but all the activities necessary to manufacture the product (or realize the service).
The product does not consume directly money, but consumes resources. The performance of activities needs a series of resources, and these resources cause costs.

An activity can be defined as a work unit: product development, or part inspection, for example.
A cost object is the reason for performing an activity; it can be either a product or a service, and activity based costing has the goal of measuring the real unit cost, resulting from the sum of all the activities that are necessary to produce a good or service.
Also important is the definition of "drivers", to measure accurately the use of different activities. The driver is a measure through which cost is allocated to the product; a measurement unit of a driver can be, for example, the number of hours worked, the number of parts produced, etc.

The main innovation in activity based costing is that the information given is not only financial information. For this purpose, a series of attributes are defined, in relation to each activity considered.
Activity attibutes (for instance, first piece inspections, indicated by number of hours of inspections) can be, first of all, cost drivers: the number of moving parts of the dye, the number of colours in printed parts. A cost driver represents the causal factor: why do we do this activity in this way? This is important because the more complex the product, the bigger the effort (and therefore the cost) required.
A second attribute can be the presence (or absence) of value added. What is the contribution of the activity analysed to the creation of value added? Is it an essential activity, from the client's point of view, or an activity that can be eliminated with a better organization of other existing activities?
The third attribute is the identifier, to measure, for instance, quality. A last information given by activity based costing are performance indicators. A performance indicator can be represented by the number of pieces refused by the client, or by the number of pieces not accepted after a quality inspection. They are measures of how well we perform the activity.

Activity based costing methodologies have not only the goal of giving an accurate assignment of costs, being thus a decision making support tool.
In addition to accurate informations on costs, more recent elaborations of activity based costing tend to be used to obtain information to make changes in business practices and introduce performance improvement programs.
Activity based costing can be used for pricing, inventory assessment, strategic management, activity management and, more generally, for a better understanding of the company.

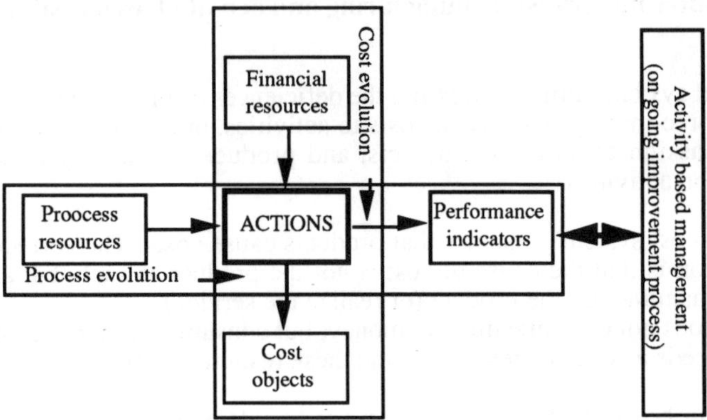

This chart shows the double dimension of the ABC model.

From the cost evolution point of view, you consider the cost of the resources such as salaries equipment, technology, which you have to diaggregate to the point of reaching the "basic work units", that is to say the activities, grouped by process or subprocess, and from these activities you are able to determine the cost of the products.
But ABC methodology give also other informations on the work performed.
In the horizontal dimension, we have cost drivers and performance indicators associated to activities. Cost drivers are identified and measured in order to understand why we do the work, which is the primary cause of the effort.
The second aspect of the process evolution is performance measurement, that takes into account time, cost, service, productivity, quality and tells us how well the work has been performed.
At this point we can perform a benchmarking action to understand, in comparison to other companies' performances, the level of performance of our work.

This second dimension, the horizontal process evolution, allows us to understand why money is spent, why work absorbs a given quantity of time, why we use given resources, why do they require a specific effort. Substantially, why do we perform work in the way we do, and how well do we perform it.
There are the non quantitative informations also useful for interpreting cost informations. The process evolution, that shifts the attention from the link between reasources and activities, is often represented using the matrix of this chart, taken from a U.S. case study. It is clear that activity based costing needs a complete knowledge of all the activities that take place in the process.
Consequently, the capacity of a CIM system to support activity based costing can be considered the indicator of the level of integration of the CIM system that supports activity based costing, in a CIM that works at its best. The latter can be considered as a benchmark for a CIM system to be improved. This is true even if the two systems are part of completely different processes.
In fact, a way to assess the good performance of a CIM system, and therefore choose the best CIM system, can be that of measuring the gap between the cost of the product or service calculated in line with activity based costing supported by CIM and the costs calculated on the basis of all the process resources used for a relatively long period of production.

References

[bon]   C. Bonini, *Simulation of information and decision systems in the firm*. Prentice-Hall, 1963.

[cal]   Joseph Cavinato, *How to Benchmark Logistics Operations,* Distribution 87, N° 8, (aug. 1988), pagg. 93-96.

[cam]   Robert C. Camp, *Benchmarking. the search for industry best practices that lead to superior performance*. ASQC quality pres, 1989.

[dro]   Thomas R. Drozdowski, *GTE Uses Benchmarking to measure Purchasing*, Purchasing 94, N° 6 (march 1983), pagg. 21-24.

[fur]   Timothy R. Furey, *Benchmarking: the Key to Developing Competitive Advantage in Mature Markets,* Planning review 15, n°5 (sept./oct. 1987), pagg. 30-32.

[hub]   G.Huber, *A theory of the effects of advanced information technologies on organizational design, intelligence, and decision making*. Academy of Management Review, 15, 1, pp.47-71, 1990.

[jh]   G. Jacobson e J.Hillkirk, *Xerox:American samurai,* New York, Macmillan Publishing, 1986

[jk1]   H.Thomas Johnson e R.S.Kaplan, *Relevance Lost: the rise and fall of management accounting,* Boston: Harvard Business School Press, 1987.

[jk2]   H.Thomas Johnson e R.S.Kaplan, *The rise and fall of management accounting*, Engineering management review, vol. 5, n. 3, autumn 1987.

[kan]   Edward J. Kane, *IBM's Quality Focus on the Business Process,* Quality Progress 19, N°4, (april 1986) pagg. 24-33.

[kap]   Robert S.Kaplan (ed.), *Measures for manufacturing excellence,* Harvard Business School Series in Accounting and Control, 1990.

[kp]   A.Kumar, P.S.Ow, M.J.Prietula, *Organizational simulation and information systems design: an operations level example*. Management Science, 39, 2, pp.218-240, 1993.

[lc1]   Byron C.Lewis, Albert E.Crews, *The Evolution of Benchmarking as a Computer Performance Evaluation Technique*, MIS Quaterly, 9, n°1 (march 1985), pagg.8-16.

[lc2]   Richard L.Lynch e Kelvin F.Cross, *Measure up: yardsticks for continuous improvement*. Basil Blackwell inc., 1991.

[lt1]   Harold A.Linstone e M.Turoff, *The Delphi Method: Techniques and Applications*, Redding, Mass. Addison-Wesley, 1975.

218

[lt2]   Mario Lucertini, Daniela Telmon, *Le tecnologie di gestione. I processi decisionali nelle organizzazioni integrate.* Franco Angeli, 1993.

[jai]   Ramchandran Jaikumar:, *An architecture for process control costing systems.* In: R.S.Kaplan (Ed.) *Measures for manufacturing excellence*, Harvard Business school series in accounting and control, 1990.

[mar]   J.March (Ed.), *Handbook of organizations.* Rand McNally, 1965.

[mas]   Brian Maskell, *Performance Measurement for World Class Manufacturing: a Model for american companies,* Productivity Press, 1992.

[ms]   T.Malone, S.Smith, *Modelling the performance of organizational structures.* Operations Research, 36, 3, pp.421-436, 1988.

[ods]   V.K.Omachonu, E.M.Davis, P.A.Solo, *Productivity measurement in contract oriented service organizations,* Int. J. of Technology management, vol 5, n. 6, pagg. 703-719, 1990.

[pet]   Tom Peters, *Thriving on caos,.* Excel/A California lim., 1987.

[pip]   Frank J.Pipp, *Management Commitment to Quality,* Quality Progress 16, N° 8 (august 1983), pag. 12-17.

[sch]   Richard J.Schonberger, *World class manufacturing. The lessons of semplicity applied*, The Free Press, 1986.

[spe]   Micheal J. Spendolini, *The benchmarking book*, AMACOM, 1992.

[tzc]   Fancis G. Tucker, Seymour M. Zivan e R. C. Camp, *How to measure Yourself against the best,* Harvard Business Review, jan.feb. 1987, pagg.2-4.

[tz]   Fancis G.Tucker, Seymour M.Zivan, *A Xerox Cost Center Imitates a Profit Center,* Harvard Business Review, may-june 1985, pagg. 2-4.

Issues in Enterprise Modelling

Mark S. Fox
Department of Industrial Engineering
University of Toronto
4 Taddle Creek Road
Toronto, Ontario M5S 1A4 CANADA
tel: 1-416-978-6823; fax: 1-416-971-1373; internet: msf@ie.utoronto.ca

Abstract

Computerization of enterprises continues unabated and so does the cost of software. The availability of a generic, common-sense enterprise model is necessary if we are to reign in costs. But in order to construct useful Generic Enterprise Models (GEM) there are a number of issues that have to be addressed. In this paper we explore the following issues: Is there such a thing as a generic enterprise model? Can the terminology be precisely defined? Does all knowledge need to be explicit? Need there be a single, shared enterprise model? How can we determine which is a better enterprise model? Can an enterprise model be consistent? Can an enterprise model be created and kept current? Will the organization accept an enterprise-wide model? We then briefly describe the TOVE project, which attempts to address many of these issues.

1.0 Introduction

As described in a recent report on Agile Manufacturing [Nagel et al. 91], if an industrial organization is to compete in the coming decade, they must produce products that are: of consistently high quality throughout the product's life, customised to local market needs, open in that they may be integrated with other products, environmentally benign, and technically advanced. The key to achieving these capabilities is "agility". Agility implies the ability to: continuously monitor market demand, quickly respond by providing new products, services and information, quickly introduce new technologies, and quickly modify business methods. But achieving agility requires far greater *integration* of functions within the enterprise, and between enterprises, than has ever been achieved; enterprises must be task oriented as opposed to organisation oriented; expertise must flow freely across the enterprise to where it is needed.

Integration is a step along the road to agility. Yet it contradicts decades of management science teachings. We have been taught that in order to cope with the complexity of enterprises, we have to decompose them into manageable pieces; each piece having minimal interaction with the others. But the decomposition impedes the free flow of information and knowledge, and the coordination of actions. In order to break-down these organizational barriers, Hansen [91] has identified five principles of integration:

S. Y. Nof (ed.), Information and Collaboration Models of Integration, 219–234.

1. "When people understand the vision, or larger task, of an enterprise and are given the right information, the resources, and the responsibility, they will 'do the right thing'."

2. "Empowered people - and with good leadership, empowered groups - will have not only the ability but also the desire to participate in the decision process."

3. "The existence of a comprehensive and effective communications network ... This network must distribute knowledge and information widely, embracing the openness and trust that allow the individual to feel empowered to affect the 'real' problems."

4. "The democratization and dissemination of information throughout the network in all directions irrespective of organizational position ... ensures that the Integrated Enterprise is truly integrated."

5. "Information freely shared with empowered people who are motivated to make decisions will naturally distribute the decision-making process throughout the entire organization."

These principles focus on two major issues: 1) how to motivate employees, and 2) how to provide employees with the right information to do their job. But in achieving the latter, there is a limit to how many meetings you can attend, memos you can read, and trips you can make! The question then is how can technology aid integration?

Over the last 10 years there has been a shift in how we view the operations of an enterprise. Rather than view the enterprise as being hierarchical in both structure and control, a distributed view where enterprise units communicate and cooperate in both problem solving and action has evolved [Fox 81]. To achieve integration it is necessary that units of the enterprise, be they human or machine based, be able to understand each other. Therefore the requirement exists for a representation in which enterprise knowledge can be expressed. Minimally the representation provides a language for communicating among units, such as design, manufacturing, marketing, field service, etc. Maximally the representation provides a means for storing knowledge and employing it within the enterprise, such as in computer-aided design, production control, etc.

The problem that we face today, is that the legacy systems to support enterprise functions were independently created, consequently they do not share the same representations. This has led to different representations of the same enterprise knowledge and the inability of these functions to share knowledge. We call this the *Correspondence Problem*: What is the relationship among concepts that denote the same thing but have different names? It is common for enterprises, especially those that are geographically dispersed to use different names to refer to the same concept. No matter how rationale the idea of renaming them is, organisational barriers impede it.

Secondly, these representations lack an adequate specification of what the terminology means (aka semantics). This leads to inconsistent interpretations and uses of the knowledge. Lastly, the cost of designing, building and maintaining a data model of enterprise knowledge is large. Each tends to be unique to the enterprise; terminology is enterprise specific.

As a solution to this problem, there has been an increasing interest in Generic Enterprise Models (GEM). A GEM is a data dictionary that defines the classes of entities (or objects) that are generic across a type of enterprise, such as manufacturing, and can be employed (aka instantiation) in defining a specific enterprise. It is believed that if one starts with a GEM, the time and cost of producing an instantiation of the model will be reduced significantly. Though much work has gone into the creation of GEMs, few have reflected upon the issues that arise in their creation and use. In the following, we explore a number of issues surrounding the creation of a Generic Enterprise Model.

2.0 Enterprise Modelling Issues

2.1 Is there such a thing as a generic enterprise model?

Yes! There exists significant amounts of knowledge that is generic across many applications. The identification and formalization of generic knowledge has come to be called "Ontological Engineering" [Gruber 93]. An ontology is a formal description of entities and their properties, relationships, constraints, behaviours. Entities are classified into one or more taxonomies.

In trying to construct an ontology that spans enterprise knowledge, the first question is where to start. Brachman provides a stratification of representations [Brachman 79]:

> **Implementation:** physical representation of data

> **Logical:** logical interpretation of the physical representation.

> **Conceptual (aka Epistemological):** primitives for representing the components of a concept: properties, structure, relations, generalization, association.

> **Generic:** domain independent concepts such as time, causality, action, space, etc.

> **Application (aka Lexical):** primitives are application dependent and may change meaning as knowledge grows.

The following diagram depicts the last three levels with examples of the type of knowledge that is represented at each. Note that the application level is re-labeled the enterprise level. Secondly, the division between levels is somewhat artificial in that each level may be further stratified. Determining what concepts should be in the generic level versus the enterprise level is based on their generality.

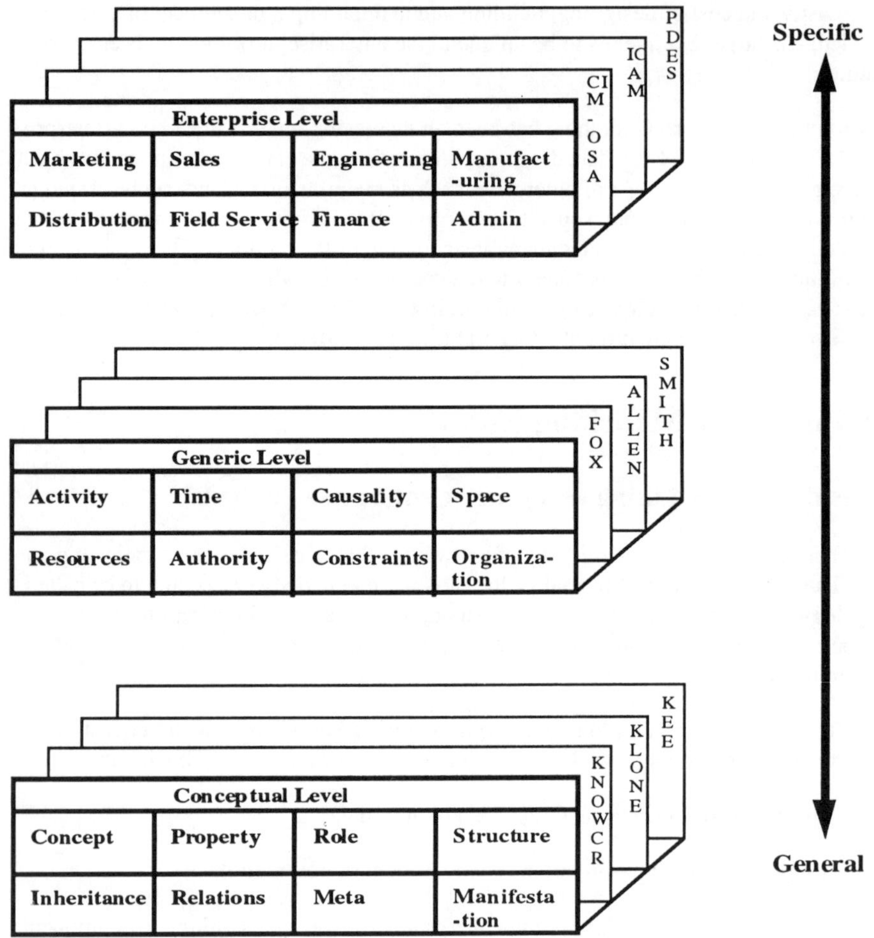

The *conceptual level* provides the building blocks for defining concepts. The basic unit of representation is an *object* for which is defined:

- **Properties:** Cardinality, Type

- **Relationships:** Range restrictions

- **Generalization/Specialization** hierarchies

- **Classification:** Prototypical descriptions vs. instances

The conceptual level received much attention in the 1970s, with the development of knowledge representation languages such as FRL [Roberts & Goldstein 77], KLONE [Brachman 77], KRL [Bobrow & Winograd 77], NETL [Fahlman 77], and SRL [Fox 79],. Many of the concepts investigated in these system have formed the basis of semantic data modelling in databases. More recently, the conceptual level has been formalized as what is now called "terminological logic" [Brachman & Schmolze 85].

The *generic level* provides ontologies for concepts common across many domains. Generic level representations include concepts such as:

- •Time [Allen 83],

- •Causality [Rieger & Grinberg 77], [Bobrow 85],

- •Activity [Sathi et al. 85],

- •Resources [Fadel 93], and

- •Constraints [Fox 83] [Davis 87].

Consider the representation of time. Time is represented by points, periods and relations. A time-point lies within an interval $\{<tmin, tmax> \mid tmin \le tmax, tmin, tmax \in N\}$. A time-period is bounded by a start and end time-point $\{<TP1, TP2> \mid tmin1 \le tmax2, TP1, TP2 \in TP\}$. We use Allen's [83] temporal relations to describe the relationships between time-points and/or time-periods. We present the thirteen possible temporal relationships and refer to Allen's paper for the transitivity table of these temporal relations.

Relation	Symbol	Symbol for Inverse	Pictorial Example	Relation	Symbol	Symbol for Inverse	Pictorial Example
X over-laps Y	o	oi		X dur-ing Y	d	di	
X same Y	=	=		X starts Y	s	si	
X meets Y	m	mi		X ends Y	e	ei	
X before Y	<	>					

One of the largest efforts underway to create an integrated set of generic representations is the CYC project at MCC [Lenat & Guha 90].

The *enterprise level* provides a data dictionary of concepts (aka reference model) that are common across various enterprises, such a products, materials, personnel, orders, departments, etc. At the enterprise level, various efforts exist in standardizing representations. For example, since the 1960's IBM's COPIC's Manufacturing Resource Planning (MRP) system has had a shared enterprise model. In fact, any MRP product contains an enterprise model. Recently, several efforts have been underway to create more comprehensive enterprise model, including:

CAMI: A US-based non-profit group of industrial organizations for creating manufacturing software and modelling standards.

ICAM: A project run by the Materials Lab. of the US Air Force [Davis et al. 83] [Martin et al. 83] [Martin & Smith 83] [Smith et al. 83].

IWI: A reference model developed at the Institut fur Wirtschaftsinformatik, Universitat des Saarlandes, Germany [Scheer 89].

The following are the basic relations and objects in their range defined for the "part" concept in the ICAM model from the *design* perspective [Martin et al. 83] [Martin & Smith 83]:

- **IS CHANGED BY:** Part Change (105) (also shown as "is modified by")
- **APPEARS AS:** Next Assembly usage item (119) (also shown as "is referenced as").
- **HAS:** Replacement part (143).
- **HAS SUBTYPE (IS):** Parts list item (118), Replacement part (143).
- **IS USED AS:** Next Assembly Usage (40), Advance material notice item part (144), Configuration list item (170).
- **IS TOTALLY DEFINED BY:** Drawing (1).
- **IS LISTED BY (LISTS):** Configuration list (84).
- **IS USED IN:** Effectivity (125).
- **IS FRABRICATED FROM:** Authorized material (145).

The following ar e the basic relations and objects they are linked to for a "part" from a *manufacturing* perspective [Smith et al. 83]:

- **HAS:** N.C. Program (318), Material issue (89), Component part (299), Alternative part (301), Part/process specification use (255), Material receipt (87), Work package (380), Part tool requirement (340), Part requirement for material (397), Standard routing use (254), Image part (300), Part drawing (181).
- **IS ASSIGNED TO (HAS ASSIGNED TO IT):** Index (351).
- **IS DEFINED BY (DEFINES):** Released engineering drawing (12).
- **IS SUBJECT OF:** Quote request (90), Supplier quote (91).
- **IS TRANSPORTED BY:** Approved part carrier (180).
- **IS RECEIVED AS:** Supplier del lot (309).
- **APPEARS AS:** Part lot (93), Ordered part (188), Serialized part instance (147), Scheduled part (409), Requested purchase part (175).
- **CONFORMS TO:** Part specification (120).

- **IS INVERSE:** Component part (299), Alternate part (301), Section (363), End item (5), Configured item (367), Image part (300).

- **IS USED AS:** Component part callout (230), Process plan material callout (74).

- **IS SUPPLIED BY:** Approved part source (177).

- **MANUFACTURE IS DESCRIBED BY:** Process plan (415).

- **SATIFIES:** End item requirement for part (227).

- **IS REQUESTED BY:** Manufacturing request (88).

- **IS STORED AT:** Stock location use for part (227).

- **IS SPECIFIED BY:** BOM Item (68).

This is only the tip of the iceberg. If one were to develop a complete GEM at the enterprise level, its sheer size would overwhelm the abilities of any database manager or knowledge engineer. There is a point at which further elaboration tends to obfuscate rather than enhance the model. On the other hand, if there is not enough detail, then its value may be limited. We will revisit this issue in section 2.7.

2.2 Can the terminology be precisely defined?

> "It is certainly praiseworthy to try to make clear to oneself as far as possible the sense one associates with a word. But here we must not forget that not everything can be defined" Gottlob Frege.

Prior to the advent of GEMs, an application's data model was defined in a database system's data dictionary. The creation of the data dictionary was and continues to be the responsibility of the database administrator who works with the end users. In the worst case, definitions for each of the objects, attributes and relations in the dictionary are not available, and their interpretation can only be derived by looking at how an application used the information. Better managed data dictionaries include definitions, usually written in a natural language such as english. Due to the inherent ambiguities of natural language, even these definitions may be interpreted differently by each user. If we are to create truly sharable GEMs, we need the ability to precisely state the meaning of each object, attribute and relation.

Precise definitions can be constructed. Through the use of logic, we can define more precisely the meaning of each object, attribute and relation as needed. Definitions may be hierarchical and circular. Hierarchical in the sense that enterprise level concept are defined in terms of generic level concepts. Circular in that enterprise level concepts are defined in terms of other concepts at the same level, and vice versa! Many, if not most, definitions can be represented using first order logic. Some definitions may require high order languages, but it is probably the case most things can exist in a first order language.

Consider the temporal relations introduced in the previous section. The following are definitions of two variations of the *before* relation:

$$\text{TimePoint1 is } \textit{possibly before} \text{ TimePoint2 IF } tmin1 < tmax2 \qquad \text{(EQ 1)}$$

$$\text{TimePoint1 is } \textit{strictly before} \text{ TimePoint2 IF tmax1} < \text{tmin2} \qquad \text{(EQ 2)}$$

Tmin1 and tmax1 bound the interval in which time point 1 is located. The first axiom states that for TimePoint1 to be possibly before TimePoint2, there must exist at least one point in time in TimePoint1's associated interval that is less than some point in time in TimePoint2's associated interval. This is true iff tmin1 < tmax2.

2.3 Does all knowledge need to be explicit?

The usefulness of an instantiated GEM is determined by the queries it can answer. Consider a model with an SQL interface. Knowledge is explicitly represented if it can be retrieved using a simple SELECT. That is, the knowledge is represented explicitly and only needs to be retrieved. Knowledge is represented implicitly if it requires a more complex query to retrieve it. For example, it may require one or more JOINs combined with SELECTs. This is equivalent to performing deduction. For example, if the model contains a 'works-for' relation and it is explicitly represented that Joe 'works-for' Fred, and that Fred 'works-for' John, then the obvious deduction that Joe 'works-for' John (indirectly) is not represented explicitly in the model but must be deduced.

We distinguish between a model that includes axioms that support deduction, versus a model without axioms where deductions are specified by the query. In the former case, the model would be able to deduce that Joe works-for John in response to a query asking who does Joe work for. In the latter case, the user would have to specify a complex query which would include as many joins as necessary to travel along the works-for relation. Since the user does not know at the outset the depth of the works-for path, they may not get the information they were looking for. We call a model which includes axioms an Axiomatised Enterprise Model (AEM). An AEM that includes a deduction engine (i.e., theorem prover) has been called either a knowledge base or a deductive database. We will refer to it as a Deductive Enterprise Model (DEM). The lack of a deductive capability forces users to spend significant resources on programming each new report or function that is required.

So far we have discussed the deductive capability of a model without reference to the nature of the axioms or rules used in performing the deductions. We say a DEM possesses Common-Sense (DEM_{CS}) if its axioms define the meaning of the terms in the ontology. By Common-Sense, we mean that the axioms enable the model to deduce answers to questions that one would normally assume can be answered if one has a "commons-sense" understanding of the enterprise.

In summary, the design, creation and maintenance of software is fast becoming the dominant cost of automation. A significant portion of these costs is for software that provides answers deduced from the contents of the enterprise model. Many of these questions could be answered automatically if the enterprise model had the "common sense" to answer them!

2.4 Need there be a single, shared enterprise model?

Not all knowledge has to be represented generically, only that which is shared among units of the enterprise, and that too may be specialized. Units of an enterprise evolve representations and procedures that are tailored to their roles and goals. The tailoring is usually necessary to achieve higher degrees of productivity and quality. Consequently, formalized models maximally affect what is communicated among enterprise units, and minimally affect how information/knowledge is represented within units.

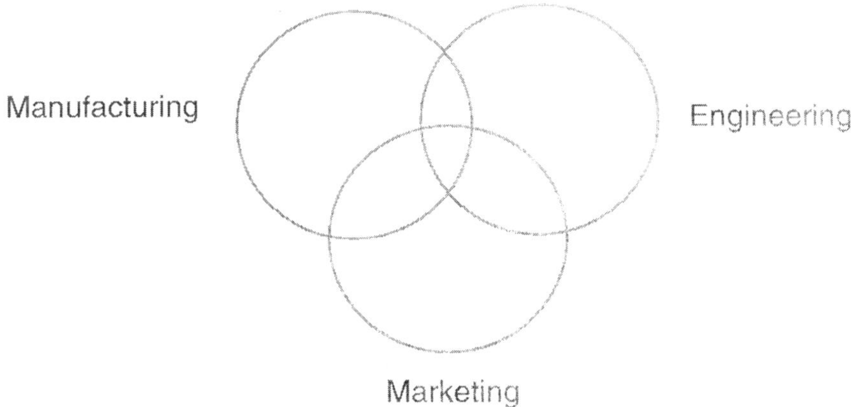

Even the interchanges among units in the enterprise neither require nor desire a single integrated model as a basis of communication. As shown in the figure above, there may be one language using for communication between engineering and manufacturing, and a different one for engineering and marketing. But all units will share some core language. Though the artificiality of the enterprise implies the possibility of an integrated model, reality tends to differ. Integrated models are really a lattice of models that are specialized to the needs of subsets of enterprise units.

2.5 How can we determine which is a better enterprise model?

Given the many efforts seeking to create a GEM, there has never been a well defined set of criteria with which these efforts could be evaluated! In fact, there is no objective means by which one can compare one GEM with another. Following are what we believe should be the characteristics of a representation:

Generality: To what degree is the representation shared between diverse activities such as design and troubleshooting, or even design and marketing?

Competence: How well does it support problem solving? That is, what questions can the representation answer or what tasks can it support?

Efficiency: Space and inference. Does the representation support efficient reasoning, or does it require some type of transformation?

Perspicuity: Is the representation easily understood by the users? Does the representation "document itself?"

Transformability: Can the representation be easily transformed into another more appropriate for a particular decision problem?

Extensibility: Is there a core set of ontological primitives that are partitionable or do they overlap in denotation? Can the representation be extended to encompass new concepts?

Granularity: Does the representation support reasoning at various levels of abstraction and detail?

Scalability: Does the representation scale to support large applications?

These criteria bring to light a number of important issues and risks. For any set of functions, how can we determine if the integrating model is *functionally complete*? A model is functionally complete if it contains the types of information necessary for a function to perform its task. Are functionally complete models specifiable? One way of specifying a model's functional requirements is as a set of questions that the model must be able to answer. We call this the *competency* of a model.

Another problem is where the representation ends and inference begins? Consider the competence criterion. The obvious way to demonstrate competence is to define a set of questions that can be answered by the representation. If no inference capability is to be assumed, then question answering is strictly reducible to "looking up" an answer that is represented explicitly. In contrast, Artificial Intelligence representations have assumed at least inheritance as a deduction mechanism. In defining a shared representation, a key question then becomes: should we be restricted to just an terminology? Should the terminology assume an inheritance mechanism at the conceptual level, or some type of theorem proving capability as provided, say, in a logic programming language with axioms restricted to Horne clauses (i.e., Prolog)? What is the *deductive capability* that is to be assumed by a reusable representation?

The efficiency criterion is also problematic. Experience has demonstrated that there is more than one way to represent the same knowledge, and each representation does not have the same complexity when answering a specific class of questions. Consequently, we cannot assume that a representation will partition the space of concepts, but there will exist overlapping representations that are more efficient in answering certain questions. Furthermore, the deductive capability provided with the representation affects the store vs. compute trade-off. If the deduction mechanisms are taken advantage of, certain concepts can be computed on demand rather than stored explicitly.

The ability to validate a proposed representation is critical to this effort. The question is: how are the criteria described above operationalised? The *competence* of a representation is concerned with the span of questions that it can answer. We propose that for each category of knowledge, a set of questions be defined that the representation can answer. Given a conceptual level representation and an accompanying theorem prover (perhaps Prolog), questions can be posed in the form of queries to be answered by the theorem prover. Given that a theorem prover is the deduction mechanism used to answer questions, the *efficiency* of a representation can be defined by the number of LIPS (Logical Inferences Per Second) required to answer a query. Validating *generality* is more problematic. This can be determined only by a representation's consistent use in a variety of applications. Obviously, at the generic level we strive for wide use across many distinct applications, whereas at the application level, we are striving for wide use within an application.

2.6 Can an enterprise model be consistent?

The assumption that enterprise knowledge can be globally consistent is ridiculous. By definition, an information system based on a distributed architecture will abound in inconsistent information. Tailoring and local context leads to ambiguities and inconsistencies in the content of what is stored and communicated. How to manage inconsistency so that it does not adversely affect operations is the problem that has to be solved.

One way of approaching this is to identify subsets of knowledge that must remain consistent among a set of "consenting" agents in the information network. Changes to this knowledge must be managed so that inconsistencies do not arise.

2.7 Can an enterprise model be created and kept current?

Enterprises are dynamic and undergo continuous change. Consequently, a process for managing the evolution of the model is required. Since the competence of a model is specified by the activities that use it, it follows that model management is an activity-based process. The information requirements of activities determine data spheres and their contents. A data sphere is a set of information that is shared by functionally-related agents. Groupings of activities lead to data spheres whose model is a point in the model lattice.

Since enterprise activities are the result of enterprise design, model specification is the outcome of enterprise design. Without an adequate process - and possibly a theory - of enterprise design, the construction of an integrated model will be either expensive or impossible. Emerging methods for enterprise analysis and possibly design include:

- GRAI: Universite de Bordeaux.
- CIM-OSA: A reference model being developed by the ACIME group of ESPRIT in Europe [Esprit 90].
- PERA: Purdue Enterprise Reference Architecture [Williams 91].

2.8 Will the organization accept an enterprise-wide model?

There is a belief that an integrated model cannot be superimposed upon an enterprise. Enterprises are both artificial and natural. Artificial in that formal structures and systems exist within the enterprise by design. Natural in that systems evolve in response to the inadequacies of the design due to changing market conditions, technologies, knowledge, etc. The artificiality of an enterprise admits the specification and utilization of an integrated model. Its adoption is imposed by the enterprise's formal structures.

3.0 TOVE: TOronto Virtual Enterprise

In the Enterprise Integration Laboratory at the University of Toronto, we have been investigating the creation of a Common-Sense Deductive Enterprise Model (DEM$_{CS}$). The goal of the TOVE project is to create a generic enterprise model that has the following characteristics: 1) provides a shared terminology for the enterprise that each agent can jointly understand and use, 2) defines the meaning of each term (aka semantics) in a precise and as unambiguous manner as possible, 3) implements the semantics in a set of axioms that will enable TOVE to automatically deduce the answer to many "common sense" questions about the enterprise, and 4) defines a symbology for depicting a term or the concept constructed thereof in a graphical context.

We approach these goals by identifying the different types of knowledge we wish to represent at the generic level. Generic concepts include representations of Time [Allen 83], Causality [Rieger & Grinberg 77] [Bobrow 85], Activity [Sathi et al. 85], and Constraints [Fox 83][Davis 87]. For each type of knowledge, we first define the competency requirements. We then define an ontology that will support the specified competency. We approach the second and third goals by defining a set of axioms (aka rules) that define common-sense meanings for the ontology, in first order logic and implemented in Prolog.

TOVE is not only an ontology but a testbed. TOVE has been used to define a *virtual company* whose purpose is to provide a testbed for research into enterprise integration. TOVE is implemented in C++ using the ROCK@+[TM] knowledge representation tool from Carnegie Group. Axiom are implemented in Quintus Prolog which is integrated with ROCK. TOVE operates "virtually" by means of knowledge-based simulation [Fox et al. 89].

4.0 Conclusion

Computerization of enterprises continues unabated. The amount of software is increasing while its cost is not decreasing. The availability of a generic, common-sense enterprise model is necessary if we are to reign in costs. But in order to construct useful Generic Enterprise Models there are a number of issues that have to be addressed. Foremost is the transition of the efforts from poorly principled data modelling into principled engineering.

The TOVE project is our attempt at creating such a model. It's goals are 1) to create a shared terminology (aka ontology) of the enterprise that each agent can jointly understand and use, 2) define the meaning of each term (aka semantics), 3) implement the semantics as a set of axioms that will enable TOVE to automatically deduce the answer to many "common sense" questions about the enterprise, and 4) define a symbology for depicting terms and concepts in a graphical context. We are approaching these goals by defining a three level representation: application, generic and conceptual.

5.0 Acknowledgments

This research is supported in part by an NSERC Industrial Research Chair in Enterprise Integration, Carnegie Group Inc., Digital Equipment Corp., Micro Electronics and Computer Research Corp., Quintus Corp., and Spar Aerospace Ltd.

6.0 References

[Allen 83]   Allen, J.F.
Maintaining Knowledge about Temporal Intervals.
Communications of the ACM. 26(11):832-843, 1983.

[Bobrow 85]   Bobrow, D.G.
Qualitative Reasoning About Physical Systems. MIT Press, 1985.

[Bobrow & Winograd 77]
Bobrow, D., and Winograd, T.
KRL: Knowledge Representation Language.
Cognitive Science. 1(1), 1977.

[Brachman 77]   Brachman, R.J.
A Structural Paradigm for Representing Knowledge.
PhD thesis, Harvard University, 1977.

[Brachman & Schmolze 85]
Brachman, R.J., and Schmolze, J.G.
An Overview of the KL-ONE Knowledge Representation Systems.
Cognitive Science. 9(2), 1985.

[Davis 87]   Davis, E.
Constraint Propagation with Interval Labels.
Artificial Intelligence. 3, 281-331, 1987.

[Davis et al. 83] Davis, B.R., Smith, S., Davies, M., and St. John, W.
Integrated Computer-aided Manufacturing (ICAM) Architecture Part III/Volume III: Composite Function Model of "Design Product" (DES0).
Technical Report AFWAL-TR-82-4063 Volume III, Materials Laboratory, Air Force Wright Aeronautical Laboratories, Air Force Systems Command, Wright-Patterson Air Force Base, Ohio 45433, 1983.

[Esprit 90]   ESPRIT-AMICE.
CIM-OSA - A Vendor Independent CIM Architecture.
Proceedings of CINCOM 90, pages 177-196.
National Institute for Standards and Technology, 1990.

[Falhman 77] Fahlman, S.E.
A System for Representing and Using Real-World Knowledge.
PhD thesis, Massachusetts Institute of Technology, 1977.

[Fadel 93] Fadel, F.
A Micro-theory for Resources
Technical Report, Enterprise Integration Laboratory, Department of
Industrial Engineering, University of Toronto, to appear.

[Fox 79] Fox, M.S.
On Inheritance in Knowledge Representation.
*Proceedings of the International Joint Conference on Artificial
Intelligence.*
Morgan Kaufmann Pub. Co.95 First St., Los Altos, CA 94022, 1979.

[Fox 81] Fox, M.S. An
Organizational View of Distributed Systems.
IEEE Transactions on Systems, Man, and Cybernetics. SMC-
11(1):70-80, 1981.

[Fox 83] Fox, M.S.
Constraint-Directed Search: A Case Study of Job-Shop Scheduling.
PhD thesis, Carnegie Mellon University, 1983. CMU-RI-TR-85-7,
Intelligent Systems Laboratory, The Robotics Institute,
Pittsburgh.

[Fox et al. 89] Fox, M.S., Reddy, Y.V., Husain, N., McRoberts, M.
Knowledge Based Simulation: An Artificial Intelligence Approach
to System Modeling and Automating the Simulation Life Cycle.
Artificial Intelligence, Simulation and Modeling. In Widman, L.E.,
John Wiley & Sons, 1989.

[Gruber 93] Gruber, T.R.
Toward Principles for the Design of Ontologies Used for
Knowledge Sharing.
Technical Report, Knowledge Systems Laboratory, Stanford
University, 1993.

[Hansen 91] Hansen, W.C.
The Integrated Enterprise. In *Foundations of World-Class
Manufacturing Systems: Symposium Papers.*
National Academy of Engineering, 2101 Constitution Ave, N.W.,
Washington DC, 1991.

[Lenat & Guha 90] Lenat, D., and Guha, R.V.
                   Building Large Knowledge Based Systems: Representation and Inference in the CYC Project.
                   Addison Wesley Pub. Co., 1990.

[Martin & Smith 83]Martin, C., and Smith, S.
                   Integrated Computer-aided Manufacturing (ICAM) Architecture Part III/Volume IV: Composite Information Model of "Design Product" (DES1).
                   Technical Report AFWAL-TR-82-4063 Volume IV, Materials Laboratory, Air Force Wright Aeronautical Laboratories, Air Force Systems Command, Wright-Patterson Air Force Base, Ohio 45433, 1983.

[Martin et al. 83] Martin, C., Nowlin, A., St. John, W., Smith, S., Ruegsegger, T., and Small, A.
                   Integrated Computer-aided Manufacturing (ICAM) Architecture Part III/Volume VI: Composite Information Model of "Manufacture Product" (MFG1).
                   Technical Report AFWAL-TR-82-4063 Voluem VI, Materials Laboratory, Air Force Wright Aeronautical Laboratories, Air Force Systems Command, Wright-Patterson Air Force Base, Ohio 45433, 1983.

[Nagel et al. 91] Nagel, R.N. et al.
                   21st Century Manufacturing Enterprise Strategy: An Industry Led View.
                   Technical Report, Iacocca Institute, Lehigh University, Bethlehem PA, 1991.

[Rieger & Grinberg 77]
                   Rieger, C., and Grinberg, M.
                   The Causal Representation and Simulation of Physical Mechanisms.
                   Technical Report TR-495, Dept. of Computer Science, University of Maryland, 1977.

[Roberts & Goldstein 77]
                   Roberts, R.B., and Goldstein, I.P.
                   The FRL Manual.
                   Technical Report MIT AI Lab Memo 409, Massachusetts Institute of Technology, 1977.

234

[Sathi et al. 85]   Sathi, A., Fox, M.S., and Greenberg, M.
Representation of Activity Knowledge for Project Management.
IEEE Transactions on Pattern Analysis and Machine Intelligence.
PAMI-7(5):531-552, September, 1985.

[Scheer 89]   Scheer, A-W.
Enterprise-Wide Data Modelling: Information Systems in Industry.
Springer-Verlag, 1989.

[Smith et al. 83]   Smith, S., Ruegsegger, T., and St. John, W.
Integrated Computer-aided Manufacturing (ICAM) Architecture Part III/Volume V: Composite Function Model of "Manufacture Product" (MFG0).
Technical Report AFWAL-TR-82-4063 Volume V, Materials Laboratory, Air Force Wright Aeronautical Laboratories, Air Force Systems Command, Wright-Patterson Air Force Base, Ohio 45433, 1983.

[Williams 91]   Williams, T.J., and the Members, Industry-Purdue University Consortium for CIM.
The PURDUE Enterprise Reference Architecture.
Technical Report Number 154, Purdue Laboratory for Applied Industrial Control, Prudue University, West Lafayette, IN 47907, 1991.

IV. Interaction and Collaborative Work

INFORMATION AND COLLABORATION
FROM A SOCIAL/ORGANIZATIONAL PERSPECTIVE

Les Gasser
Computational Organization Design Lab
Institute of Safety and Systems Management
USC, Los Angeles, CA 90089-0021 USA
(213) 740-4046
gasser@usc.edu

ABSTRACT. Building computer-based information systems that collaborate or that support collaboration requires several types of theory, including collaboration theories and theories of modeling and implementation. These theories can be developed from numerous perspectives. This paper treats two such perspectives: Distributed Artificial Intelligence (DAI) and social/organizational perspectives. The DAI perspective focuses on simple, cognitively-motivated agent structures, shared communication languages, well-defined agent, process, and task boundaries, and a common semantic basis for interpreting environments and interactions. This contrasts with the social organizational perspective, which considers a different problematic: the structures and boundaries of agents (e.g. work groups, organizations), the languages they use, the nature of their tasks, and their working interpretations are all matters of continuing negotiation and evolution---they are the subject matter of collaboration, and not just its framework. The paper investigates how each perspective conceptualizes the problems and important issues of coordination and modeling/implementation, and suggests several integrating directions.

1. INTRODUCTION

Fields such as distributed artificial intelligence have developed interesting and useful methods to model information and collaborative activity. Most of these methods focus on simple, cognitively-motivated agent structures, shared communication languages, well-defined agent, process, and task boundaries, and a common semantic basis for interpreting environments and interactions. Detailed empirical social studies of collaborative work find that most of these conceptualizations are problematic in practice: the structures and boundaries of agents (e.g. work groups, organizations) the languages they use, the nature of their tasks, and their working interpretations are all matters of continuing negotiation and evolution are the subject matter of work, not just its framework.

Most theoretical and experimental approaches in DAI give only a partial account of multi-agent collaboration, because they have not grappled with several basic issues, including the practical bases of shared knowledge, semantics, and assumptions, reasoning under inconsistency or incompatible representations, etc. [Gasser 91]. Virtually all current DAI approaches to these problems depend upon common interagent semantics with at most one or two reflective or contextual levels, correspondence theories of representation and belief, global measures of coherence, and the individual agent as the unit of analysis

237

S. Y. Nof (ed.), Information and Collaboration Models of Integration, 237–261.
© 1994 *Kluwer Academic Publishers. Printed in the Netherlands.*

and interaction. Most DAI experiments and theories depend upon closed-system assumptions such as common communication protocols, a shared global means of assessing coherent behavior, some ultimate commensurability of knowledge, or some boundary to a system. These current theories are inadequate for supporting the integration of heterogeneous systems with possibly-incompatible internal semantics (e.g. Hewitt's "microtheories" [Hewitt 91]). Moreover, approaches based on common interagent semantics are incomplete from a theoretical standpoint [Gasser 91, Gasser 92a]. All of these issues have impacts on the structure and capabilities of implementation platforms and development methodologies for DAI.

Findings like these motivate alternate perspectives on information and collaboration, termed social-organizational perspectives. This paper contrasts the standard DAI models of information and collaboration with alternative conceptions based on more fluid and dynamic notions of organizations, information, and collaboration.

1.1. COLLABORATION, INTERACTION, AND IMPLEMENTATION THEORIES

Thinking about information and collaboration from the standpoint of implementing computational systems that act collaboratively or support collaboration requires three types of theory. First, we need a basic *theory of information and collaboration*. This theory would

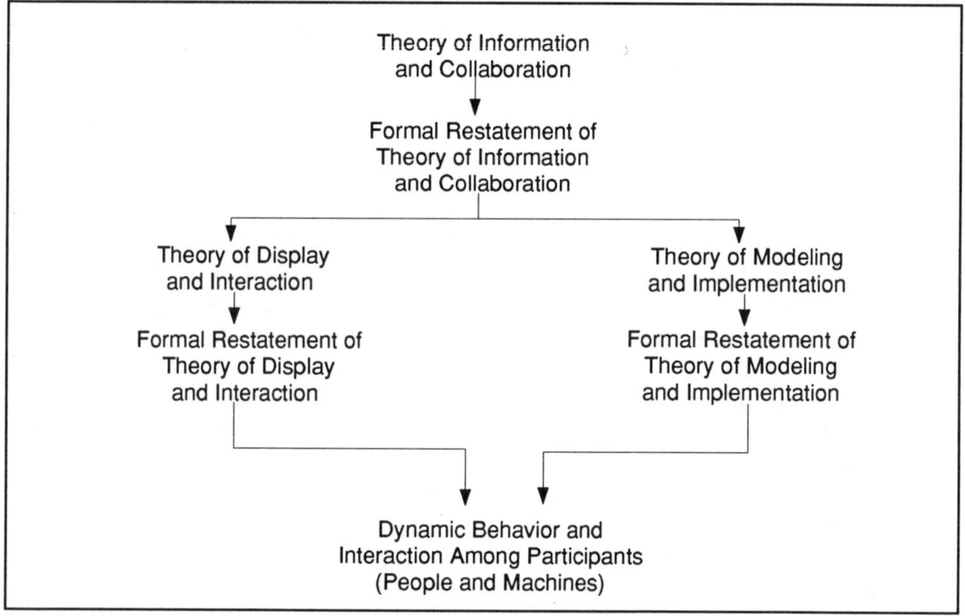

Figure 1: Three Types of Theory

include a set of objects, concepts and structures that capture and describe relationships among important aspects of information and collaboration. Second, we need a *theory of display and interaction*, that describes how to present and manipulate the objects, concepts, and structures of the collaboration theory. Third, we need a *theory of modeling and implementation* that can be used to bridge the gap between the conceptual machinery of the information/collaboration theory, the interaction theory, and a concrete, dynamic computational realization that puts those theories to work. A theory of modeling and implementation describes how to address each of the conceptual elements of information, collaboration, display, and interaction with specific display and programming mechanisms. Figure 1 shows relationships among these three types of theory.

Theories of information and collaboration address conceptual modeling and control issues, while driving requirements for theories of display/interaction (i.e. specifying things to be displayed and interacted with) and of modeling and implementation (i.e. specifying what must be modeled and implemented dynamically). From a computational perspective, theories of modeling and implementation present representational structures and limitations for stating theories of information, collaboration, display, and interaction.

1.2. THEORY

By *theory* we mean a set of objects representing important concepts and a set of particular relationships among those objects. The set of concepts is the theory's *conceptual ontology*. For purposes of this paper, we can consider each of the theories we are interested in to have a formal and an informal component. Ideally, the informal component serves as a general statement of intent, and the formal component specifies precisely (that is, precise in relation to the language of a virtual machine that will interpret the theory) the important objects and relationships of the theory. If this is done, dynamic behavior and interaction may be flexibly achieved with a virtual machine that integrates and interprets the three theories. When the theory is varied, the corresponding formal part of the theory can be re-interpreted by the virtual machine, leading to new dynamic behavior and interaction. Such an approach has been taken in the ACTION organization analysis/design system, in which an organization theory, an interface theory, and a design process theory are integrated with a virtual machine, to allow theory-driven organization design and analysis, and to allow for easy integration of revised organization, interaction, or process theories [Gasser *et al.*, 93].

1.3. THEORETICAL PERSPECTIVES

In this paper we are primarily concerned with collaboration theories and modeling/implementation theories. If a theory includes a conceptual ontology, then different ontologies provide different (and possibly incommensurate) theories. In the arenas of information and collaboration, the substantive domains of interest for this paper, there are many alternative theoretical perspectives available, both for capturing the basic concepts of information and collaboration, and for the modeling and implementation theories which organize the operationalization of these concepts to create useful computational systems and dynamic behavior.

This paper is organized around two general categories of perspective on the theories of interest here: DAI-oriented perspectives and social/organizational perspectives. The paper investigates each of the theories (collaboration and modeling theories) from each of the two perspectives (DAI and social/organizational), to develop insights into the following basic questions:

- •What is the conceptual machinery that can be brought to bear to understand information and collaboration for computational systems?

- •How can this understanding be formalized, modeled, and implemented computationally?

- •What are the present limits of conceptualization, formalization, and implementation?

2. INFORMATION/COLLABORATION THEORIES FROM A DAI PERSPECTIVE

Distributed AI has focused on the problems of representing and reasoning about collective, coordinated activity among groups of large-grain agents. Typically, both agents and the community of agents as a whole embody processing architectures that reflect either a descriptive theory of human problem-solving and cognition, or a normative high-level approach to problem-solving. That is, DAI is concerned with representing and modeling processes of problem-solving at both individual and social levels, using computational methods. DAI researchers are especially interested in modeling problem-solving in which control decisions (decisions about what reasoning or problem-solving actions to take next) and representation decisions (decisions about how to form, stabilize, and utilize significant concepts) are taken dynamically by a community of agents itself, rather than by programmers. To do this, DAI research has addressed five basic issues [Bond and Gasser 88]. These five problems are inherent to the design and implementation of any system of coordinated problem solvers, and thus provide in part a set of foundational questions for DAI implementation and modeling theories. The problems include:

- How to formulate, describe, decompose, and allocate problems and synthesize results among a group of intelligent agents.

- How to enable agents to communicate and interact: what communication languages or protocols to use, and what and when to communicate.

- How to insure that agents act coherently in making decisions or taking action, accommodating the non-local effects of local decisions and avoiding harmful interactions.

- How to enable individual agents to represent and reason about the actions, plans, and knowledge of other agents in order to coordinate with them; how to reason about the state of their coordinated process (e.g., initiation and termination).

- How to recognize and reconcile disparate viewpoints and conflicting intentions among a collection of agents trying to coordinate their actions.

The primary focus of DAI has been to model coordination and collaboration among artificial agents.

Coordination among problem-solving activities has been a concern of computing for decades. Recently, Gelernter and Carriero have noted the ubiquity of computing ensembles collections of (possibly-asynchronous) activities including people, computational processes, and other ensembles. They have argued that computation (within ensembles) and coordination (between ensembles) are separate and orthogonal dimensions of all useful computing, and have proposed coordination languages as a class of tools for managing interaction. Moreover, "A computation language by itself is useless. A computation must communicate with its environment or it serves no purpose" [Gelernter and Carriero 92]. The environment of a computation clearly may include other computations, or people, and the shape of this interaction is the province of coordination, and the focus for coordination languages. They go on to define many common processes in computing (e.g. operating systems) as fundamentally coordination-centered activities.

DAI research has generated several approaches to achieving well-coordinated collaborative systems. These include treating coordination as distributed search, as settled and unsettled questions, and as organization design.

2.1. COORDINATION AS SEARCH

Lesser and his colleagues have viewed coordination in a cooperative distributed problem solving system as a matter of effective control of distributed search [Lesser 91]. Distributed problem solving can be viewed as distributed search, with attendant control and coordination problems, as follows:

- The space of alternative problem states can be seen as a large search space investigated by a number of problem-solvers.

- Each problem solver has to make local control decisions: Each problem solver has to make its own local decisions about what areas of the search space to explore, given the information it has at hand when the decision is made.

These local control decisions have impacts on the overall effort expended be the collection of problem-solvers. The local decisions, taken together across the entire group of problem solvers, focus the overall problem-solving effort through regions of the global search space. To the extent these regions are overlapping, some search has been duplicated and coordination has been suboptimal. To the extent the regions are larger than the optimal the search has been inefficient.

But what is control? We can define *control decisions* as decisions about what action to take next in a problem-solving process. We may term any knowledge that informs these decisions *control knowledge*. Each control choice is the outcome of an overall control regime that includes knowledge about 1) what are the control alternatives (what is the range of options from which to choose, along with a specification of the granularity of action---that is, what is the degree of change between control choice points), 2) what are the decision criteria used to choose among alternatives, and 3) what is the decision procedure that applies the criteria to the decision alternatives to make the control choice.

When we view coordination in cooperative distributed problem solving as a control problem, control choice becomes more complicated to the degree that control decisions are more:

- •Numerous (there is a greater number of control decisions being made simultaneously)

- •Asynchronous (there is less temporal coordination among control decisions)

- •Decentralized (control decisions are made in different places, which may lead to control-knowledge uncertainty or incompatibility across problem-solvers.)

Based on our definition of control as next-action choice, we can define control decision uncertainty as ambiguity in that choice. That is, the greater the ambiguity in control choice, the greater the control decision uncertainty. According to Lesser and his colleagues, distributed problem solving systems need two kinds of control.

Network control or cooperative control comprises decision procedures that lead to good overall performance by the problem-solving network as a whole, and that are based on network-level information. One type of network control involves the allocation of search-space regions to problem solvers, because this is allocation impacts the set of alternatives to consider in individual control decisions.

Local control refers to decision procedures that lead to good local decisions, and that are based on local information only. Local information concerns the status and progress of a single node in its own local environment and its own local search-space region.

To achieve network and local control, most current coordination approaches involve combinations of two strategies: reducing the degree and/or reducing the impact of uncertainty in local-level or network-level decisions. The impact of control uncertainty is felt in the arbitrariness of control decisions, and its effects are related to the density of goals in the search space. On average, for constant goal density, greater control uncertainty would be expected to increase search effort.

The impact of control uncertainty can be reduced by reducing coupling between the activities of separate problem-solvers. Coupling can be reduced by reducing common dependencies that problem solvers share, such as logical dependencies or and resource dependencies. Dependencies influence control decisions. For example, because one agent's set of viable control alternatives depends on another agent's actions, uncertainty may be increased or decreased by the control choices and actions of the other agent. Obviously, communication plays an important role in establishing the actual degree of control uncertainty, by increasing contextual decisionmaking information.

Within this distributed search formulation, there is a range of approaches to coordinating a collection of semi-autonomous problem solvers.

2.1.1. *Organization* is a network-level coordination mechanism. In most DAI research, a particular organizational arrangement usually specifies the distribution of specializations among problem solvers in a collection. In effect, organization specifies which classes of subproblem each agent will agree to address, and which classes of subproblem each agent will forego. Viewed this way, organization is a precise way of dividing the problem space without specifying particular problem subtrees. Instead, agents are associated with problem types, and problem instances circulate to the agents which are responsible for instances of that type. The distribution of problem types may be along a variety of axes, such as functionally-oriented or product-oriented dimensions (cf. [Malone, 1987]).

2.1.2. *Exchanging metalevel information* is another way that problem solvers can improve their coordination. Metalevel information is control-level information about the current priorities and focus of the problem-solver. Metalevel information exchange is imprecise, because it is aggregated and because it is indirect. It does not specify which goals an agent will or will not consider. Rather, it specifies on which goal types an agent will focus. It is also a moderate-time-horizon source of information, and as such, it reduces effective agent reactivity when it is the sole source for coordination information.

2.1.3. *Local and multiagent planning* are also useful coordination techniques, with a somewhat shorter potential time horizon. When agents generate, exchange, and synchronize explicit plans of action to coordinate their joint activity, they arrange a priori precisely which tasks each agent will take on. Plans specify completely a sequence of actions for each agent; they reduce control uncertainty to 1 at each choice point for the duration of the plan. This has the effect of temporarily converting a search process into an algorithmic process. Viewed another way, planning trades control uncertainty for reactivity by increasing the granularity of action from individual actions to entire plans.

2.2. COMMITMENT

There is a particular assumption of temporal consistency of actions underlying each of these coordination mechanisms, that we can term *commitment*. In effect, each coordination mechanism sets up a relationship between agents wherein one agent can "count on" the actions of another. In this way, all coordination mechanisms are founded on inter-agent commitments. When agents use explicit and fine-grained synchronization, each dependent agent carries out a blocking procedure that prevents it from acting while the decisionmaking agent is choosing. In effect, this blocking procedure is a commitment not to act on the part of the dependent agent. This blocking procedure is a procedure carried out by an operating system, but that is simply the mechanism by which the commitment is enforced. When agents generate and exchange plans or meta-level information, they are in effect making commitments to act in accordance with those plans or that information. When agents establish an organization structure of goal-type allocations, they are making implicit commitments about what types of activities they will pursue, that all agents can use to reduce their control uncertainty. Without the force of these commitments, each mechanism's coordinating impacts would be lost. The coordination value of plan exchange springs from the fact that the receiving agent can count on the sending agent following its plan---else why send it? The coordination value of organization structuring follows from fact that every agent abides by the organization structure. Agents do take on goals for which they are structurally responsible, and they do not take on other goals. Knowledge of this is what allows for reductions in control uncertainty. Unfortunately, further treatment of sources and mechanisms of commitment are beyond the scope of this paper, (see [Bond 90, Cohen and Levesque 90, Gasser 91, Shoham 91]) but it is an active area of contemporary DAI research.

2.3. EFFECTIVENESS OF COORDINATION

Coordination may refer to the process of structuring decisions so as to maximize overall effectiveness of a collection of problem-solvers. Or it may refer to the outcomes of a collection of control decisions.In either case, a collection of problem solvers is more coordinated if overall it performs better on some measure of performance (such as efficiency, solution quality, timeliness, or some combination.) For example, Durfee, Lesser and their colleagues have conceptualized degree of coordination as a kind of minimal wasted effort, in measuring the performance of their Distributed Vehicle Monitoring Testbed (DVMT) system [Durfee *et al.* 87]. They measure performance by counting the total number of knowledge sources (KSs) activated by all nodes in the collection as the system reaches a previously-known, optimal solution. The closer the collection comes to the known optimal number of knowledge-source activations, the less the wasted effort, and the better the performance.)

2.4. COORDINATION AS PATTERNS OF SETTLED AND UNSETTLED QUESTIONS

An intelligent agent faces a variety of choices about what to believe, what knowledge is relevant, and what actions to pursue. At any moment, some of these choices are settled, and some are open or unsettled [Dewey 38]. The settled issues need no further attention because the agent itself, some other agent, or perhaps a designer has removed choice. However, any settled question---e.g., a solution to a coordination or control problem--- may have to be unsettled in a new situation; the problem may have to be re-solved in a new way for the new situation (see also [Dewey16, Dewey38]). As above, settled questions provide a basis of commitment for agents to take action. They are a "loan of certainty" [Dewey 38] that allow for coordinated action, and that reduce uncertainty just as metalevel communication or other notions of organization treated above serve to reduce uncertainty.

For example, there are several levels of settlement in the well-known DAI pursuit problem [Gasser *et al.* 89] in which some number of "blue" agents trying to surround a "red" agent to keep it from moving on a 2D grid. These nested layers of settled questions include: problem instance existence, prey location, team constituents, team member roles, and local actions. Higher-level questions provide context for lower-level ones.

We can view organization itself from this perspective, if we view an organization as a particular set of settled and unsettled questions about belief and action through which agents view other agents. Said another way, an organization should not be conceived as a structural relationship among a collection of agents or as a set of externally-defined limitations to their activities. Instead, to achieve the simultaneous aims of decentralized control, no global viewpoints or sharing, and fidelity to observable human action, we can locate the concept of organization in the beliefs, expectations, and commitments of agents themselves. When agents experience "organization," they do not see global structures or fixed constraints on action instead they see interlocking webs of commitment (e.g., to prior but changeable settlements and patterns of action (e.g. routines of others being played out). Organizational change means opening and/or settling some different set of these questions in a different way, giving agents new sets of control decisions and new levels of certainty about them.

This viewpoint on organizational behavior and problem solving introduces additional constraint relationships among agents, beyond what have been discussed above. These include the constraints imposed on an agent by

- Its commitments to others, which reduce their uncertainty by settling some of their questions

- Its commitments to itself and to its own future actions

- Its beliefs about the commitments of others

In effect, any agent can use its expectations of the routine actions of other agents as fixed points with which to reason. The more organized a group of agents, the more their action can be based on these expectations or defaults; the expectations or defaults are the organization. Agents which participate in highly organized collectives have highly constrained actions, because most choices have already been made in the enactment of routines and encoded as default beliefs and the concomitant expectations of others' behavior.

To summarize, a collection of locally-settled questions provides a set of default expectations and commitments which constrain an agent's actions and provide a set of fixed points or (when taken together across agents) routines, which can be used for coordination. The fixed points can be called an organization, or possibly a coordination framework for the agents.

2.5. COORDINATION AS DESIGN

Coordination activities and coordination frameworks are as much a part of human organizations as of computational ones. In modern flexible manufacturing enterprises, for example, configuration of coordination for ensembles of people and automated production machinery is a key issue during planning, design, implementation, and operation [Majchrzak and Gasser 92]. Recognizing the difficulty of reasoning about coordination[1] in human-machine aggregates, the HITOP-A (Highly Integrated Technology, Organizations, and People-Automated) decision-support system has been implemented. HITOP-A is an automated knowledge-based design, decision support, and simulation system, developed to aid in the formulation and analysis of human infrastructures for computer-based manufacturing systems. HITOP-A incorporates a large collection of specific decision rules and heuristics, drawn from subject matter experts, the current theoretical literature, current best-practice approaches, and formal analyses. These rules predict in detail, for a wide range of contexts, technologies, and management values, a number of aspects of the human infrastructure needed to support a proposed technology. The experimental domain

---

1. As well as other issues such as job design, skill requirements, and performance management systems.....

for HITOP-A has been the human infrastructure supporting flexible manufacturing cells (FMCs), but initial investigations support the possibility that the domain knowledge embodied in HITOP-A generalizes to other types of technologies and business processes. In particular, HITOP-A coordination knowledge may be abstracted to reasoning about coordination (and other aspects) of DAI systems.

HITOP-A models coordination as forms and styles of communicative interaction between people within an interdependent workgroup (lateral), with a supervisor (vertical), and with units outside the immediate workgroup (external lateral or vertical). It is analyzed along dimensions of formality (whether coordination should involve formal and standard interactions or whether it should be flexible and adaptive), interactiveness (whether coordination should be based on multidirectional dialogues among members of a group, or whether it should be simply a unidirectional information flow), and speed (whether coordination should be reactive, real-time, and responsive to circumstances, or whether it should be periodic and "batched"). Several of the predictor variables for determining appropriate coordination structures include:

- *Uncertainty of tasks.* Task uncertainty affects the degree of the degree to which coordination practices can be standardized, and the attributes of coordination needed. HITOP-A designs more flexible and dynamic coordination structures as uncertainty increases.

- *Degree of workflow coupling.* The degree of coupling, can be measured as the amount of buffering allowable between tasks, and as other mutually constraining features. It affects the need for and type of coordination, as well as the breadth of knowledge needed for articulation.

- *Reciprocal interactiveness of workflow.* Reciprocal interactiveness, measured as the degree to which work is performed in a concurrent and interactive fashion (as reflected in a graph of tasks, goals and interdependencies), affects the degree to which work and coordination procedures can be standardized and the degree to which coordination regimes can be standardized and decoupled from the actual process of work.

- *Degree of decisionmaking discretion.* Discretion refers to the allowable decisionmaking latitude for a set of tasks. The needs decisionmaking discretion with people doing other jobs impacts how much coordination must occur and how tightly interdependent are the activities the coordination must support.

To compute coordination requirements, decision rules can be applied to sets of aggregated features of the organization and its task and information structure. These rules analyze the predictor variables to generate coordination attributes. For example, HITOP-A contains a rule that states that if cross-job coupling is loose and cross-job interactiveness is complex, then coordination should be informal, not standardized. Once coordination attributes are generated, HITOP-A can recommend mechanisms for achieving the appropriate coordination outcomes, via lookup tables. For example, the lookup table for coordination specifies a variety of ways to encourage different attributes of coordination. Interactive computer conferencing, initial co-location of team members, and frequent and informal meetings are all ways in which informal, immediate, and interactive coordination can be facilitated by management

Coordination is a key topic in DAI research. It can be approached from a number of different directions. Treating coordination as a problem of control in distributed search allows us to reason about mechanisms to reduce the degree and impacts of control uncertainty as a way of improving coordination. Viewing coordination as multilevel settlements helps to capture the context-dependency of coordination and to understand the requirements and scope of flexible coordination structures. Systems like HITOP-A hold promise for supporting the design and analysis of coordination structures in both automated (DAI) systems and in human organizations in which technology plays a role.

3. DAI IMPLEMENTATION THEORIES

Each of these control and coordination mechanisms requires and implies some mechanisms for modeling and implementation. Researchers have evolved a range of mechanisms for coordination representation and problem-solving. These can be arranged along a spectrum of control, autonomy and flexibility, as shown in Figure 1 (adapted from [Gasser 92b]). Regimes such as master-slave procedure calls are inflexible and centralized coordination regimes, specified by designers and enacted by computational processes by passing control threads. Approaches such as semaphores coordinate multiple-process access to common resources. They are temporally adaptive and flexible, specified by designers, and enacted by processes with a centralized arbiter. Moving down the spectrum of autonomy, we move from "designed in" coordination to coordination mechanisms that are opportunistically both designed and enacted by processes during their joint activities, and hence more responsive in both form and content to the character of those joint activities. In addition, as the distinctions between coordination activities and computation become blurred higher in the spectrum, the process boundaries and definitions of processes and process-ensembles become flexible, defined by the ongoing computations and interactions together.

Coordination Type	Degrees of Control and Adaptation
Explicit central control; Procedure calls (Master/Slave)	Explicit constraints; Centralized; common language constraints; Minimally adaptive under programmer specification.
Explicit synchronization and communication (Semaphores, Monitors, etc.)	Interaction constraints; Semi-centralized; common language constraints; Adaptive to temporal uncertainty

Table 1: Coordination Mechanisms and Degrees of Control

Coordination Type	Degrees of Control and Adaptation
Shared-data abstractions (Linda; Concurrent prolog)	Locally-centralized; Common language constraints; Adaptive temporally and spatially.
Functionally-accurate/cooperative (FA/C) approach; Triangulation and convergence on results	Opportunistic control; Fixed interactions; Adaptive to some semantic and temporal uncertainty; Locally centralized; Common language constraints.
Reasoned Control: Agents use knowledge of selves and others to build and revise coordination frameworks.	Predictions and adaptive interaction; Adaptive to more semantic, temporal, and interactional uncertainty. Minimal sharing; decentralized.
Evolving Interactions: Agents evolve during interaction; No shared semantics. Coordination is an emergent property of interaction patterns	Decentralized; Pragmatic (non-shared) semantics; Fully adaptive to semantic, temporal, and interactional uncertainty; Flexible balance of adaptation and stability.

Table 1: Coordination Mechanisms and Degrees of Control

Coordination in DAI most often refers to the process of control decisionmaking that guides the overall behavior and performance of a collection of problem-solvers. Coordination mechanisms are typically centered in the middle ranges of Table 1.

Typical DAI systems have used four approaches to capture and implement mechanisms of coordination and collaboration. Each of these modeling and implementation approaches stresses a different aspect of the information/collaboration problem space. They include:

- *Object-Based Concurrent Computation* (OBCP) which introduces concurrency, multiple control threads, and object autonomy into object-based computing paradigms. OBCP focuses on the definition and identity of objects and interactions among them, and less directly on control and coordination.

- *Blackboard and Distributed Blackboard Architectures* which provide interaction and control structures as well as shared, structured data areas for interaction

- *Integrative Systems* which allow for connection of heterogeneous and sometimes multi-grained problem-solving and conceptual modeling processes, but focus less on the definition of specific objects or specific modes of problem-solving coordination.

- *Experimental Testbeds* which focus on measurement, display, and experimentation aspects of coordination.

The most basic DAI modeling and implementation theories are based on the OBCP paradigm; others can easily be generated as special cases of that [Ferber and Briot 88, Ferber and Carle, 91, Gasser and Briot 92]. For handling representation and coordination issues, OBCP-based implementation theories grapple with a number of modeling and implementation problems, including:

3.0.1. *Identity.* One key problem is the problem of object identity, which can be further analyzed as three issues: the need for distributed object representations, the need for dynamic object representations, and the need for situated object representations.

3.0.2. *Communication/Interaction.* In OBCP all communication is uniformly done through message passing. Any kind of message may be sent. This provides uniformity of interaction but does not give any explicit information about the intention of the communication, nor does it support a theory of communication and interaction linked to a theory of individual or joint problem-solving activities. In the general OBCP formulation, it is the programmer who determines the semantics of messages and the shape of protocols, not the community of agents.

3.0.3. *Activity.* OBCP models activity by message acceptance and reaction. Primary issues are at what grain to implement activity---within an object or among objects, how to make the grain flexible, and how to link the grain to the identity of objects.

3.0.4. *Organization.* Composition and inheritance are the basic mechanisms used to structure a collection of objects. When the number of objects becomes large, it becomes necessary to organize them in larger entities because individual referencing of objects tends to be less manageable. Such grouping mechanisms have been proposed for some specific needs such as debugging [Honda and Yonezawa 88]. Configurations, and concurrent aggregates [Chien and Dally 88] have also been proposed as constructs to abstract groups of objects. However these simple organization mechanisms do not take into account the various articulations between objects inside a same group and between groups (e.g., master-slave) to structure coordination between them to achieve a common task. Most current conceptions of groups employ a representative or object that serves as a surrogate for the group.

4. INFORMATION/COLLABORATION THEORIES FROM A SOCIAL/ORGANIZATIONAL PERSPECTIVE

One point of entry into the study of aggregated multiagent systems is through the study of human work organizations, especially the study of workplaces in which people work together with machines. These work situations are interesting because they incorporate both automated and non-automated agents in collaborative activity, allowing us to consider both the adaptive complexity of real human behavior and the constrained performance of automated processes. There are many such workplaces. Several of the more interesting are scientific workplaces, in which people use fairly complex analytical and experimental tools as part of their work, and design/manufacturing workplaces, where technologies of varying complexity and rotundity are used.

Over the past fifteen years or so, I have studied several sets of problems that arise in scientific workplaces, in design/manufacturing workplaces, and in DAI systems, and that seem to be closely related [Gasser 86, Gasser *et al.* 87, Ishida *et al.* 92]:

- •How computing and human work integrate over long periods of time under approximately routine organizational conditions.

- •How processes of scientific inquiry unfold, generating robust ways of doing things over time despite mistakes, failures, uncertainty, and disagreement.

- •How collections of automated problem-solvers can organize and reorganize themselves as they adapt in changing circumstances.

These three problem sets seem to me to be related in that they all involve the following aspects [Gasser 86]:

- •Relatively persistent structures of knowledge and action, in relatively dynamic environment.

- • Active redefinition of conventional meanings (by which I mean regularized responses) and adaptive reshaping of action structures such as standard operating procedures.

- • Active integration of anomalies and unexpected events.

In addition, researchers are beginning to confront conceptual problems with current DAI approaches when trying to address some of the social aspects of agent behavior, for example:

- • When they consider individual agents' cognition situated in group and social contexts [Rogoff and Lave 84]. How does the group context impact the individual cognition?

- • When they consider groups (e.g., organizations) as loci of action or knowledge (cf. [Hutchins and Klausen 92, Weick and Roberts 92]).

- • When we consider knowledge and activity that is distributed over space, time, semantics, etc. [Bond and Gasser 88, Gasser 91, Gasser 92a]

- • When we consider open versus closed systems questions, such as creativity and the generation of fundamentally new forms of interaction and knowledge.

The socially-oriented conceptual model that we consider here derives from symbolic-interactionist sociology (e.g., [Blumer 69, Charon 79], and modern social studies of science and technology (e.g., [Latour 87]). The basic ideas of symbolic interactionism are that:

- • The primary units of analysis are interactions, not individuals.

- • Symbols are both the means and content of important interactions. The significance and meaning of symbols are in turn established through interaction.

- Individuals are dynamic and evolving agents with many components, not stable, structured, unitary personalities.

- The many context-dependent, continuously-evolving selves of an individual are shaped via dynamic selection and interpretation (as symbols) of stimuli in interaction.

- Actions are guided (but not determined) by perspectives, which are learned through communication.

- Society comprises individuals in processes of patterned interactions, embedded in, and always (re-)creating, perspectives.

The basic concepts addressed by social/organizational theories include facts and social facts, interactions, and organizations.

4.1. FACTS AND SOCIAL FACTS

There are several kinds of alternatives to the relatively conventional (in AI) conceptualizations view that facts are statements about the world known to be true in all contexts (e.g., possible worlds semantics of knowledge and belief). These include:

- the treatment of facts as continuously reinterpreted statements with dynamic "facticity"--statements repeatedly transformed, reinforced, and re-valued as they are incorporated in ongoing discourses (cf.[Latour and Woolgar 79]) so that their stable or reified character is a product of action, rather than a basis of action, and

- Durkheim's notion of "social facts." Durkheim described social facts as "ways of acting, thinking, and feeling" that exist outside individual consciousness, that are diffused widely within a group, and that exert "a coercive power" over the activities of individuals, "recognizable by the resistance that it offers any individual action that would violate it." He points out that when taking on certain social commitments, "I perform obligations which are defined outside myself and my actions...we are ignorant of the details of the obligations we must assume, and...to know them we have to consult the legal code and its authorized interpreters...the above statements will apply [to] each member of a society in turn" [Thompson 85], pp. 68-71. The point is that social facts reside in collectivities, not in individuals: "The determining cause of a social fact must be sought among antecedent social facts, and not among states of individual consciousness" [Thompson 85], pp. 86.

Individuals (as in individual statements or facts, individual actions, individual agents). Individual knowledge, performance and achievement has long been the focus of AI (cf. [Bond and Gasser88, Bobrow 91]). But much has been taken for granted. What is the nature of the individual agent? In what sense is it possible to conceive of an individual, carving one out of a continuous web of social interaction and involvement? Said another way, what aspects of individuals are not social facts? What is the boundary of any individual, in terms of action, time, knowledge, perception, etc., and how is our knowledge of these boundaries constituted socially or non-socially?[2] How do stable individuals emerge in the collective action of societies and organizations?[3] Gerson has presented a simple and cogent conception of individual as "something for which nothing else will substitute for each and every purpose" [Gerson 91] pg 1. He points out that any conception of an individual thing depends on a recognizer (who assesses the substitutability and differentiation of the individual thing), and that the ongoing process of recognition is subject to mistakes. We discover and correct these mistakes, in general, due to the restrictions on action that they entail. Suppose a medical-diagnosis knowledgebase is mistakenly loaded into a circuit-diagnosis system, and doesn't substitute. The diagnosis system does *something*--- maybe it beeps, crashes, or emits an error message---but it doesn't cooperate with its user in diagnosing circuits. Trying to treat a painter like a car mechanic won't work, because we depend on the painter's participation in fixing the car, and it's not forthcoming. In Gerson's words, "in specific local circumstances we live in a world of alliances which corrects mis-identifications" [Gerson 91] pg. 1. (cf. Durkheim's concept of the coercive power of social facts, mentioned above).

Thus, says Gerson, non-substitutability "is a function of cooperation and response from others; there is no single thing in general...We can reliably recognize something as an individual and as the same individual only if there is equivalence of criteria across recognizers [over time and place]. This is achievable only for very narrow purposes and for relatively short periods of time." Moreover, "some things [e.g., people] can actively manipulate the process of recognizing [by how they] anticipate and negotiate the criteria which others use to recognize them. They can insist on some criteria and rule out others....they can decide to be another individual, or [to be] individual in another way. When this happens we have things recognizing or constructing each other as individuals, the identity of each being dependent upon its cooperation with the identity of the other. In this situation, things demand recognition of their identities on their own terms as the price of cooperation" [Gerson 91] pgs. 2-4.

4.2. INTERACTIONS.

What is the nature of interaction among individuals? Do we need a clear and delineated conception of the individual in order to conceive of interaction? For example, once we have located the very nature of individual agenthood in social processes---once agents become social facts---against what ground are we to give semantics to messages which travel across time and place between agents? What are the boundaries of interaction? For instance, suppose a sending agent gives notice to a receiving agent that "a proposal will arrive in a following message." Where and when does the interpretation of the proposal

---

2. Once you think you have a clear answer to the boundary question, consider individuals as aggregates of parts---as (de)composable systems---and see if your answer holds up! See below.

3. For the uninitiated, the notion of individuals emerging in collective activity may seem strange, but is very real. Two examples: stable software processes built and maintained by software teams are very clearly individuals that continuously (re)emerge in social processes. Similarly, people are products of collective action in very physical ways---food, clothing, shelter, health care, etc. are all continuously and collectively (re)arranged, and the knowledge involved in these activities is no less so.

message begin and end? Does it begin with the notice message? Does it begin with prior messages, activities, and world states that over time generated the internal and external structures that allowed for interpretation of the notice message and subsequent assimilation of the proposal message? How do we separate the interpretation of a message from the activity and structures that establish the context in which it is interpreted? (cf. Gerson's note on the ways agents can influence their own substitutability and identity). It is certainly possible to set up very complex interpretation structures beforehand and to reduce interaction to sending a very small set of tokens, or even to sending none [Genesereth *et al.* 84]! It would seem that the nature of the boundaries of interpretation, hence of the meaning of "message," is contingent on the socially-emergent definition of "message."

4.3. ORGANIZATIONS AND GROUPS AS ACTIVE, COGNIZING, PERCEIVING, REMEMBERING ENTITIES.

How can concepts such as action, cognition, perception, and memory be conceived at an organizational level of analysis (cf. [Hutchins and Klausen 92,Weick and Roberts 92])? For each of these concepts, where is its locus, and what gives it its stability or pattern? How does organization emerge along with the collective action of individuals?

First, let us consider the issue of aggregation: how to "put together" collections such as knowledge-based processes, "agents," and/or people into an organized whole, and how to have them act together in response to some higher-order phenomena--that is a phenomenon at the level of the whole, not at the level of the components. What would this look like?

We can think of aggregation a having the following four aspects. First, there must be some identifiable entities that are put together. Second, these entities must be individually responsive to some environmental circumstances on their own---there must be a way of talking about them as individuals, with respect to some class of environmental stimuli and substitutability criteria. Third, there must be some mechanism or process that welds them together into an ongoing unit that exhibits some routineness, stability, or pattern. Fourth, this higher-order unit must itself respond in some patterned way to some qualitatively different class of stimuli, such that the overall response of the aggregate is different from the response of the individual units. That is, the group of individuals will not be substitutable for the aggregate with respect to the ongoing aggregate-level environment.

From this description we can see there is some relationship between the interconnecting process and the class of higher-order stimulus that defines the nature of the aggregation. We can also see that the defining characteristic of an aggregate is:

- •that it is a higher-order patterned response, which means that the interactions among parts must also be patterned to some degree, and

- •that the identifying character of the aggregate is determined in part by the character (and level) of the stimulus and response; in effect, the environment has a hand in defining whether something is an aggregate or not. (This is in line with the previous discussion of individuality as non-substitutability.)

Note that nowhere have we spoken of the members of the aggregate having any sort of (common) goals or intentions. To be identified as an aggregate vis-a-vis some observer or interactor, it is sufficient that there is an overall pattern to the members' collective activity in response to a class of stimuli, and we need not attribute to that pattern any notions of "cooperation" or "working together."

Overall, for thinking about socially-constituted knowledge and action, we would like to avoid the notions of goal and intention because we want to deal with multilevel aggregates at multiple and arbitrary levels of aggregation. In such structures, concepts such as goals and intention become problematic, because we don't have a clear idea of where to situate responsibilities (e.g. of parties for goals, when parties are aggregates) or how to allocate action (e.g., for achieving goals) when action is distributed and simultaneous.

5. SOCIAL/ORGANIZATIONAL IMPLEMENTATION THEORIES

Most DAI research works from the premise that some stable set of agents with stable architectural boundaries come together and coordinate their activities in the solution of joint problems. That is, a stable society of agents emerges from the constructive interactions of multiple pre-existing members. The primary problem, then, is how to design the individuals so that they can effectively coordinate when enlisted in joint problem-solving situations. Social roles and social-level effects are founded in individual action and knowledge.

There can be other points of view on the individual-society relationship, however. For example, we could imagine society as a collection of interactions, from which individual actors or agents emerge in response to social circumstance. The individual-society design problem then becomes 1) how to describe and manipulate the boundaries of agents, 2) how to flexibly aggregate and disaggregate agents in response to changing conditions in the society, and 3) how to give agents stable identities when their natures are changing.

In joint research with Toru Ishida and Makoto Yokoo, we designed and experimented with just such a system for organization self design [Ishida *et al.* 92]. In this system, a problem-solving society was conceived as a set of interactions among packets of knowledge. Each packet of knowledge was represented as a standard OPS-5 production rule, and interactions among them were represented as Working Memory Elements (WMEs) flowing from rule to rule. Rules had interdependencies with other rules that supplied or consumed WMEs. Rules also had interferences with rules that led to possibly-conflicting conclusions. An agent in this system was simply a mapping of some set of rules and WMEs to a locus of action - a production-system interpreter. By creating or deleting interpreters and changing this mapping, the population of agents could adaptively respond to changing environmental demands for solution quality or timeliness. In particular, the mapping was changed as agents either chose to compose or to decompose, as environmental conditions changed. Decomposition split an agent into two new agents, partitioning the knowledge and dependencies (a form of disaggregation), while composition joined two preexisting agents together into a new agent, combining their knowledge and dependencies (a form of aggregation).

This research demonstrated that flexible aggregation of agents with manipulable boundaries and dynamic identity could be effective in adaptive problem-solving. It opened a host of new questions about the epistemological and methodological status of agent boundaries, (dis)aggregation processes, and agent identity. It also raised the possibility of defining knowledge as organization---as the ability for a collection of processes to adaptively arrange themselves so as to accomplish some end---rather than as correspondences between statements in a language and states of a world or sets of possible worlds.

The key social/organizational ideas on which we have begun to base DAI system structure and implementation are the following.

- The notion of modeling other agents as the key integrating mechanism for organizational and coordination processes. This idea was drawn from Mead's theory of the unification of self and society through the processes of "taking the role of the other" and reflexive self-identity [Mead 34]. It was reflected in the MACE system as the acquaintance structure that was a part of all agents [Gasser *et al.* 87a].

- The idea that DAI systems should be multi-agent, possibly human-machine aggregates at all levels, which led to the approach of using the help of a community of system agents for constructing and interacting with a DAI system [Gasser *et al.* 87a].

- The idea that long-term problem-solving proceeded through a series of "frozen accidents" that became embedded, reified, and aggregated into stable structures of action and interaction [Gasser *et al.* 89].

- The idea of organizations as stable, overlapping, and nested patterns of action and knowledge---as patterns of settled and unsettled questions---rather than organizations as fixed structures of responsibility, communication, or control. For example, the MACE notion of organization was captured in the knowledge agents had about each other. The boundary of an organization was simply the boundary of knowledge about how and when to include other agents in particular problem-solving processes. The MACE representation of a single node of a contract net, for example was a four-agent aggregate in which each agent had highly restricted knowledge of with whom to interact [Gasser *et al.* 87b]

- The observation that social activity is inherently a multi-perspective process, and that disparities can occur at any level of description or context from many different points of view simultaneously. Hence, processes in practical DAI systems will be subject to observational and organizational dynamism and incongruity [Gasser 91].

- The ideas that the loci of action and knowledge in social systems are dynamically aggregated units, that aggregation must be modeled across numerous levels and from numerous perspectives simultaneously, and that in effect, agents can be construed as emerging from interactions, rather than vice-versa [Gasser 91, Ishida *et al.* 92].

- The importance of both prescriptive (i.e., targeted for explicit organization design goals) and exploratory (or emergent) approaches to generating organizational form and structure [Gasser *et al.* 93, Majchrzak and Gasser 92].

6. KEY ISSUES FOR INTEGRATING DAI AND SOCIAL/ORGANIZATIONAL VIEWS

There are a number of key issues for beginning to integrate social/organizational and DAI viewpoints. Here we focus on the modeling and implementation theory layers, since they place the primary limitations on thinking computationally about social/organizational collaboration theory.

6.0.1. *Encapsulation and Local Control.* Both to manage complexity in applications, and to provide a natural fit to the problems and domain requirements of DAI systems, object-based paradigms are common. Encapsulation is directly related to the conceptual issues of object boundaries, identity, and referencing, which become significant problems in the context of aggregated objects.

Fixed boundaries for objects do not reflect either the theoretical positions emerging in the DAI world or the reality of multi-level aggregations of action and knowledge (e.g. multi-level groups or organizations). From these viewpoints, agents are dynamically defined by reference to their changing position in a community, and by reference to their position in a frame of reference---the "same" computational units (which may be individuals or groups, depending on levels of analysis) may participate in different "agents" simultaneously. Finally, when agents are aggregated into dynamic groups, the issue of continuity of identity emerges---what maintains the stable identity of an "agent" when it is composed of ever-changing object definitions and patterns of interaction? This can be an important issue; Hogg and Huberman have shown the potential for chaotic behavior among agents with particular decision making procedures, and has illustrated how chaos can be ameliorated by making agents better predictors of each others' behavior [Hogg and Huberman 90]. However, such prediction depends upon the continuous identity of agents.

6.0.2. *Message-Based Communication.* Truly distributed systems require message-based communications for interaction. Message-based communication is related to encapsulation and to identity and referencing of objects, since messages must be directed to a receiver, and must be acted upon by an interpreter, which may itself be an aggregate.

6.0.3. *Heterogeneous Multi-Grain Objects.* Agents or objects may exist at different levels and types of granularity in a complex system. implementation platform must address the issues of mapping objects, for expansion/contraction of objects, and for flexible object composition implementation. Heterogeneity and multi-granularity are also problems of referencing, activity, identity, and composition, since heterogeneity may arise from alternative compositions or aggregations of objects.

6.0.4. *Language Support for Flexible Organization and Interaction Structures.* To maximally exploit concurrency, as well as to provide for flexible reconfiguration of the interactions among objects in an information system, higher-level modules may be composed of lower-level objects which are themselves capable of local coordination. The approach to design may be either reductionist/decompositional---explaining and implementing high-level behaviors in terms of lower-level concurrency by constraining the behaviors of lower-level components---or constructionist/compositional---aggregating lower-level concurrent modules into emergent higher-level structures with emergent behaviors, which may not be directly derivable from the descriptions of the lower-level modules. Flexible organization and interaction are also problems of referencing, activity, identity, and composition, since flexible organizations accommodate dynamic aggregation and thus encounter dynamic referencing, identity, and boundary issues.

6.0.5. *Agent Modeling*. Adaptive coordination, and especially coordination with limited communication, requires the ability to model the behavior and knowledge of other agents. Agent modeling involves problems of referencing, activity, identity, and composition, since aggregate agents may refer to models of other agents which are themselves dynamic aggregates. This of course raises issues of dynamic referencing, identity, and boundaries.

6.0.6. *Reusable Shells*. Practical information systems engineering will require the ability to reuse abstract descriptions of system components, and to reuse knowledgebases and coordination structures. Object-Based Concurrent Programming (OBCP) languages already provide useful structures for concept and code reuse, and to the extent these can be elaborated into high-level integrative shells, the construction of more robust and complex systems will be enhanced.

6.0.7. *Testbed and Measurement Tools*. Practical construction of information system systems also demands high-level simulation, measurement, and control environments, and development tools to build and manage large-scale systems. The difficult problems of shell and testbed construction involve how to incorporate support for dynamic analysis of object identity, composition and aggregation, boundaries, concurrent object activity, etc.

6.0.8. *Composition/Decomposition Transformations*. For purposes of load balancing, adaptive reorganization to fit new problem circumstances, (including adaptive clustering of processes, changes in data sources, etc.), and flexible control, it may be desirable to change the grain size of agents, and the control and knowledge relationships among agents, during the course of problem solving. This is only possible if the agents themselves are decomposable or composable. Thus, description languages must be available which allow for composition of agent-groups from individual agents, and treating agent groups as single agents at higher levels.

It is also important to be able to view the activities of individual problem solvers in a group at several levels of analysis. Just as social theorists conceptualize social action in numerous levels, such as individual, small group, organization, society, and so on, it may be useful to be able to analyze an intelligent distributed problem-solving process at varying levels in vitro. This can only be done using tools and analytic techniques which treat and conceive the problem-solving agents at differing levels of granularity, using observational mechanisms of decomposition and aggregation. Much more research needs to be done to address these compositional, decompositional, and representational mechanisms, but several existing approaches can provide suggestive frameworks for implementation.

6.0.9. *Meta-Level Architectures and Reflection.* Meta-level (also called reflexive or reflective) architectures---those unified architectures in which some part of the architecture objectifies, reasons about, and influences some other part---are advocated in many information systems contexts. This issue is deeply related to the primary social/organizational problems of modeling other agents as a basis for social coherence.

6.0.10. *Dynamic Interpretation and Language.* We would like to see computational systems that modify both their knowledge and their activity structure at all levels of analysis---i.e., communities of programs that evolve the languages in which they are written. FOr example, we might define and demonstrate social mechanisms of dynamic category formulation, classification, and concomitant reification---the active formation of agreed-upon basic concepts and their use in joint interpretation and discourse processes. Such social mechanisms would be those in which categories, classification activities, reifications, structures, etc. were subject to Durkheimian social coercion processes.

7. SUMMARY AND CONCLUSIONS

This paper has treated problems of information and collaboration from the standpoint of a collection of interacting theory types and a collection of interacting theoretical perspectives:. The theory types considered are first, theories of information and collaboration, and second, theories of modeling and implementation. We have left untreated the aspect of developing specific theories of display and interaction (but see, e.g., [Gasser and Korner 90]). Each of these theory types has been investigated from two perspectives:

- •The perspective of distributed artificial intelligence, which treats information as facts and propositions, and treats coordination as distributed search, commitment, settled/ unsettled questions, and as a design problem.

- •A social/organizational perspective, which treats information as structured patterns of activity, and which treats collaboration and coordination as issues of continuous symbolization and negotiation.

Theories of implementation and modeling for DAI theories and social/organizational theories have been compared, and some suggestions made for structuring more flexible and socially-oriented implementation theories to support more socially-dynamic approaches to information and collaboration.

The argument in this paper has been first that we might need to incorporate greater sociability into machines, and second, to propose some more directly social angles for thinking about the machine/human ensembles that we do work with. I'd like to suggest that with computers as partners, we have several opportunities to explore alternative theoretical models of sociability and culture, namely, the varieties of society and culture that emerge among collections of semi-autonomous machines and people-machine ensembles. I suggest seriously treating these as alternative, model cultures and societies, to learn more about how far our current conceptualizations of computation, as well as of culture and society, go.

8. ACKNOWLEDGEMENTS

The ideas in this paper are the product of several years of discourse in an informal, distributed community of people. Phil Agre, Jean-Pierre Briot, Elih Gerson, Ann Majchrzak, and Leigh Star, have been particularly influential. In addition, I thank Shimon Nof and the organizers and funders of the NATO ARW on Information and Collaboration models for their motivation and support of this paper.

9. REFERENCES

[Blumer 69] H. Blumer, *Symbolic Interactionism: Perspective and Method*, Prentice Hall, Englewood Cliffs, NJ, 1969.

[Bobrow 91] Daniel G. Bobrow, "Dimensions of Interaction," *AI Magazine*, Fall, 1991.

[Bond and Gasser 88] Alan H. Bond and Les Gasser, "An Analysis of Problems and Research in Distributed Artificial Intelligence" in Alan H. Bond and Les Gasser (eds.), *Readings in Distributed Artificial Intelligence*, Morgan Kaufmann Publishers, San Mateo, CA, 1988.

[Bond 90] Alan H. Bond, "Commitment: A Computational Model for Organizations of Copperating Intelligent Agents," in *Proceedings of the 1990 ACM Conference on Office Information Systems*, Cambridge, MA., April, 1990.

[Charon 79] Joel Charon, *Symbolic Interactionism: An Interpretation, Evaluation, and Critique*, Prentice-Hall, Englewood-Cliffs, NJ, 1979.

[Chien and Dally 90] A.A. Chien and W.J. Dally, "Concurrent Aggregates," *Symposium on Principles and Practice of Parallel Programming*, March 1990.

[Cohen and Levesque, 1990] P.R. Cohen and H. Levesque, "Intention is Choice with Commitment" *Artificial Intelligence*, 42:3, 1990.

[Dewey 16] John Dewey, *Essays in Experimental Logic*, Dover Publications, New York, 1916.

[Dewey 38] John Dewey, *Logic: The Theory of Inquiry.* Henry Holt and Company, New York, 1938.

[Durfee *et al.* 87] E.H. Durfee, V.R. Lesser, and D.D. Corkill, "Coherent Cooperation Among Communicating Problem Solvers," *IEEE Transactions on Computers*, C-36, pages 1275--1291, 1987.

[Ferber and Briot 88] J. Ferber and J.-P. Briot, "Design of a Concurrent Language for Distributed Artificial Intelligence," *International Conference on Fifth Generation Computer Systems (FGCS'88)*, Vol. 2, pages 755--762, Icot, Tokyo, Japan, November-December 1988.

[Ferber and Carle 92] J. Ferber and P. Carle, "Actors and Agents as Reflective Concurrent Objects: a Mering-IV Perspective," *IEEE Transactions on Systems, Man, and Cybernetics*, 21(6), November/December 1991.

[Gasser 86] Les Gasser, "The Integration of Computing and Routine Work," *ACM Transactions on Office Information Systems*, Vol 4:3, July 1986, pp. 225-250.

[Gasser *et al.* 87a] Les Gasser, Carl Braganza, and Nava Herman, "MACE: A Flexible Testbed for Distributed AI Research" in M.N.Huhns, Ed., *Distributed Artificial Intelligence, Pitman*, pages 119--152, 1987.

[Gasser *et al.* 87b] Les Gasser, Carl Braganza and Nava Herman, "Implementing Distributed Artificial Intelligence Systems Using MACE," *Proceedings of the Third IEEE Conference on Artificial Intelligence Applications*, pages 315--320, 1987.

[Gasser *et al.* 89] Les Gasser, Nicholas Rouquette, Randall Hill, and Jon Lieb. "Representing and Using Organizational Knowledge in DAI Systems," in L. Gasser and M.N. Huhns, eds., *Distributed Artificial Intelligence, Volume 2*, Pitman, 1989.

[Gasser and Korner 90] Les Gasser and Kim M. Korner, "Human Interfaces and Distributed Intelligent Systems," *Proceedings of the FRIEND21 Conference on Next Generation Human Interfaces*, Oiso, Japan, October, 1990.

[Gasser 91] Les Gasser "Social Conceptions of Knowledge and Action," *Artificial Intelligence*, 47, January/February 1991.

[Gasser 92a] Les Gasser "Boundaries, Aggregation, and Identity: Plurality Issues in Multi-Agent Systems," in Y. Demazeau and J-P. Muller, eds., *Decentralized AI 3*, North Holland, New York, 1992.

[Gasser 92b] Les Gasser, "DAI Approaches to Coordination," in N.M. Avouris and L. Gasser, eds., *Distributed Artificial Intelligence: Theory and Praxis*, Kluwer, 1992.

[Gasser and Briot 92] Les Gasser and Jean-Pierre Briot, "Object-Based Concurrent Programming and Distributed AI," in N.M. Avouris and L. Gasser, eds., *Distributed Artificial Intelligence: Theory and Praxis*, Kluwer, 1992.

[Gasser *et al.*, 93] Les Gasser, Ingemar Hulthage, Brian Leverich, Jon Lieb and Ann Majchrzak, "ACTION: Computational Analysis and Design of Organizations", COD Research Memo 23, Computational Organization Design Lab, Institute for Safety and Systems Management, USC, 1993 (submitted).

[Gelernter and Carriero 92] David Gelernter and Nicholas Carriero, "Coordination LAnguages," *Communications of the ACM*, 35:2, February, 1992.

[Genesereth, et al. 84] Michael Genesereth, Matthew Ginsberg, and Jeffrey S. Rosenschein, "Cooperation Without Communication" Stanford University Computer Science Department Technical Report HPP-84-36, September, 1984.

[Gerson 91] Elihu M. Gerson, "Individuals," Paper presented at the AAAI Spring Symposium on Composite Systems Design, Stanford, CA, 27 March 1991. Available from Tremont Research Institute, 458 29th Street, San Francisco, CA. 94131 (tremont@ucsfvm.ucsf.edu).

[Hewitt 91] C.E. Hewitt, "Open Information Systems Semantics for Distributed Artificial Intelligence," *Artificial Intelligence*, January 1991.

[Hogg and Huberman 90] T. Hogg and B.A. Huberman, "Controlling Chaos in Distributed Systems," Technical Report SSL--90--52, Dynamics of Computation Group, Xerox Palo Alto Research Center, Palo Alto, CA, 1990.

[Honda and Yonezawa 88] Y. Honda and A. Yonezawa, "Debugging Concurrent Systems Based on Object Groups," *ECOOP'88*, LNCS, No 322, Springer-Verlag, August 1988.

[Hutchins and Klausen 92] E. Hutchins and T. Klausen, "Distributed Cognition in an Airline Cockpit," in D. Middleton and Y. Engestrom, eds., *Communication and Cognition at Work*, Cambridge University Press, 1992.

[Ishida *et al.* 92] T. Ishida, L. Gasser, and M. Yokoo, "Organization Self-Design of Distributed Production Systems," *IEEE Transactions on Data and Knowledge Engineering*, 4(2), pages 123--134, 1992.

[Latour and Woolgar 79] Bruno Latour and Steve Woolgar, *Laboratory Life: The Social Construction of Scientific Facts*, Sage, Beverly Hills, CA., 1979.

[Latour 87] Bruno Latour, *Science in Action*. Harvard University Press, 1987.

[Lesser 91] Victor R. Lesser, "A Retrospective View of FA/C Distributed Problem Solving," *IEEE Transactions on Systems, Man, and Cybernetics*, 21:6, pp 1347-1362, November/December, 1991.

[Majchrzak and Gasser 92] A. Majchrzak and L. Gasser, "HITOP-A: A Tool To Facilitate Interdisciplinary Manufacturing System Design," *International Journal of Human Factors in Manufacturing*, 2(3), pages 255--276, 1992.

[Malone 87] T. Malone, "Modeling Coordination in Organizations and Markets," *Management Science*, 33:10, 1987.

[Mead 34] George H. Mead, *Mind, Self and Society*, University of Chicago Press, 1934.

[Rogoff and Lave 84] B. Rogoff and J. Lave (eds.) *Everyday Cognition: Its Development in Social Context*, Harvard University Press, Cambridge, MA, 1984.

[Shoham 91] Y. Shoham, "Agent0: An Agent-Oriented Language and its Interpreter," *Proceedings of the National Conference on AI* (AAAI-91), pages 704--709, 1991.

[Thompson 85] Kenneth Thompson, *Readings from Emile Durkheim*, New York: Tavistock Publications Ltd., 1985.

[Weick and Roberts 92] Karl E. Weick and Karlene H. Roberts, "Organization Mind and Organizational Reliability: The Case of Flight Operations on a Carrier Deck," working paper presented at the Workshop on Adaptive Processes and Organizations: Models and Data, Santa Fe Institute, February, 1992.

PROFESSIONAL WORK, COOPERATIVE WORK, MEETING SOFTWARE:
A PRACTICAL VIEW[1]

MICHAEL SHARPSTON, The World Bank, Washington, D.C. USA

Introduction: The Nature of Professional Work

To decide how we can aid professional work with new technologies, we first need to consider the true nature of that work. In the context of an office for professional work (which could be any organization — military, commercial, or industrial — but excluding organizations doing straight volume-processing such as paying insurance claims), the single most important process is probably professional interaction. It is through professional interaction (often cross-disciplinary, increasingly often multi-cultural) that complex assignments largely acquire their professional value added. An office is very much a *social* environment, and sociolinguistic and ethnographic research shows that "social maintenance work" — greetings, incidental chat, informal discussions — and the transmission of substantive information are often inextricably interwoven, even in the same utterance.[2] Staff are always engaged in the joint production of meaning, especially with those belonging to another group as they work to understand each other's points of view and even the words they seem to have in common (for example, words such as "file" and "archive" mean significantly different things to a computer specialist and a records management specialist or archivist).

Indeed, office studies [3] show that the higher up an organization you go, the less time is spent in writing and the more in direct, interactive communication (for example, telephone conversations, conferring with a secretary, scheduled and unscheduled meetings): for upper management this can be 40 to 60 percent of total working time. The ongoing adoption by senior managers of electronic mail and voice mail could slightly lower these percentages for interactive communication, but because face-to-face communication is the richest communication medium in terms of total (verbal and non-verbal) information conveyed in a time period, it is likely to remain a favourite with senior managers.[4]

Professional interaction requires "hard" communication (the creation and interchange of facts, data, information, and ideas, and their subsequent production as projects or publications) as well as "soft" communication: information on how to get things done, for example, or who has which attitude on what matter and why. The quality of communication determines how effective these interactions — formal and informal, social and professional — will be at conveying information, building an

---

[1]   I am deeply grateful to Ms. Laura Goodin for her substantive editing work, and I should also like to thank Ms. Karen McGraw of Ventana Corporation for her help in tracking down some difficult references.

[2]   For a beautiful analysis of how this is true even in an apparently unsophisticated working environment, see Eleanor Wynn, *Office Conversation as a Communication Medium*, Ph.D. Thesis, University of California at Berkeley, 1979.

[3]   There are surprisingly few good studies of how time is spent in offices, (which as suggested in the text may also be changing because of new technologies). Two of the most frequently cited studies are Harvey L. Poppel, "Who needs the office of the future?" in Harvard Business Review, November-December 1982, and G. Engel, J. Groppuso, R. Lowenstein, and W. Traub, "An Office Communication System", IBM Systems Journal, Vol. 18, No. 3, 1979. For a good review article, see Raymond R. Panko, "Managerial Communication Patterns", in Journal of Organizational Computing, 2(1), 95-122, 1992.

[4]   Mintzberg is associated with some of the best work on the real nature of a manager's work. See for example Henry Mintzberg, "The Manager's Job: Folklore and Fact", originally published in the Harvard Business Review, July-August, 1975, reprinted in James Brian Quinn, Henry Mintzberg, Robert M. James, *The Strategy Process: Concepts, Contexts, & Cases*, Prentice Hall, 1982, pp.22-31.

263

S. Y. Nof (ed.), Information and Collaboration Models of Integration, 263–268.
© 1994 *Kluwer Academic Publishers. Printed in the Netherlands.*

institutional culture and an institutional memory, and allowing staff to benefit from their colleagues' experiences and insights.

Trends in Professional Work

In the management literature, there is general agreement that organizations need to become flatter, with ad hoc working groups and/ or the more permanent work "clusters" that work together more flexibly as needed, supported by the appropriate information technology.[5] Ideally, one can even build technology infrastructure to support new patterns of communication: thus Claudio Ciborra is aiming to facilitate peer-to-peer communication internationally between research, development, and marketing units in a multinational pharmaceutical firm, in an attempt to shorten the total product innovation and development/marketing cycle.[6] This would go along very well with the evidence that it is the informal organization, at least as much as the formal organizational structure, which is responsible for innovation.[7]

In fact, the successful introduction of groupware — that is, computer-supported cooperative working tools — is no easy task. Unfortunately, Charles Grantham is probably right when he quotes a groupware designer thus, "Show me a group of people who are working in a cooperative way first, then we will build some computer technology to support that work. Don't do it the other way round."[8] Creating cooperative working practices may well involve changes in organizational cultures: both intangible factors such as staff members' perception of the power of organizational boundaries, and such tangible personnel devices as reward systems. Indeed, one writer argues that a major advantage of investment banks over their larger competitors, commercial banks, is their experience in network management, including the associated reward system which supports flexible working together.[9]

Truly cooperative work results from staff interacting through all the means they have at their disposal: scheduled, unplanned, formal, informal, one-to-many, one-on-one, technology-assisted, face-to-face. Each form of interaction serves a different function: an organization and its staff can derive benefit not only from scheduled meetings but unscheduled as well.[10] (Indeed, the Chief Executive of Scandinavian Airline Systems had a new corporate headquarters designed primarily to foster professional interaction, including serendipitous meetings. It includes a commons area for each

---

[5]   See for example Peter F. Drucker, "The Coming of the New Organization", in <u>Harvard Business Review</u>, January-February 1988, or Henry Mintzberg, "The Adhocracy" in James Brian Quinn, Henry Mintzberg, Robert M. James, *The Strategy Process: Concepts, Contexts, & Cases*, Prentice Hall, 1988, pp. 607-626.

[6]   For his general philosophy, see Claudio Ciborra, *Tecnologie di coordinamento: Informatica, telematica e istituzioni economiche*, FrancoAngeli, 1989.

[7]   Among the most interesting articles on this topic is D.M. Muslim-Zade, TE.E. Imamliev, and CH.M. Gajiev, "On the Theory of Fundamentally New Organizations (Some Sociopsychological and Organizational Aspects)," *Soviet Journal of Psychology*, 1986, Vo. 7(2), p. 245-256. The article is in English.

[8]   Charles E. Grantham, "Can Groupware Get Serious?", in <u>Research Highlights</u>, February 11, 1991 (featuring Intelligent Document Management), New Science Associates, pp. 765-770.

[9]   Robert G. Eccles and Dwight B. Crane, "Managing Through Networks in Investment Banking", in <u>California Management Review</u>, Fall 1987, Vol. XXX, No. 1.

[10]   Personal communication, Dr. Tora Bikson, Senior Research Scientist, RAND.

particular set of offices, and a covered "Main Street" with a pavement cafe between office buildings.) Yet the scheduled meeting remains the most recognized means of professional interaction in many organizations, and, as such, can be a useful place to begin exploring the benefits of cooperative work and the role of technology in assisting such collaboration.

Scheduled Meetings

Scheduled meetings can be of many different kinds: getting to know each other, information dissemination, persuasion/negotiation, team-building, conflict resolution. Indeed, the same meeting can serve different functions at different moments. Various attempts have been made to modify the meeting process so that participants can submit and consider a broader range of suggestions, reactions, and other contributions. With "nominal group technique"[11] for example, each participant writes down an answer to the issue before the group: this is intended to foster independent thinking, and keep a few dominant speakers from carrying the flow of discussion in a particular direction before others have had a chance to contribute their own ideas. If we are writing down our ideas separately but simultaneously, I cannot censor myself in order to avoid setting out an opposing idea to my boss. If all the ideas, including contradictory ones, are then set out on a flip chart, it becomes acceptable to consider a wide range of possibly contradictory ideas. The ideas are "over there" on the flip chart, less attached to the originator; an idea is more readily considered on its own merits rather than the status or personality of its author, and a criticism of them is less a criticism of him or her.

Meeting Software and Its Practical Impact

The advent of computer technology has introduced a number of ways to assist participants in accomplishing their goals during meetings. The benefit of information technology can start even before the meeting begins: electronic mail can facilitate establishing an agenda and ensuring that staff come to the meeting suitably prepared. But the potential for better meetings extends far beyond the rather limited interaction that electronic mail provides:

- With electronic brainstorming the scope for parallel input from all participants permits a much more rapid accumulation of ideas, and can keep good ideas from getting lost in a flow of discussion.

- The computer can also offer anonymity — perhaps not perfect if participants can recognize each other's style (although it may also be a disastrous gambit to claim to know who has authored a remark when in fact it was authored by another). Anonymity can be useful as a disinhibitor, allowing participants to say something overtly which everyone knows but has been reluctant to discuss.

- Other tools such as electronic voting can have important uses that are not immediately obvious: even in organizations where consensus is the rule at meetings, voting could be used to ascertain on which aspects of a complex issue there was already de facto agreement, thus

---

[11] The classic work on nominal group technique, (also published in 1975), is Andre L. Delbecq, Andrew H. Van de Ven, David H. Gustafson, *Group Techniques for Program Planning: A Guide to Nominal Group & Delphi Processes*, Green Briar Press, 1986.

saving time and effort for discussion of the remaining aspects. The voting tool in this case would serve to open and organize discussion, not to close it.

- An important benefit of the use of meeting software is that there is a record immediately available of the meeting, which can serve also to document decisions and the allocation of assignments.

There is some presupposition behind the use of such meeting software that attention *should* be paid to ideas separately from their originators, and that a more free flow of ideas is better than a less free flow. But reality in an organization is typically that some protagonists are more powerful than others; also, a "good" idea about how I should perform the function with which I am charged may not prove so "good" if I am sufficiently hostile to it. Even the disinhibition effect of anonymity could also prove counterproductive. However, there is scope to use meeting software in a way which allows for power and role realities. The phase of "topic analysis", when ideas originally brainstormed are consolidated, in any case tends to respond to more conventional meeting pressures such as hierarchical relationships among participants. And it is possible to use meeting software to allow wider input without necessarily moving to "democratic" decision-making: provided these "ground rules" are made clear in advance the process is honest and may be very beneficial.

It should be noted that using meeting software does not remove the importance of skills in facilitating a meeting; rather, it augments the options available to a good facilitator. For example, if electronic brainstorming has just brought out some uncomfortable truths previously not admitted to open discussion, the facilitator might be the one to recognize and suggest that this might be the moment to break for drinks and very informal chat for group mutual reassurance before proceeding further — rather than continuing the meeting electronically at all.

Judging the Benefits of Meeting Software

An important issue is how to judge the "success" or impact of meeting software, beyond an immediate novelty effect. It's difficult to write software that can orchestrate the many functions a meeting serves; the electronic brainstorming tool is probably the aspect of meeting software which offers the easiest and most evident benefit. Among the items one would like to judge would be:

(a) SPEED AND QUALITY OF DECISION-MAKING

The empirical evidence on the effects of meeting software are curiously mixed. Experimental research produces very ambiguous results, but actual users in real-life situations tend to be much more positive, often attributing to meeting software very high savings in time, as well as improved decisions.[12] On the whole, it seems more appropriate to go with the results of field use in

---

[12] One excellent source, with a first-class bibliography, is Alan R. Dennis, Joey F. George, Len M. Jessup, Jay F. Nunamaker, Jr, Douglas R. Vogel, "Information Technology to Support Electronic Meetings", *MIS Quarterly*, December 1988.

For an enthusiastic experience, see T.I. Dallavalle, Alicia Esposito, Steve Lang, (Bellcore), "Groupware — One Experience", presented at The Fifth Conference on Corporate Communication, Fairleigh Dickinson University, Madison, New Jersey, May 20-21, 1992. Another is Ron Grohowski, Chris McGoff, "Implementing Electronic Meeting Systems at IBM: Lessons Learned and Success Factors", in *MIS Quarterly*, December 1990. A third is Brad Quinn Post, (Boeing), "Building the Business Case for Group Support Technology", *Proceedings of the 25th International Conference of the System Sciences, January 1992, Vol. 4 pp.34-45.*

corporations: in this, as in other matters to do with communication, students probably are a rather different species to executives in middle life. Their expectations, motivations, and, hence, reactions may well be different. On the particular issue of decision speed, one might note that there is a possible interaction between the time to reach decisions and the speed of — and commitment to — implementation (for example, Japanese corporations are notoriously slow on the first and good at the second), so that speed in decision-making itself needs to be put in a broader context (discussed below).

(b)   EFFECT ON GROUP RELATIONSHIPS

For groups in which participants continue to have power and influence for a long time, the effect on group relationships is clearly important. Indeed, in an organization such as the World Bank, where individuals typically continue to run across each other for years in successive different roles, effects on relationships between individuals can have an importance that outlives even a particular work group.

(c)   EFFECTS ON IMPLEMENTATION

It is all too easy to assume that decision-making automatically implies corresponding implementation. However, computer-assisted meeting software is on the whole likely to have a favourable effect on implementation, since it can permit more people (perhaps the next echelon of managers) to participate — or at least to be present — when a decision is taken, which tends to increase at least comprehension, and optimally, commitment as well.

On the whole the World Bank team evaluating software for computer-assisted meetings judged that the software might expand the range of ideas considered, facilitate frankness in discussion, improve record-keeping at meetings, and tend to improve commitment to implementation.

For the future, there are two clear trends which meeting software needs to follow. First, software needs to be responsive to the fact that a meeting is typically only an episode in a office task, not a stand-alone event: office members are usually involved with this same task both before and after a meeting. Second, we can expect meetings increasingly where some of the participants are present in the flesh, others there on a video wall or similar device. We need software and devices which can aid this new kind of meeting. At a more mundane level, there is also a need for better integration with presentation tools, including multi-media.

Because the variety of meeting situations is so great, the complexity of attempting to design useful hardware and software to support meetings is formidable (indeed, a good vendor offers a palette of tools appropriate to different settings).

The Broader Picture

It is probably unrealistic to have exaggerated hopes merely because it is increasingly feasible technologically to devise the means for computer-supported cooperative work. It is however true that appropriate use of technology *can* act as a catalyst for organizational change, if one is aware of the limitations that an ambient organizational culture enforces at a given moment. In many organizations,

---

There are also very positive newspapers accounts; for example, Claudia H. Deutsch, "Business Meetings by Keyboard", *The New York Times*, Sunday, October 21, 1990, featuring Greyhound Financial and the U.S. Army among other organizations. Jim Bartimo, "At These Shouting Matches, No One Says a Word", in *Business Week*, June 11, 1990; and Chris Kraul, "Anonymity Makes Electronic Boardroom Work" in *Los Angeles Times*, Tuesday, November 6, 1990, cover similar ground.

electronic mail has had a considerable impact in breaking down barriers to horizontal communication, for example, and facilitating informal communication.[13] It could well prove that meeting software will work out in retrospect to have been a rather good way to introduce an organization to more extensive use of groupware of many kinds: meeting software may have both more immediate appeal and a lower "entry" requirement than some other computer-supported cooperative work tools, yet also serve as an introduction to this way of working.

---

[13] Electronic mail has already had a substantial impact on the way work is done in the World Bank. See Tora Bikson, Sally Ann Law, and Michael Sharpston, *Electronic Mail Use at the Bank: A Survey and Recommendations*, ITF Staff Paper Number 9, Information, Technology and Facilities Department, The World Bank, May 1992.

CM^3, Looking into the Third and Fourth Dimensions of GDSS

Gavish, B., Gerdes, Jr., J., Sridhar, S.
Owen Graduate School of Management
Vanderbilt University
Nashville, TN, 37203

Abstract: The area of Group Decision Support System (GDSS) is a relatively new field. To date, most research has focused on small group activity using a Decision Room concept. 'Computer Mediated Meeting Management' (CM^3) is a new GDSS system designed to address the challenges of meetings decentralized in time and space. Design principles, operational characteristics and experiences gained from sessions using the CM^3 system are discussed. We also discuss the fundamental issues related to decentralized meetings including mechanisms to aid in identifying conversational threads in brainstorming sessions. Unique features of the CM^3 brainstorming and voting modules are presented.

1 Introduction

The area of Group Decision Support System (GDSS)[1] emerged over ten years ago. Today, hundreds of GDSS facilities exist both in academia and industry. IBM, for example, has over 30 rooms installed on its premises. Current GDSS rooms require significant capital investment, ranging from hundreds of thousands to millions of dollars per decision room.

To date most research and development effort has focused on synchronous, single site, small group activity using the Decision Room concept [4, 17, 26]. In contrast, the areas of decentralized, asynchronous, and very large group GDSS have not been as fully investigated. *Computer Mediated Meeting Management* (CM^3) is a new GDSS designed to address the challenges introduced by these additional dimensions.

In Section **2** we present an overview of Group Decision Support Systems – classification schemes, implications of decentralized and asynchronous GDSS, issues related to the role of the facilitator and aspects concerning anonymity. Section **3** provides an introduction to the Computer Mediated Meeting Management System. Sections **4** and **5** look at the brainstorming and voting modules as implemented in CM^3. Section **6** gives a summary and planned research directions

---

[1] For those unfamiliar with the concepts behind Group Decision Support Systems, DeSanctis and Gallupe [6], Dennis, et. al. [3] , Kraemer and King [23], and Nunamaker, et. al. [26] provide a good introduction.

269

S. Y. Nof (ed.), Information and Collaboration Models of Integration, 269–299.
© 1994 *Kluwer Academic Publishers. Printed in the Netherlands.*

2   Issues Relating to Group Decision Systems

DeSanctis and Gallupe [6] define Group Decision Support Systems as "an interactive computer-based system that facilitates the solution of unstructured problems by a set of decision makers working together as a group." In the following sections we look at various issues related to GDSS. Section 2.1 focuses on the various taxonomies proposed to categorize the GDSS field. Sections 2.2, 2.3, and 2.4 address the consequences of extending GDSS to support space and time dispersed meetings. Section 2.5 delves into the role of the facilitator and how this role is affected by extended GDSS systems. Section 2.6 looks at the implications of anonymity on participant performance. Finally we present in Section 2.7 the problem and issues of rewarding individuals for their contributions during a GDSS session .

2.1   GDSS TAXONOMIES

Several useful taxonomies have been proposed to categorize both the body of research addressing GDSS as well as the capabilities and focus of the systems themselves. Dennis et. al. [3] proposed a taxonomy which segmented the study into three dimensions – Group Size, Group Proximity, and Time Dispersion (see Figure 1). Instead of time dispersion, DeSanctis and Gallupe [7] focused on the task confronting the group in their taxonomy (see Figure 2).

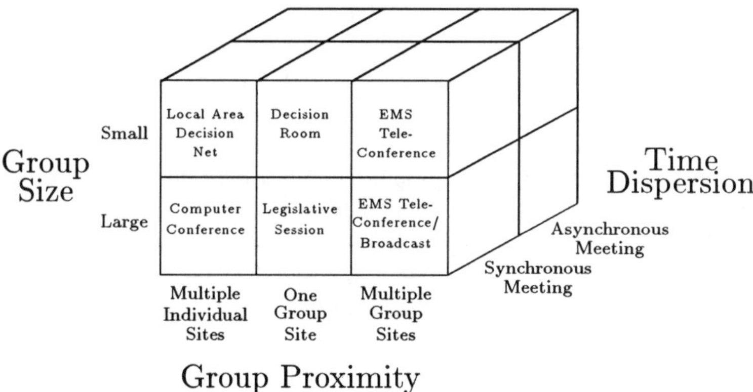

Figure 1: GDSS Taxonomy relating Group Size, Group Proximity, and Time Dispersion (Dennis et.al. [3])

DeSanctis and Gallupe [7] define three Levels of information exchange support: *Level 1* GDSSs provide mechanisms aimed at removing the barriers of communications.

Level 2 systems incorporate structured support techniques, such as mathematical models. *Level 3* systems integrate protocol and meeting-management-oriented expert systems into the GDSS.

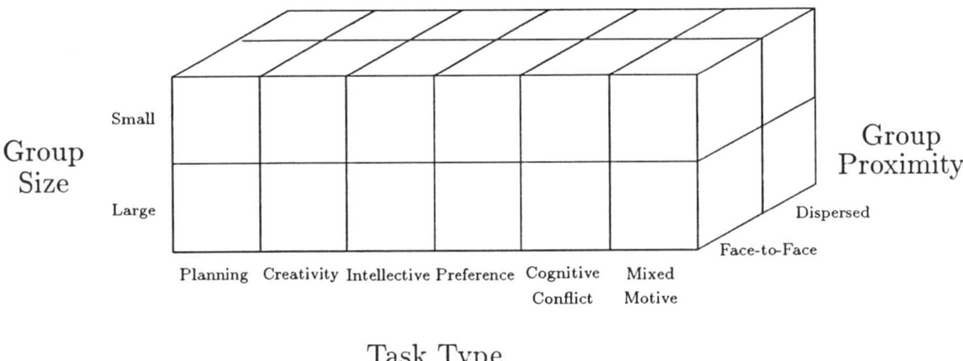

Figure 2: GDSS Taxonomy relating Group Size, Group Proximity, and Task Type (DeSanctis [7])

Over the past ten years researchers have studied various aspects of computer supported GDSS systems, often with contradictory results (see Dennis et. al. [4] for a review of recent GDSS research). The cause for so much contradictory evidence could be that these studies did not consider all of the important independent variables. We feel the study of group decision support systems must be expanded to include at least nine dimensions as discussed below. Note that proposed dimensions contain those included in both Dennis et. al.'s [3] and DeSanctis and Gallupe's [7] taxonomies.

I) Group Size: Existing taxonomies consider *Small* meeting as those involving less than 10 participants, and *Large* meetings involving 10 to 30 participants. This view of group size is restrictive, for it does not consider the needs of meetings involving hundreds or even thousands of participants.

II) Task Type: In addition to those tasks identified by DeSanctis [7], we include tasks which involve multiple decisions, surfacing of emotions, and linked decisions and actions, where decision (action) impacts other decisions (actions).

III) Group Proximity: Meetings can be fully decentralized where each participant is at a separate location; partial decentralized using multiple, linked meetings rooms;

or centralized, where all participants are at one physical location (see discussion in Section 2.2).

IV) Time Dispersion: Independent of the groups proximity, a meeting can be either synchronous or asynchronous (see discussion in Section 2.3).

V) Group Structure and Composition: The structure of the group (i.e., Peer, Hierarchical) can influence individual interaction and group activity. A group might be composed of highly educated individuals, those of less educational background, or a mixture of the two. Groups can consist of multi-national and multi-cultural participants.

VI) Time Horizon: Different mechanisms are needed for decision processes that take hours versus processes which take days, weeks or months. Although related to the time dispersion dimension it does represent a different aspect of a system or problem. The time horizon for the meeting could be a one month time period, but the protocol may require all participants to meet in a synchronous session. Similarly, the time horizon could be relatively short, just a few hours, but participants can have the flexibility to interact when it is convenient for them.

VII) Anonymity: A system can support different levels of anonymity In addition to Full anonymity and No anonymity, a spectrum of limited anonymity is possible. Under limited anonymity, participants may be identified as belonging to a specific subgroup. Under this scenario, the system needs a mechanism which allows a user to prove a claim of authorship while mainta ning an anonymous environment (see Section 2.7). Anonymity can be influenced by e familiarity of participants within the group. Word selection, spelling, gramma nd comment temperament can all give clues to the identity of the author. These ues are intensified the more familiar participants are with each other.

VIII) Meeting Control: Some mechanism is needed to control the administrative functions of the meeting. This can be accomplished either through a central facilitator, group of facilitators, or distributing these functions to the group.

IX) Task Complexity: Depending on the difficulty of a particular task, different group performance can be expected. Easy problems will tend to garner a high level of consensus rather quickly, whereas difficult problems will typically yield more disagreement. Also, different tools will be needed depending on the complexity of the task.

2.2 DECENTRALIZED GDSS – THIRD DIMENSION

The ability to hold decentralized meetings, where participants are located at multiple distinct sites, yields a number of benefits. Many of these benefits relate to the added flexibility and improved anonymity gained with decentralized meetings. Studies indicate a reduction in participant satisfaction with decentralized meetings [12, 19]. However, there are situations where it is not possible to hold centralized meetings due to the disperse nature of the participants.

We distinguish between decentralized GDSS and a computer conferencing system [21, 22]. Although computer conferencing can support many of GDSS features, it can't support multiple conversations to the degree supported in GDSS. The spontaneity gained with GDSS systems allow for immediate group reaction to the evolving meeting. This dynamic aspect of group decision support systems is illustrated in Section 4.1 paper with an example.

The various issues related to decentralized GDSS are discussed below.

Elimination of meeting related travel time and expense [25]. Remote participation reduces the expense of attending meetings. It essentially eliminates both monetary outlay (air travel, hotel and food expenses) and the non-productive travel time and fatigue normally associated with distant meetings. Remote participants do incur additional communication expense, but this is much smaller than conventional travel expenses. This lower cost (measured in both time and money) may make it feasible for some individuals to attend a meeting which otherwise is cost prohibitive.

Easier scheduling of meetings. The ability to schedule a meeting depends on the availability of all participants. As the number of participants increases, the potential number of conflicts increases exponentially. This is especially true when the meeting involves individuals at the higher levels of the organization. The higher echelons of the company (i.e., the president and vice presidents) have more time commitments than do the lower echelons (engineers and managers). This makes it much harder to schedule a conflict-free meeting involving the higher levels of the company. Allowing decentralized meetings decreases the number of constraints which must be satisfied (i.e., all participants need not be at the same site for the meeting).

Greater Participation. Participation in a centralized meeting is constrained by the meeting facility's size. Most existing GDSS's are designed to support centralized meetings, often in a Decision Room setting (see [4]). As a result, the size of the Decision Room limits participation, with most Decision Rooms supporting less than 30 participants. This constraint is eliminated in a decentralized meeting. Meeting size is now limited by the communication backbone. The technology of Wide Area Networks currently enables meetings involving large numbers of simultaneous, worldwide users.

Enhanced Anonymity. In single site meetings it is difficult to ensure anonymity. Keyboard activity, verbal communication (laughter, comments, tone of voice), and body language can give clues to the source of *anonymous* input. In decentralized meetings these non-verbal cues are not present.

Increased anonymity could be significant when cultural differences exist among the participants. Individuals from aggressive cultures may inhibit the participation of those from less aggressive cultures (see discussion in Session 2.6). Similarly, when involved in single site meetings, participants can be influenced by the appearance or traits (sex, race, age, etc.) of other participants. For example, a single site meeting involving a paraplegic would likely influence the tenor of a session addressing health care or handicapped workplace issues. The meeting may progress quite differently if all participants are attending remotely. Similarly, personal biases could emerge in meetings involving multiple ethnic groups, or religious backgrounds. When participants are physically dispersed, participant's personal appearance and physical traits no longer influence the group activity.

Reduced impact of influential participants. The presence of an influential participant in a centralized meeting can inhibit interactions of other participants. This effect was observed during a CM^3 brainstorming session involving a group of employees and a member of upper management. While the manager was in the room, the employee group was reserved in their comments. At our request the manager left the room and joined the meeting from a remote site. Although the employees were aware that the manager was still interacting in the meeting, the tone became more open and free flowing with more substantive issues being addressed.

Verification of identity. Ensuring only authorized attendance in centralized meetings is relatively easy through the use of passwords and an isolated communication network. When the meeting is decentralized, added security is needed to prevent unauthorized access the meeting. There is the potential of an individual masquerading as either a valid participant or as the system host. Some mechanism is needed to prove the identity both the user and remote host when absolute security or anonymity is required.

Greater need for system self-containment. Centralized meetings can make use of additional support mechanisms. For example, large format displays [11, 25, 26], flip charts, multi-media presentations and verbal communication are all possible. Other mechanisms must be utilized when all participants are not at the same location. Ideas and information normally conveyed through these tools must now be communicated through the GDSS system.

Increased chance of partitioned meetings. A remote participant (or group of participants) can become disconnected from the rest of the meetings. Depending on the activity, it may be desirable to permit the disconnected participants to continue work on the focus problem and later integrate and resolve differences between independent sessions.

Partitioning of participants may be consciously done. Within a large meeting, subgroups may develop to address specific tasks. For example an election committee could split into Platform, Strategy, and Fund raising committees. The results of each committee will be later be used together, so it is desirable that they are associated, but their immediate tasks are unrelated. Partitioning the committees would allow control over participation and focus attention within each committee on their assigned task.

The issues above represent many of the positive aspects of decentralized meetings. There are limitations imposed by this approach. System security and participant anonymity is harder to ensure. Even it is possible to provide a secure, anonymous system, it may be difficult to convince users that their comments are not being monitored or can not be intercepted. An interesting problem is when a member of the group is asked to leave a session. In a centralized session, that individual can simply leave the room, but in a decentralized session, how do you ensure that individual is not involved in the session when anonymity issues prevent the system from identifying the actions of specific users? These are all important research issues.

2.3 ASYNCHRONOUS GDSS – FOURTH DIMENSION

Conventional meetings are held *synchronously* – all participants attending at the same time. Although meetings may span extended time periods with fluctuating attendance, in general there is a high percentage of participants involved during any given time in the meeting.

Computer and technology mediated meetings enable *asynchronous* sessions where each participant can 'attend' a meeting on a schedule which is most convenient to that individual. The various issues related to asynchronous GDSS are discussed below.

Easier scheduling of meetings. Similar to space dispersed GDSS systems, the ability to schedule a meeting with flexible attendance times will reduce scheduling conflict making it easier to schedule meetings. However in an asynchronous meeting, what is meant by a schedule? Meeting schedules can now be flexible, with participation allowed over a span of hours, days or weeks.

Effects on Meeting efficiency. The immediacy of feedback and the dynamics of debate will be affected when all participants are not simultaneously present in the meeting. Immediate reaction reduces filtering and participant inhibition.

Prioritized Meetings. Participants will necessarily make time in a busy schedule for high priority meetings, but it may be difficult to schedule low priority meetings. An extended, asynchronous meeting would allow participants to interact whenever individual schedules permit. Several CM^3 sessions extended over a multi-week time period. Participants entered comments at various times over that period, at their convenience (sometimes at night and on weekends).

Flexible meeting attendance. During a single site meeting, some participants will exhaust their comments sooner than others. Either some individuals wait for the rest of the group finishes their input, or the activity is terminated thereby losing the input of those not yet finished. In asynchronous meetings it is up to each participant to determine the value of his or her continued attendance.

Facilitator function a bottleneck. In same room meetings, facilitator plays an active role in facilitating a meeting. Requiring a facilitator adds an additional level of abstraction to the decision making process. This can have either a positive or negative effect. The facilitator can offer procedural guidance to the meeting, with the quality of that guidance directly impacting the meeting effectiveness. On the other hand the facilitator can become an inhibitor by restricting the free flow of activities and ideas. In asynchronous meetings, either a separate facilitator or a distribution of the facilitator's duties to the participants is required. By providing the required tools, it is hoped that the facilitator's duties can be assumed by the group, giving the group more direct control over the meeting. (see further discussion in Section 2.5).

Capturing time phase information. For some activities, the time phasing of participant's activity may have significance. For example, a multi-media presentation may trigger comments by the participants. Capturing the temporal relationship (the time phasing) between the comment and the multi-media presentation may have added significance to the comment, for it places it in context. To capture this information, it may be useful to enable quasi-real time interaction for asynchronous meetings.

Time phased meeting. To simulate a synchronous meeting it would be possible to time phase an asynchronous meeting. In such a protocol, all user interaction is time phased relative to the beginning of the meeting. Whenever a participant enters the session, all previous input is replayed capturing the temporal relationship of the meeting. The new user can interact with the system, add new comments and see *new* comments being added just as if all participants were simultaneously interacting with the system. In a decentralized environment, the remote participant would theoretically be unable to distinguish this time phased meeting from a synchronous meeting.

Minimum group sizes. In conventional, synchronous meetings, some minimum number of participants is required. Each participant is aware of not only the number of participants, but also the identity of all participants. This may not be true with asynchronous meetings. In decentralized GDSS meetings, participants have the flexibility to enter and leave the session, and so the specific participants involved will change over time. Meeting dynamics will necessarily be different when only one individual is active in the meeting at any given time rather than a group of individuals.

The issues above represent the benefits of asynchronous meetings, but there is an obvious drawback. There is less continuity and spontaneous interaction when all participants and not involved simultaneously. But the ability to have an asynchronous meeting makes it possible to have meetings where otherwise it is not possible. Take for example a meeting involving personnel assigned to some critical function, like power plant operators. Since there always must be one shift on duty, it is not possible to have a synchronous meeting involving all plant operators.

2.4 DECENTRALIZED, ASYNCHRONOUS GDSS

Additional issues emerge when considering a GDSS system supporting meetings occurring over an extended time period. Such a system is needed to support an electronic global conference, possibly on thinning of the atmospheric ozone layer, providing disaster relief, or addressing some critical problem within an international company.

Many of the problems and issues identified for Time Dispersed and Space Dispersed GDSSs are magnified when both are supported. Time phasing becomes more difficult due to communication delays. decentralized meetings will tend to improve the anonymity, but the timing (time when entered) of some user's responses may give a clue to the input's source. For example comments entered very late in the evening could indicate the author is in a different time zone.

Meetings distributed across time zones. In the limit, distributed meetings can be global in nature, involving participants from multiple time zones. Holding synchronous meetings would be difficult due to the local time differences. Allowing asynchronous meetings reduces the impact of time conflicts, making it easier for individuals to participate.

Handling Communication Delays. How should comments be time phased when participants are exposed to different subsets of meeting input? This is similar to the problem experienced in distributed database systems when they become partitioned. Continued processing within a partitioned environment can cause data inconsistencies. Mechanisms are then needed to resolve the differences in the partitioned data. Similar inconsistencies can result a GDSS environment when different users enter comments based on different base states making it impossible to generate a partial ordering of the comments.

Ordering of comments is not a problem in fully synchronized meetings. Comment order in this environment is maintained by providing a consistent view of the meeting to each participant.

2.5   THE ROLE OF THE FACILITATOR

In a GDSS system, the facilitator performs the administrative duties of the meeting chairperson. The facilitator's role is to dynamically gauge the progress of the meeting, keep the meeting focused on the stated objectives, address the group's procedural problems, and in general keep the group moving forward in the analysis of the target problem. As new issues are raised, the facilitator issues new topics to elicit comments on these issues. The facilitator administers the meeting protocols, including how much time is spent on a particular activity, and the choice of specific protocols. Some protocols may force all participants to address a specific topic while other protocols may allow participants to freely choose their own topic area. The facilitator need not be knowledgeable (and generally is not) in the specific topic area being addressed, to perform this administrative function.

So, by definition, the facilitator is charged with administering the meeting protocols to expedite the decision process. Unfortunately, the facilitator may control the meeting agenda in a nonproductive way. In general, the facilitator will not be an expert on the meeting's subject area. While an expert in the subject area would recognize substantive interaction on a topic, the facilitator would not. As a result, the facilitator's control of the session is based on incomplete knowledge, and thus may not be very productive. In many cases the facilitator becomes an inhibitor instead of a facilitator.

The need for a facilitator in synchronous GDSS sessions (where participants attend at the same time) does not generally present a problem, although it does represent additional overhead. As GDSS systems move toward decentralized, asynchronous meetings, the requirement for a physical facilitator becomes a major system limitation. GDSS technology enable extended meetings, lasting days, weeks or even months. In such an environment participants can interact when it is convenient for them. If the facilitator is needed to provide full, system functionality, the support of extended meetings requires continuous staffing of the facilitator position.

2.6 IMPACT OF ANONYMITY

Most group decision support systems ensure participant's anonymity as part of the meeting protocol. However, anonymity can be a double edged sword. Anonymity can promote increased interaction allowing fuller investigation of the issues. Unfortunately, participants can hide behind the anonymity and act irresponsibly without fear of retribution.

By reducing potential costs of expressing unpopular opinions, anonymity can promote increased interaction. It negates external factors, such as the stature or position of participants, allowing each comment to be evaluated and debated on its own merits. Some studies have shown that participants may be more inclined to express critical comments than they would in a public forum [1, 18].

Anonymity can have a detrimental effect on group activity because it removes some of the checks and balances inherent in conventional meetings. It allows participants to act irresponsibly without fear of retribution [28]. In a brainstorming environment, it is possible for participants to enter false or inflammatory remarks without having to justify those remarks. Under some meeting protocols, these comments may not be discovered during the meeting. This is true when each participant is allowed to work on different tasks. Subgroups focus on a subset of the discussion questions and never see the comments entered under another discussion area. These comments will eventually be detected when the meeting summary is printed and analyzed, but this maybe some time after the meeting has completed.

In a non-anonymous meeting, participants may be less open and free flowing due to potential repercussions. Any and all statements made can be directly attributed to specific individuals. Libelous or inflammatory statements would likely be immediately challenged with requests for justification. These safe guards are not present in an anonymous system.

Both of these aspects of anonymity (increase in comments generated and irresponsible behavior) have been observed with the CM^3 system. During one particular session focusing on working condition issues, participants were very critical of some specific management practices. What was interesting was that upper management was not aware that there was employee concern regarding some of these practices prior to the meeting.

In this case the anonymity permitted participants to take part in what could be viewed as a politically risky activity (i.e., criticizing current management practices which had to that point not been aired). In another meeting, a group of participants executed a scathing attack on another participant, the full scope of which was not realized until the meeting summary was printed and reviewed. This graphically illustrated the risks of an anonymous system.

Other aspects of anonymity affect system performance. Due to anonymity, participants are not associated with comments and stances established during the session. This may make it easier for them to objectively evaluate alternative approaches to an issue without fear of *losing face* [25] or appearing indecisive. Also in the absence of anonymity, some individuals resist making comments for fear of appearing ignorant. System anonymity may diminish this fear, thereby promoting greater group interaction. Anonymity makes it easier for individuals to change their initial positions on an issue. Their early stance is not tied to them personally, thus making it easier to refine and pick an alternate position later in the meeting.

Group dynamics can be affected by the individuals which constitute the group – strong personalities tend to dominate their environment which may influence individuals with less aggressive personalities. This effect can be negated through decentralized GDSS meetings which enable both anonymous attendance as well as anonymous comment entry (see Section 2.2).

2.7 REWARD ALLOCATION PROBLEM

Anonymity does not necessarily increase discussion or the number of unique ideas generated. There is often value associated with good ideas. This could be in terms of recognition for the individual putting forth an idea, or in terms of actual monetary rewards. Participants may not be as open with their ideas if there is no way to obtain credit for their intellectual contribution. This is not a problem encountered in conventional meeting forums. In non-anonymous meetings, credit for an important idea is easily attributable to specific individuals.

In a GDSS setting, it is possible to identify the author of each comment by attaching some identification tag to each comment, thereby eliminating benefits of anonymity. Attaching a tag and simply suppressing its display ultimately defeats anonymity and thus we also deem this approach unacceptable. The challenge is to develop a mechanism which will allow participants to prove a claim of authorship without losing the benefits of anonymous interactions.

Any system devised must be able to prove claims of authorship. There may be multiple claims placed on a specific comment or set of comments. The reward process (and thus the process of claiming involvement) may take place long after the meeting has finished. As a result, it may be unclear in the participant's minds which comments they did make. It is also not enough to take a single claimant's word as to authorship just because other individuals do not contradict his claim. The person making a false claim may want to be associated with the comment, whereas the actual author may not wish to acknowledge his or her authorship. By contradicting the false claim, the actual author has to expose that which he or she does not want to expose. An even greater problem is that there are now two or more people claiming authorship. Without some mechanism to prove authorship, it is unclear who if anyone will receive the reward.

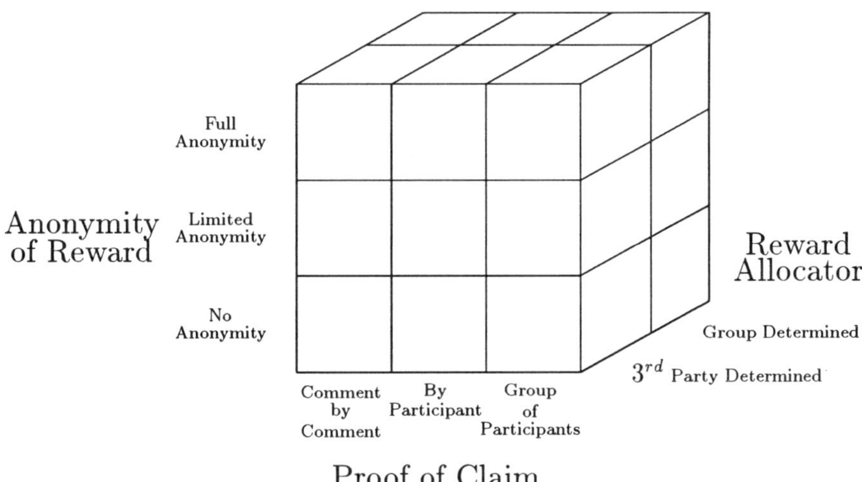

Figure 3: Reward Taxonomy relating: Reward Allocater, Proof of Claim, Anonymity of Reward

We have identified three dimensions required of any reward allocation mechanism: How are claims proven; Who allocates rewards; and What is the level of anonymity afforded individuals when collecting a reward (see Figure 3). The full scope of the reward issue is beyond the scope of this paper, it is addressed in a forthcoming paper [8].

3   Computer Mediated Meeting Management (CM^3)

Computer Mediated Meeting Management $(CM^3)^2$ is a term used to describe a computer and telecommunication based decision support environment designed to facilitate group consensus formation and enhance the group decision-making processes. The objective of the CM^3 project is to apply computing and telecommunication technologies to "value add" to a group decision process, not simply recreate the group's natural decision process within a computer setting. The CM^3 system is a prototype system that attempts to address many of the issues raised in Section **2.**

The current CM^3 system focuses on facilitating intergroup communication, thereby falling under Gallupe and DeSanctis's Level 1 classification [7, 11]. Unlike systems designed around the Decision Room concept [11, 25], CM^3 was designed to support the unique demands of decentralized meetings. Although the CM^3 system has successfully been used in single site meetings, it does not depend on facilities typically found in a Decision Room setting, such as overhead projectors, flip charts, and projection screens. Because it is based on a decentralized paradigm it is not constrained by the limitations of a physical meeting room.

One key CM^3 feature is the ability to support anonymous input of suggestions and comments. Laboratory and empirical experience suggest a leveling of a group's hierarchy results from anonymity, stimulating the generation of new ideas [1, 15, 18]. This significantly improves brainstorming sessions and, in many cases, improves the quality of decisions made by the group [1, 18]. Another benefit of using CM^3 is the strong emphasis on group activities and group reward and incentive systems, eliminating many of the pitfalls stemming from a conventional decision-making process.

CM^3 is a group decision support system designed to support both decentralized and asynchronous meetings. A prototype demonstration system has been developed and is operational at the Owen Graduate School of Management, Vanderbilt University. Over the past two years it has been used in over 200 sessions including Government Agencies, Civic organizations, Military, Academic, and Industrial groups. The system supports decision making processes that are very difficult to support by other methods. CM^3 is especially helpful for policy formation in complicated and fuzzy situations, for politically delicate questions, and for group consensus formation. The methodology supports all of the group processes of conventional meeting and at the same time, allows multiple, one-on-one exchanges of ideas, data, and opinions between meeting participants.

---

[2]The name CM^3 was coined by Ron Grohowski, then with IBM, after using the prototype decision support system at Owen Graduate School of Management.

CM^3 meetings are characterized by very high input intensities from meeting participants. The reason for the high input intensity stems from the fact that all participants can "speak" (through their keyboard) at the same time, while observing on their display all inputs provided by other participants. The participant can also move from topic to topic, focusing on those which are of interest to him or her. As the number of participants in a meeting increase, the system's parallel nature increases the overall input rate of new ideas, feedback, and issue analysis. Empirical experiments indicate that as the number of participants in a meeting increase, so does the relative advantages of the system: shorter meeting duration, greater exchange of ideas, and a higher quality of decisions.

3.1 CM^3 DESIGN PRINCIPLES

The CM^3 system was designed around three basic principles:

Support GDSS Dispersed in Space and Time
- The system should support sessions which are decentralized in time, space or a combination of the two.

Flexibility
- Move meeting control into the hands of the group, rather than depending on a centralized facilitator.
- Each participant should have the flexibility to control his or her local environment. This extends to the determination and scheduling of activities.

Anonymity
- Protocols should be available to support anonymous activities, even in a decentralized, asynchronous environment.

The ability to adequately support sessions decentralized in time and space greatly broadens the scope and potential benefits of GDSS. This goal prompted the user interface to be more monitor driven than some systems which depend on projection systems [3, 5, 24]. In a decentralized environment, the only means of communicating information is through the computer's display monitor (and possibly video and audio links), and so the support systems must be designed with this in mind. Because the participants are decentralized, users need to have the capability to generate any session related output locally whenever it is necessary.

The impact of a decentralized system extends far beyond the user interface. As discussed in Sections 2.2 and 2.3, a number of new issues, opportunities, and problems emerge when decentralized sessions are considered. New protocols and support systems are needed to address these issues and take advantage of the opportunities that a decentralized system affords.

3.2 CM^3 FEATURES

The CM^3 system supports many of the features found in other GDSS systems. These include: Support for centralized meeting (i.e., Decision Rooms); Anonymity of participants, reducing effect of shy and dominant personalities; High input intensity stemming from participants working in parallel; Triggering of new thought processes; Group memory provides for fully documented meetings, and the ability to support extended meetings. The merits and implications of these GDSS features have been discussed in many of the earlier papers [2, 3, 7, 11, 13, 15, 23, 26, 27, 32]. However, CM^3 contains a number of features not found in other systems. These aspects are introduced below.

- **Dynamic formation of subgroups within meetings.** Some existing GDSS systems support *breakout sessions* to address subtopics within the main session. In addition to this process, CM^3 supports natural, dynamic subgroup formation within the main session. Users individually determine which issues to address at any given time. This creates multiple conversational threads within the session. Since interaction is interest driven, the involvement in any conversational thread will be dynamic with members entering and leaving at their discretion. Compared to *breakout sessions*, anonymity is enhanced since the participants are not identified.

  Figure 4 illustrates the dynamic formation of subgroups within meetings. In this session, a participant interjected a comment off of the stated subject area, but was of immediate interest – *the location of a good place to eat.* A conversational thread quickly emerged and about 20 comments within two minutes. It was not a major disruption. There was a flurry of interest, the issue was discussed, and people moved on to other issues. Note that in Figure 4 , the comment numbers are not consecutive, indicating that other threads were simultaneously being addressed in the same meeting.

- **Real-time user feedback to the developers.** It is important to capture user feedback in real time, otherwise these immediate impressions may be lost. For example, a new user may have a problem activating a particular function within the system, however once he is shown the required key strokes, it becomes somewhat natural. A good example is the key to access lotus type menus (i.e., the backslash key). We have seen that if the user is asked for problem areas at the end of the session, these initial operational difficulties are often omitted. It is important to capture these initial impressions as they happen, rather than to wait until the end of the session.

The CM^3 system is designed to support electronic group meetings via a wide area network, eliminating the need for all participants to physically come to the same location.

Q17: In your opinion, what will be the major issues in telecommunications
in the next 5 years ?

. . .

24: ISDN and standard protocols

25: Managing high bandwidth networks

   29: Agree, but what are the major problems?

28: Anyone found a good place to eat around here at a reasonable price ... something different

    36: Try the sushi place next door, you can find anything you like around. Simply tell me
what type of food you like.

      38: Junk Food

        41: I agree

    42: I hear the South Street Smokehouse is good and not too expensive

    43: If you like a college town flavor, try Satco across the street.

      44: Regarding Satco across the street. Which Street?

        51: Do they serve brews? By the way, Eckerds is 3-4 blocks down on West End has a
wide variety of Imported Beers!

          54: Do miss Nashville's own Market Street ale – not bad at all!
            56: Where can I find Market Street Ale? On Market Street? Far from here?

          58: No, just look for a place with a good selection of beer.

    49: Granite Falls is pretty good. It is about a block from here.

. . .

Figure 4: Example of a spontaneous generation of a conversational thread. Comment 28 interjected an issue of immediate interest to the group, and was quickly address in parallel to other discussions.

3.3 DISTRIBUTING THE FACILITATOR ROLE

The CM^3 system attempts to distribute the facilitator function to the meeting participants. Consider a brainstorming session. At the meetings onset, the chairperson generates a set of topic areas for discussion. As individuals join into the meeting, they control their own resources. They have full control over the topics displayed on their monitor with the ability to move at will between Subject areas and topics within Subject areas. They also have the ability to generate any of the session printouts locally at any time during the session. This reallocates many of the facilitators duties to the individual user.

However, the facilitator role has not been totally eliminated. The facilitator is still the only individual who can modify the meeting protocol – i.e., issue new topic areas for discussion, force discussion on a particular topic, and terminate a particular activity. We are currently investigating ways of distributing these remaining functions either to the group as a whole, or to a subset of the group.

3.4 INVESTIGATING THE IMPACT OF ANONYMITY

Past research on the impact of anonymity has lead to contradictory result s. Connolly et. al. found an increase in performance with anonymous groups [1], whereas other researchers [14, 18, 20] found no difference.

Empirically we have observed the reported trend [1, 18] of participants expressing more critical comments in an anonymous forum than in a public forum. To get further insight into this phenomenon, a follow-up questionnaire was sent to each participant of anonymous CM^3 sessions. These sessions focused on different issues within civic, educational, and commercial enterprises. Each session contained a variety of issues, ranging from the mundane to very sensitive, complicated issues. The brainstorming sessions involved approximately 20 participants, generating in excess of 40 pages of comments. The follow-up form recounted all comments made during the session. Each participant was asked to rank the tenor of the comments, ranging from neutral to highly controversial. In addition they were asked to identify comments that he or she had made during the session. Accompanying the form was a letter explaining that the purpose of the questionnaire was strictly research related and assured each participant that the authorship of the various comments would not be divulged.

When analyzing the responses we noted that many comments remained unclaimed. Further analysis revealed that the set of claimed comments were limited to those comments designated by the group as having little controversial content. In all cases tested, there **was not one** instance in which someone identified himself or herself as the author of a comment ranked either controversial or highly controversial. Participants were only willing to claim authorship of less critical comments. The process of identifying authorship and ranking of comments was done after the CM^3 sessions had been completed. When making their own evaluations, participants did not know how the group would classify the comments. Thus, participants used their own judgement of the comments at the time they were indicating comment authorship. The decision to *not* claim comments ultimately viewed as controversial by the group was a personal, voluntary decision.

Based on the unwillingness of individuals to be credited with the more controversial comments, we concluded that many of these comments would not have been made had the interaction not been anonymous. Moreover, this limited study points to the fact that the importance of anonymity is content dependent. It is very important for charged issues, and less important for run-of-the-mill issues. This might explain the contradictory results in earlier studies. A more detailed study of this issue is in progress.

4 CM^3 Brainstorming Process

The Brainstorming Module is the central module in the CM^3 system, and is the focus of a forthcoming paper [9]. It provides an anonymous forum for a group to generate ideas, discuss issues, and develop alternative courses of action. The meeting organizers or coordinators prepare a short list of thought-provoking questions. They also provide the list of attendees which will participate in a meeting, and the initial meeting protocols to be used during the meeting. A meeting participant logs onto the system and is presented three windows on his display (see Figure 5).

- **A Question Window** in which a participant is presented a questions, one either from the set of question prepared by the meeting organizer or coordinator, or from questions interjected by other participants during the meeting. The participant is expected to respond to the presented question.

- **A Response Window** in which a meeting participant types his response to the questions. A participant can and is encouraged to provide multiple responses to the same question. He is also encouraged to comment on responses made by other participants. The referenced comments need not be visible to comment on them. Functions are available that allow the individual to present new questions to be analyzed by the meeting participants.

- **A Board Window** in which participants can scroll through and view earlier responses to the questions under consideration. The fact that the participant can see earlier responses reduces the number of redundant or duplicate responses. Comments provided by earlier respondents can trigger a participant to explore new possibilities that were suggested by others. The participant enters a response if he has something to add to the subject area. New responses are appended to the board in real time as they are generated by meeting participants.

An attempt has been made to make effective use of color throughout the CM^3 system. Color is used to differentiate aspects of the system from one another and give visual prompts to the user. There are plans to investigate the use of color to indicate content information. For example, a user profile could indicate interest areas. The system could automatically highlight comments which match the individual user's interest profile.

When a participant feels that he had enough with a given question, he can switch to a different question and repeat the commenting process. A user has the flexibility of scrolling through and reviewing previous comments, entering a response on the topic question, or commenting on a previous response. Each individual can move back

```
                    *  U S E R    S T A T I O N  *
 ┌────────────────────────────────────────────────────────┐
 │  1: Cellular communications will be big.                 │
 │  2: Expansion of available networks                      │
 │  3: Network Management system development                 │
 │  4: Managing High Bandwidth Netorks - Reconfigurable networks
 │  5: What do you mean by big (with respect to global traffic volume)?
 │  6: How do you think Clinton/Gore's proposal for the data highway will
 │     affect this?                                          │
 │  7: Organizational & Legal: more open and competitive markets
 │  8: What will be major research issues?                   │
 │  9: What type of legal implications: Security of transmissions?      BOARD WINDOW
 │ 10: It will delay investment and development of high bandwidth
 │     projects.  If I know that Big Daddy is going to invest in something
 │ 11: I meant regulatory                                    │
 │ 12: Squeezing the maximum capacity out of limited spectrum.
 │ 13: Why?                                                  │
 │ 14: What did Clinton/Gore propose?                        │
 │ 15: Have you got any suggested approaches for achieving this?
 │ 16: Creating the infrastructure for integrated services (i.e. multimedia)
 │ 17: If downstream service has a fairly stable frequency distribution,
 │     you can get substantial improvement by methods I've developed
 │ 18: Just balancing loads among routing options buys you something
 └────────────────────────────────────────────────────────┘

 ┌────────────────────────────────────────────────────────┐
 │ Q 1:   In your opinion, what will be the major issues in telecommunications in
 │        the next 5 years ?                                      QUESTION WINDOW
 │                                                          │
 └────────────────────────────────────────────────────────┘

 ┌────────────────────────────────────────────────────────┐
 │ Please enter your response (upto 3 lines) to the above question ;
 │ press ENTER after you complete the response ....              RESPONSE WINDOW
 │                                                          │
 └────────────────────────────────────────────────────────┘
  F1-Help   F2-Next Q   F3-Prev Q   F4-View Rsp   F6-Utility    F9-Quit  F10-Comment
```

Figure 5: The brainstorming screen is divided into three windows. The middle window gives the Question Statement. The top window gives the responses pertaining to the current Question. The Bottom window is used to enter new comments.

and forth between questions, and always can see how others responded to any given question, but never knows the identity of any respondent.

The system also supports many other functions. Direct communications between participants (either anonymously or not anonymously) allows them to form subsets within an ongoing session. It also supports communication between a participant and the facilitator. This can be used to suggesting new questions, initiating changes in the meeting protocol, or asking the facilitator for help.

During or at the end of the meeting a user can invoke a printout of all questions and responses generated by the meeting participants. Their printout is used both as a documentation device and as an aid for further study and analysis by the participants.

4.1 FINDING ORDER IN CHAOS

By its very design, the brainstorming process generates a large set[3] of comments having little natural internal order. In a GDSS environment, multiple participants

---

[3]Frequently, within a short period of time, 40 to 100 densely filled pages of comments are generated.

can simultaneously enter ideas and comments. Each participant will have been exposed to a different subset of meeting's comments. Due to difference in individual's interests and perspectives, brainstorming sessions quickly tend to develop multiple conversational threads. The initial discussion typically focuses on a particular issue, but when an interesting or controversial idea is put forth, a certain subset of the responses begin to address this new aspect, in parallel to the original conversational thread.

This triggering affect is an important aspect of GDSS systems, for it improves the depth with which a subject is investigated. This added depth, linked with the unstructured nature of the comment stream, can create a problem when analyzing the session output. Multiple threads can be difficult to identify. Also, without knowing to what it refers, the meaning of a comment may be misinterpreted. When a participant enters a comment, it may be in response to an entry which had just been read, in which case the associated comment is currently displayed. Since other participants are making comments simultaneously, the respondent's comment will likely be separated from its associated comment thereby losing the natural association obtained from a sequential listing. In general it is not possible to generate such an absolute ordering. Consider two independently generated comments Y and Z both triggered from comment X. If comment ordering is considered significant, two orderings are possible: X-Y-Z or X-Z-Y. Without some external linkage mechanism, there is no way of indicating that the last comment is associated with comment X and not with the second comment listed.

CM^3 uses three mechanisms to provide structure during and after of the brainstorming session. The first provides Subject Classifications to guide the discussion. A session can contain multiple Subject classes. Under each of these general topic areas are placed multiple discussion statements used to initiate the meeting dialog. This provides only a very coarse control of the meeting, because of eventual formation of multiple conversational threads under each Subject class and discussion statements.

The second mechanism allows the user to explicitly indicate any association with previous comments. Each comment is sequentially numbered. When a participant's comment relates to a previous comment, a pointer is prepended to the comment indicating the association. For example, the Figures 6 and 7 give a sample of a session dialog. In Figure 6 the comments are listed chronologically with the comment number separated from the comment by a colon. The initial comments are disjoint, reflecting the independent thought processes of the group. As the session progress however, the previous comments have triggered responses from the group. Comments 5 & 8 raise questions to clarify the statements expressed in earlier comments. In turn, a whole discussion arose based on the original statements. Simultaneously it appears

Q17: In your opinion, what will be the major issues in telecommunications in the next 5 years ?

1: Cellular communications will be big.

2: Expansion of available networks

3: Network Management system development

4: Managing High Bandwidth Networks - Reconfigurable networks

5: What do you mean by big (with respect to global traffic volume)?

6: How do you think Clinton/Gore's proposal for the data highway will affect this?

7: Organizational & Legal: more open and competitive markets

8: What will be major research issues?

9: What type of legal implications: Security of transmissions?

10: It will delay investment and development of high bandwith projects. If I know that Big Daddy is going to invest in something

11: I meant regulatory

12: Squeezing the maximum capacity out of limited spectrum

13: Why?

14: What did Clinton/Gore propose?

15: Have you got any suggested approaches for achieving this?

16: Creating the infrastructure for integrated services (e.g., multimedia)

17: If downstream service has a fairly stable frequency distribution, you can get developed

18: Just balancing loads among routing options, with a threshold control on each, buys you something.

Figure 6: Chronological listing of comments. Although comment 5 was made with reference to comment 1, this is not captured. It is difficult to discern conversational threads.

that other conversational threads take place. It is impossible to find sense in the material presented in Figure 6.

The chronological organization of the comments and responses captures the temporal relationship between statements and the various conversational threads. The disjoint nature of the comments comprising a thread makes it difficult to ascertain the whole meaning of each topic being discussed. The pointer system allows the comments to be organized in topical order, which captures the relationship between the session's comments (see Figure 7).

The third mechanism CM^3 uses to provide order to a brainstorming session is the ability to *spin-off* evolving threads into new discussion areas. Some conversational threads grow to the point that they become major topics of discussion. When this happens it is advantageous to extract this thread and create a new subject area, thereby improving the continuity of both discussion areas.

<table>
<tr><td colspan="2">Q17: In your opinion, what will be the major issues in telecommunications in the next 5 years ?</td></tr>
</table>

> 1: Cellular communications will be big.
>> 5: What do you mean by big (with respect to global traffic volume)?
>> 8: What will be major research issues?
>>> 12: Squeezing the maximum capacity out of limited spectrum
>>>> 15: Have you got any suggested approaches for achieving this?
>>>>> 17: If downstream service has a fairly stable frequency distribution, you can get substantial improvement by methods I've developed.
>>>>> 18: Just balancing loads among routing options, with a threshold control on each, buys you something.
>
> 2: Expansion of available networks
> 3: Network Management system development
> 4: Managing High Bandwidth Networks - Reconfigurable networks
>> 6: How do you think Clinton/Gore's proposal for the data highway will affect this?
>>> 10: It will delay investment and development of high bandwith projects. If I know that Big Daddy is going to invest in something
>>> 13: Why?
>>> 14: What did Clinton/Gore propose?
>
> 7: Organizational & Legal: more open and competitive markets
>> 9: What type of legal implications: Security of transmissions?
>> 11: I meant regulatory
>
> 16: Creating the infrastructure for integrated services (e.g., multimedia)

Figure 7: Topical session listing. Utilizing explicit references to previous responses, comments are rearranged making it easy to follow the conversational threads. Comment 5 is located under comment 1.

An alternate approach to *spinning off* threads would be to reorder comments topically when displayed to the user. A drawback of viewing the session in a topical format rather than chronologically is that participants loose the temporal dimension and may not be exposed to new comments. New comments are appended to the session log. If the display is chronological, the participant will be exposed to new comments from all threads. However, if the display is arranged topically, new comments are displayed with their associated thread. If an individual focuses on a particular thread, comments pertaining to other threads will be missed.

5 CM^3 Voting Module

In brainstorming sessions, ideas and concepts are freely generated, but there comes a time when discussion must end and a decision made. A mechanism is needed to permit participants to express their views in a more structured manner and converge on a plan of action. This is the purpose of the CM^3's Voting Module [10]. It provides a mechanism with which the group can evaluate and rank alternative solutions proposed during brainstorming sessions. To do this efficiently, the users can select from CM^3's various voting protocols. The voting module provides a framework within which the group can quickly reach a consensus on difficult issues.

- Referendum – Simple yes/no/abstain vote

- Multiple Choice, Select Multiple – Given a set of n options, select k

- Ranking – Rank the order of preference for a set of options

- Chip Allocation – Allocate available number of chips, or tokens, among multiple options.

The basic premise of the voting module is common to many of the similar modules found in other GDSS systems [13, 23, 26]. As a result, only the unique features found in CM^3 are discussed in the following sections.

5.1 CM^3's UNIQUE VOTING FEATURES

The Voting Module incorporates a few unique features. Anonymity is central to the concept of a secret ballet. The way it has been implemented in the past may be overly restrictive. Conventional protocols not only keep the votes of individuals secret, but they hide the intermediate group vote until the end of the voting cycle, supposedly so the voters are not unduly biased by the voting of others. This has been an issue in presidential elections with network news groups releasing east coast voting results before west coast states have closed their polls.

Unlike a brainstorming activity where one can argue on both sides of an issue without taking a stand, a voting activity requires the participant to commit to some position. In a referendum vote, that position can either be in favor of, opposed to, or abstain from the referendum question. If a participant is forced to state his or her position publicly, as is the case in non-anonymous votes, there is a tendency for that individual to maintain that position so as not to seem indecisive, or wavering. In an anonymous setting, the individual does not have to expose his position, and thus may be more willing to alter it as the issues become more clear. For example, assume there are eleven participants involved in a CM^3 session, each allocating 100 chips among 7 alternatives (see Table 1). Because of differences in individual utility functions, all the alternatives valued by Voter A ended up with the lowest ranking. Without some mechanism to change his initial vote, the conclusion reached from this single vote would probably not yield an optimum solution, for none of the top three options are valued by Voter A. In addition, his most valued option, *Vacation* was last in its overall group ranking as a result of little overall group support. In this case the group's utility functions are such that even if Voter A's placed all of his 100 chips on *Vacation,* none of the rankings would change.

Which of the following components in the companies benefit package should be improved? Allocate your 100 chips between the following alternatives.							
	INITIAL VOTE				REVOTE		
	Voter A	Voters B - K	Total Vote	Rank Order	Voter A	Voters B - K	Total Vote	Rank Order
Vacation	40	10	50	7	–	10	10	7
Salary	30	115	145	6	60	115	175	2
Health Insurance	20	130	150	5	40	130	170	3
Life Insurance	–	160	160	3	–	160	160	5
Pension Plan	–	260	260	1	–	260	260	1
Savings Plan	–	170	170	2	–	170	170	3
Job Security	10	155	165	4	10	155	165	6

Table 1: In results of initial voting, none of Voter A's valued alternatives are in the group's top three alternatives. Voter A affected the group results by reallocating his chips.

If the participants have the opportunity to change their votes Voter A could reallocate his chips placed on items with low overall group value (e.g., *Vacation* and *Job Security*). Concentrating these chips on other items valued by Voter A (e.g., *Salary* and *Health Insurance*) will affect the outcome of the vote (see Table 1). *Salary*, Voter A's second most valued option changed from sixth to second, and *Health Insurance* changed from fifth to third. This leads to a better consensus compared to a single voting cycle.

The re-voting process is dynamic. As one person changes his or her vote, it can affect the voting of others in the group. For example, in Table 1, Vacation still received 10 chips which if reallocated could effect the group ranking once more. Because of these factors, the interactive voting and re-vote process tend to provide rapid group convergence, even when group members started from opposing positions.

CM^3 incorporates two mechanisms to provide this re-vote function. The first allows users to see the intermediate voting results during the voting period. After the user casts his or her vote, and a minimum threshold of participants have cast their votes,

294

the intermediate voting results are presented. These results are only presented to those individuals who have voted, thus encouraging participation. During the voting period, each user has the option to alter his vote, whereby the intermediate votes are updated. Only the last vote cast by the individual is reflected by the group vote, so individuals may change their votes as often as desired.

The second mechanism is the ability to request a revote on the issue. Once the voting period has ended, the votes are finalized and can not be changed. It is possible that there might be a flurry of votes entered at the very end of the voting cycle, not giving the other participants an opportunity to evaluate the interim results and modify their vote accordingly. After seeing the final results, individuals want to change their vote. This is addressed through an anonymous request for a re-vote on the same issue. Pressing a function key updates a counter in the system indicating that some participants want a revote. Based on the protocols in force, and the decision of the group, the revote request can be accepted or denied.

If intermediate results are displayed, some mechanism is needed to maintain voter anonymity. If two people are the only ones who have voted, it is trivial for these individuals to determine the other's vote. For this reason, intermediate results are withheld until a minimum threshold of voters have cast their votes.

These three features (only showing intermediate votes to those who vote, allowing revotes, and the voting threshold) enhance the power of the voting process. The initial votes are unbiased but the process can benefit from the increased knowledge afforded with the intermediate results.

As with the Brainstorming Module, it is possible to printout any of the session results at participant's local site. This is important when participants are not centrally located. More details on the voting mechanisms, their uses and implications can be found in /cite GGvote.

5.2   CHIP ALLOCATION – A POWERFUL VOTING METHOD

Of the voting modules available in CM^3, the Chip Allocation protocol is preferred by most users. Each user is allocated a given set of *chips* or tokens. He allocates his set of chips among the set of alternatives presented. After making original allocation and vote, (and number of voters exceeds threshold) the intermediate voting results are reported (see Figure 8). Based on this added information, the participant has the option of changing his or her vote (see Section 5.1) for discussion on implications of changing one's vote).

```
 7:34:41                        Voting Module              Chip Allocation

 ┌──────────────────────────────────────────────────────────────────────┐
 │ 2) Which of the following components in the company's benefit package  │
 │    should be improved?                                                 │
 └──────────────────────────────────────────────────────────────────────┘

 ┌─ Allocate Chips between Alternatives ───────Your Vote─Group Vote─┐
 │ 1) Vacation                                    40  │      4.5    │
 │ 2) Salary                                      30  │     13.2    │
 │ 3) Health Insurance                            20  │     13.6    │
 │ 4) Life Insurance                                  │     14.5    │
 │ 5) Pension Plan                                    │     23.6    │
 │ 6) Savings Plan                                    │     15.4    │
 │ 7) Job Security                                10  │     15.0    │
 ├────────────────────────────────────────────────────────────────┤
 │ Total Chips Remaining =                         0  │ Total      │
 │                                                    │ Voted   11 │
 └────────────────────────────────────────────────────────────────┘


     Scroll - cursor  Ctrl-Enter - Send Vote  F4-Req. Cyc.  F9-Quit  F10-Comment
```

Figure 8: The Chip Allocation voting protocol aggregates participant's votes and reports intermediate results.

Chip Allocation provides both a ranking and a measure of preference intensity. A ranking protocol can order a set of alternatives, but can not indicate the degree of preference. The Chip Allocation method captures both metrics. A large differential between successively ranked items indicates a rapidly decaying utility function, whereas small differentials indicates little preference among the options. based on usage of this protocol, the increased information gleaned from a Chip Allocation voting session appears to far outweigh the marginal increase in complexity of this voting protocol.

The Resource Allocation protocol extends the chip allocation protocol to multiple dimensions. As the name implies, this protocol is useful in polling the groups opinions regarding a resource allocation problem. In the example given in Figure 9, manpower is being allocated among various positions. For each position (horizontal row) there is an allotment of chips (100 in this example) to be allocated among the options listed at the head of each column. Two different protocols could be used – allow chips to be placed on only one option; allow chips to be allocated among the options. As with the other voting protocols, intermediate results can be made available, as illustrated with the values shown in brackets. For complicated issues, optimization models can be initiated using these results as input.

6   Planned Research Directions

We have raised many interesting research issues in this paper. Some have been addressed and solved, research is in progress on others.

```
7:47:22                    Voting Module              Resource Allocation

5) Given our limited budget and manpower, where should we allocate our
   personnel?
```

Chips Remaining	Chips Left	FDL	RWH	SEL	MKM	LG	DMG
New Business Development	0	65 [46.7]	35 [22.5]	[30.8]			
Mgr. of House Accounts	0		[9.2]			75 [80.5]	25 [10.3]
Mgr. of Domestic Sales Force	0				25 [9.2]	75 [73.5]	[17.3]
Mgr. of Foreign Sales Force	0				50 [72.2]	20 [11.3]	30 [17.5]
Mgr. of R & D	0			100 [87.3]	[12.3]		
Plant Manager	0		75 [53.5]			25 [56.5]	
Accounting	0					[14.8]	100 [85.2]
Eng. and Drafting	0	100 [76.2]			[23.8]		

```
Scroll - cursor  Ctrl-Enter - Send Vote  F4-Req. Cyc.  F9-Quit  F10-Comment
```

Figure 9: Participants allocate chips in a row-wise manner in the Resource Allocation protocol. Intermediate voting results are displayed within brackets.

There are many open and interesting research questions associated with group decision support systems. There appears to be a trend toward decentralized meetings which will support larger participation. A consequence of the parallel process capability of GDSS system and larger meeting size is the real potential for information overload [16]. Within the brainstorming environment, establishing linkages between comments using pointers is a modest beginning at managing this potential flood of information. We are looking at additional ways to consolidate information in a meaningful way.

To assist groups in their decision-making tasks, we are investigating the incorporation of various structured support techniques, such as computerized mathematical models, thus moving CM^3 from a Level 1 to a Level 2 system [7, 11]. The focus of the research is to find ways to effectively introduce modeling techniques and tools in a decentralized, asynchronous environment. Many empirical studies are needed to understand the full implications of the enhanced capabilities that CM^3 provides to the individual user and to the group. Several of these studies are underway and will be reported, once completed.

References

[1] Connolly, T., Jessup, L. M., and Valacich, J. S., *Idea Generation in a GDSS: Effects of Anonymity and Evaluative Tone*, Working Paper, University of Arizona, Tucson, AZ, 1988.

[2] Connolly, T., Jessup, L. M., and Valacich, J. S., *Effects of Anonymity and Evaluative Tone on Idea Generation in Computer-Mediated Groups*, **Management Science**, Vol. 36, No. 6, June 1990, pp. 689-703.

[3] Dennis, A., and Nunamaker, J. et. al., *Information Technology to Support Electronic Meetings*, **MIS Quarterly**, Dec. 1988, pp. 590-619.

[4] Dennis, A., and Nunamaker, Jr., J., and Vogel, D. R., *A Comparison of Laboratory and Field Research in the Study of Electronic Meeting Systems*, **JMIS**,Vol. 7, No. 3, Winter 1990-91, pp. 107-135.

[5] Dennis, A., and Nunamaker, Jr., J., and Paranka, D., *Supporting the Search for Competitive Advantage*, **JMIS**, Vol. 8, No. 1, Summer 1991, pp. 5-36.

[6] DeSanctis, G., and Gallupe, R. B., *Group Decision Support Systems: A New Frontier*, **Database**, Winter, 1985, pp. 3-10.

[7] DeSanctis, G., and Gallupe, R. B., *A Foundation for the Study of Group Decision Support Systems*, Management Science , 1987, Vol. 33, No. 5, pp. 589-609.

[8] Gavish, B., and Gerdes, Jr., J., *Challenges of Reward Systems in Anonymous GDSS and their Solutions*, forthcoming.

[9] Gavish, B., and Sridhar, S., *Brainstorming in a Decentralized Environment using the CM^3 System*, forthcoming

[10] Gavish, B., and Gerdes, Jr., J., *CM^3, Voting Mechanisms and their implications*, forthcoming

[11] Gallupe, R., and DeSanctis, G., *Computer Based Support for Group Problem Finding: An Experimental Investigation*, **MIS Quarterly**, June 1988, pp. 277-296.

[12] Gallupe, R., and McKeen, J. D., *Beyond Computer-Mediated Communication: An Experimental Study into the Use of a Group Decision Support System for Face-to-Face versus Remote Meetings*, **Proceedings of the ASAC 1988 Conference**, Halifax, Nova Scotia, pp. 103-116.

[13] George, J., Nunamaker, J., and Vogel, D., *Group Decision Support Systems and Their Implications for Designers and Managers: The Arizona Experience,* **Transactions of the Eighth International Conference on Decision Support Systems**, Boston, MA, June 1988, pp. 13-25.

[14] George, J. F., Easton, G. K., , Nunamaker, J., and Northcraft, G. B., *A Study of Collaborative Group Work With and Without Computer Based Support,* **Inf. Sys. Res.**, Vol. 1, No. 4, 1990, pp. 394-415. *Group Decision Support Systems and Their Implications for Designers and Managers: The Arizona Experience,* **Transactions of the Eighth International Conference on Decision Support Systems**, Boston, MA, June 1988, pp. 13-25.

[15] Grohowski, R., McGoff, C., Vogel, D., Martz, B., and Nunamaker, Jr., J., *Implementing Electronic Meeting Systems at IBM: Lessons Learned and Success Factors,* **MIS Quarterly**, Dec. 1990, pp. 368-383.

[16] Hiltz, S. R., and Turoff, M., *Structuring Computer-Mediated Communication Systems to Avoid Information Overload,* **CACM** , July 1985, p.680-689.

[17] Huber, G., *Issues in the Design of Group Decision Support Systems,* **MIS Quarterly**, Sept. 1984, pp. 195-204.

[18] Jessup, L. M., *Group Decision Support Systems: A Need for Behavioral Research,* **International Journal of Small Group Research**, September 1987, pp. 139-158.

[19] Jessup, L. M., Tansik, D. A., and Laase, T. D., *Group Problem Solving in an Automated Environment: The Effects of Anonymity and Proximity on Group Process and Outcome with a Group Decision Support System,* **Proceedings of the 1988 Annual Meeting of the Academy of Management**, Anaheim, CA, August 1988.

[20] Jessup, L. M., Connolly, T., and Galegher, J., *The Effects of Anonymity on Group Process in an Idea-Generating Task,* **MIS Quarterly**, September, 1990, pp. 313-321.

[21] Johansen, R., **Teleconferencing and Beyond**, McGraw-Hill, New York, 1984.

[22] Kerr, E. B., and Hiltz, S. R., **Computer-Mediated Communication Systems: Status and Evaluation**, Academic Press, New York, 1982.

[23] Kraemer, K. L. & King, J. L., *Computer-Based Systems for Co-operative Work & Group Decision Making,* **ACM Computing Surveys**, Vol. 20, No. 2, June 1988, p.115-146.

[24] Nunamaker, J. F., Applegate, L. M., and Konsynski, B. R., *Computer-aided Deliberation: Model Management and Group Decision Support*, **Journal of Operations Research**, Vol. 36, 1988, pp. 826-848.

[25] Nunamaker, J. F., Vogel, D., and Konsynski, B. R., *Interaction of Task and Technology to Support Large Groups*, **Decision Support Systems**, 1989, pp. 139-152.

[26] Nunamaker, J. F., Dennis, A. R., Valacich, J. S., Vogel, D., and George, J. F., *Electronic Meeting Systems to Support Group Work*, **Communications of the ACM**, Vol. 34, No. 7, July 1991, pp. 40-61.

[27] Seibold, D. R., *Making Meetings more Successful: Plans, Formats, and Procedures for Group Problem-Solving*, **Journal of Business Communication**, Vol. 16, 1979, pp. 5-20.

[28] Siegel, J., Dubrovsky, V., Kiesler, S., and McGuire, T. W., *Group Processes in Computer Mediated Communication*, **Organizational Behavior and Human Decision Processes** Vol. 37, 1986, pp. 157-187.

[29] Smith, J. Y., and Vanecek, M. T., *Dispersed Group Decision Making Using Non-simultaneous Computer Conferencing: A Report of Research*, **JMIS**, vol. 7, no. 2, Fall 1990, p.71-92.

[30] Steeb, R., and Johnston, S. C., *A Computer Based Interactive System for Group Decision-making*, **IEEE Transactions on Systems, Man, and Cybernetics**, SMC-11:8), August, 1981, pp. 544-552.

[31] Watson, R., DeSanctis, G., and Poole, M. S. *Using a GDSS to Facilitates Group Consensus: Some Intended and Unintended Consequences*, **MIS Quarterly**, Vol. 12, No. 3, September, 1988, pp. 287-478.

[32] Zigurs, I., Poole, M., and DeSanctis, G., *A Study of Influence in Computer Mediated Group Decision Making*, **MIS Quarterly**, Dec. 1988, pp. 625-644.

PICTORIAL AND TEXT EDITORS FOR THE COLLABORATIVE WORK ENVIRONMENT

JACK W. POSEY
School of Industrial Engineering
1287 Grissom Hall
Purdue University
West Lafayette, IN 47907-1287
USA

ABSTRACT. A collaborative work environment implemented on computer workstations for cross-functional teams of participants must be adaptable to the different characteristics of each participant. The characteristics that can affect the outcome of collaborative activity are listed. Different modes of interaction, alone or combined, are one way to provide the necessary adaptability. Experiments with a pictorial editor and a text editor for expert system rules are described. The findings of the research are applied to the adaptability issues of collaborative work environments. Interfaces for different types of participants are proposed.

1. Introduction

The participants in a collaborative activity may have very different personal characteristics. For groups of people who have similar backgrounds and experience, individual traits and tendencies may significantly affect the outcome of collaborative work. When members of a collaborative group have widely different backgrounds and experience, the differences may be so large that, if not productively harnessed, they could defeat the good intentions of the group and cause the collaboration to be halted before any worthwhile outcome could be accomplished. The effects of the differences also depend on the purpose of the collaboration or the type of task that is the object of the collaboration.

Here is a list of personal characteristics that are likely to have an impact on collaboration:

1) Language skills and fluency.
2) Education level, area of study, and chronology.
3) Cognitive attributes.
4) Motivation and goals
5) Personal and career experience.
6) Position in organization.
7) Duration since last collaborative participation.

When the collaborative activity takes place over a distributed network of computer workstations as described in Abdel-Wahab (1990) and Abdel-Wahab and Feit (1991), the human-computer interface becomes an important factor for the execution and the outcome

S. Y. Nof (ed.), Information and Collaboration Models of Integration, 301–320.
© 1994 *Kluwer Academic Publishers. Printed in the Netherlands.*

of the collaboration. Due to the differences in personal characteristics, different interface styles for different participants may be necessary for the collaboration to be successful.

The research reported in this paper provides insight into the issue of using different interfaces for different participants during a collaborative activity. The origins of the research come from the desire to determine the effect of the use of direct manipulation interfaces on mental models of acquired knowledge. The research consisted of experiments in which undergraduate engineering students used a text editor or a pictorial editor for expert system rules to acquire knowledge about a domain. The students then used the knowledge to perform several tasks. In the final experiment, the two editors were used in four different combinations to perform the same task. The experimental results suggested that the pictorial interface was most useful for tasks that require knowledge transfer and the solution of novel problems whereas the text interface was most useful for tasks that require sequential, repetitive cognitive processing.

An extensive literature review is presented in section two. The research is described in limited detail in section three. Results of the research are presented and discussed in section four. In section five, the applications of the research results to collaboration technology are described. Four ways to present multiple interface modes to collaboration participants are presented in section six. Section seven has conclusions about the use of collaboration for integration and comments on the potential for using different interfaces during collaborative work.

2. Pictorial Editor Background

2.1. PROBLEMS OF KNOWLEDGE REPRESENTATION

McDermott (1982) described the development and the use of the R1 expert system for configuring VAX computer systems. He said that a major shortcoming of the OPS4 production system language used to program R1 was the need to have the programmer act as a translator to explain the production rules. In a highly regarded landmark text on expert systems, Buchanan and Shortliffe (1984) stated that one of the needs for future developments in knowledge representation was a more flexible style to make the knowledge engineer's and the domain expert's roles easier and more successful. They said that knowledge representation techniques as used in their programs forced the knowledge engineer to be the translator between the expert's expressions of knowledge and the syntax of the expert system language knowledge representation.

Some methods of knowledge representation were essentially useless to experts. Tsuji and Shortliffe (1983) stated that displays of text expert system rules were poorly suited for showing relations between rules, outcomes of the execution of a series of rules, and the broad, overall problem solving approach embodied in the knowledge. Clancey (1983a) found that the MYCIN knowledge base, a text format rule knowledge base, could not provide an overall concept of the medical diagnosis task it was designed to perform. Clancey also said that this caused difficulty when modifying the rule base for anyone other than the original rule authors. Musen et al. (1986) expressed surprise at how little use had been made by knowledge engineers of graphics and non-keyboard information entry devices for knowledge acquisition. They said that the limited use was disturbing because often the overwhelming burden of constructing an expert system came from "communication difficulties" between the domain expert and the knowledge engineer. Lee (1989) stated that complex, computer science-oriented languages for knowledge representation and unfriendly user interfaces were not attractive to most experts. Lee et al.

(1990) warned against using formal rules and frames for knowledge representation because they did not represent the natural expressions of experts.

Buchanan et al. (1983) defined "representation mismatch" as the problem of having to describe the knowledge as the expert thinks of it in an expert system language syntax. Gruber and Cohen (1987) said that representation mismatch was a "fundamental obstacle" to knowledge acquisition. Kidd and Cooper (1985) called the expert system knowledge representation in a non-expert's style an "alien format." Te'eni (1990) described knowledge representation that was composed of abstract rules in unfamiliar syntax to be "irritating."

Several authors have called for consideration of human cognitive capabilities in the knowledge acquisition process. Basden (1983) criticized expert system interfaces for their lack of a "well-engineered man-machine interface of high quality which conveys maximum useful information." He added that such an interface "...does not necessarily imply natural language input." In their paper on design for supervisory control systems, Murphy and Mitchell (1986) called for the consideration of human cognitive facilities in the presentation of information in order to avoid misinterpreting or even missing the information. Hoffman (1987) criticized authors of knowledge representation papers for only considering the format and style of knowledge representation, and ignoring the act of extracting the knowledge from the expert. Wexelblat (1989) reflected on many years of personal experience in developing knowledge-based systems. He objected to the reflective character of most research papers on expert system interface issues and the great lack of empirical data. He also criticized expert systems that had a user's interface different from the interface used by the knowledge engineer to construct the expert system.

2.2. DESIRED FEATURES OF KNOWLEDGE REPRESENTATION

The need for knowledge representation to be complete and realistic has been addressed. Buchanan and Shortliffe (1984) stated that debugging in context was the best way to perform knowledge base refinement because it was easier to look at specific cases than to review abstract rules for completely and correctly stating the knowledge.

Different kinds of knowledge must be accommodated by the knowledge representation. Gruber and Cohen (1987) called for explicit procedural knowledge. They said this was especially important if an expert system was created by multiple experts. Diederich et al. (1987) recommended multiple knowledge representations in order to accommodate different types of knowledge from one or more sources. Two examples were procedural and declarative knowledge.

The need for the expert to understand and be able to customize the knowledge representation has been recognized. Kidd and Cooper (1985) called for "intelligibility of representation," defined as compatibility between the user's model and system's model of the domain, so that the knowledge representation and the problem solution could be understood by the user. They said this was necessary for competent and confident use of the expert system. Murphy and Mitchell (1986) recommended that the user be given the capability to alter the display of information to remove inconsistencies and ambiguities in the knowledge representation.

Lee et al. (1990) stated that the knowledge representation used by the expert during knowledge acquisition by the expert system should be different from the representation used by the expert system language for solving problems. This was because the expert should be using a domain-dependent representation and the problem solving program should be using a domain-independent representation.

In a paper addressing issues in human-computer collaboration, Silverman (1992) described a "creativity system." This was software that would engage a user in a way that stimulated the user's creativity and encouraged the user to express ideas in a natural fashion.

2.3. ASSESSMENT AND DEVELOPMENT TOOLS FOR KNOWLEDGE REPRESENTATION

Five paradigms from psychology, cognitive science, and computer science are helpful tools to direct and assess developments in knowledge representation. Mental imagery is the paradigm from psychology. The cognitive science paradigms are mental models and stimulus-central processing-response compatibility. The computer science paradigms are interactive graphics interfaces and direct manipulation. The paradigms served as guides to the development of the pictorial expert system rule editor and influenced the experiments in the research.

2.3.1. Mental Imagery. In a summary review of research by imagists and propositionalists, Anderson (1978) critiqued their theories and research. Drawing all of his findings together, Anderson espoused a "tricode theory" of information. He said that there are three kinds of information – visual-spatial, verbal-sequential, and abstract-propositional – and different representations for each. Anderson noted that "translations" between representations are assumed so that the first choice of representation for information does not mean another representation that is better suited for another process can't be used. The tone of Anderson's overall appraisal was that humans will use a representation that allows the most "efficient" execution of a task. The development of a robust theory of representation is not served by attempting to restrict the representation to only one type.

One part of imagery theory that Anderson did agree with was that the context in which an image was presented was important to remembering the image. Corollary to this was the notion that pictures that can be "meaningfully interpreted" are the best-remembered pictures.

The dual-coding theory of Paivio (1986) views mental representation as consisting of a verbal part and a nonverbal part, with representational units of logogens and imagens, respectively. The parts are separate but partly interconnected. Logogens that are abstract and imagens that are unnamed are not directly interconnected. Processing may use just one of the representations or both, depending on the stimuli from the context of the problem, the mode and task of the instructions, and individual differences. The mode that the information is presented in has an influence on which code is used to store the information. Thus, pictures as stimuli are represented with the image code, and text is represented with the verbal code.

Forbus (1980) emphasized the ability of humans to apply their visual capabilities to reasoning about the motion of objects and the relations between objects. He gave the use of diagrams as an example of this application. Forbus said that humans apply their visual capabilities in order to think about the objects in diagrams as they would think about the actual objects. Hollan (1984) attributed the value of graphics in an interface to the natural skill developed by humans for dealing with the spatial relations of physical objects. Eberts (1984) said that the relationships and object behaviors were internalized by mentally imaging the objects and performing spatial reasoning on the image.

Two concepts that depend on a user's mental imagery capabilities and spatial reasoning for their utility were discussed in Clancey's (1983b) paper on the use of personal workstations for education. One concept was "graphical envisioning" to allow a student to interactively create a process and investigate its behavior. The other concept was showing graphical results of mathematics calculations so that users would learn how to apply mathematical concepts and formulas as a prelude to learning the details of the mathematics.

2.3.2. *Stimulus—Central Processing—Response Compatibility.* Wickens et al. (1983) proposed the concept of stimulus-central processing-response (SCR) compatibility. It was based on the concept of stimulus-response compatibility developed by Fitts and Seeger (1953) that arose from research during and after World War II conducted by A. M. Small (1990) on stimulus-stimulus compatibility. Wickens et al. showed that in addition to the compatibility of stimulus and response, the compatibility of the stimulus and response with the cognitive processing resources also had an effect on reaction time. They found that consistency in all three produced the fastest times.

Before Wickens et al. published their work on SCR compatibility two other papers described results related to the concept. Carroll et al. (1980) conducted two experiments to study the effects of no representations and different representations to subjects to use to perform isomorphic problems. They concluded that a graphic representation helped subjects organize intermediate solutions for later use, possibly encouraged subjects to spend more time trying to solve the problems, and might make a problem easier to solve but not easier to understand. Bastick (1982) hypothesized that moving the sprite around the interface may promote holistic processing.

Changes in mental models over time were studied in the context of SCR compatibility in Eberts (1983). The results indicated that without sufficient development time, mental models created under consistent training conditions would not provide superior performance. Eberts and Schneider (1985) found that consistency during the development of a mental model for a second-order tracking task resulted in better performance on a transfer task.

2.3.3. *Mental Models.* The field of mental models developed from a blend of parts of cognitive psychology and artificial intelligence (Stevens and Gentner, 1983).

Williams et al. (1983) said that people used mental models as a memory aid and for predicting, justifying, and explaining physical system behavior by qualitatively modeling system changes. Rasmussen (1983) said one of the uses of mental models was to perform "internal" experiments. Other uses were prediction, explanation, and understanding. He also stated that multiple mental models were a way that a human solves problems. Rumelhart and Ortony (1977) stated that the context in which a stimulus was presented would affect the resulting mental model.

Larkin (1983) said that novices and experts had different types of mental models. Novices constructed mental models from developments in real time and the models were composed of objects in a "naive" problem representation that was runnable. Experts had an additional mental model development approach that was based on a physical representation based on formal physics. Larkin stated that these two representations were different from a mathematical representation that was formed from equations of physical principles. Larkin said that this two-model dichotomy was useful for describing the difference in behavior between novices and experts.

Hollan et al. (1980) used computer graphics to teach steam propulsion plant operation to trainees. They said that a mental model did not provide quantitative results like a mathematical model but instead gave a person a "feel" for the system and enabled the human to respond rapidly and, if not optimally, at least reasonably well to unanticipated system changes. Eberts (1984) performed research on mental models used for prediction. He stated that when it was not practical for a person to manipulate actual physical objects, a computer graphics display of the objects could be manipulated and provide the interaction to enable the construction of mental models of the objects.

Rouse and Morris (1986) produced a comprehensive survey of mental model research and applications of mental models. Their work confirmed the wide appeal of the mental model concept and explained the tendency of different disciplines to assign different definitions to mental models. From their survey they concluded that the generally accepted uses for mental models were description of purpose and form, explanation of function and state, and prediction of state.

2.3.4. *Interactive Graphics Interfaces.* Tsuji and Shortliffe (1983) listed three features that gave the graphics interface its power in expert system development and management. The first was user-friendly control over the level of detail shown on the display screen. The second was the wide variety of representations which enabled a user to select the most understandable representation and also helped the user see the overall system better than with a text representation. The third feature was the use of windows to control the presentation of information and allow the user to readily "compare and contrast" information.

2.3.5. *Direct Manipulation.* Shneiderman (1983) described direct manipulation as a powerful technique for human-computer interaction. In direct manipulation, a user controls a pointing device to manipulate objects on the display. The power of the computer is utilized to provide real-time or near-real-time response. Dale (1969) described a "cone of experience" that listed 11 levels of experience ranging from the tip of the cone, named verbal symbols (least concrete), to the base of the cone, named direct purposeful experience (the actual experience). The level just above the base was named contrived experiences. Direct manipulation is an example of experience at this level.

The semantic/syntactic model of knowledge was used by Shneiderman to explain the appeal of direct manipulation to users. This model divides human memory into two parts – semantic (largely system independent) and syntactic (system dependent). He said that direct manipulation lets users manipulate objects corresponding to a high level in the semantic knowledge. The dependence on the syntactic knowledge for specifics of interface syntax was greatly reduced thus "reducing operator problem-solving load and stress."

2.4. PREVIOUS DEVELOPMENTS IN PICTORIAL KNOWLEDGE REPRESENTATION

A number of researchers have proposed or implemented ways to improve knowledge representation and the knowledge acquisition process. Some contributions were made in other research pursuits but are relevant or illuminating nonetheless.

Bijl and Szalapaj (1985) created the MOLE system for practitioners in experience-based disciplines. MOLE was designed to use a combination of text and graphics in the interface. However, pictorial graphic displays themselves were not stored as descriptions per se, they were only a means of telling a user about the model stored as text descriptions in the representation structure. A "graphical grammar" was used to permit text-to-graphics translation and vice-versa. The goal of Bijl and Szalapaj was to allow the maximum possible freedom to the user to mix graphics with text in the interface.

Freiling and Alexander (1984) integrated pointing at electronic schematic diagrams and circuit board maps on a display screen with a structured interface language to rapidly develop expert system families for troubleshooting electronic devices. The expert systems had 60% to 80% correct diagnosis which was satisfactory for the intended applications. The pointing feature was used for knowledge acquisition and diagnosis. The natural language rules were translated into an internal rule format for use in an expert system program written in Prolog.

Barker and Najah (1985) developed a technique in which a user touched a stylus to a picture on a high-resolution digitizer pad to edit a database or perform database operations. A commercial database program provided the database functions. The use of concept keyboards and paper-based images for user interaction with expert systems was described in Barker and Manji (1987). They said that this was just one area of research of user interfaces based on paper forms overlaid on a digitizer for stylus pointing or a concept keyboard for finger pointing.

A knowledge representation structure called a "picture-frame" was described in Strothotte (1989). The picture-frame structure was used to enable incremental solution presentation in pictorial form. The picture-frame was based on the semantic frame, with graphic slots used to make changes or additions to the pictorial solution.

3. Experiments with the Pictorial and Text Expert System Rule Editors

3.1. RESEARCH PLAN

The overall intent of the research was to utilize the concept of mental model development and use by humans to measure the effects of the two different rule editors. The experimental procedures caused each subject to develop, expand, and revise mental models of a simple chemical processing system. Tests were then administered to measure the prediction, understanding, explanation, and transfer properties of each subject's mental models. Posey (1992) contains a complete, detailed report on the experiments.

The development and assessment tools described in section 2.3 were utilized in different ways. The pictorial rule editor incorporated the direct manipulation and interactive interfaces paradigms and promoted the use of mental imagery by the subjects to think about the chemical processing system. Since the chemical processing system was an arrangement of objects with specific spatial relationships, the pictorial rule editor provided SCR compatibility. Subjects developed, expanded, and revised their mental models of the chemical processing system as they were trained to use the interface and then as they performed the tasks of the experiments.

Subjects who were trained on the text rule editor and used it to perform experimental tasks were working with the same knowledge domain as pictorial editor subjects. However, the text format of the rules was not compatible with the spatial nature of the chemical processing system. The text subjects' mental models were developed in a context that was different from the pictorial subjects' context.

The last experiment was the rule correctness experiment. This was the only experiment that crossed the training groups with the non-trained rule editor. This provided opportunities to observe the effects of transfer between rule editor modes, pictorial and propositional memory, and SCR compatibility.

The subjects used for the research were undergraduate engineering students but none were chemical engineering majors. The subjects had a variety of computer experience but all had used computers regularly for programming, text editing, or other purposes.

3.2. EXPERIMENTAL PROCEDURE

A pictorial expert system rule editor interface and a text expert system rule editor interface were developed with the X Window System and the Athena Widget Set. The C programming language was used for the source code. Each interface as it first appears on the workstation display is shown in Figure 1 (pictorial interface) and Figure 2 (text interface).

A simple chemical processing system was the knowledge domain used for the experiments. The system parts and operating characteristics were conceived after reviewing texts on polymer processing and chemical engineering. This domain is suitable for pictorial rules because it is composed of physical objects that are spatially oriented to one another.

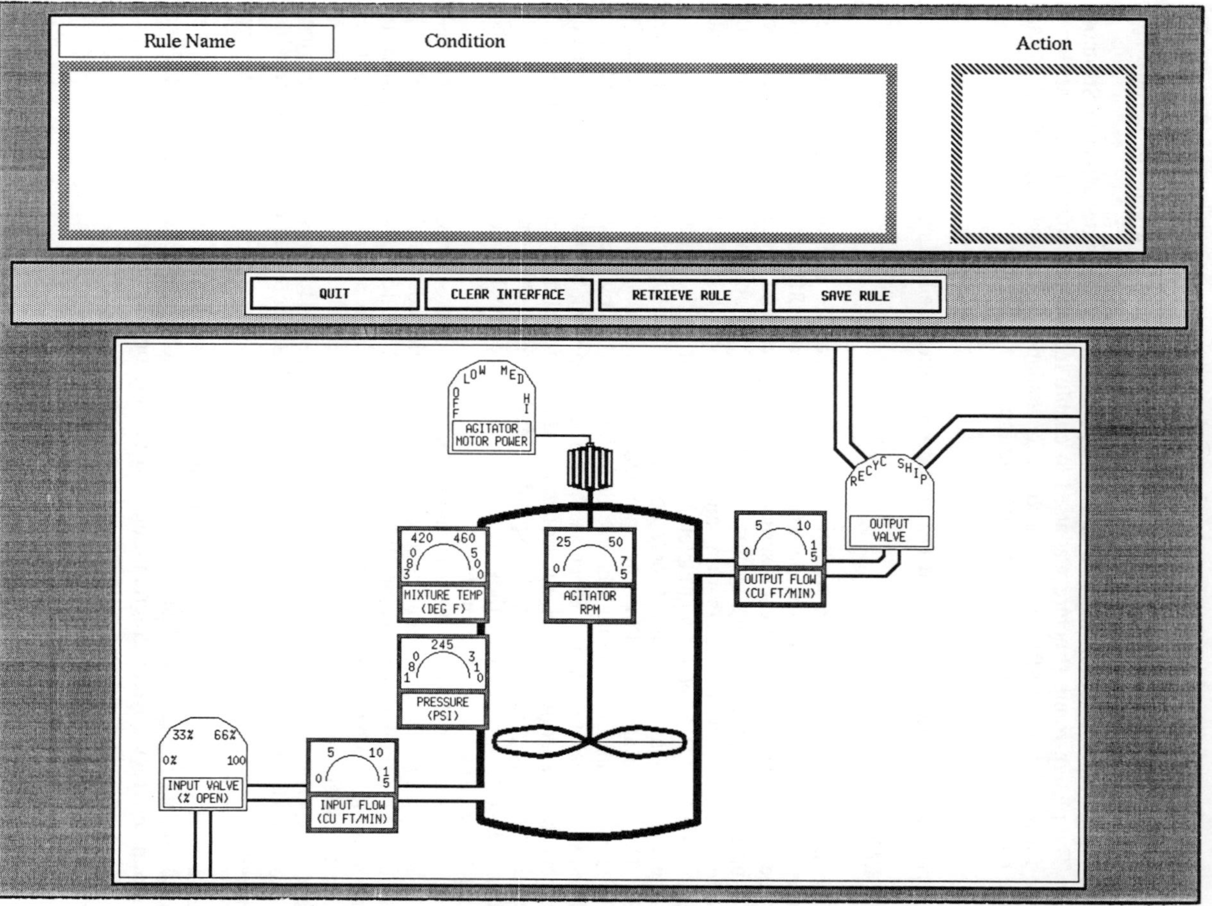

Figure 1. Pictorial mode expert system rule editor interface before any rule editing interaction.

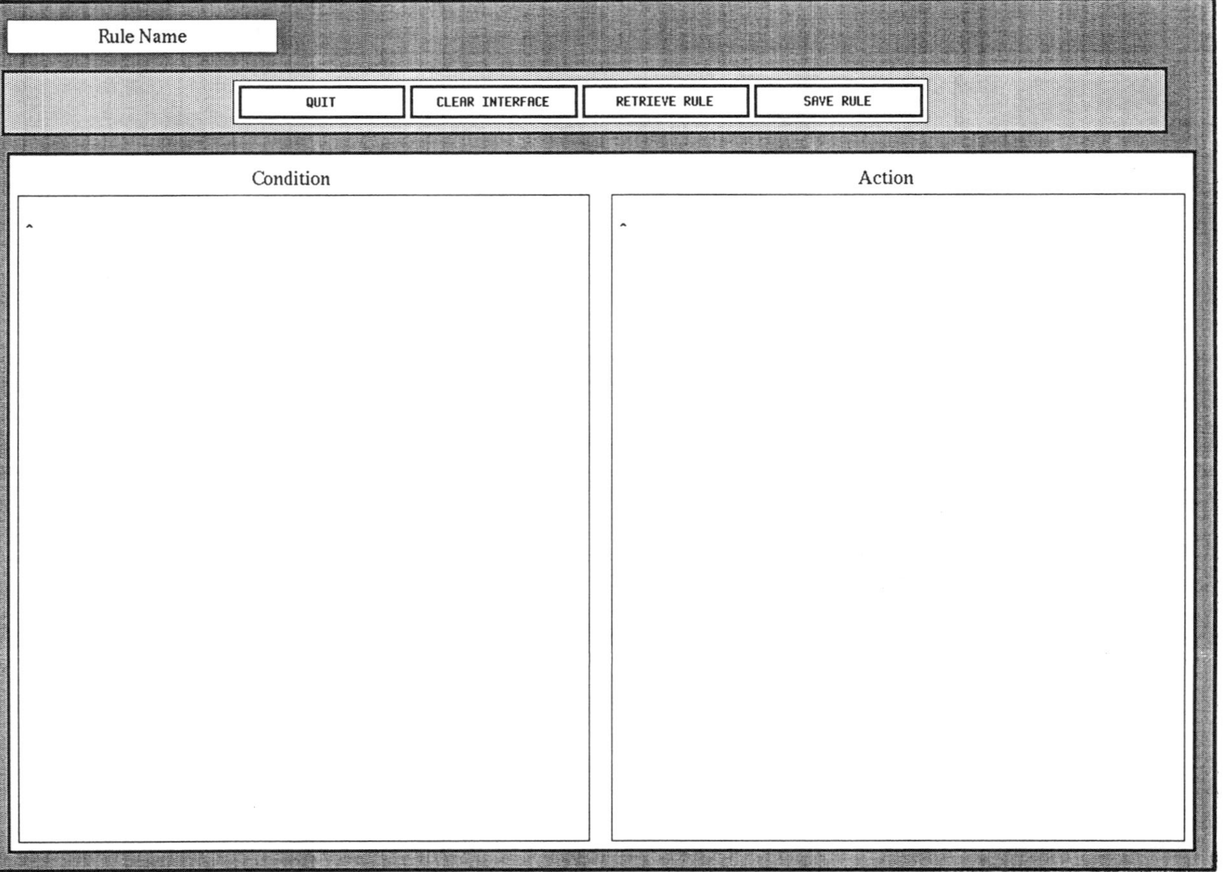

Figure 2. Text mode expert system rule editor interface before any rule editing interaction.

Twenty-one procedures useful for operating the system were defined and a rule set was developed for each procedure. The procedures were for starting the system, shutting down the system, and adjusting the system after flow rate changes. A rule from a sequence of rules that describe system startup is shown on the pictorial interface in Figure 3 and in the text interface in Figure 4.

Each subject individually performed the experiments across five sessions. The experiments are listed in Table 1. The first session consisted of training in the use of the interface. The first part of session two was spent on reviewing the interface training and learning about the chemical processing system. The rest of session two and the third, fourth, and first part of the fifth sessions were experimental tasks. The last part of the fifth session was for the administration of cognitive tests.

Each subject learned about the chemical processing system by reading a description and performing two sets of exercises. In the written description the parts of the system, their spatial and functional relationships, and the purpose and use of the system were presented. The two sets of exercises contained questions about the components of the system, the operation of the components, and the operating states of the system. Each subject had to correctly answer all of the questions before beginning the first experimental task.

4. Experimental Results

4.1 MENTAL MODELS

Table 2 shows the experiments divided into groups based on type of cognitive activity. Both procedure memorization experiments, Experiment 1 and Experiment 5, are included in procedure memorization in Table 2. Three of the experiments represent processing-dominant work and the remainder represent thinking-dominant work. A checkmark (✓) indicates significantly better performance. This dichotomy of the experiments shows that the text group is superior for processing tasks and the pictorial group is superior for thinking tasks.

Rasmussen's (1983) theory of three levels of models of human performance provides a framework for understanding the dichotomy of tasks. Rasmussen states that three levels of models and their respective information units are skills and signals, rules and signs, and knowledge and symbols. Skills are at the lowest level, rules the middle level, and knowledge the highest level. Skills represent automatic behavior, rules represent learned behavior, and knowledge represents behavior developed to respond to new situations. A person cannot consciously report on skills-level behavior control, but can report on rules-based and knowledge-based behavior.

The rule matching, rule firing, and rule correctness exercises require behavior at the rules model level. This is because the subject would be expected to be able to report on the steps performed to solve each exercise, but the subject did not have to synthesize new knowledge from existing knowledge, or integrate new knowledge, to perform the exercises. The six other experiments required performance at the knowledge level.

For some of the nine tasks, the performance of the two groups was not significantly different. For tasks where the text group was superior, the work was at the rules level. For tasks where the pictorial group was superior, the work was at the knowledge level.

The pictorial subjects developed good macro procedure proposals faster than the text subjects. This concurs with the findings of Eberts and Schneider (1985). They showed that subjects trained with consistent augmentation developed an accurate internal model of a system, whereas subjects with no or inconsistent augmentation did not. Eberts and Schneider

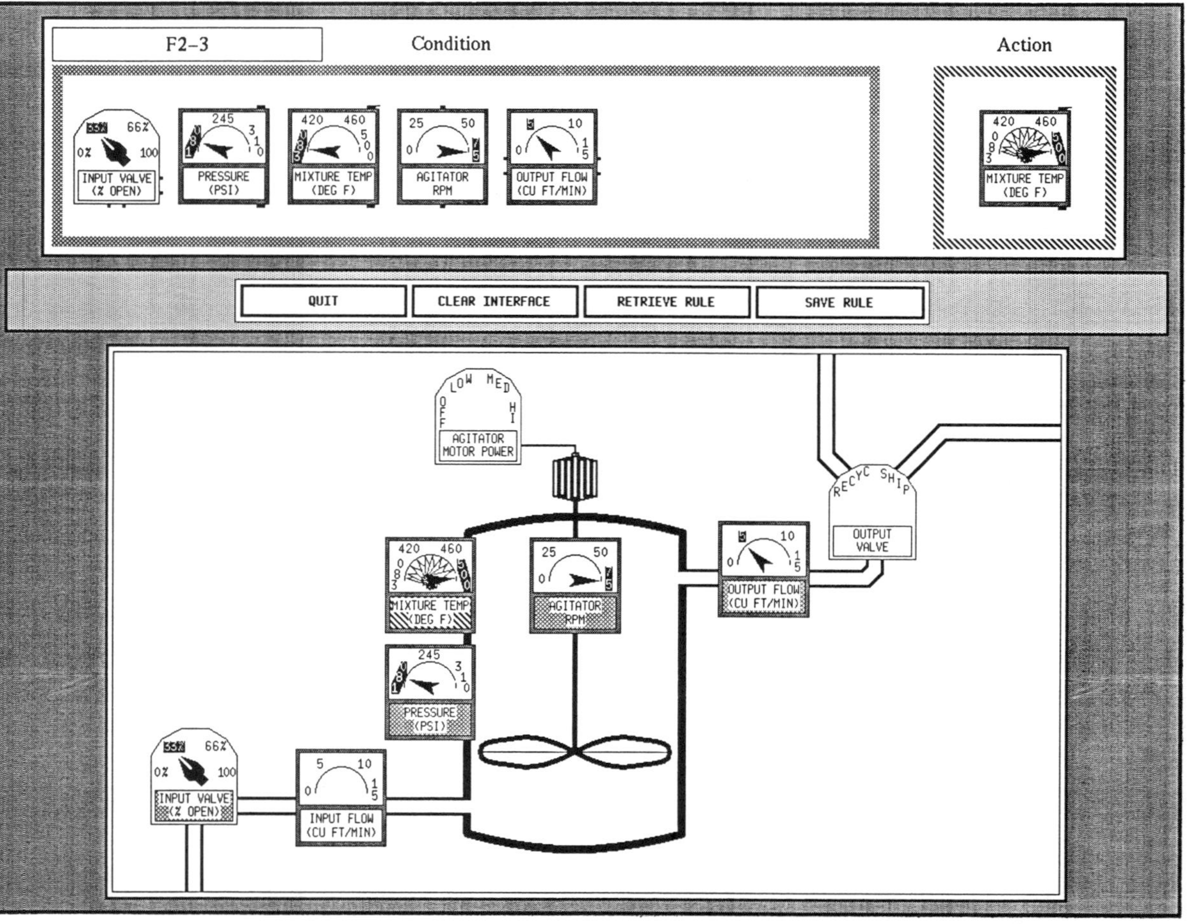

Figure 3. A rule from a processing system startup sequence shown on the pictorial mode interface.

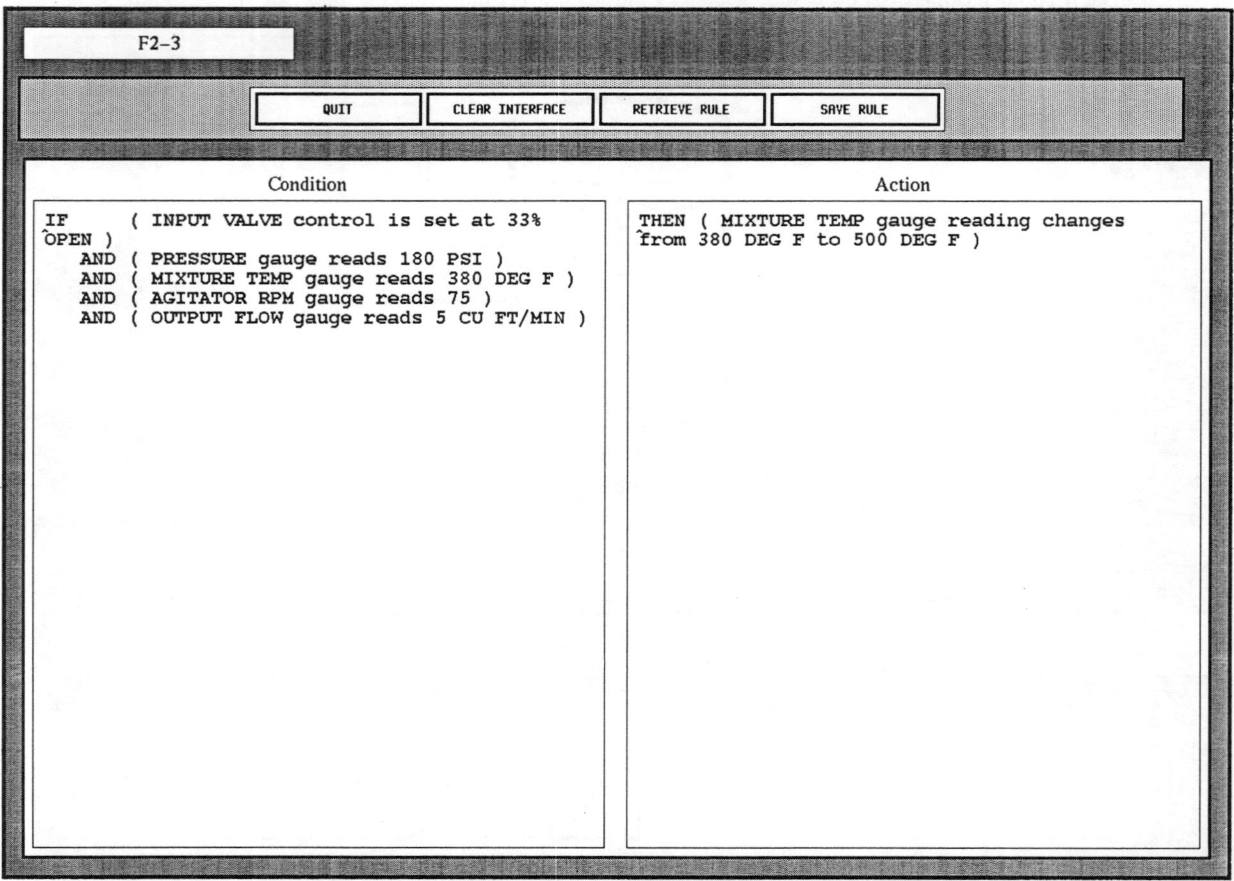

Figure 4. The same rule from Figure 3 on the text mode interface.

Table 1. Description and Purpose of Experiments.

Experiment	Session	Purpose	Description
Procedure Memorization	2	Learn procedure	Memorize rules, test memory
Procedure Interpretation	2	Show system knowledge	Write interpretations
Rule Matching	2	Test rule retention, system knowledge	Decide if rule is from procedures
Rule Firing	3	Test memory and cognitive processes	Fire rules to change system
Micro Procedure Memorization	4	Learn procedure	Memorize rules, test memory
Rule Completion	4	Test understanding of operating procedures	Add missing parts of rules
System Operation Guidelines	4	Test understanding of operating procedures	Write guideline of procedure
State Change Prediction	4	Test prediction ability	Predict changes in system
Macro Procedure Proposal	4	Test transfer of knowledge	Write proposal to solve new operating problem
Rule Correctness	5	Test compatibility of interface modes	Decide if rule on interface is same as rule specification

Table 2. Classification of Cognitive Activity of Experiments and Superior Training Group for Experiments (✔ indicates superior performance).

Experiment	Cognitive Activity	Pictorial Group	Text Group
Rule Matching	Processing		✔
Rule Firing	Processing		✔
Rule Correctness	Processing		
Procedure Memorization	Thinking	✔	
Procedure Interpretation	Thinking		
Rule Completion	Thinking		
System Operation Guidelines	Thinking	✔	
State Change Prediction	Thinking		
Macro Procedure Proposal	Thinking	✔	

found that subjects from the consistent augmentation group (accurate internal model) were better at solving transfer problems. Eberts and Schneider concluded that this indicated the subjects with the accurate internal model performed spatial manipulations in order to transform their knowledge to solve the transfer problem. They said that the subjects who had been trained with no or inconsistent augmentation used simplifying rules, and that some of the rules were incorrect, which resulted in significantly poorer performance on the transfer tasks. Eberts and Schneider attributed the ability to manipulate the consistent internal model to practice with an external display of the system on which the consistent augmentation was provided.

The pictorial rule editor provides a consistent representation of the system and through direct manipulation causes a user to practice with the system during rule creation and rule firing. The text rule editor does not provide a consistent representation of the system or practice with the system, only rules about the system. Based on the conclusions of Eberts and Schneider, pictorial editor users would be expected to perform better than text editor users on a transfer task. The results of the macro procedure proposal experiment did indeed show this; the good proposals were developed in significantly less time by the pictorial rule editor subjects. This shows that the pictorial group subjects' model of the system is superior to the system model developed by the text group. A more consistent model would require less cognitive processing to develop a solution to a transfer problem, and this was the case for the macro procedure proposal experiment.

4.2. STIMULUS—CENTRAL PROCESSING—RESPONSE COMPATIBILITY

For the system operation guidelines experiment, the difference in performance was due to quality of subject responses. For all the other experiments both groups performed similarly in terms of number of correct responses or quality of response. For the other four experiments where there was superior performance by one of the two groups, it was due to less time. The SCR compatibility theory of Wickens et al. (1983) combined with Rasmussen's (1983) levels of performance shows which group has a model at the rules level and which group has a model at the knowledge model and explains the differences in performance.

In general, the best SCR compatibility gives the shortest time for performing a task. Since the text training group has the shortest times for tasks most appropriate to solution with the rules level model, the text group's model is probably a rules model. Similarly, since the pictorial group has the shortest times for tasks most appropriate to solution with the knowledge level model, the pictorial group's model is probably a knowledge model. The better performance by the pictorial group on the system operation guidelines also supports the Rasmussen theory. This is because the task requires reformulation of knowledge, which is best done with a model at the knowledge level. Since both groups were presented with the same information about the system, in the same order, but in different modes, the modes must be the cause for the differences in each group's mental model of the system.

4.3. DIRECT MANIPULATION

Te'eni (1990) performed experiments to determine if feedback from direct manipulation interaction resulted in better performance than feedback provided as information separate from the user's interaction. He stated that his research was founded in part on Paivio's (1986) dual coding theory. Te'eni concluded that his results showed that subjects who used feedback through direct manipulation were faster, required less cognitive effort, and committed fewer errors than subjects using only distinct feedback. He found these effects stronger for the more complex of two tasks. In addition, Te'eni believed that the direct manipulation subjects also benefited from additive effects of the distinct text feedback provided on the display and the direct manipulation interaction.

The pictorial and text rule editor results of the research are similar to Te'eni's findings. The tasks where the pictorial group performed better (procedure memorization, system operation guidelines, and macro procedure proposal) were more complex than the tasks in which the text group performed better (rule matching, rule firing). Furthermore, the rule correctness experiment showed strong additive effects of dual coding.

4.4. IMPLICATIONS FOR EXPERT SYSTEM INTERFACE DESIGN

The procedure memorization outcomes in experiment one and experiment five showed the pictorial editor could be used to learn operating procedures in less time. Spatial, concrete presentation of knowledge and direct manipulation of the knowledge is a useful, promising approach for managing the knowledge representation of an expert system. From the experiments that followed the memorizations it is evident that the mental model of the knowledge developed with the pictorial rule editor was different from the mental model developed from use of the text rule editor. Depending on the type of task in each experiment, the pictorial group performed better, the same, or worse than the text group. The implication of this result is that the end use of the expert system should be considered when selecting the type of interface and knowledge representation method to use to develop and implement the expert system.

The results of the macro procedure proposal experiment indicated that pictorial-developed mental models can transfer better to a design task for the domain than text-developed mental models. The pictorial mode can be useful for dealing with different types of knowledge besides simple, shallow, declarative knowledge. The consistency of mental model development and multiple levels of mental models is the likely reason for this. The implication of this result is similar to the previous implication although it is more specific. In this case, the implication is an end use of the expert system for facilitating the discovery of new procedures for the existing knowledge domain.

The interactions in the rule correctness experiment showed that the combination of type of problem and method of presentation may produce effects not attainable for unmixed interface and rule specification modes. The almost non-existent effect due to training mode suggests the transfer between text and pictorial modes was successful for both groups. This may be a positive example of additive processing from dual coding of memory representation. The implication is that a mixed mode of interface and representation may provide the highest expert system performance metrics, and the level of performance may not be predictable from experience with single-mode representation methods.

The dichotomy of cognitive activity that is shown in Table 2 leads to another implication. It is that text-mode rules should be used to explain to humans how a computer processes expert system rules, but pictorial-mode rules should be used to comprehend the meaning of the rules with respect to the knowledge domain.

5. Application of Experimental Results to Collaboration Technology

The results of the research on the pictorial and text expert system rule editors provide three insights into solutions for the problem of accommodating differences in personal characteristics of collaboration participants. First, the demonstration that interface users were able to acquire knowledge about a domain that they were not familiar with. Second, the indication that some tasks were better suited to one interface mode than the other. Third, for one type of task, the two interfaces combined resulted in better performance than using just one interface.

All of the experimental subjects were engineering students representing several different engineering majors but none were majoring in chemical engineering. Every subject was able to acquire knowledge of the chemical processing domain through experience with an interface. This suggests that collaboration participants who are not familiar with a specific technical domain, but who are familiar with or experienced in the general principles that the specific domain is founded on, will be able to participate in collaboration about that domain.

The superiority of one mode over the other for different tasks indicates that different interfaces should be used for different tasks. In a collaborative activity, this means that if there is more than one type of task there may need to be different interfaces. This could arise in at least two ways. One way is when there is a sequence of tasks performed to accomplish the collaboration. The second way is for multiple tasks to be ongoing concurrently, perhaps with different participants performing different tasks, or one or more participants involved in more than one task at the same time.

The results from the correctness task show that an interaction may occur from combining two interfaces. When the interaction results in better performance than the performance with a single mode, the combined modes should be used. For collaboration activity, combined interfaces may result in better performance of some types of tasks by all participants. In other instances, some participants in the collaboration may be more comfortable or productive using combined modes and other participants may do better with a single mode.

6. Multiple Interface Modes for Collaboration

There are different ways to implement multiple interfaces for collaboration. Four ways are proposed in this section.

The first method for implementing multiple interfaces is a single mode used by all collaboration participants. The mode can be changed, perhaps by a facilitator, but at any time during the collaboration all participants are using the same type of interface.

A second method is to provide a menu of modes on each user's screen. A participant can then select any interface at any time. This method enables each participant to use the interface that best suits his or her needs during the collaboration and to change the interface at will.

The third method is a mixed-mode interface that is an extension of the second method. Here mixed-mode means side-by-side presentation of two or more interface modes, each mode presenting the same information but in different representations. This method can be used for productive interactions between modes.

The fourth method is a hybrid mode interface. In this method, each participant can interweave different interfaces into a customized representation of information. This provides tremendous flexibility. A challenge is to develop menus or other interaction techniques for participants to use to create hybrid modes in a pleasant, efficient way. An alternative to a participant directly constructing a hybrid mode is to embed the hybrid mode construction capability in an adaptive interface.

7. Conclusions

Silverman (1992) poses the question "What type of users relate well to which types of knowledge and under what circumstances?" This research poses a more specific question: What type of users relate well to which types of knowledge *display* and under what circumstances? The research results presented in this paper show that different interface modes can be used to meet the needs of collaboration participants with different personal characteristics. Providing participants with the option of selecting an interface mode would have a positive impact on the outcome of any collaboration. Each participant would be able to select the interface mode most useful for any task or objective of the collaboration. One of the objectives of organizational integration is the empowerment of all members of the organization (Little, 1992). The use of different interface modes directly strives for this objective by allowing for many differences in collaboration participants.

In section six, only four possible interface mode selection methods were described. This should not be taken to mean that these are the only four ways. Instead, this should be considered to be just one view of the ways that interface modes could be offered to collaboration participants.

One of the objectives of using different interface modes is to promote consistency in the mental models of all of the collaboration activity participants. The mental model developed by a participant from using an interface mode is dependent on personal cognitive attributes. Different individuals may develop similar mental models even if each individual uses a different interface mode because of the individual's cognitive attributes and other personal characteristics.

Collaboration between participants distributed geographically over large or small distances can be done today. Bringelson (1993) describes research in which readily available hardware, software, and networks are regularly used. The author of this paper and several colleagues regularly used a collaborative software tool to hold weekly research meetings with a major advisor on sabbatical at a university 2,000 miles distant.

An example of a collaborative activity that nicely draws together collaborative work and expert system development is provided by Gasser (1991). He lists modular knowledge acquisition.as one of the reasons for having distributed artificial intelligence.

Different interface modes also may prove useful in extending collaboration to include teaching or training objectives. In this instance, interface modes that were understandable to novices, such as students or trainees, would be used. The result would provide the novice with a way to quickly become involved and interested in the collaboration. The novice would learn more from the collaboration experience and would be more likely to look upon collaboration in a favorable light.

In the research described in this paper only two interface modes were used. Upon reflection of the tremendous number of differences that there would be for a collaboration involving a number of participants, the task of providing enough interface modes to meet all participants' demands may seem a very large assignment. However, the results of research by Silverman (1992), which investigated one human–one intelligent agent collaboration, suggest that just two modes may achieve a very high rate of fulfilling needs of collaboration participants. His results indicate that just a novice level and an expert level of representation may provide enough flexibility for most users of a collaborative system. This means that for relatively little development effort most of the gains to be had from providing different interface modes may be attained.

Integration may occur vertically, horizontally, or along progression in time within an organization or between organizations. It is reasonable to assume that the collaborative tools that are used to instantiate the integration should be different for users along these three dimensions. For example, vertical integration may result in associations among personnel from top executives to inexperienced young workers and every level in between these two groups. Horizontal integration may involve participants who have roughly similar experience, authority, and objectives, but who are in different departments of an organization. Integration along progression in time would result in temporal effects due to different cognitive attributes, amount of time since the most recent collaboration, and duration of recent collaboration participation. In addition, the time pattern of initial and subsequent collaborations could result in different learning and experience effects. By providing knowledge representation in a format that is most useful to each user that is affected by integration, multiple interface modes can help provide the flexibility that is essential to the success of integration.

Acknowledgements

This research was partially funded through a Presidential Young Investigator award to Ray E. Eberts sponsored by the Division of Design, Manufacturing, and Computer Integrated Engineering of the National Science Foundation and Digital Equipment Corp.

References

Abdel-Wahab, H. M., (1990). Multiuser tools architecture for group collaboration in computer networks. Computer Communications, **13**(3), 165-169.

Abdel-Wahab, H. M., and Feit, M. A., (1991). XTV: A framework for sharing X Window clients in remote synchronous collaboration. Proceedings of TRICOMM '91, IEEE Conference on Communications Software: Communications for Distributed Applications and Systems, 159-167.

Anderson, J. R., (1978, July). Arguments concerning representations for mental imagery. Psychological Review, **85**(4): 249-277.

Barker, P. G., and Manji, K. A., (1987). Pictorial knowledge bases. P. of the Third Conference of the British Computer Society, Human-Computer Interaction Specialist Group. Cambridge: Cambridge University Press, pp. 163-173.

Barker, P. G., and Najah, M., (1985). Pictorial interfaces to data bases. International Journal of Man-Machine Studies, **23**, 423-442.

Basden, A., (1983). On the application of expert systems. International Journal of Man-Machine Studies, **19**, 461-477.

Bastick, T., (1982). Intuition: How we think and act. Chichester: John Wiley, 1982.

Bijl, A., and Szalapaj, P., (1985). Saying what you want with words and pictures. In Shackel, B. (Ed.), Proceedings of Human-Computer Interaction – INTERACT '84. Amsterdam: Elsevier, 275-280.

Bringelson, L. S., (1993). Human factors and group decision support systems: An integrative perspective to an integrated environment. Proceedings of 2nd Industrial Engineering Research Conference. In press.

Buchanan, B. G., Barstow, D. K., Bechtel, R., Bennett, J., Clancey, W., Kulikowski, C., Mitchell, T., and Waterman, D. A., (1983). Constructing an expert system. In Hayes-Roth, F., Waterman, D. A., and Lenat, D. B., (Eds.), Building expert systems. Reading, MA: Addison-Wesley, 127-167.

Buchanan, B. G., and Shortliffe, E. H., (Eds.), (1984). Rule-based expert systems: The MYCIN experiments of the Stanford heuristic programming project. Reading, MA: Addison-Wesley.

Carroll, J. M., Thomas, J. C., and Malhotra, A., (1980). Presentation and representation in design problem-solving. British Journal of Psychology, **71**, 143-153.

Clancey, W. J., (1983a). The epistemology of a rule-based expert system – a framework for explanation. Artificial Intelligence, **20**, 215-251.

Clancey, W. J., (1983b). Communication, simulation, and intelligent agents: Implications of personal intelligent machines for medical education. Proceedings of American Association for Medical Systems and Informatics Congress 83, 556-560.

Dale, E., (1969). Audiovisual methods in teaching, 3rd edition. NY: Holt, Rinehart, and Winston.

Diederich, J., Ruhmann, I., and May, M., (1987). KRITON: A knowledge-acquisition tool for expert systems. International Journal of Man-Machine Studies, 26, 29-40.

Eberts, R. E., (1983). The effects of an inaccurate internal model on subsequent learning. Proceedings of the Human Factors Society 27th Annual Meeting, 156-160.

Eberts, R. E., (1984). Augmented displays for problem solving. In Salvendy, G. (Ed.), Human-Computer Interaction. Amsterdam: Elsevier, 261-265.

Eberts, R. E., and Schneider, W., (1985). Internalizing the system dynamics for a second order system. Human Factors, 27, 371-393.

Fitts, P. M., and Seeger, C. M., (1953). S-R Compatibility: Spatial characteristics of stimulus and response codes. Journal of Experimental Psychology, 46, 199-210.

Forbus, K. D., (1980). Spatial and qualitative aspects of reasoning about motion. Proceedings of National Conference on Artificial Intelligence, 170-173.

Freiling, M. J. and Alexander, J. H., (1984). Diagrams and grammars: Tools for mass producing expert systems. Proceedings of the First Conference on Artificial Intelligence Applications, Denver, Colorado, 537-543.

Gasser, L., (1991). Social conceptions of knowledge and action: DAI foundations and open systems semantics. Artificial Intelligence, 47, 107-138.

Gentner, D., and Stevens, A. L., (Eds.), (1983). Mental models. Hillsdale, NJ: Erlbaum.

Gruber, T. R., and Cohen, P. H., (1987). Design for acquisition: Principles of knowledge-system design to facilitate knowledge acquisition. International Journal of Man-Machine Studies, 26, 143-159.

Hoffman, R. R., (1987, Summer). The problem of extracting the knowledge of experts from the perspective of experimental psychology. AI Magazine, 53-67.

Hollan, J. D., (1984). Intelligent object-based graphical interfaces. In Salvendy, G. (Ed.), Human-Computer Interaction. Amsterdam: Elsevier, 293-296.

Hollan, J., Stevens, A., and Williams, M., (1980). STEAMER: An advanced computer-assisted instruction system for propulsion engineering. Proceedings of the Summer Simulation Conference, 400-404.

Kidd, A. L., and Cooper, M. B., (1985). Man-machine interface issues in the construction and use of an expert system. International Journal of Man-Machine Studies, 22, 91-102.

Larkin, J. H., (1983). The role of problem representation in physics. In Gentner, D., and Stevens, A. L. (Eds.), Mental models. Hillsdale, NJ: Erlbaum, 75-98.

Lee, J. K., Lee, I. K., Choi, H. R., and Ahn, S. M., (1990). Automatic rule generation by the transformation of expert's diagram: LIFT. International Journal of Man-Machine Studies, 32, 275-292.

Lee, N. S., (1989). Graphical knowledge programming with KNAPS. International Journal of Man-Machine Studies, 31, 611-641.

Little, J. D. C., (1992). Tautologies, models and theories: Can we find "laws" of manufacturing? IIE Transactions, 24(3). 7-13.

McDermott, J., (1982). R1: A rule-based configurer of computer systems. Artificial Intelligence, 19, 39-88.

Murphy, E. D., and Mitchell, C. M., (1986). Cognitive attributes: Implications for display design in supervisory control systems. International Journal of Man-Machine Studies, 25, 411-438.

Musen, M. A., Fagan, L. M., and Shortliffe, E. H., (1986). Graphical specification of procedural knowledge for an expert system. Proceedings of IEEE Computer Society Workshop on Visual Languages, 167-178.

Posey, J. W., (1992). Pictorial and Text Editors for Expert System Rules. Ph. D. Dissertation, Purdue University, West Lafayette, IN.

Paivio, A., (1986). Mental representations: A dual coding approach. Oxford: Oxford University Press.

Rasmussen, J., (1983). Skills, rules, and knowledge; signals, signs, and symbols, and other distinctions in human performance models. IEEE Transactions on Systems, Man, and Cybernetics, SMC-13, 257-266.

Rouse, W. B., and Morris, N. M., (1986). On looking into the black box: Prospects and limits in the search for mental models. Psychological Bulletin, 100, 349-363.

Rumelhart, D. D., and Ortony, A., (1977). The representation of knowledge in memory. In R. C. Anderson and R. J. Spiro (Eds.), Schooling and the acquisition of knowledge. Hillsdale, NJ: Erlbaum, 99-135.

Shneiderman, B., (1983, August). Direct manipulation: A step beyond programming languages. IEEE Computer, 16(8): 57-69.

Silverman, B., (1992). Human-computer collaboration. Human-Computer Interaction, 7: 165-196.

Small, A. S., (1990). Foreword. In Proctor, R. W., and Reeve, T. G., Stimulus-response compatibility: An integrated perspective. Amsterdam: North-Holland, v-vi.

Stevens, A. L., and Gentner, D. (1983). Introduction. In Gentner, D., and Stevens, A. L. (Eds.), Mental models. Hillsdale, NJ: Erlbaum, 1-6.

Strothotte, T., (1989). Pictures in advice-giving dialog systems: From knowledge representation to the user interface. Proceedings of Graphics Interface '89, London, Ontario, 94-99.

Te'eni, D., (1990). Direct manipulation as a source of cognitive feedback: A human-computer experiment with a judgement task. International Journal of Man-Machine Studies, 33, 453-466.

Tsuji, S., and Shortliffe, E. H., (1983). Graphical access to the knowledge base of a medical consultation system. Proceedings of American Association for Medical Systems and Informatics Congress 83, 551-555.

Wexelblat, R. L., (1989, Fall). AI Magazine, 66-78.

Wickens, C. D., Sandry, D. L. and Vidulich, M., (1983). Compatibility and resource competition between modalities of input, central processing, and output. Human Factors, 25(2), 227-248.

Williams, M. D., Hollan, J. D., and Stevens, A. L., (1983). Human reasoning about a simple physical system. In Gentner, D., and Stevens, A. L., (Eds.), Mental models. Hillsdale, NJ: Erlbaum, 131-153.

NEURAL NETWORK BASED AGENTS FOR COORDINATION OF INTERACTION

Ray Eberts
School of Industrial Engineering
Purdue University
West Lafayette, IN 47907
USA

ABSTRACT. Recent research on neural network-based agents for coordinating and filtering information is presented. Neural net agents have advantages over rule-based agents because they can be easily adapted to the information needs of the user through the learning capabilities of neural networks. Three application areas for these agents are discussed: electronic bulletin boards, Group Decision Support System (GDSS) meetings, and electronic mail (e-mail). This paper reports accuracy results of the network for electronic bulletin boards and meeting messages. The agents could accurately filter messages, under optimal conditions, in the range of 85% to 99%. Interface designs for incorporating these agents in GDSS and e-mail systems were discussed and illustrated. The advantages of neural network-based agents over rule-based agents were discussed.

1. Introduction

The sources for electronic information are multiplying. Electronic mail (e-mail) is used to communicate with others, across the building or across the world, mostly through textual information. Meeting announcements, requests for information, or discussions occur through electronic bulletin boards. Interaction and coordination of group or team activities occurs through group decision support systems (GDSS) or computer supported cooperative work (CSCW). GDSS is most closely associated with brainstorming meetings where all the participants are interacting through a networked computer system so that the meeting can occur in parallel. For CSCW, tasks such as text-editing or computer programming are done concurrently through computer support of multiple group members.

These sources of information can be overwhelming as a computer user becomes swamped with information. With e-mail, as with surface mail, junk mail becomes a problem. The number of topic areas for electronic bulletin boards seems to be increasing weekly; the number of postings under the topics increases at an even faster rate. Searching through postings to find relevant information is a time-consuming process. Finally, GDSS is most effective if the number of participants is high, usually around 15 people. The number of messages which can be generated by this number of people can be quite large. Any particular participant has to read through several irrelevant messages before finding a relevant message in which to respond.

To alleviate these problems, methods are needed to search, coordinate and filter information. One successful technique is the use of multiple agents to search and filter information. Each agent can be programmed to find certain kinds of information and, instead of the human performing the

S. Y. Nof (ed.), Information and Collaboration Models of Integration, 321–346.

searching tasks, the computer-based agent can automatically search information sources such as e-mail, electronic bulletin boards, and GDSS meeting transcripts.

In the past, agent-based systems have used mostly rule-based expert systems to determine how to search, coordinate, and filter information. These systems have had some success (e.g., Fischer and Stevens, 1991), but they have inherent problems. Development time can be very high because the information needs of users have to be determined through system developer intervention, and rules must be constructed from these analyses. Programming time for rules is long because of the complexities of this type of programming. Expert systems are good for identifying invariant features of information needs but relatively poor at recognizing patterns not specified in the rules. Because recognition of information characteristics can be formulated as a pattern recognition task, this paper reports on investigations of using neural networks for filtering and coordinating information.

This paper is divided into the following sections. The first section provides background material on neural networks and some previous applications. The remaining sections discuss research and development of neural network agents for searching, coordinating, and filtering information in three application areas: electronic bulletin boards, GDSS, and e-mail. Research will be reported on the methods used to develop and train the neural network agents, tests on the accuracy of the agents, and the interface designs for supporting the agents in these application areas.

2. Neural Network Background

Searching for and filtering information can be considered a pattern recognition task. If a person is performing these tasks, he or she will look for certain patterns of words to determine if a message is important or not. In other domains, such pattern recognition problems have been attacked successfully through the use of neural net modeling.

One of the primary applications of neural networks has been in machine vision. Neocognitron (Fukushima, 1987, 1988a, 1988b; and Fukushima, Miyake, and Ito, 1983), a model of visual pattern recognition, tries to be faithful to physiological brain functions and human information processing. It can learn to recognize visual patterns and is especially adept at recognizing patterns with partial information. NETtalk (Sejnowski and Rosenberg, 1987) has been designed to learn to produce speech from English text. PARSNIP (Hanson and Kegl, 1987) is a neural network which learns natural language grammar after being exposed to natural language sentences. PARSNIP was trained syntactically on sentences of 15 words or less. Once trained, it was able to syntactically label words in the trained sentences and to successfully generalize the syntactic labels to new sentences.

A characteristic of all these neural net programs is that they have a large number of feature units which, when turned on and off, can represent input patterns that need to be recognized by the neural net. These input patterns are then mapped into a smaller number of outputs which represents the classification of the pattern. This mapping occurs by setting connection weights on the arcs connecting the nodes from one level to the next. Through a set of training examples which specifies how an input pattern is related to an output categorization, the neural network uses a learning algorithm to change the weights so that the mapping of input to output, from the training examples, is accurate. Each time the network changes the set of connection weights is called a training cycle. Several thousand cycles may be required before the net is considered to be trained to a high enough accuracy level. After the net is trained, it can generalize from the trained patterns to other new patterns not trained specifically. This generalization ability is what distinguishes the neural network from other intelligent systems such as expert systems.

Although the specific form of neural networks varies, they usually have several common features. Rumelhart and McClelland (1986) state that a certain class of neural networks, the backpropagation model, contains the following seven features.

- A set of processing units
- A state of activation
- An output function for each unit
- A pattern of connectivity among units
- A propagation rule for propagating patterns of activities through the network of connectivities
- An activation rule for combining the inputs impinging on a unit with the current state of that unit to produce a new level of activation for the unit
- A learning rule whereby patterns of connectivity are modified by experience

Rumelhart and McClelland (1986) describe these seven features as follows.

The set of processing units are the nodes in the network. These nodes are usually organized into input and hidden layers. The input layer is the representation of the pattern which needs to be recognized. Depending on the application, these units in the input layer can represent different concepts. In visual pattern recognition tasks, they could represent visual features such as letter features, letters, words, and phrases (Fukushima, 1987, 1988a, 1988b; Fukushima, Miyake, and Ito, 1983). For general programs not specific to a particular application, they represent abstract elements over which meaningful patterns can be defined. The nodes in the hidden layer have the same characteristics as those in the input layer although they may not represent a meaningful concept. The hidden layer is needed for mapping some complex conjunctions and disjunctions of input nodes into the output nodes. A node in an output layer is similar to nodes in the input and hidden layer except that an output node is not connected to any higher processing layers. All the processing in neural nets is carried out by these units with no executive control over them. The unit simply receives input from neighbors, combines all inputs using a specified function, and computes an output value to pass on to the neighboring units in the next level. Under this scheme, several units can be activated at any particular time so that parallel processing can occur.

The state of activation of the system at any time t can be represented by a vector, $a(t)$, where each of the N units in the network corresponds to a value in the vector (Rumelhart and McClelland, 1986). The pattern of activation determines what the network is representing at any particular time. The vector values can be discrete, continuous, or stochastic either bounded or unbounded. For discrete values, the vector value is usually either 0 or 1, representing on or off. Continuous values can be activated as any real number. Stochastic values could be a function of a defined probability distribution. The neural nets described in the following applications take on continuous values bounded between 0 and 1. Because values of 0 and 1 can have undesirable effects on the learning algorithms, values of 0.1 and 0.9 were used for off and on, respectively.

Each unit, except the input units, will receive activation from its neighbors in the preceding layer of the network. The units sum all the input activity and, if the value exceeds some threshold, will output signals to its neighbors. The function relating the input values to the output determines the strength of the output of the units. The functions depend on the model used and the purpose of the model. Rumelhart and McClelland (1986) identify common functions such as the identity function, step threshold functions, continuous threshold functions, and stochastic functions based upon some probability distribution.

The pattern of connectivity is important to how the information is processed. Humans usually process information in a combination of top-down and bottom-up processing. For bottom-up processing, the inputs to the network represent perceptual units such as letter features in vision or phonemes in speech recognition. These low level units are combined into larger units until a meaningful high level unit is output, such as a word. For top-down processing, the input to the network is some kind of context which is used to activate the lower level units. As an example, if the overall context was animals and the network received input of three letters with "C" and "T" the only ones which could be identified, then the middle letter "A" would be activated instead of "U" because of the top-down activation. Some neural network models use either bottom-up or top-down processing alone or a combination of the two. The pattern of connectivity is established by determining the weights or strengths associated with the connections between the units. In most models, the input to the unit is a weighted sum of the separate inputs from each of the individual units. The weights can be excitatory, a positive value, or inhibitory, a negative value. A weight of 0 would represent no connection between two units. The weights are usually represented as a matrix of weights in which the units to be connected are from the rows and columns of the matrix. The knowledge of the neural network is represented in the weights between the connections. When a neural network learns, it changes the weight values so as to produce input/output pairings corresponding to the values observed for the training examples.

The rule of propagation states how the matrix of connection weights is combined with the output values of the units to produce a net input for each type of input into a unit. Generally, the matrix of weights is multiplied by the output vector to determine the net input. Two types of inputs are usually considered, the excitatory and inhibitory inputs, so that separate net input values would be determined for each.

An activation rule states how the net inputs of each type impinging on a particular unit are combined with one another and with the current state of the unit to produce a new state of activation. Two useful activation rules are decay or saturation over time. For decay, the activation is descreased over time and for saturation the activation is increased over time.

One of the important aspects of neural networks is that the network can change according to experience by modifying the patterns of interconnectivity. Three kinds of modifications can occur: the development of new connections; the loss of existing connections; and the modification of the strengths of connections that already exist. Little work has been done on the first two but these first two can be considered a special case of the third one if a new connection is considered established when the weight becomes non-zero and a disconnection occurs when the weight becomes zero (Rumelhart and McClelland, 1986). Most of the important learning rules are based on that of Hebb (1949) who stated, simply, that when one unit receives input from another unit, and both are highly active, then the connection weight between the two units should be strengthened. In practice, this simple rule can take on many forms. The equation most approximating that of Hebb is:

$$\Delta w_{ij} = g(a_i(t), t_i(t)) h(o_j(t), w_{ij})$$

where Δw_{ij} is the change in weight for the connection from unit u_j to unit u_i at time t. The equation represents the product of two functions, $g()$ and $h()$. The $g()$ function has parameters corresponding to the activation, $a_i(t)$, of u_i, and a teaching function, $t_i(t)$. The $h()$ function has parameters corresponding to the output value, $o_j(t)$, of u_j, and the connection strength between the two units, w_{ij}. Variants of this include the following. If the teaching function is not specified, then the functions g and h are proportional to their first arguments and the change in weight is:

$$\Delta w_{ij} = \eta a_i o_j$$

where η is the constant of proportionality representing the learning rate. If the amount of learning is taken to be proportional to the difference between the actual activation achieved and the target activation provided by a teacher (this is often called the delta rule) then the equation is:

$$\Delta w_{ij} = \eta(t_i(t) - a_i(t))o_j(t)$$

A further variant is the following rule which was specified by Grossberg (1976):

$$\Delta w_{ij} = \eta a_i(t)(o_j(t) - w_{ij})$$

These are the main learning rules; others exist for more specialized situations. Through application of the Hebbian learning rules, the neural network determines the connection strengths between the nodes so that the input-output relationships in the training examples are captured. After training, test data can be propagated through the trained network so that the model can recognize input patterns by mapping them to output nodes using the existing connection weights established during learning.

The main advantage of the neural network approach constructing agents is that these agents can be trained from examples to be sensitive to certain kinds of information. For electronic bulletin boards, GDSS messages, and e-mail, the user only has to indicate which messages should be mapped to a particular category. The category could occur along several dimensions such as the importance or the content of the message. The agent can also be dynamic depending on the length of the memory and how often the agent is retrained. These advantages can be contrasted to rule-based agents in which the rules must be specified through some kind of intervention and are relatively invariant over time. Another advantage is that these agents are very good for pattern recognition; solutions can be found based upon partial information.

3. Electronic Bulletin Board Applications

As indicated in the introduction, electronic bulletin boards are becoming quite large. A user of these services could spend much unproductive time searching through irrelevant information trying to determine if any postings are valuable. Users could become more productive if multiple agents performed some of the preliminary searches to determine if any postings may be important to the user.

The following section reviews a previous research project, Fischer and Stevens' (1991) INFOSCOPE, in which rule-based agents were developed to search for relevant electronic bulletin board messages. This approach is contrasted to a neural net-based agent approach developed by Habibi and Eberts (1991). The method used to develop the agents is described in some detail, and then studies on the accuracy of the agents in filtering bulletin board postings is described.

3.1. INFOSCOPE

The INFOSCOPE filter developed by Fischer and Stevens (1991) use a cognitive science concept, Anderson's (1990) Rational Analysis of Human Memory, to make decisions about the messages which should be stored or deleted. Anderson indicates that the probability an item in memory will be needed at a later time is a function of three effects: frequency, recency, and spacing. Frequency refers to the number of times that the item has been used in the past. Recency refers to the amount of time since it has been used. Spacing refers to the distribution of usage over time. One could expect that the same three effects are important for analyzing the need to select newsgroups within a bulletin board and then to read messages from that newsgroup. A filter using these three effects could be used to prioritize messages within a newsgroup. As an example, if messages within a newsgroup have been read frequently, then the filter will assign these messages a higher priority to display on the screen. If messages within a newsgroup have been read recently, other messages from this newsgroup would also receive a high priority.

The users of INFOSCOPE do not have to construct the filters themselves. Instead, the filter is constructed and modified through the use of agents. The agents are collections of rule-based heuristics which monitor the behavior of the user in relation to the messages received by the user to construct rules. The rules allow the system to be personalized to the particular user and, in consultation with the user, can assist in the modification of the rules used for the filter. The system works as follows. First, the user reads messages in a newsgroup. Second, an agent recognizes that the user only reads 15% of the messages in a particular newsgroup. This is judged to be too low a number by the agent. Third, this first agent tells a second agent to start gathering messages which may be more interesting to the user. Fourth, once these messages have been gathered, the second agent sends a suggestion to the user indicating that more interesting messages have been found because the user had not been reading many messages. Finally, the user can either accept this suggestion or not. If it is not accepted, another agent must try to determine why it was not accepted and modify the behavior of the other agents.

Evaluating INFOSCOPE is very difficult. No evaluation of the accuracy of INFOSCOPE is offered by Fischer and Stevens. The agents work, within the confines of the programs which construct the rules, but it is difficult to determine the accuracy of the agents in finding the relevant bulletin board postings. The determination of the filtering needs of the user is based upon how well the poster of the bulletin board message chooses the appropriate newsgroup; the user of the agents has very little control over accuracy. As an example, if a posting is sent to an unusual newsgroup by the poster, then the user's agent will not be able to find it because the agent does not search by content.

3.2. Adaptive Agents for Categorizing Bulletin Board Messages

Habibi and Eberts (1991) developed neural net-based agents for categorizing messages from electronic bulletin boards. In this approach, the neural net agents are trained through training examples in which the user has specified the message and the categorization of the message. The neural net uses a dictionary of words, constructed from the messages, as inputs, and the output nodes represent the category of the message. After training, the agents are able to take any message as input, propagate the message pattern through the trained net, and then determine which category best fits the message.

The following describes the method used to develop the adaptive neural net agents. Preliminary accuracy results on categorizing bulletin board messages along several category dimensions are also discussed. To determine the most accurate method to construct the agents, first different kinds of dictionaries were tested and, second, different information sources, such as the message heading or text, were manipulated.

3.3. Research on Dictionary Types

This research addressed the issue of the best way to construct a dictionary for the neural net-based adaptive agents. Two dictionary types, to be used as the inputs for the net, were considered. A common words dictionary combined all the words that were common to the files within a subtopic. Thinking that the patterns which need to be mapped to the category outputs may not be distinguishable from one another, a difference words dictionary approach was also considered. For this method, only the words that were different among the five subtopics were used in the dictionary. This may help to differentiate the patterns to be mapped. In addition to the dictionary construction methods, dictionary size was also manipulated to determine the effect of size on net accuracy.

3.3.1. Method. Two neural net software packages were used. Initially, a shell program called NETS, developed by NASA (Baffes, 1989), was used. Next, a shell called PlaNet version 5.6 (previously called SunNet) (Miyata, 1990) was used. Both incorporate the backpropagation algorithm described in Rumelhart and McClelland (1986).

Data were taken from the rec.arts.startrek newsgroup. This newsgroup was used because it receives many postings daily. In addition, even though everyone discusses the same general topic (Star Trek), numerous branchings take place thus creating many subtopics. Postings were chosen from five of these subtopics.

Two methods for constructing a dictionary of words to be used by the agent were considered, the common and difference dictionaries. The differences between these two methods and the method for constructing the dictionaries is illustrated in Figure 1. The words in the dictionary were taken from either two or three news postings from each of the five topics to construct small and large dictionaries, respectively. The top of the figure shows the procedure for constructing the small dictionaries from the words in two news postings of subtopic 1. First, the words and characters from the postings are sent through a filter that replaces all nonalphabetic characters with a carriage return, converts the remaining characters to lower case, and sorts the words. The result is two output files. Second, the words in the output files are compared, and the words that are the same from the two files are sent to common word file 1. Up to this point, the same procedure is followed for the other four subtopics resulting in four more common word files. Finally, the words from all five common word files are sent through the last filter. This filter is used differently depending upon the type of dictionary that is constructed. For the Common Dictionary, all the words from the common word files are combined into the one dictionary. For the Difference Dictionary, only those words that are dissimilar in the five common word files are included in the dictionary.

Smaller dictionaries were constructed from the two postings as compared to the larger dictionaries for the three postings for each subtopic. For the two-posting condition, the common dictionary had a size of 172 words (using both NETS and PlaNet) and the difference dictionary had a size of 205 words (PlaNet was used for this). The larger dictionary, constructed from the three postings for each subtopic, used only a common dictionary and contained 341 words (PlaNet was also used for this).

Before training the net, the researcher read all the postings and categorized them under the five sub-topics. The dictionary words were used as inputs, 30 nodes were in the hidden layer, and the five subtopics were used as category outputs (see Figure 2). For each news posting used for training or for a test, the input nodes from the dictionary were turned on (a value of .9 in the figure)

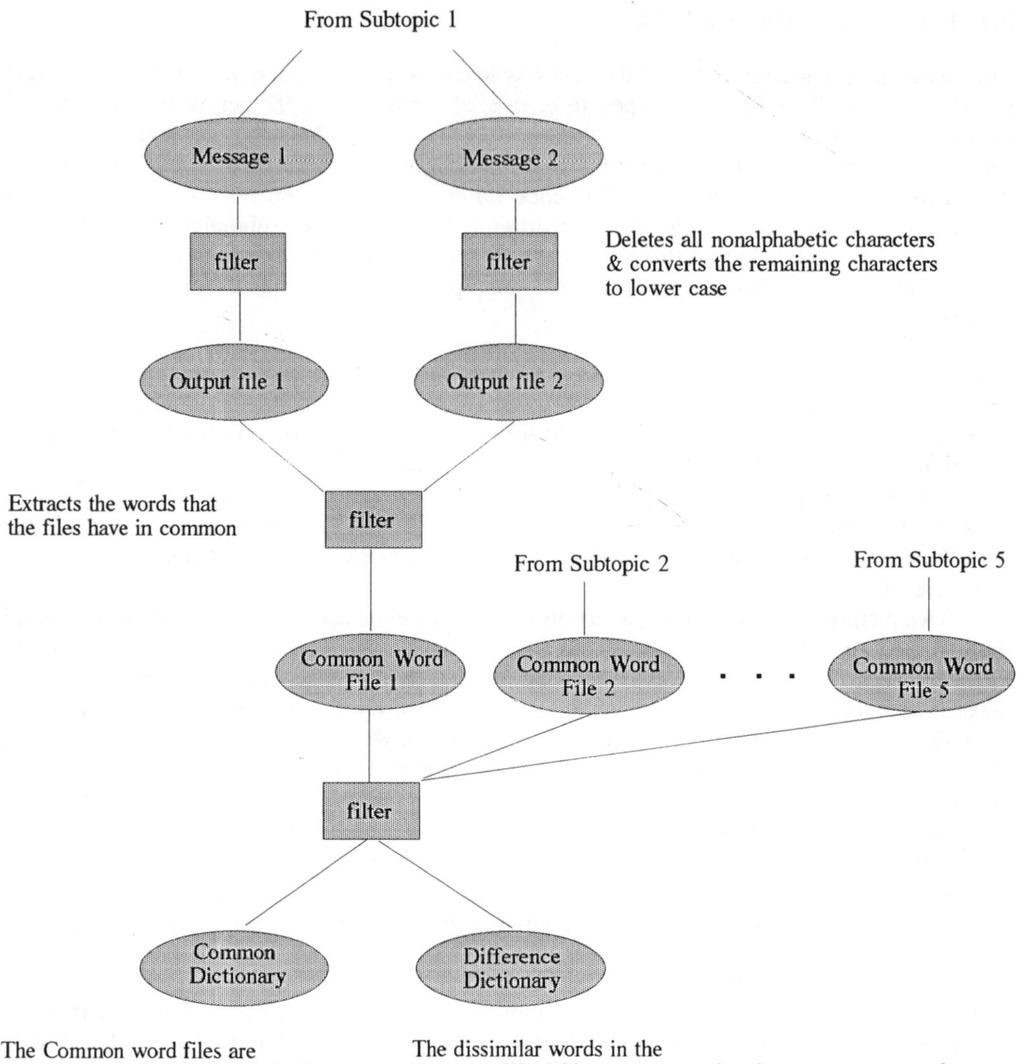

Figure 1. Illustration of the method to construct the dictionary for the neural net-based agent (from Habibi and Eberts, 1991).

if the news posting contained the word corresponding to the particular input node; the node remained off (a value of .1) if the news posting did not contain the word corresponding to the particular node. The net was trained with the news postings that were used for creating the dictionary, plus enough more to have six postings from each topic. Thus, the net learned from a total of 30 files that served as training examples. After the net learning was complete, 64 new news postings were used to test the network.

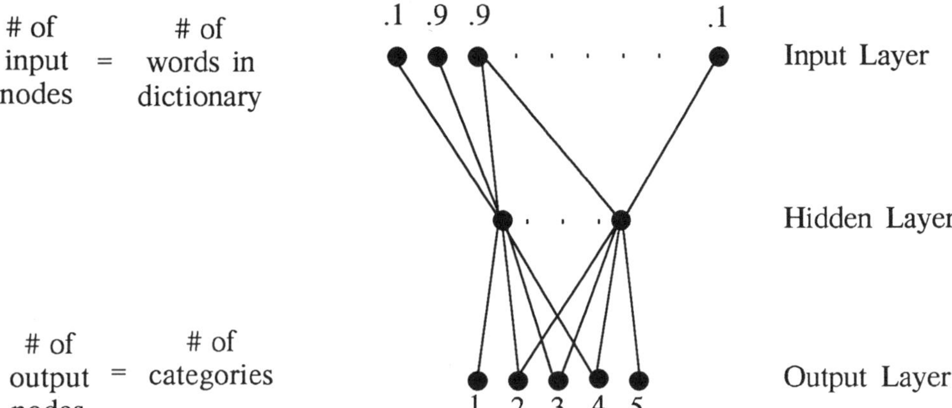

Figure 2. Depiction of the structure of the neural network used for the neural net-based agents (from Habibi and Eberts, 1991).

3.3.2. Results. Accuracy of the agent was determined in the following way. Before training the net, the experimenter classified all the messages into the five subtopics. Some of the messages were used for training the net, and the rest were used for testing the net. During the test phase after the net had been trained, accuracy was determined by counting the times that the net classified the messages the same as the experimenter had originally classified the messages.

Results for the networks on correctly classifying the 64 new news postings after training are shown in Table 1. For the 172 difference dictionary using both NETS and PlaNet, out of the 64 samples, the net failed to identify 11 of them. Best performance occurred for the 205 word common dictionary in that PlaNet failed to identify only 4 of the samples. A larger dictionary did not seem to help; the 341 word dictionary exhibited 14 errors.

TABLE 1.

Accuracy of agent filtering as a function of dictionary type, software, and size (from Habibi and Eberts, 1991)

dictionary type	*software shell used*	*size of dictionary*	*# of errors out of 64*
difference	NETS	172	11
difference	PlaNet	172	11
common	PlaNet	205	4
common	PlaNet	341	14

The results can be summarized as follows. No difference occurred between the NETS and PlaNet software in terms of classification performance. The only difference was that the learning time with NETS was extremely slow and dictionaries of size greater than 200 would not converge on a solution. The common dictionary method worked better than using the difference dictionary method. When the size of the dictionary was increased up to over 300 words, performance decreased.

3.4. Information Source

In some situations, header information is available and this can be used as the source of the text for the pattern recognition. For bulletin board postings, each message has a header which provides a topic area. When another user responds directly to a posting, this header information is carried over to this new posting, thus providing a method to track the different pathways taken by the information and responses. Most e-mail use a similar method for the header information. In other situations, however, header information may not be included, and the only source of text for the pattern recognition would be in the body of the text. This experiment determined how much information is provided by the header for categorizing the postings.

3.4.1. Method. PlaNet was used in this experiment. The method and the training procedure was much the same as previously. The variable of interest was the source of the text. In the header source condition, the body of the text was removed and the dictionary was constructed from the words in the header resulting in a dictionary of size 21. In the text body source condition, the header was deleted and only the words in the body of the text were used resulting in a dictionary of size 146. The data from these two conditions can be compared with the common dictionary of size 205 from the previous section in which both header and text body were included. The common dictionary method was used to construct the dictionary in all cases.

3.4.2. Results . Best performance for the net was the header source condition in which only one error out of 64 possible classifications occurred (see the third column of Table 2). This was even better than the best condition from the previous section using both header and text body in which only 4 errors, out of 64, occurred in classification. The text body source condition, without the header information, exhibited 12 out of 64 errors.

In examining the kinds of errors from the text body source condition, it was concluded that the classification used for training the net, set by the researcher, may be wrong. These classifications, shown in the third column of Table 2, were mostly from the header information,

TABLE 2.
Accuracy of agent filtering as a function of information source (from Habibi and Eberts, 1991)

source of text	*size of dictionary*	*# of errors out of 64 (classify by header)*	*# of errors out of 64 (classify by body)*
header & body	205	4	1
body	146	12	6
header	21	1	1

so that high performance by the net in that condition may not be surprising. In examining the text in the body of some of the postings, many of the messages deviated greatly from the header information which was automatically carried over to the new posting. In the text body condition, 6 out of 12 errors were due to deviation from the subject heading. Also, 3 out of 4 errors made in the header and text body condition were due to deviations from the subject heading. In these messages, the header was not related to the body of the message. If the messages are classified according to the text in the body of the messages, then the performance of the agent that used the header and body was equal to that of the agent that used the header only (see the far right column of Table 2).

3.5. Discussion and Conclusions

The results from this series of experiments showed that the performance of the agent is very dependent on the dictionary it uses. Best performance occurred using a common words dictionary with a medium sized dictionary. Overall, the common words dictionary method exhibited slightly fewer errors than the difference word dictionary. Increasing the size of the dictionary actually decreases the accuracy of the agent. Information source had a large effect on accuracy. Only including the header information exhibited more accurate performance than the text in the body only.

In determining the accuracy of the agent, it is difficult to determine if the experimenter or the agent is correct. Accuracy was determined by how the experimenter thought the agent should have classified the messages. The agent may have found word patterns in the messages that were not noticed by the experimenter.

This research showed that using neural net-based agents that adapt to the category specifications of a user could accurately filter and coordinate information. Unlike the INFOSCOPE agents, the neural net adaptive agents did not rely on pre-programmed rules for filtering information. Additionally, they did not rely on the message poster to place the messages in the correct newsgroup; the adaptive agents can filter according to the message content. The only possible drawback is that if the categories are not well defined, then the agent may not be able to perform the classification.

4. GDSS APPLICATIONS

When many people are participating in a GDSS session, the number of messages can be large. A participant would be likely to be overloaded with information and thus miss relevant messages. To alleviate overload when many people are participating, one solution has been to send messages randomly to a subset of the participants. This increases the likelihood that a person will miss relevant information and is an unsatisfying solution to the information overload problem.

A better solution to the overload problem is to have an agent search the GDSS messages, adapting to the information needs of a user, so that only the relevant information is sent to a user. This kind of intelligent filtering is more satisfying than the randomness employed in previous filtering attempts.

The following describes some preliminary research which has been performed on using neural net agents to filter GDSS messages. After these research results, some of the important issues still to be considered in agent-based GDSS systems are considered.

332

4.1. Filtering Meeting Messages

Before testing the neural net-based agents on GDSS messages, a preliminary study was performed on messages from traditional meetings. The source of the messages was a 1973 Federal Reserve meeting. At that time, the Federal Reserve kept transcripts of all meetings, so this provides a rich source of data to test the ability of the agents to filter information into categories. The following describes the method to produce the neural net agents, and then results from applying the agents to classsify the messages from the meeting transcript according to the content of the message are discussed.

4.1.1. Dictionary Construction. The method for constructing the agents was similar to that used for the electronic bulletin board, with some important differences. Figure 3, which is similar to Figure 1, illustrates this new method. In the previous study only two or three postings were used for training to extract the common words. With this small number of postings, extracting the words common to all the postings was possible. With a larger number of files used in training, though, the requirement that all files have a common word becomes too stringent; if one file did not have an important word, then the word would be lost. In the following research, we tested two critical value (CritVal) percentages: 50 and 75. The 50% condition extracted a word if it was present in 50% of the files from the category; the 75% condition extracted the word if in 75% of the files.

Since the previous study found that small dictionaries were better than large dictionaries, the size of the dictionary was reduced by including a second filter at the top of the figure to remove the typical words such as "a" or "the." These typical words added nothing to the ability of the agent to classify messages according to content. Finally, the previous study showed that the common words dictionary was most accurate, so this dictionary was used in the present research.

4.1.2. Method. The experimenter read through the transcripts from the Federal Reserve meeting and found that many of the messages addressed five major subtopics: economic recovery, prices, the dollar, banking, and growth. Figure 4 shows a typical message from the economic recovery subtopic. Several training sets were composed in which a message was coupled with its classification. Some of the files were not in the training set; these were reserved for the test phase to determine the accuracy of the net.

Three parameters were manipulated: CritVal (the critical value percentages for including a common word in the dictionary), the number of cycles (the length of training, either 500 or 1000 cycles), and eta (the learning rate, either a slow or fast rate). CritVal is important because if it is too high, then important words may be missed. If too low, irrelevant words, which do not differentiate the categories, may be placed in the dictionary. For the number of cycles, one would like to reduce the learning time (have fewer cycles) but, at the same time, have enough cycles so that the weights can be adjusted properly. For the adjustment of the learning rate eta, if the learning rate is too fast the solution weights will jump around and may miss a good solution. If too slow, the solutions may settle on a local minimum which is not very accurate.

The resulting network was very similar to that shown in Figure 2. Once again, the dictionary words were used as input to the net and these were mapped through a hidden layer to five subtopic categories. The major difference in this study is that, to reduce the number of connections in the network, the number of nodes in the hidden layer was reduced from 30 to 8.

4.1.3. Results. Although a systematic study was not performed on the three parameters, some observations on the performance of the agent was illuminating. As before, the agent's accuracy

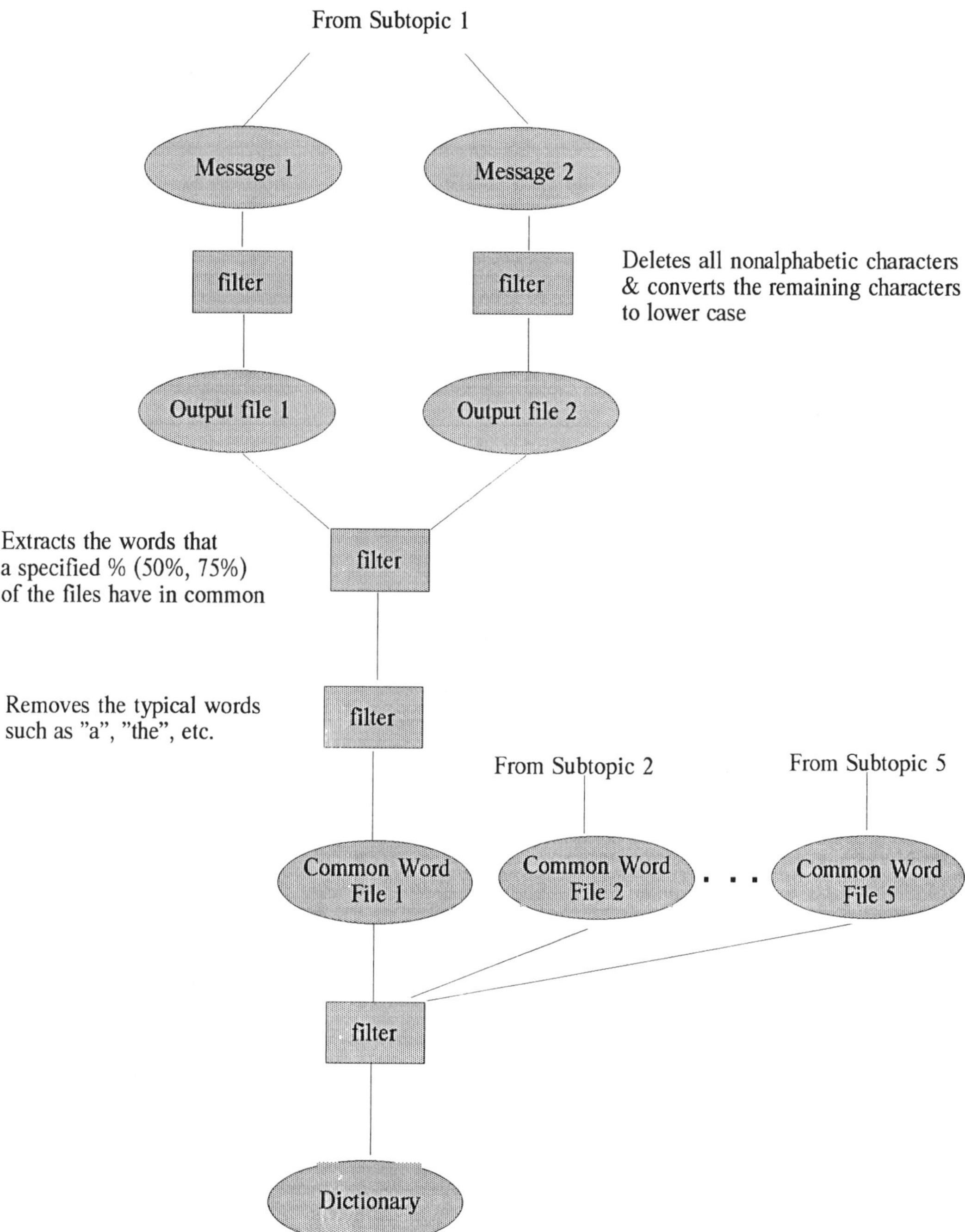

Figure 3. Illustration of the dictionary construction method for the neural net-based agents for filtering information from meeting messages.

> A serious financial disturbance could occur, and if it did, economic activity in the months ahead would be considerably weaker than projected. That possibility aside, greater weakness in the first two quarters of this year, if it developed, would most likely result from larger-than-projected declines in real expenditures for business fixed investment and more severe liquidation of inventories. On the other hand, housing starts--which had declined further in December, contrary to staff expectations--might now be at or near their low, and consumer confidence, already so weak, seem more likely to recover somewhat than to deteriorate further. In any case, the prospects for recovery beginning by summer or early fall will improve considerably if the President's fiscal proposals are enacted.

Figure 4. Example of a meeting message from the Federal Reserve meeting. This message was classified under the economic recovery subtopic.

was tested by having the experimenter categorize the test messages under the five subtopics before the agent was trained. Table 3 shows a summary of the results from the tests. The accuracy ranged from a low of 24% correct to 85% correct. The parameter which had the most effect on this variability was CritVal. Performance was poor when CritVal was 50% (accuracy ranged from 24% to 73%), but improved markedly when CritVal was 75% (a range of 70% to 85%). Overall, a slower learning rate for eta was preferable. More learning cycles improved performance marginally.

4.1.4. Conclusions. The accuracy of the agent could be quite high if the correct parameters are chosen. The ranges of accuracy seen in this study indicates that a participant in a GDSS could train and use an adaptive agent to search through and filter the incoming messages. Use of the agent in this way is certainly preferable to the standard practice of routing messages to the participants randomly. Agent accuracy does not have to be 100%; the 85% level seen in the best condition could certainly be used to reduce the participants' information overload in a more effective way than the random method of message routing. This research also showed that CritVal had the most

TABLE 3.

Accuracy of the agents as a function of critical value, learning rate and training cycles for categorizing the meeting transcripts

CritVal	eta	cycles	Average% correct
0.50	0.20	500	46
0.50	0.60	500	24
0.50	0.05	1000	73
0.50	0.20	1000	55
0.75	0.05	500	82
0.75	0.20	500	82
0.75	0.05	1000	85
0.75	0.20	1000	82
0.75	0.60	1000	70

effect on agent performance. If CritVal was too low, then too many irrelevant words were passed to the dictionary and the agent had difficulty mapping the message words to the category subtopic.

The accuracy of the agents for meeting messages was close to the accuracy of the text body condition of the electronic bulletin boards. Meeting messages do not typically have subject headings, so that the content of the text body is the only information source which could be utilized.

4.2. Incorporating the Agents in GDSS

The results from the meeting transcripts showed that the neural net-based agents could be useful for filtering and then routing messages in a GDSS session. For those participants interested in certain topic areas, the user could adapt the agent through the neural net training procedure so that it looks for certain kinds of information. After training, the agent could be turned on so that it would read all the incoming messages, prioritizing the messages according to the information needs of the user. The information overload would be reduced, and the participant could concentrate more on the important meeting topics rather than the mass of unfiltered information coursing through the GDSS network.

The GDSS interface has already been developed and has been used in other contexts. Figure 5 shows a screen dump of this interface. It has been developed on Sun workstations using X Windows and the Motif toolkit. By knowing the address of the workstations, the interface can be displayed on any workstation, at any site throughout the world, as long as network access to this machine is available.

To use the interface, the facilitator first connects all the workstations by specifying the addresses of the participants. Once started, the user has several windows in this particular version. The window at the top left (My Thoughts) is where the user composes the messages that will be sent to the group. After the message is completed, the user has two buttons which can be clicked in order to send the message: Group or Decisions. If the Group button is clicked, then the message is sent to the large lower window on the left (Group Discussion). This is used by the group to discuss the group process, such as ground rules, protocols, and procedures for how the group should function. Discussion topics can also include any items which need to be discussed before a final decision is reached. If the Decisions button is clicked, then the messages are sent to the Group Decisions Window at the lower right. This window displays the important output from the group; the critical decisions that are made in response to the group topics.

The agent window can be seen at the bottom right of the interface. The agent has been trained, at this point, and is waiting to be activated to filter the information for this participant. Each participant would have an agent, or several agents, which have been adapted to the information needs of the individual. The agent would read the messages which are being sent to the Group Discussion window; all participants should see the final decisions in the other window.

The software and interface has already been developed. Research is planned to test the interface in the near future.

4.3. Future Enhancements

A GDSS in the future will have to be more sensitive to team support issues than they have in the past. When a company assigns a team to solve problems, the dynamics within the team are different from those dynamics observed for small, loosely formed collaborative groups such as occurs in GDSS sessions. In particular, each member has a relatively defined and fixed position on the team (Baker and Salas, 1992). Team members must also be able to know the weaknesses

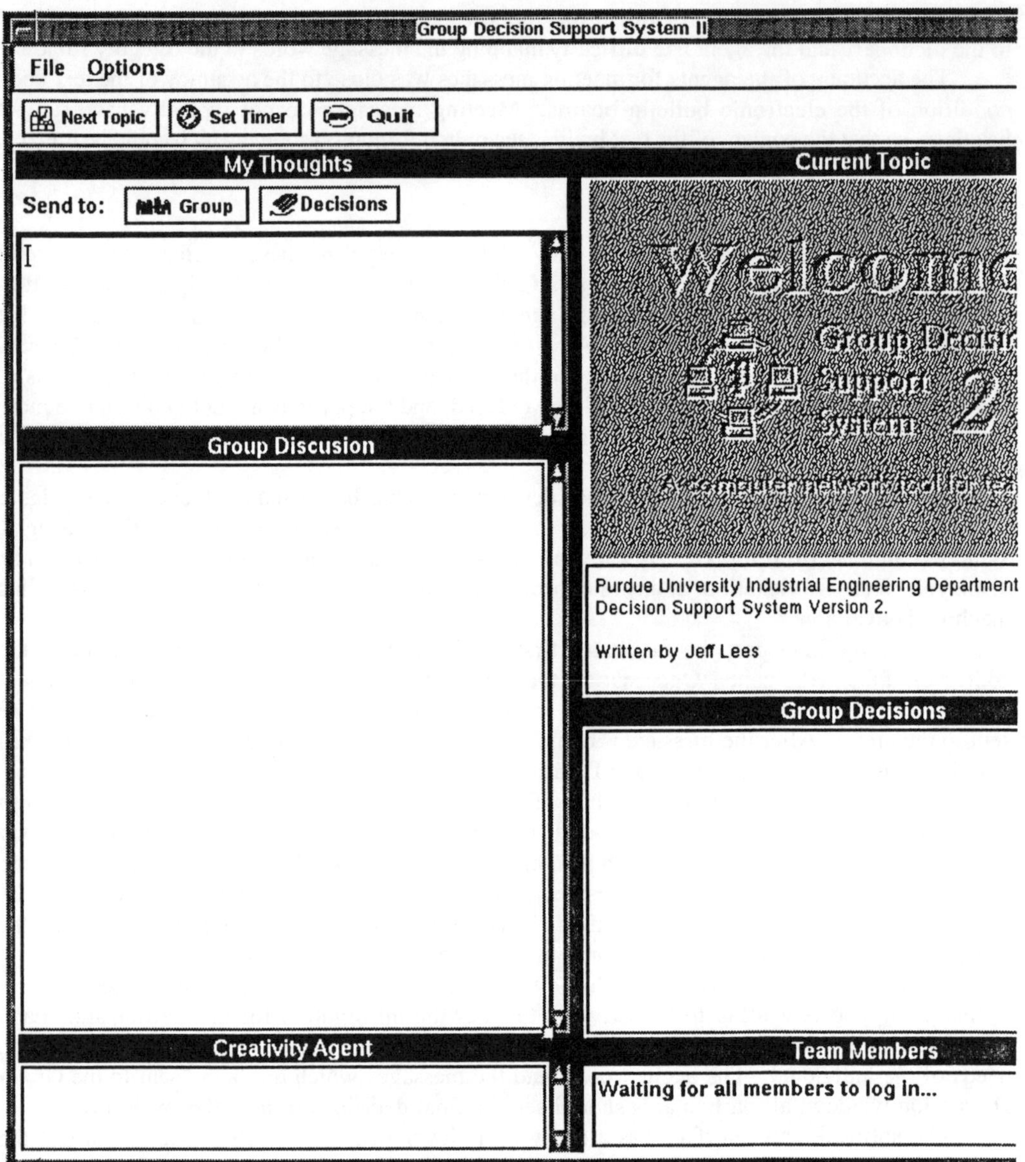

Figure 5. Screen dump of the interface design for the GDSS sessions.

and strengths of fellow team members. Parker (1990) finds that effective teams have a clear mission and plan; have excellent communication, openness, and trust among team members; and have a blend of people, each contributing a special talent. Many GDSS environments have been designed to support loosely-structured groups but not the structured teams so important to current business practices.

In order to support GDSS teams, we must understand the characteristics of effective teams. Two aspects of teamwork should be important to a GDSS environment: team player roles and team stages. In a non-GDSS situation, a team is identified by those participants who meet together, exchange information and ideas, and make decisions or solve problems. In the GDSS environment, teams can be formed, in similar ways, by routing messages within a set of members so that these participants can communicate with each other to accomplish the goals of the session.

4.3.1. Team Player Roles. An effective team requires a mix of team members who exhibit a variety of teamwork styles. Parker (1990) has researched the roles played by members in teams and has determined that effective teams need the following role players:

- Contributor - provides the group with good technical information and data

- Collaborator - defines the big picture or the vision of the team

- Communicator - listens to the other team members and facilitates the team by resolving conflicts and providing a comfortable environment where ideas can be presented freely

- Challenger - critiques the ideas of the team and encourages the team to take well-conceived risks

In simple terms, the contributor is concerned with the task, the collaborator is concerned with the goal, the communicator is concerned with the process, and the challenger is concerned with the criticism.

4.3.2. Team Stages. The team players may assume different roles as a team progresses through stages. Tuckman (1965) determined that teams must progress through four stages:

- Forming - team members cautiously grope for guidance, determining the roles of the other team members, and defining the task itself

- Storming - characterized by conflict among team members and resistance to the task while the team continues to determine its structure

- Norming - team members express a willingness to make the team work and develop norms of behavior and member interaction

- Performing - the team is ready to tackle the task with a clearly defined structure and purpose.

All of the stages are useful for developing a team identity. The challenge within a GDSS environment is to allow team members to float between the teams while at the same time understanding the team by being aware of the stages that the team has already gone through.

Future plans are to develop the agents so that they can support teamwork. It is expected that a team will form as a subset of participants within the total GDSS session. The following specialized agents are planned and will be developed in a similar manner to the content agents discussed in the previous sections.

- *Team Content Agent.* These agents will be similar to the content agent discussed above, except they will be trained on team content instead of individual content. The agent will be trained on previous messages from a team. Once trained, the agent can be sent out to read the messages from other teams in the GDSS. Those messages that correspond to the trained categories will be identified and accessed by the group. With this tool, a team should be able to determine if other relevant information is being discussed by the other teams, and how this information may affect their own team decisions.

- *Team Stages Agent.* This agent will be pretrained to identify the different stages in the team process: forming, norming, storming, and performing. This agent will be used to monitor the progress of the teams, to offer intelligent advice about how to proceed through the stages, and to help incorporate new team members into a core group. As an example of these functions, if a new team is found to skip one of the stages, intelligent advice will be provided to the team about the functions that should be discussed by the team in this particular stage before moving on to another stage. If a new team member is to be incorporated into a core team, then the team agent can be used to update the new member about the current stage of the team.

- *Role Type Agent.* This agent will be pretrained to identify the team member type. If a team reaches a stage, and the Role Type Agent determines that the team is not role balanced, then one of two things can happen: 1) the software tool will provide intelligent advice to remind those originally identified to play the role that they have to perform more of the functions of their member role; or 2) a new team member could be added to balance out the teams.

4.4. Discussion and Conclusions

This section has discussed how neural net-based agents can be used in a GDSS environment. The results from the meeting transcripts of the Federal Reserve showed that an agent can be trained to search and filter information according to individual content areas. A GDSS interface has been developed, and the agents have been incorporated into the software for this system. Research on these GDSS agents will occur in the near future.

The challenge of GDSS in the future is to support teamwork within a larger set of GDSS participants. It is envisioned that subsets of GDSS participants will form as teams, and that these teams will be structured and develop similar to traditional teams without computer support. Ideas were presented for how agents can help support teams in a GDSS environment.

In conclusion, a neural network approach to the development of the agents has several desirable features when incorporated in a GDSS:

- Agent memory can be lengthened or shortened by specifying the number of recent messages which will be used in the training set.

- Agents can recognize different dimensions of information from the same message by changing the mappings of messages to category.

- Recent unpublished work indicates that team player roles and team stages are two dimensions which can be recognized by neural network agents.

- Team agents can be trained on-line and can monitor information in real-time (Eberts, Villegas, Phillips, and Eberts, 1992).

5. E-MAIL APPLICATIONS

The final application area for nerual net-based agents is e-mail. Similar to the other application areas, users can be over-run with junk mail. The following describes typical e-mail systems, a previous attempt to filter e-mail messages (the Information Lens System), and a new neural net-based agent design for e-mail, called PostMaster.

5.1. Early E-Mail Systems

An e-mail message has to be structured so that the sender of the message indicates to whom the message will be sent. This requirement can be made more usable by incorporating a feature which allows the message to be addressed to groups of people, usually by a predefined single name, so that when the message is sent all the members of the predefined group will receive the message.

Another usability feature is that each message typically has a topic which is specified by the sender. When received by another person, messages will be tagged by this header information so the words used in the header could provide important context for the receiver. After the header, the sender enters the text of the message. Once received by another person, this user has several options after reading the message. The message could be deleted so that a copy is no longer available. Some systems provide an automatic means to save the message. The receiver of the message could be expected to reply to the sender on some aspect of the message. Some systems have a method to reply easily, such that the address is entered automatically, the header information is retained from the previous message, and the user has only to enter the new text.

One of the important issues with e-mail is how to facilitate transfer of information among the users without overloading the user (Malone, Grant, Lai, and Turbak 1988). A problem is that some users may be inundated with messages that can be classified as "junk mail." One method for performing this filtering is to use distribution lists. In a distribution list, users with similar topic interests are placed on a list and whenever a question arises on this topic, or information sharing is needed, the e-mail message is sent to all members on the distribution. This kind of filtering is left to the members of the distribution list to ensure that only messages relevant to the topic area will be distributed. Many of the advances in e-mail systems have occurred by trying to address this e-mail information filtering problem.

5.2. Malone's Information Lens System

Malone (1987) was one of the first to recognize the information filtering problems for e-mail applications and developed a method to filter information on e-mail which he termed the Information Lens (e.g., Malone et al., 1988). The Information Lens works upon the principle of semi-structured messages which are of an identifiable type and contain fields in which text can be entered. Different kinds of e-mail messages may have different structures and different fields. As an example, a meeting announcement message has a certain structure; a memo message has a different structure. The e-mail user potentially has many different structures to choose from when sending a message. Malone et al. (1988) has provided some capability for the user to define the structures for e-mail messages.

Different kinds of messages will have different kinds of structures and fields; these patterns are called message templates. When sending a message, the user must first select a template, specific to the kind of message sent, and then fill in the fields on the template. To be successful, the user must be aware of the different kinds of messages and the templates which are available for these messages. If the wrong template is chosen, the message information will not fit into the structure.

The structure of the message is not used to filter the information directly; the filtering is performed on the receiver's end. Since the structures of the messages have been defined and confined through the templates, rules can be developed for searching the fields and routing messages depending on the information in the fields. As an example, the "From" field indicates who the message is sent from. A simple rule for the information in this field would be if the message is from your boss, then assign a high priority to the message and read it quickly. Malone et al. (1988) have developed other rules also. The rules can be modified somewhat by the user to enhance the message routing and classification capabilities.

Malone et al. (1988) suggest that this Lens system can be incorporated into a larger system, such as a computerized calendar. For meeting announcements, the time of the meeting could be entered automatically on the user's computerized calendar. If the time slot on the calendar had already been filled by another appointment, then a message could be sent back automatically to the sender requesting a different time.

A problem with the Information Lens system, as argued by Fischer and Stevens (1991), is that it has placed a good deal of the filtering burden on the sender of the message instead of the receiver. The receiver of the message is responsible for constructing the filter that is used to classify and prioritize the messages, but the filter is useless unless the sender used the templates to construct the e-mail messages. Fischer and Stevens argue that the senders may not be motivated to go to the extra effort to use the templates, because the benefit is to the receiver. Only the receiver of the messages would be likely to pay the cost to achieve the filtering benefits.

5.3. Neural Network-Based Agents for E-Mail: PostMaster

PostMaster has been developed from the earlier Habibi and Eberts (1991) research to use neural networks to learn, from the user, how to sort e-mail messages into categories as the messages are received. The interface design for PostMaster, shown in Figure 6, has been developed on a Sun workstation using the Motif tool kit on top of the X Windows functions.

The following describes the features of the interface design. A window at the top right represents a pigeon hole where unsorted e-mail messages can be placed. In the figure, several messages have been received and are ready for sorting. At this point, the user has two choices: either read the messages in this box personally or have the agent read and sort the messages. The agent is waiting at the bottom of the window, and is poised for action when needed. If the user decides for the agent to read and sort the messages, then the mouse pointer is moved to the agent and clicked.

Once set in action, the agent moves to the top of the screen, towards the unsorted mail pigeon hole, and is depicted as reading a message (see Figure 7). When a message is read, the letters depicted in the pigeon hole are reduced by one. The three pigeon holes at the upper left of the figure are for sorted mail. The three boxes represent important mail to the left, medium important mail in the middle, and junk mail to the right. After the agent reads the mail, and the neural network classifies the mail based upon previous learning, the mail piece would be placed in one of the three boxes. At this point, a mail piece depiction would appear in the box. The agent continues reading the messages, sorting, and placing them in boxes until all the messages are read.

The user can now read the mail in any of the boxes. In each of the boxes, the mail has been prioritized by perceived importance, not by when the message was received. So, on the first click

Figure 6. A screen dump of the interface design for PostMaster. Twelve messages are in the unsorted box. The agent is ready to be activated for filtering the messages.

Figure 7. At this point for PostMaster, the user has clicked the agent button and the agent is reading and classifying the messages into the three boxes.

of the important mail box, the most important message from all the possibilities in the box would appear in the large window in the middle of PostMaster.

The learning for the neural network takes place in two ways. First, if PostMaster has not been used in the past, then all messages remain in the unsorted box and the agent cannot be activated. The reader must read the messages from the unsorted box and, when finished, PostMaster asks the user to sort the message by clicking on one of the boxes. PostMaster then saves the words for the dictionary representing that category. The neural net can then learn how to set the mapping between the input words and the three classifications.

The second way for PostMaster to learn is actually a relearning process. It is possible that a message may be classified wrongly from the previous learning. After reading the message, the user has the option to reclassify the message by clicking on a Reclassify button (see Figure 8). PostMaster takes this new information, retrains the neural network to account for it, and will use the new weights to classify any further messages.

The accuracy of PostMaster depends on how well the user trains the neural network (thus the importance of the master-apprentice relationship). If the user has a good categorization scheme, then PostMaster can learn this scheme. If the categorization scheme of the user is random, this will appear in the categorizations by PostMaster.

Several learning options have been parameterized, and tests arebeing performed currently to determine optimal values for these. It is possible that the user may change criteria for importance over time. The speed of adaptation to these changes can be set by changing the memory of the agent through the number of messages which are kept for each category. If the memory is high (such as 20 items), then PostMaster will be slow to adapt to changes in categorizations by the user. If the memory is low (such as 5 items) then PostMaster may change too quickly.

5.4. Discussion and Conclusions

Unlike Malone's Information Lens, PostMaster places the burden of categorization of messages on the receiver instead of the sender. The receiver is more likely to spend mental effort training the neural network, and, once trained, the agent can become a valuable assistant in increasing the productivity of the user. PostMaster has a possible advantage over the Fischer and Stevens (1991) INFOSCOPE method for categorization in that PostMaster relies less on pre-programmed rules for agent behavior, and the learning model is simpler. Similar to INFOSCOPE, which is based upon a cognitive theory of human memory functioning, PostMaster is based upon a cognitive theory of how humans learn to recognize patterns. With INFOSCOPE the emphasis is on memory, with PostMaster the emphasis is more on learning.

Several enhancements are planned for PostMaster. If the user already has folders for storing e-mail messages according to category, then the neural net can learn the contents of the folders. In line with this, the user should have some capability to increase the number of pigeon holes to correspond to specific user categories, not just importance of the mail. The e-mail system should also have the capability to read the messages in the background as they are received and indicate to the user the importance of the message through a pop-up window. The user can then decide whether or not to suspend current operations and look at the mail.

6. Overall Conclusions

The previous has discussed how neural net-based agents can be used to filter and coordinate information. The plan for the future is to integrate the agents for these separate applications into

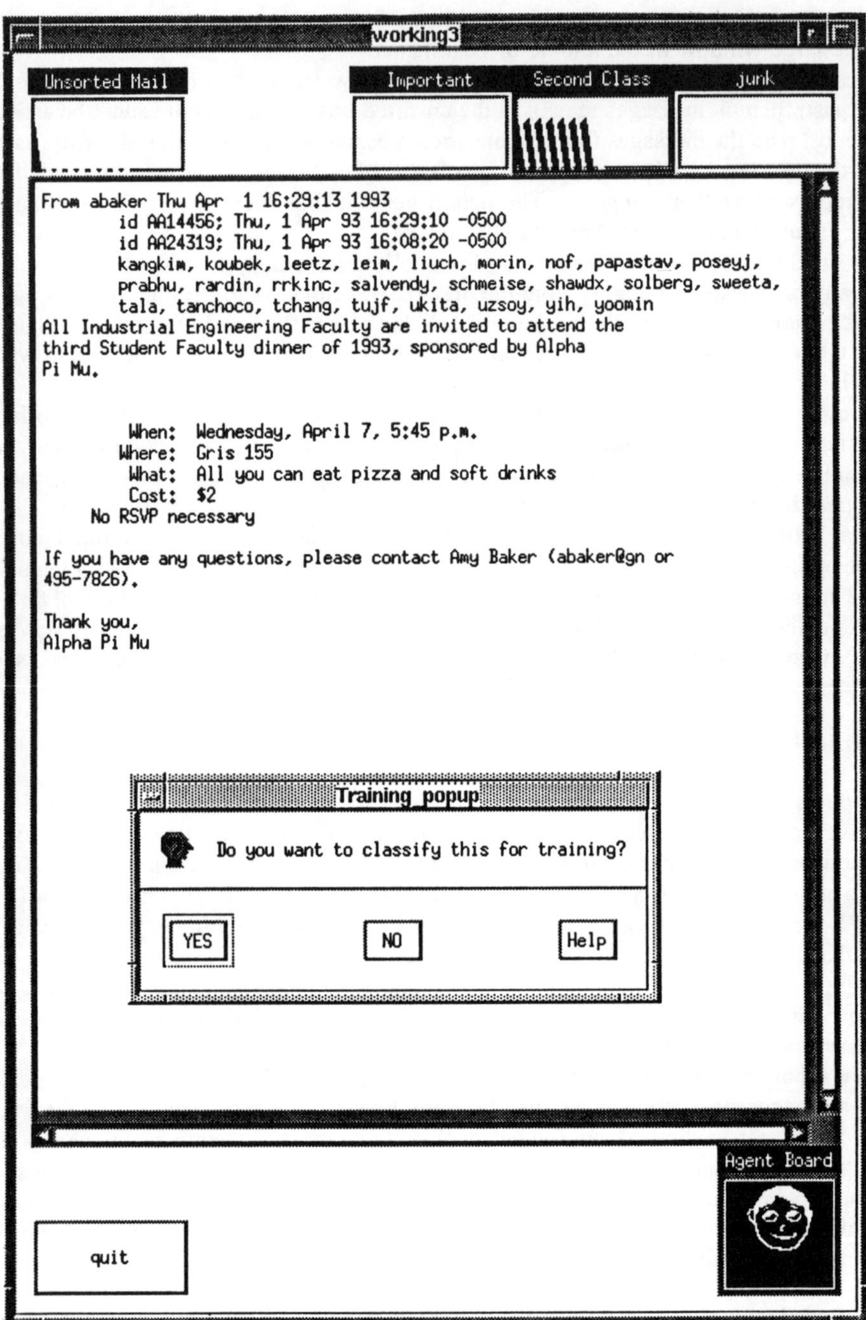

Figure 8. The user can read any of the messages which have been placed in the boxes. If the user thinks that this particular message was classified in the wrong way by the agent, then it can be reclassified. The agent will then learn the reclassification at a later time.

one system. In particular, the same agent developed for searching and filtering electronic bulletin board messages could be used in GDSS meetings or e-mail systems. It is envisioned that a particular user will have several agents trained to perform different tasks. The different agents could have different icon depictions depending on the characteristics. For example, an agent with a long memory could look different from an agent with a short memory. An agent searching for certain kinds of information could have certain physical features. This integration is planned for the near future.

7. References

Anderson, J. R. (1990). *The adaptive character of thought.* Hillsdale, NJ: Erlbaum.

Baffes, P. T. (1989). *NETS user's guide* Johnson Space Center: NASA.

Baker, D. P., and Salas, E. (1992). Principles for measuring teamwork skills. *Human Factors, 34,* 469-475.

Eberts, R. E., Villegas, L., Phillips, C., and Eberts, C. (1992). Using neural nets for user assistance in HCI tasks. *International Journal of Human-Computer Interaction., 4,* 59-77.

Fischer, G., and Stevens, C. (1991). Information access in complex, poorly structured information spaces. *Human Factors in Computing Systems Conference Proceedings, CHI'91* (pp. 63-70). New York: ACM.

Fukushima, K. (1987). Neural network model for selective attention in visual pattern recognition and associative recall. *Applied Optics, 26,* 4985-4992.

Fukushima, K. (1988a, March). A neural network for visual pattern recognition, *IEEE Computer,* 65-75.

Fukushima, K. (1988b). Neocognitron: A hierarchial neural network capable of visual pattern recognition. *Neural Networks, 1,* 119-130.

Fukushima, K., Miyake, S., and Ito, T. (1983). Neocognitron: A Neural network model for a mechanism of visual pattern recognition. *IEEE Transactions on Systems, Man, and Cybernetics, SMC-13,* 826-834.

Grossberg, S. (1976). Adaptive pattern classification and universal recoding: Part I. Parallel development and coding of neural feature detectors. *Biological Cybernetics, 23,* 121-134.

Habibi, S., and Eberts, R. (1991). Using neural networks to categorize bulletin board messages. In C. H. Dagli, S. R. T. Kumara, Y. C. Shin (Eds.), *Intelligent engineering systems through artificial neural networks* (pp. 947-952). New York: ASME Press.

Hanson, S. J., and Kegl, J. (1987). PARSNIP: A connectionist network that learns natural language grammar from exposure to natural language sentences. *Proceedings of the Ninth Annual Conference of the Cognitive Science Society* (pp. 106-119).

Hebb, D. O. (1949). *The organization of behavior.* New York: Wiley.

Malone, T. W. (1987). Computer support for organizations: Toward an organizational science. In J. M. Carroll (Ed.), *Interfacing Thought: Cognitive Aspects of Human-Computer Interaction* (pp. 294-323). Cambridge, MA: MIT.

Malone, T. W., Grant, K. R., Lai, K-Y., Rao, R., and Rosenblitt, D. (1988). Semistructured messages are surprisingly useful for computer-supported coordination. In I. Greif (Ed.), *Computer-Supported Cooperative Work: A Book of Readings* (pp. 311-331). San Mateo, CA: Kaufmann.

Miyata, Y. (1990). *A user's guide to SunNet version 5.6.* Computer Science Department, University of Colorado, Boulder.

Parker, G. M. (1990). *Team players and teamwork.* San Francisco: Jossey-Bass.

Rumelhart, D. E., and McClelland, J. L. (Eds.). (1986). *Parallel Distributed Processing: Explorations in the Microstructure of Cognition, Volume One: Foundations.* Cambridge, MA: Bradford Books/MIT Press.

Sejnowski, T. J., and Rosenberg, C. R. (1987). Parallel networks that learn to pronounce English text. *Complex Systems, 1,* 145-168.

Tuckman (1965). Development of sequences in small groups. *Psychological Bulletin, 63,* 384-399.

8. Acknowledgments

Partial support of this research has been provided by the Division of Design, Manufacturing, and Computer Integrated Engineering of the National Science Foundation, Award Number 8657590-DMC to Ray Eberts under the Presidential Young Investigator program. This work has benefitted from the collaborations of many people. Manoel Tenorio provided many helpful comments and suggestions during the course of the work on electronic bulletin boards. Shidan Habibi performed the test of the electronic bulletin boards. Jeffrey Hudson performed the test of the meeting messages. Jason Priebe programmed an early version of e-mail system and Jeffrey Lees programmed the GDSS interface. Finally, the computer team support ideas were developed in collaboration with Shimon Nof.

V. Information Models, Software, and Theories of Integration

NEW APPROACHES TO MULTI-AGENT PLANNING

Jeffrey S. Rosenschein
Eithan Ephrati
Computer Science Department
Hebrew University
Givat Ram
91904 Jerusalem
Israel

ABSTRACT. Coordination of actions by a group of agents corresponds to a group planning process. A multi-agent planning technique is presented that makes use of a dynamic, iterative search procedure. Through a process of group constraint aggregation, agents incrementally construct a plan that brings the group to a state maximizing social welfare. At each step, agents vote about the *next* joint action in the group plan (i.e., what the next transition state will be in the emerging plan). Using this technique agents need not fully reveal their preferences, and the set of alternative final states need not be generated in advance of a vote. With a minor variation, the entire procedure can be made resistant to untruthful agents.

1. Introduction

The term *planning*, in classical Artificial Intelligence (AI), usually refers to the generation of sequences of actions leading from an initial state to a goal state. Most planning research in AI has historically been concerned with a single agent, but more recently the term *multi-agent planning* has come to be used for the generation of coordinated sequences of actions for groups of agents.

The field of Distributed Artificial Intelligence (DAI) is concerned with coordinated, efficient activity by groups of autonomous agents. Given this characterization of the field, it is clear that one of the central missions of DAI is to explore multi-agent planning. The term, however, can be used in many different ways, and DAI researchers have chosen "planning" as a catch-all phrase that characterizes very different streams of work.

For example, one focus of DAI planning research has been that of "planning for multiple agents," which considers issues inherent in centrally directed multi-agent execution. Smith's Contract Net [30, 31] falls into this category, as does other DAI work such as [28, 26, 19]. A second focus for research has been "distributed planning," where multiple agents all participate in coordinating and deciding upon their actions [4, 29, 5, 37, 36, 6, 27].

The question of whether the group activity is fashioned centrally or in a distributed manner is only one axis of comparison. Another important issue that distinguishes DAI research in multi-agent planning is whether the goals themselves need to be adjusted, that is, whether there may be any fundamental conflicts among different agents' goals. Thus, for

349

S. Y. Nof (ed.), Information and Collaboration Models of Integration, 349–364.

example, Georgeff's early work on multi-agent planning assumed that there was no basic conflict among agent goals, and that coordination was all that was necessary to guarantee success [14, 15, 32]. Similarly, planning in the context of Lesser, Corkill, and Durfee's research often involves coordination of activities (e.g., sensor network computations) among agents who have no inherent conflict with one another (though surface conflict may exist). "Planning" here means avoidance of redundant or distracting activity, efficient exploration of the search space, etc.

Another important issue is the relationship that agents have to one another, e.g., the degree to which they are willing to compromise their goals for one another (assuming that such compromise is necessary). *Benevolent Agents* are those that, by design, are willing to accomodate one another; they have been built to be cooperative, to share information, and to coordinate in pursuit of some (at least implicit) notion of global utility. These benevolent agents for the most part characterize research in the DAI subfield of Cooperative Problem Solving. In contrast, research in the subfield of Multi-Agent Systems posits the existence of self-motivated agents, perhaps built by different designers, with purely local notions of utility. Each will cooperate only when it is in its best interest to do so [38]. Still another potential relationship among agents is a modified master-slave relationship, called a "supervisor-supervised" relationship, where non-absolute control is exerted by one agent over another [7, 9].

The synthesis, synchronization, or adjustment process for multiple agent plans thus constitute some of the (varied) foci of DAI planning research. Synchronization through conflict avoidance [14, 15, 32], distribution of a single-agent planner among multiple agents [3], and the use of a centralized multi-agent planner [28], have all been explored. Other recent work includes [33, 34, 35, 22, 17, 20].

In this paper, we present a new approach to deriving multi-agent plans (this work has also been reported on in [12]). We consider how agents could reach consensus about what multi-agent plan to carry out using a voting procedure, without having to reveal full goals and preferences (unless that is actually necessary for consensus to be reached). Our technique also does away with the need to generate final alternatives ahead of time (instead, candidate states arise at each step as a natural consequence of the emerging plan). The agents iteratively converge to a plan that brings the group to a state maximizing social welfare.

2. The Scenario

Our scenario involves a group of n agents operating in a world currently in the state s_0. Each agent has its own private goal. Since all agents operate in the same environment and may share resources, it is desirable that they agree on a joint plan for the group that will transform the world to an agreed-upon final state.

Each agent is assumed to have a value, a *worth*, that it associates with states that satisfy its goal. This worth can be used to assign a utility to any state, goal or not, that is reached by a particular plan (the techniques for doing this assignment are discussed below).

Given the existence of such a worth function, we want to give the agents a method for choosing a group plan. The agents will all participate in the choice process, as well as in carrying out the resulting multi-agent plan. The group plan should then result in a compromise state that is in consensus.[1]

2.1. AN EXAMPLE

Consider a simple scenario in the slotted blocks world. There are four slots (a, b, c, d), five blocks $(1, 2, 3, 4, 5)$, and the world is described by the relations: **On**(Obj_1, Obj_2)—Obj_1 is stacked onto Obj_2; **Clear**(Obj)—there is no object on Obj; and **At**$(Obj, Slot)$—Obj is located at $Slot$. The function **loc**(Obj) returns the location (slot) of Obj. Slots themselves function as (stationary) objects (e.g., block 1 in slot b could be described by $On(1, b)$).

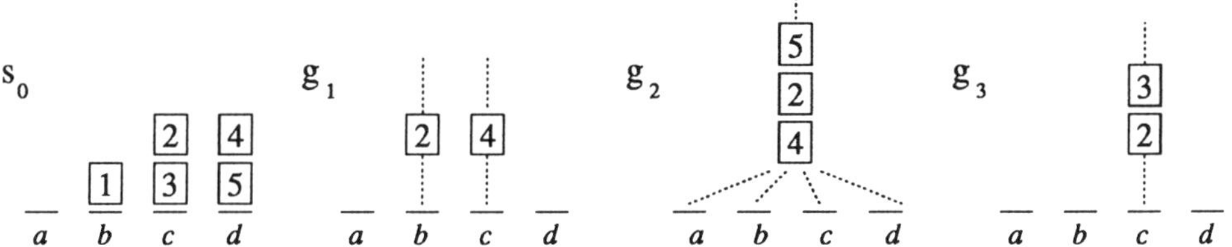

Figure 1: A Blocks World Example

There are three agents operating in the world. The start state is shown at the far left of Figure 1. As further represented in that figure, these agents have (respectively) the following goals:
$g_1 = \{At(4, c), At(2, b)\}, g_2 = \{On(2, 4), On(5, 2)\}, g_3 = \{On(3, 2), At(2, c)\}$, with respectively the the individual worths: 8, 12 and 16.

There is only one available operator: **Move**(Obj_1, Obj_2)—place Obj_1 onto Obj_2. This operator can be characterized by the following STRIPS-like lists:
$[Prec: Clear(Obj_1), Clear(Obj_2), On(Obj_1, Obj_x)]$,
$[Del: On(Obj_1, Obj_x), Clear(Obj_2), At(Obj_1, loc(Obj_1))]$,
$[Add: On(Obj_1, Obj_2), At(Obj_1, loc(Obj_2)))]$.
Assume that when a single agent performs the $Move$ operation there is a cost of 4, while if two agents $Move$ an object together the operation costs a total of 3 (1.5 each).

We want the agents to jointly choose a plan that results in a compromise state.

3. General Overview of the Process

The algorithm enables a group of agents to find the state that maximizes their social welfare function f^U. There are many possible global utility functions, such as taking the product of individual agent utilities, or taking their median, or taking their sum. Our procedure here is not dependent on any particular function, and is equally suitable regardless of the one chosen. Given the set of individual utility functions of the agents in the decision group, we define the Social Welfare/Utility $U(E)$ (of any set of constraints E) to be that function f^U of all the individual worth functions, minus the cost of achieving the set from s_0 (that is, $U(E)$ equals f^U minus the cost of achieving E).

---

[1]The decision procedure, to be presented below, may also serve in a single-agent planning scenario, where an agent is trying to integrate a group of disparate goals.

The underlying idea is the *dynamic generation of alternatives* that locates the most desirable state for the society. At each step, all agents reveal additional information about their private goals. The current set of candidate states is then expanded and (possibly) pruned to comprise the new set of candidate states. The process continues until all newly formed sets of constraints have lower social utility than their ancestor sets. Note that our primary concern here is not with the complexity of the planning process but rather with the resulting state of the multi-agent plan.

The search procedure needs an accurate value for the social utility of each candidate state in order to proceed correctly (i.e., the search space of alternatives is dynamically pruned by the social welfare criterion). To provide this value, the agents vote over the set of candidates at each step.

The search method combines aspects of hill-climbing with breadth-first search. For example, parallel searches are carried out in each promising direction (where social utility is growing or constant within the gap bound $\nabla_f u$[2]). When the search encounters a direction where social utility decreases beyond this limit, however, the search is terminated (reminiscent of hill-climbing).

The procedure has the following advantages: **(a)** alternatives are generated by the entire group dynamically (allowing the procedure to be distributed [11]); **(b)** utilities will be calculated and submitted only for "feasible" alternatives (utilities of infeasible alternatives need not be revealed, reducing the choice procedure's computational complexity, and also respecting agent privacy when possible [10]); **(c)** agents are required to submit only the minimally "conflict-sufficient" information about their goals (described below), further maintaining their privacy.

3.1. DEFINITIONS

- $e(g_i)$ is the set of absolutely necessary constraints needed for any optimal plan to achieve the goal g_i, starting at the initial state s_0. In accordance with the partial order over these constraints,[3] we divide $e(g)$ into subsets of constraints. Each such subset within $e(g)$ comprises all the constraints that can be satisfied within j (optimal) steps, and are necessary at some subsequent step after j. The total number of operators (the length of the plan) that satisfies the set of constraints $e(g)$ is denoted by $l(e(g))$.

  We denote $e(g)$'s components by $\bigcup_j E_g^j$, such that E^j includes all the constraints that can be satisfied within j steps, and are necessary at some step $\geq j$. For any $j \geq l(e(g))$, we define E^j to be the description of the goal g.

  Example: In the blocks world scenario from Section 2.1., $e(g_1)$ for Agent 1 would be: $E_1 = [C(2), C(4)] (= E_1^1) \cup [C(4), A(2,b)] (= E_1^2) \cup [A(4,c), At(2,b)] (= E_1^3)$

- A constraint $I \in E^j$ is said to be *temporary* if later in the plan there is a constraint $\bar{I} \in E^k$ (where $k > j$) that denies it ($I \wedge \bar{I} \models F_{alse}$). We say that a set of apparently conflicting constraints E is semi-consistent ($E \not\models_{tmp} F_{alse}$) if the removal of temporary

---

[2] Given f^U and the agents' utility functions, we define the gap bound $\nabla_f u$ to be the maximal gap between any local maximum of f^U and any local minimum that follows it.

[3] Constraints are temporally (partially) ordered sets of the domain's predicates associated with the appropriate limitations on their codesignation.

constraints makes it consistent (see Section 4.1. for an example). Given a semi-consistent E we define $r(E)$ to be the set of all maximal consistent subsets of the predicates in E.

- $P(E)$ denotes the set of the cheapest "grounded" (or "complete" [1]) plans that achieve the final subset of E. We denote these states by $s(E)$
  $(= \{s \mid (s \models E^{l(E)}) \wedge c(s_0 \rightsquigarrow s) = \min_{k \models E^{l(E)}} c(s_0 \rightsquigarrow k)\}$).

- $F_{ollow}(E)$ is defined to be the set of constraints that can be satisfied by invoking at most one operator, given the initial set of constraints E ($F_{ollow}(E) = \{I \mid \exists op \exists P[op(P(E)) \models I]\}$).

4. The Algorithm

This section describes the algorithm in more detail, along with a running example. At each step of the procedure, agents try to impose more of their private constraints on the group's aggregated set of sets of constraints. Since agents want to maximize their own utility, they will impose as many constraints as they can at each step. The set of all non-pruned aggregated sets of constraints at step k is denoted by $\mathcal{A}^k$ (its constituent sets will be denoted by A_j^k, where j is simply an index over those sets). $\mathcal{A}^{k+}$ denotes the set $\mathcal{A}^k$ before it is pruned (similarly, its non-pruned components are denoted by A_j^{k+}).

As an example, assume a simple scenario of the slotted blocks world (we use the same operators and predicates as described in Section 2.1.). There are 3 blocks $(1,2,3)$ and two agents (a_1, a_2). The initial state is $\{On(1, b), On(3, 1), On(2, c)\}$. The agent's goals are (respectively) $g_1 = \{On(1, 2)\}$ and $g_2 = \{On(2, 3)\}$.[4] The exact procedure is defined as follows:

1. At step 0 each agent i finds $e(g_i)$—the individual temporally ordered set of constraints that achieves the goal g_i, starting from the initial state. The virtual set of alternatives is initialized to be the empty set ($\mathcal{A}^0 = \emptyset$).

   In our example, we have:[5] $e(g_1) = \{[\texttt{C}(2)] \cup [\texttt{C}(2), C(1)] \cup [O(1, 2)]\}$ (this ordered set induces the plan $\langle M(3, \bar{2}), M(1, 2) \rangle$) and $e(g_2) = \{[\texttt{C}(3), C(2)] \cup [O(2, 3)]\}$ (inducing the plan $\langle M(2, 3) \rangle$). Note that $\texttt{C}(2)$ and $\texttt{C}(3)$ are temporary since they are denied later in the plan.

2. At step k, each agent may declare $E_i^{A_j^k} \subseteq E_i^{l+1}$ only if any $k \leq l$ E_i^k was already declared and accepted by the group, and the declaration is "feasible," i.e., it can be reached by invoking one operator on some set of constraints that was reached by the procedure during the previous step: $\exists A_j^k [(A_j^k \in \mathcal{A}^k) \wedge (\bigcup_{n=1}^l E_i^l \subseteq A_j^k) \wedge (E_i^{A_j^k} \subseteq F_{ollow}(A_j^k))]$. i can try to impose elements of his "next" private subset of constraints on the group decision only if they are still relevant and his previous constraints were accepted by the group.

---

[4]This is reminiscent of Sussman's Anomaly in the single-agent planning scenario—where the plan to achieve one sub-goal obstructs the plan that achieves the other. The example assumes that although the final state that satisfies both agents' goals costs 9 to reach, it is the state that meets the social welfare criterion.

[5]We will use the first letter to denote the full operator predicate, and $\bar{y}$ to denote any location excluding y's. We also use a typewriter font to denote temporary constraints.

At the first step, each agent i may declare $E^1_{g_i}$. In our example, this will be $E^1_{g_1} = [\mathtt{C}(2)]$ and $E^1_{g_2} = [\mathtt{C}(3), C(2)]$. At the second step, a_1 declares $E^{A^1_1}_1$, which in this example equals $E^2_{g_1} = [\mathtt{C}(2), C(1)]$. Similarly, a_2 declares $E^2_{g_2} = [O(2,3)]$. (Both are in $F_{ollow}(\mathcal{A}^1)$, which contains only one subset.) At the third step, a_1 declares $[O(1,2)]$ and a_2 declares $[O(2,3)]$ (his final goal, which is already in $\mathcal{A}^2$).

3. For each set of constraints $A^k_j \in \mathcal{A}^k$, we generate all the maximal consistent or semi-consistent extensions $\{A^{(k+1)+}_j\}$ with elements of $\bigcup_i E^{A^{(k+1)+}_j}_i$ (i.e., each element in $A^{(k+1)+}_j$ is defined as $\{A^k_j \cup \{I \mid (I \in \bigcup_i E^{A^{(k+1)+}_j}_i) \wedge ((I \cup A^{(k+1)+}_j) \not\models_{tmp} F_{alse})\}\}$).

   At the first step, the aggregated set of constraints is $\{\mathtt{C}(3), C(2)\}$. At the second step, both declarations may coexist consistently, and there is therefore only one successor to the previous set of constraints: $\{\mathtt{C}(2), C(1), O(2,3)\}$. At the third step, the aggregated set is $\{O(1,2), O(2,3)\}$, and it satisfied both agents' goals.

4. At this stage, all extensions are evaluated so as to enable the pruning of sets that reduce social utility. Each agent declares the utility it associates with each newly-formed set of aggregated constraints.

   The first set of aggregated constraints is satisfied by the initial state, and thus induces the null plan (with cost of zero). The value given to that first set is the value that agents assign to the initial state. The second set can be achieved by the plan $\langle M(3,a) \rangle$, with cost of 3. In order for the set not to be pruned, it must be the case that the agents value the set by at least $3 + \nabla_{f^U}$ more than the initial set. The third set is the set that satisfies the social welfare criterion (by our assumption above), and therefore (by definition) will have higher value than previous steps (and not be pruned).

5. The next set of sets of constraints is pruned so that it contains only sets that do not decrease the social utility (with respect to their ancestor set) by more than the gap bound ∇_{f^U}. Formally, $\mathcal{A}^{k+1} = \{A^{k+1}_j \mid A^{k+1}_j \in \mathcal{A}^{(k+1)+}_j \wedge \forall l \leq k[U(A^{k+1}_j) \geq U(A^l_j) + \nabla_{f^U}]\}$.

   In this simple example, all the individual extensions are mutually consistent. Therefore, there is only one aggregated extension at each step. Since this set brings the agent closer to the set that satisfies the social welfare criterion, it is not pruned.

6. The process ends when $\mathcal{A}^{k+1} = \mathcal{A}^k$. The values assigned by each agent to each state in $\mathcal{A}^k$ are then compared to find the consistent sets that maximize social utility ($\hat{\mathcal{A}}^k = \{A \mid A \in \mathcal{A}^k \wedge \forall A^*[A^* \in \mathcal{A}^k \Rightarrow U(A^*) \leq U(A)]\}$). Among these equivalent sets, one is randomly chosen.

   In the example there is only one such set—the final set which is determined in the fourth step.

Theorem 1 *Given any social welfare function f^U, and an appropriate gap bound ∇_{f^U}, this mechanism finds all sets that maximize the Social Welfare. Consensus will be reached after at most $O(\max_A l(P(A)))$ steps such that $P(A)$ results in a state that maximizes the group's social welfare.*

Proof. The proof of this theorem, and the one below, appear in [8] with slightly different notations. □

4.1. CONSTRAINTS

There are several important aspects of, and requirements for, the procedure above to succeed, which we discuss in this section. Figure 2 shows three simple scenarios in the slotted blocks world that will help us explain the issues involved. In these examples, two agents can achieve a consensus state that fully satisfies both agents' goals. There are three slots, and several blocks in each world. Only the *Move* operator is available, and it costs 2 under all circumstances.

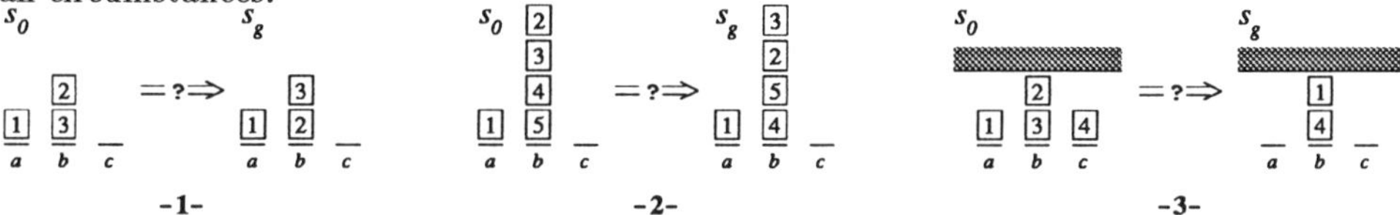

Figure 2: Three scenarios in the slotted blocks world

4.1.1. Specification of Constraints.

An important requirement for the success of the procedure is that each agent identify and declare only the absolutely necessary constraints needed for its individual plan to succeed. As an example, consider the first scenario in Figure 2. Two agents want to achieve the following goals: $g_1 = \{A(1,a), C(1)\}$ and $g_2 = \{O(3,2), A(2,b)\}$. These two goals can co-exist, as shown in the final goal state. To achieve his goal, a_1 need not take any action. a_2, on the other hand, has to carry out $\langle M(2,x), M(3,\bar{2}), M(2,b), M(3,b)\rangle$. The only way for the second step of this plan to be completed is by stacking either block 3 or block 2 onto block 1. By including either of these into his set of constraints, a_2 would encounter a_1's opposition, thus (perhaps) preventing his own goal from being achieved.

However, the purpose of the second *Move* operator is just to achieve $A(3,\bar{b})$. If a_2 declares his set of constraints to be $[C(2)]\cup[C(2),C(3)]\cup$
$[C(2),C(3),A(3,\bar{b})]\cup[C(2),C(3),A(2,b)]\cup[A(2,b),O(3,2)]$, the conflict is avoided.

4.1.2. Aggregation of Temporary Constraints.

In the second scenario, a_1's goal is the same, but a_2's goal is now
$\{O(5,4), O(3,2), A(2,b), A(4,b)\}$. As in the previous example, there is no way for him to achieve his goal without trying to temporarily violate a_1's goal. But in contrast to that example, there is no way to specify a_2's constraints in a way that would avoid the conflict. Here, for example, $A(2,\bar{b})$, although temporary, is crucial for a_2's plan's success. Were the process to consider sets of constraints to be relevant only as long as they were consistent, the process might stop after two steps, avoiding the goal state from even being considered.

It is therefore necessary that the *temporary* conflicts between the agents' plans not cause deadlocks (that is, if there is a temporal order that can later resolve them). This phenomena is achieved by allowing the existence of semi-consistent sets of constraints, and having agents express preferences over these sets. For that reason, agents should recognize what their temporary constraints are (terms that will be violated by their own future actions). Identifying the temporary constraints of his own plan, a_2's set of constraints (E_2) would then become:

$[C(2)] \cup [C(2), A(2,\bar{b})] \cup [A(2,\bar{b}),C(4), A(3,\bar{\bar{2b}})] \cup [A(2,\bar{b}),$
$C(4), A(3,\bar{\bar{2b}}), C(5), A(4,\bar{b})] \cup [A(2,\bar{b}),C(4), A(3,\bar{\bar{2b}}),C(5),$
$A(5,\bar{\bar{4b}})] \cup [A(2,\bar{b}), C(4), A(3,\bar{\bar{2b}}), C(5),A(5,\bar{\bar{4b}}), C(2),A(4,$
$b)]\cup[A(3,\bar{\bar{2b}}),C(2), A(4,b),O(5,4),C(3)]\cup[C(2),A(4,$
$b),O(5,4),C(3), A(2,b)]\cup[A(4,b),O(5,4),O(3,2),A(2,b)]$

As can be seen in this specification, all the constraints that contradict a_1's goal are temporary. Therefore, even though a_2's plan actually violates a_1's goal, the mutual goal state is reachable.

4.1.3. Aggregation of Functional Constraints.

The generation of consensus sets of constraints is based on the aggregation of the individual sets of constraints. As the third example in Figure 2 shows, this is not always trivial. Here, there is a ceiling that makes it impossible for more than two blocks to be stacked. This time a_1's goal is $A(1,b)$ while $g_2 = A(4,b)$. We assume that the function $h(b)$ returns the height (in number of blocks) at slot b. As in all our scenarios, s_0 satisfies E_i^1 of both agents.[6] Following the second step of his plan, each of the agents has as temporary constraints $h(b) \leq 1$ and $C(x)$, where x is the block he wants to move onto slot b. These constraints enable each agent to move "his" block to slot b after removing block 2.

The aggregation of these constraints requires careful analysis. First, it must be recognized when the aggregated constraints of two identical terms such as $h(x) \leq 1$ will be $h(x) \leq 2$, or $h(x) \leq 1$, or even $h(x) \leq 0$. Had block 4 been located on block 1 in the initial state of our example, the third solution would be appropriate. Using it in the given scenario, however, would yield as the aggregated set of constraints in the second step $[C(1),C(4),h(b) \leq 0]$. The induced states of this set cost 10 move operators, while the actual plan that achieves the mutual goal costs only 5. The problem here is that both constraints are *temporary* (each agent needs a free space for one block only momentarily). Thus, in this case, the aggregated height should stay 1, leading to a state that is only three *Move* steps distant from the initial state. Unfortunately, it not clear how in general this subtle analysis is to be done.

4.1.4. Evaluation of Sets of Constraints.

Evaluation of sets of constraints plays an important role in the search procedure. One straightforward worth function for an arbitrary set A might be built by taking the worth of a goal state (assumed to be available), subtracting the cost of the single-agent plan from A to the goal, then subtracting the agent's share of the cost of the multi-agent plan to get from start state s_0 to A.

---

[6] $E_1^1 = [C(1), C(2), h(c) \geq 1]$ and $E_2^1 =[C(4),C(2),h(a) \geq 1]$. Note that by its nature, the height constraint is temporary, since it always is a precondition of an action that violates it!

Note, however, that using the above equation the worth of s_0 for an agent would simply be the worth of his goal, minus the cost of his one-agent plan to reach that goal (in a one-agent scenario this would be true for every set). Thus, since the evaluation function does not capture the notion of progress in the plan, the agent has no motivation to carry out his plan at all.

There are several ways to refine the worth function so as to solve this problem. One way is by making the "future-cost" (i.e., $w_i(g_i) - c_i(A \rightsquigarrow g_i)$) more sensitive to the progress of the plan. A simple approach is to take into consideration only that fraction of the goal's worth which reflects the amount of work already done to achieve it ($\approx w_i(g_i) \times [l(P(s_0 \rightsquigarrow A))/l(P(A \rightsquigarrow g_i))]$, which is meaningful only if $w_i(g_i) > c_i(s_0 \rightsquigarrow g_i)$). Another way is to give greater weight to the cost of operators that are located further along in the plan ($\approx w_i(g_i) - \sum_1^k k \times C(op_k)$). Or, assuming that each operator has a probability ($pr(op_k)$) associated with its success, we could use $\approx (\prod_1^k pr(op_k) \times w_i(g_i)) - c_i(A \rightsquigarrow g_i)$. These evaluations may be further refined by having weighted costs and/or probability of success associated with each of the constraints that needs to be achieved in order to transform the given set into the goal set (see [16] and [18] for richer probabilistic approaches).

Note that instead of assigning worth to sets of constraints, it may sometimes be more natural to evaluate their induced states ($s(A)$ instead of A). In any case, the worth associated with all states induced from a single set of constraints will be equivalent. In addition, note that for many variations of the above worth functions, it will be sufficient to take the gap bound $\nabla_f v$ to be zero (what was called a progressive worth function in [8]). For example, it would be sufficient to assume above that $\forall ic(A_1 \rightsquigarrow A_2) \leq c_i(s_1 \rightsquigarrow s_2)$).

5. Manipulative/Insincere Agents

In choosing a state that maximizes social welfare, it is critical that agents, at each step, express their true worth values. However, if our group consists of autonomous, self-motivated agents, each concerned with its own utility (and not the group's welfare), they might be tempted to express false worth values, in an attempt to manipulate the group choice procedure. This is a classic problem in voting theory: the expression of worth values at each step can be seen as an (iterative) cardinal voting procedure, and we are interested in a non-manipulable voting scheme so that the agents will be kept honest. We have investigated other aspects of this problem in previous work [10].

Fortunately, there do exist solutions to this problem, such that the above plan choice mechanism can be used even when the agents are not necessarily benevolent and honest. If the social welfare function is taken to be the (weighted) sum f_{sum}^U (or average) of the individual utilities, it is possible to ensure that all agents will vote honestly. This is done by minor changes to the procedure of Section 4., that allow it to use a variant of the Clarke Tax mechanism (CTm).

In [6] we proposed the CTm as a plausible group decision procedure. The basic idea of the mechanism is to make sure that each voter has only one dominant strategy, telling the truth. This phenomenon is established by choosing the alternative that scores the highest sum of bids/votes and then taxing some agents. The tax (if any) equals the portion of the agent's bid for the winning alternative that made a difference to the outcome. Given this scheme, revealing true preferences is the dominant strategy.

In our procedure, the agents are participating in many intermediate votes, and alternatives are generated dynamically (as a consequence of the intermediate votes). Therefore, the original version of the CTm cannot be used efficiently. Instead we use an iterative variation of the CTm; at each step, the tax is defined with respect to all previous steps, but is actually levied only at the final step. We use the following definitions for the stepwise Clarke tax mechanism:

- The function $v_i : \mathcal{A} \to \mathbb{R}$, returns the *true* worth (to a_i) of each aggregated set A. Similarly, the function $d_i^k(j)$ returns the *declared* worth of the set A_j by agent a_i at step k. $\vec{d_i^k}$ denotes the vector $\langle d_i^k(1), \ldots, d_i^k(m) \rangle$, the agent's declared worth over all alternatives and $\vec{v_i^k}$ denotes the true value.

- The profile of preferences declared by all agents at step k is denoted by D_n^k, where $D_{\neg i}^k$ denotes this set excluding i's preferences, such that $D_n^k = (D_{\neg i}^k, \vec{d_i^k})$.

- The choice function $f : D_n^k \times \mathcal{A} \to \mathcal{A}$ returns the state that is the *maximizer* of $\sum_{i=1}^n d_i^k(A)$;

- The tax imposed on i at step k is $t_i^k(f(D_n^k)) = \sum_{j \neq i} d_j^k(f(D_{\neg i}^k)) - \sum_{j \neq i} d_j^k(f(D_{\neg i}^k, \vec{d_i^k}))$, if this value is positive. Otherwise, t_i^k will be zero. Therefore, the utility $u_i^k(f(D_n^k))$ of agent i with respect to the chosen alternative is $w_i(f(D_n^k)) - t_i^k(f(D_n^k))$.

The planning algorithm itself should also be updated in two ways. First, since each intermediate vote is only over a subset of candidates, there is the possibility that an agent will "shift" his vote by a constant, keeping a single round's preferences accurate while undermining inter-vote comparisons. To maintain truth telling as a dominant strategy, it is necessary that artificial shifting does not occur. Therefore, we will require that all votes be relative to some "benchmark": we include A^0 (the empty set) in the set of alternatives at every step. If each agent is motivated to give his true preferences over the other states relative to A^0 ($v_i(A)$), then the score of each state s in the vote is exactly $f_{sum}^U(A)$.

Second, the tax is calculated with respect to the final choice. Knowing that a *semi-consistent* set cannot be chosen, an agent might give it an artificial value in order to change the final outcome. We therefore allow agents to vote only over consistent sets. Step 4 of the algorithm is therefore changed as follows:

- For each $A \in \mathcal{A}^{(k+1)+} \setminus \mathcal{A}^k$ find $r(A)$, its maximal consistent subsets. Each agent gives its vote regarding each state in $r(A) \cup A^0$. The worth of a consistent set is simply the sum of individual worths given to that state (note that for any consistent set A, $r(A) = A$). The worth of each semi-consistent set in $\mathcal{A}^{(k+1)+}$ is computed as follows: for each agent and semi-consistent set A, there is a consistent set with maximal worth in $r(A)$. The worth of a semi-consistent set is taken to be the sum of these maximal worth sets in $r(A)$, over all agents ($\sum_i \max_{E \in r(A)} w_i(E)$).

At the and of the process, each agent is fined the Clarke Tax with respect to the final group choice.

Theorem 2 *At any step k of the procedure, i's best strategy is to vote over the alternatives at that step (A^k) according to his true preferences $\vec{v_i}$.*

5.1. USING THE PROCEDURE ON OUR EXAMPLE

We now use the iterative procedure with its CTm to solve the first problem presented in Section 2.1.. We assume that the cost of reaching a state is divided equally among the agents (by side-payments if necessary), and that each agent i uses the worth function $w_i(A) = w_i(g_i) - c_i(s(A) \rightsquigarrow g_i)$ (the goal's worth minus the cost of the work needed to transform a state induced by the set to the goal state). From the agents' individual goals, we get the following constraints:

$E_1 = [C(2), C(4)](= E_1^1) \cup [C(4), A(2, b)](= E_1^2) \cup [A(4, c), At(2, b)](= E_1^3)^7$

$E_2 = [\mathtt{C}(2), \mathtt{C}(4)] \cup [\mathtt{C}(4), \mathtt{C}(2), C(5)] \cup [C(2), C(5), O(2, 4)] \cup [O(2, 4), O(5, 2)]$

$E_3 = [\mathtt{C}(2)] \cup [\mathtt{C}(2), C(3)] \cup [C(2), C(3), \mathtt{A}(3, \bar{c})] \cup [\mathtt{C}(2), C(3), A(2, c)] \cup [A(2, c), O(3, 2)]$

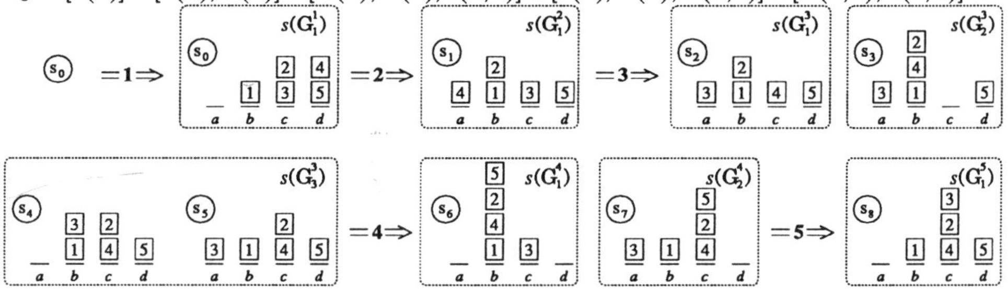

Figure 3: Induced States of the Five Steps

Figure 3 presents the induced states at each step (in this example, all the generated sets are consistent). At the first step, each agent declares E_i^1. $\mathcal{A}^1$, the set that includes all possible consensus sets of constraints, has only one member: $A_1^1 = \{[C(2), C(4)]\}$. $s(A_1^1)$ also has only one member, s_0. Agents then vote on this state, and it receives a score of 0 (for example, a_2 goal's worth is 12, and to achieve it the agent would perform $\langle M(4, \bar{c}), M(2, 4), M(5, 2) \rangle$ at a cost of 12; therefore, he values s_0 as 0).

At the second step, each agent hands in E_i^2 (which is in $F_{ollow}(A_1^1)$ for each i). Since all these constraints coexist consistently, $\mathcal{A}^2 = [C(2), C(4), C(5), C(3), A(2, b)]$. This set induces the single state $s_1 (= s(A_1^2))$ as described in Figure 3. Note that there are many other states that could satisfy this set of constraints, but s_1 has the *minimal cost*. This state can be achieved by $\langle Move(4, a), Move(2, 1) \rangle$; therefore, the state costs 6. Subtracting this cost from the worth values given by each agent (4 in this case by all three agents), the state scores 6. Since this score is greater than that of the preceding state s_0, the process continues.

At the third step, the new added constraints generate 3 possible maximally consistent extensions:

$A_1^3 = [A(2, b), A(4, c), C(2), C(5), C(3), A(3, \bar{c})]$ inducing s_2,

$A_2^3 = [O(2, 4), A(2, b), C(2), C(5), C(3), A(3, \bar{c})]$ inducing s_3,

and $A_3^3 = [O(2, 4), A(4, c), C(2), C(5), C(3), A(3, \bar{c})]$ inducing s_4 and s_5.

---

[7] a_1's plan might be $\langle Move(2, 1), Move(4, 3) \rangle$. The first operation is enabled since the constraint $C(2)$ is satisfied. Th $C(2)$ constraint is satisfied by s_0, which is included in E_1^1. $C(4)$ is needed for a future operator, but it is also satisfied by s_0, and therefore it too is included in E_1^1. $A(2, b)$ can be satisfied within one move, and is necessary at all future times for the plan to succeed, so it is included in any future set of constraints.

These induced states respectively score 11, 3, 12, and 12; A_2^3, which decreases the social utility, is therefore pruned.

At the fourth step, the two remaining extensions are extended further; a_1 hands in $\{A(2,b), A(4,c)\}$ (which is in $F_{ollow}(A_1^3)$), a_2 hands in $\{O(2,4), O(5,2)\}$ and a_3 $\{C(3), C(2), A(2,c)\}$ (both in $F_{ollow}(A_3^3)$). These constraints yield six different extensions that again induce the states s_2, s_3, s_4, s_5 and the new states s_6 and s_7 (that score respectively -5 and 5). Therefore, only the sets that induce s_4 and s_5 can be further extended (by a_3) to induce s_8. Although s_8 fully satisfies a_3's goal, it scores only 5 and the process ends.

All intermediate votes are now gathered for the final vote. Both s_4 and s_5 maximize the social welfare utility (both are one operation distant from each of the agents' goals). Both a_2 and a_3 are taxed 2. a_2 improves its utility by 2 and a_3 by 6 (a_1's utility is not improved with respect to s_0). The group's social utility is therefore improved by 8.

6. Related Work

There are a number of artificial intelligence researchers whose work relates to the approach we have been discussing above. Some of this work follows in the footsteps of Korf [21], who showed that the planning search space can be reduced if the final goal can be decomposed into several sub-goals, and the plans that achieve these sub-goals can be combined to achieve the original goal. This result suggests that the multi-agent planning algorithm presented in this paper can also serve to reduce the search space in a single-agent planning scenario if a non-optimal solution is acceptable.

A similar approach is taken in [25] to find an optimal plan. It is shown there how planning for multiple goals can be done by first generating several plans for each subgoal and then merging these plans. Finding the solution is guaranteed (under several restrictions) only if a sufficient number of alternative plans is generated for each sub-goal. Our approach does away with the need for several plans for each subgoal by using constraints instead of grounded plans (a level of abstraction that represents all possible grounded plans).

In [13] it is shown how to merge *grounded linear plans* (as opposed to aggregating constraints) in a dynamic fashion. To achieve an optimal final plan it takes that algorithm $O(\prod_{i=1}^n l(P(g_i)))$, while the approximation algorithm that is presented there takes polynomial time.

Our approach also resembles the GEMPLAN system [23]. There, the search space of the global plan is divided into "regions" of activity. Planning in each region is done separately, but an important part of the planning process within a region is the updating of its overlapping regions (in our terms, all individual plans are generated and aggregated simultaneously). This model served as a basis for the DCONSA system [27] where agents were not assumed to have complete information about their local environments. The combination of local plans was done through "interaction constraints" that were pre-specified.

The concept of solution that our algorithm employs (maximization of social welfare) also resembles the approach taken in CONSENSUS [2] where several expert systems "elect" a plan that scores the highest rating with respect to the individual points of view. There, however, the election refers to different complete global plans that each expert generates.

Another advantage of our proposed process is that it can easily be modified to deal with dynamic priorities. Since the search is guided by the vote taken at each step, it is possible to allow the agents to change their "tastes" or priorities over time (for example, due to environmental changes). As an example, in the Multi-Fireboss Phoenix system [24] planning (the actions needed to assess and contain fires) is performed by several spatially distributed agents. The system addresses, through a sophisticated negotiation protocol, the dynamic allocation of resources. Our algorithm would solve this problem in a direct manner, without negotiation. At each time interval, the agents would vote over the possible relevant distributions (one step of the algorithm per time interval). Given the individual utilities, the accurate distribution of resources would be chosen that maximizes the social utility (minimizes the damage according to the group's perspective). In addition (as mentioned in Section 5.), there is no need to assume that the agents are benevolent.

7. Conclusions

We have introduced a dynamic, iterative voting procedure. It enables a group of agents to construct a joint plan that results in a final state that maximizes social welfare for the group. The technique is more direct and formally specified than other consensus procedures that have been proposed, and maintains agent privacy more effectively. Techniques such as these provide a natural method for the coordination of multi-agent activity. Conflicts among agents are then not "negotiated" away, but are rather incrementally dealt with. Agents iteratively search for a final state that maximizes the entire group's utility, incrementally constructing a plan to achieve that state. The search can be constructed so that any manipulation by an untruthful agent will harm the agent more than it helps him.

Acknowledgments

This research was partially supported by the Israeli Ministry of Science and Technology (Grant 032-8284).

References

[1] D. Chapman. Planning for conjuctive goals. *Artificial Intelligence*, 32:333–377, 1987.

[2] R. Clark, C. Grossner, and T. Radhakrishnan. ONSENSUS: a planning protocol for cooperating expert systems. In *Proceedings of the Eleventh International Workshop on Distributed Artificial Intelligence*, pages 77–94, Glen Arbor, Michigan, February 1992.

[3] D. Corkill. Hierarchical planning in a distributed environment. In *Proceedings of the Sixth International Joint Conference on Artificial Intelligence*, pages 168–175, Tokyo, August 1979.

[4] Daniel D. Corkill. *A Framework for Organizational Self-Design in Distributed Problem-Solving Networks*. PhD thesis, University of Massachusetts, Amherst, Massachusetts, 1982. Also published as COINS Technical Report 82-33, Computer and Information Science, University of Massachusetts, Amherst, Massachusetts, December 1982.

[5] Edmund H. Durfee, Victor R. Lesser, and Daniel D. Corkill. Cooperaton through communication in a distributed problem solving network. In Michael N. Huhns, editor, *Distributed Artificial Intelligence*, chapter 2, pages 29–58. Morgan Kaufmann Publishers, Inc., Los Altos, California, 1987.

[6] E. Ephrati and J. S. Rosenschein. The Clarke Tax as a consensus mechanism among automated agents. In *Proceedings of the Ninth National Conference on Artificial Intelligence*, pages 173–178, Anaheim, California, July 1991.

[7] E. Ephrati and J. S. Rosenschein. Constrained intelligent action: Planning under the influence of a master agent. In *Proceedings of the Tenth National Conference on Artificial Intelligence*, pages 263–268, San Jose, California, July 1992.

[8] E. Ephrati and J. S. Rosenschein. Multi-agent planning as search for a consensus that maximizes social welfare. In *Pre-Proceedings of the Fourth European Workshop on Modeling Autonomous Agents in a Multi-Agent World*, Rome, Italy, July 1992.

[9] E. Ephrati and J. S. Rosenschein. Planning to please: Planning while constrained by a master agent. In *Proceedings of the Eleventh International Workshop on Distributed Artificial Intelligence*, pages 77–94, Glen Arbor, Michigan, February 1992.

[10] E. Ephrati and J. S. Rosenschein. Reaching agreement through partial revelation of preferences. In *Proceedings of the Tenth European Conference on Artificial Intelligence*, pages 229–233, Vienna, Austria, August 1992.

[11] E. Ephrati and J. S. Rosenschein. Distributed consensus mechanisms for self-interested heterogeneous agents. In *First International Conference on Intelligent and Cooperative Information Systems*, Rotterdam, May 1993. To appear.

[12] E. Ephrati and J. S. Rosenschein. Multi-agent planning as a dynamic search for social consensus. In *Proceedings of the Thirteenth International Joint Conference on Artificial Intelligence*, Chambery, France, August 1993. To appear.

[13] D. E. Foulser, M. Li, and Q. Yang. Theory and algorithms for plan merging. *Artificial Intelligence*, 57:143–181, 1992.

[14] M. Georgeff. Communication and interaction in multi-agent planning. In *Proceedings of the National Conference on Artificial Intelligence*, pages 125–129, Washington, D.C., August 1983.

[15] M. Georgeff. A theory of action for multi-agent planning. In *Proceedings of the National Conference on Artificial Intelligence*, pages 121–125, Austin, Texas, August 1984.

[16] P. Haddawy and S. Hanks. Issues in descision-theoretic planning: Symbolic goals and numeric utilities. Technical report, University of Illinois at Urbana-Champaign, IL, 1990.

[17] M. Kamel and A. Syed. An object-oriented multiple agent planning system. In Les Gasser and Michael N. Huhns, editors, *Distributed Artificial Intelligence, Volume II*, pages 259–290. Pitman Publishing/Morgan Kaufman Publishers, San Mateo, CA, 1989.

[18] K. Kanazawa and T. Dean. A model for projection and action. In *Proceedings of the Eleventh International Joint Conference on Artificial Intelligence*, 1989.

[19] Matthew Katz and Jeffrey S. Rosenschein. Verifying plans for multiple agents. *Journal of Experimental and Theoretical Artificial Intelligence*, 1993. To appear.

[20] D. Kinny, M. Ljungberg, A. Rao, E. Sonenberg, G. Tidhar, and E. Werner. Planned team activity. In *Pre-Proceedings of the Fourth European Workshop on Modeling Autonomous Agents in a Multi-Agent World*, Rome, Italy, July 1992.

[21] R. E. Korf. Planning as search: A quantitative approach. *Artificial Intelligence*, 33:65–88, 1987.

[22] Thomas Kreifelts and Frank von Martial. A negotiation framework for autonomous agents. In *Proceedings of the Second European Workshop on Modeling Autonomous Agents in a Multi-Agent World*, pages 169–182, Saint-Quentin en Yvelines, France, August 1990.

[23] A. L. Lansky. Localized search for controlling automated reasoning. In *Proceedings of the Workshop on Innovative Approachess to Planning, Scheduling and Control*, pages 115–125, San Diego, California, November 1990.

[24] T. Moehlman and V. Lesser. Cooperative planning and decentralized negotiation in Multi-Fireboss Phoenix. In *Proceedings of the Workshop on Innovative Approaches to Planning, Scheduling and Control*, pages 144–159, San Diego, November 1990.

[25] D. S. Nau, Q. Yang, and J. Hendler. Optimization of multiple-goal plans with limited interaction. In *Proceedings of the Workshop on Innovative Approaches to Planning, Scheduling and Control*, pages 160–165, San Diego, California, November 1990.

[26] Edwin P. D. Pednault. Formulating multiagent, dynamic-world problems in the classical planning framework. In Michael P. Georgeff and Amy L. Lansky, editors, *Reasoning About Actions & Plans*, pages 47–82. Morgan Kaufmann Publishers, Inc., Los Altos, California, 1987.

[27] R. Pope, S. Conry, and R. Meyer. Distributing the planning process in a dynamic environment. In *Proceedings of the Eleventh International Workshop on Distributed Artificial Intelligence*, pages 317–331, Glen Arbor, Michigan, February 1992.

[28] J. S. Rosenschein. Synchronization of multi-agent plans. In *Proceedings of the National Conference on Artificial Intelligence*, pages 115–119, Pittsburgh, Pennsylvania, August 1982.

[29] J. S. Rosenschein and M. R. Genesereth. Deals among rational agents. In *Proceedings of the Ninth International Joint Conference on Artificial Intelligence*, pages 91–99, Los Angeles, August 1985.

[30] Reid G. Smith. *A Framework for Problem Solving in a Distributed Processing Environment*. PhD thesis, Stanford University, 1978.

[31] Reid G. Smith. The contract net protocol: High-level communication and control in a distributed problem solver. *IEEE Transactions on Computers*, C-29(12):1104–1113, December 1980.

[32] Christopher Stuart. An implementation of a multi-agent plan synchronizer. In *Proceedings of the Ninth International Joint Conference on Artificial Intelligence*, pages 1031–1033, Los Angeles, California, August 1985.

[33] Frank von Martial. Multiagent plan relationship. In *Proceedings of the Ninth International Workshop on Distributed Artificial Intelligence*, pages 59–72, Rosario Resort, Eastsound, Washington, September 1989.

[34] Frank von Martial. Coordination of plans in multiagent worlds by taking advantage of the favor relation. In *Proceedings of the Tenth International Workshop on Distributed Artificial Intelligence*, Bandera, Texas, October 1990.

[35] Frank von Martial. Coordination by negotiation based on a connection of dialogue states with actions. In *Proceedings of the Eleventh International Workshop on Distributed Artificial Intelligence*, pages 227–246, Glen Arbor, Michigan, February 1992.

[36] Gilad Zlotkin and Jeffrey S. Rosenschein. Cooperation and conflict resolution via negotiation among autonomous agents in noncooperative domains. *IEEE Transactions on Systems, Man, and Cybernetics*, 21(6):1317–1324, December 1991.

[37] Gilad Zlotkin and Jeffrey S. Rosenschein. Incomplete information and deception in multi-agent negotiation. In *Proceedings of the Twelfth International Joint Conference on Artificial Intelligence*, pages 225–231, Sydney, Australia, August 1991.

[38] Gilad Zlotkin and Jeffrey S. Rosenschein. A domain theory for task oriented negotiation. In *Proceedings of the Thirteenth International Joint Conference on Artificial Intelligence*, Chambery, France, August 1993. To appear.

MODELING AND PROTOTYPING COLLABORATIVE SOFTWARE PROCESSES

P. DAVID STOTTS
Department of Computer Science
University of North Carolina
Chapel Hill, NC 27599-3175
USA

RICHARD FURUTA
Department of Computer Science
University of Maryland
College Park, MD 20742
USA

ABSTRACT. The correct and timely creation of systems for coordination of group work depends on the ability to express, analyze, and experiment with protocols for managing multiple work threads. We present an evolution of the Trellis model that provides a formal basis for prototyping the coordination structure of a collaboration system. Like its predecessor, the new Trellis model has the nicely exploitable duality of being both graph formalism and parallel automaton. The automaton semantics provide dynamic information about the interactions of agents in a collaboration; the graph structure forms the basis for the static link structure of a hyperdocument. We give several analysis techniques for the model, and demonstrate its use by expressing the interaction structure of some common forms of collaborative system.

This work is partially supported by the National Science Foundation under grant numbers IRI–9007746 and IRI–9015439, and by the Software Engineering Research Center (University of Florida and Purdue University).

1 Introduction

The Trellis project [SF89, SF90b] has investigated for the past several years the structure and semantics of human computer interaction in the context of hypertext/hypermedia systems, program browsers, visual programming notations, and software process models. Our design work has been guided since the early projects by a *simplicity-over-all* principle; this means we develop as simple a model as practical at first, and study how far towards a general solution it will take us before we add more capability, or "features" to the formalism. As a result, our interaction models strike a balance between fully-programmable/non-analyzable (like Apple's Hypercard product) and fully-analyzable/non-programmable (static directed graphs).

In this report we will refer to an information structure in Trellis as a *hyperprogram*. Due to the unique features combined in the Trellis model, a hyperprogram integrates user-manipulatable information (the hypertext) with user-directed execution behavior (the process). We say that a hyperprogram *integrates task with information*.

When using Trellis in a CSCW context, the net structure serves several functions with a single notational framework: its structures shared applications; its synchronizes loosely coupled parallel executing applications; it provides a repository for application information and historical data; and it provides mechanisms for joint decision making and action. Semantic

365

S. Y. Nof (ed.), Information and Collaboration Models of Integration, 365–390.
© 1994 *Kluwer Academic Publishers. Printed in the Netherlands.*

nets and link typing may be as useful for pure hypertext description. Object-based message passing languages are probably as appropriate for expressing parallel threads. Production systems are probably as useful for specifying group interactions. However, Trellis provides a single formalism for all these aspects of a collaboration support framework.

Due to the heavy interpretation as hypertext, Trellis hyperprograms are especially useful for processes in which human direction is an important aspect of the control flow. An example is the software development process we discuss in section 6. Such computations are referred to as being *enacted*, rather than as being executed, to distinguish the major role human input and human decisions (and for CSCW, human interactions) have in the unfolding of the actions described in the hyperprogram.

2   Formal definitions

The Trellis project is an ongoing effort to create interactive systems that have a formal basis and that support analytical techniques. The first such effort was a hypertext model [SF89], with a followup framework for highly-interactive time-based programming (termed *temporal hyperprogramming* [SF90b]). The model we present here is an extension of these earlier designs that explicitly distinguishes the various agents acting within a linked structure, and that provide an analyzable mechanism with which agents may exchange data. This new model basically follows the Trellis framework of annotating a form of *place/transition net (PT net)*, and using both graph analysis and state-space analysis to exploit the naturally-dual formalism.

The following short section outlines some of the basic concepts and terminology of PT nets, their structure, and common behaviors; readers already familiar with these notions may choose to skip it. Following that we introduce the group- and timing-specific net definitions, and finally the model of collaboration structures based on these nets.

2.1   Net theory basics

The notation used here is taken from Reisig [Rei85]. For the interested formalist, Murata [Mur89] gives a broad and thorough introduction to net theory and modeling applications. We present here just the basics required for understanding our application of this theory.

A PT net is a bipartite graph with some associated execution semantics. The two types of nodes in a net are termed *places*, represented visually as circles, and *transitions*, represented visually as bars. Activity in the net is denoted with *tokens*, drawn as dots in the places. Two nodes of the same type may not be joined by an arc. Given the arc structure of a net, the set of inputs to a node n is termed the *preset* of n, denoted $\bullet n$, and the set of output nodes is termed the *postset* of n, denoted $n\bullet$. Figure 1 shows the common representation of these PT net components (we will discuss the interpretation of this figure later); the varying patterns on tokens in this diagram represent *colors*, a mechanism for class typing discussed in detail later.

One widely used form of PT net is the *Petri net*.[1] A transition t in a Petri net is said to be enabled if each place in $\bullet t$ is *marked*, i.e., contains at least one token. Once enabled,

---

[1] We will use the general term *PT net* to describe the place and transition net syntax that is common to

a transition t may *fire*, causing a token to be removed from each place of $\bullet t$ and depositing one token in each place of $t\bullet$. A *net marking*, or *net state*, is a vector of integers, telling how many tokens reside in each place. Execution of a Petri net begins with some initial marking, and continues through a sequence of state changes caused by choosing an enabled transition in the current state and firing it to get a new state. Execution certainly terminates if a state is reached in which no transitions are enabled, but it may also be defined to terminate with any marking that has some special significance to the user of the net.

2.2 COLORED TIMED NET

The Trellis model is based primarily on a synchronously executed, transition-timed Petri net as the structure of a hyperprogram. For use in CSCW, we have employed a form of net model known generically as *high-level* nets. High-level nets have been introduced in several forms by different researchers, including predicate-transition nets [GL81], colored Petri nets [Jen81], and nets with individual tokens [Rei83].

We present our ideas in a hybrid notation. We will use the Jensen's terminology of colored nets, but the simplified syntax presented by Murata in his high-level net summary [Mur89]. All forms of high-level nets can be translated into one another, and are thus equivalent, but the simple syntax we use creates explanations that are more clear. We will discuss these other syntaxes after the examples.

In colored nets, tokens have type (color) and may carry data structure. A token of one color is distinguishable from a token of another color; within a color class, however, individual tokens cannot be distinguished from one another. The timing of the original Trellis model has been retained and combined with color to produce this model:

Definition 1 *Colored timed net structure*

A colored timed net structure CTN is a 5-tuple, $CTN = <S, T, F, \kappa, \tau>$ in which

$S = \{p_1, \ldots, p_n\}$ *is a finite set of* places *with $n \geq 0$,*

$T = \{t_1, \ldots, t_m\}$ *is a finite set of* transitions *with $m \geq 0$, and $S \cap T = \emptyset$,*

$F \subseteq (S \times T) \cup (T \times S)$ *is the* flow relation, *a mapping representing arcs between places and transitions.*

$\kappa : \{\kappa_1, \ldots, \kappa_r\}$ *is a finite set of* colors *for typing tokens, where each color is a function $\kappa_i : S \rightarrow \{0, 1, 2, \ldots\}$;*

$\tau : T \rightarrow \{0, 1, 2, \ldots\} \times \{\infty, 0, 1, 2, \ldots\}$ *is a function mapping each transition to a pair of values termed* release time *and* maximum latency *respectively. For any transition $t \in T$, we write $\tau(t) = (\tau_t^r, \tau_t^m)$ and we require that $\tau_t^r \leq \tau_t^m$.*

In this model, we have simplified the notation used in Reisig [Rei85] by assuming that the weight on each arc is 1, and that the token capacity of each place is unbounded. A net marking is a vector of token counts, with each token count being a vector of color counts; a marking provides a snapshot, at some point during execution, of how many tokens of each color reside in each place.

---

many forms of concurrent computation model. We reserve the term *Petri net* to describe a form of PT net with a specific (and familiar) execution semantics.

For a transition $t \in T$, its release time represents the number of time units that must pass once t is enabled before it can be fired; its maximum latency represents the number of time units that may pass after t is enabled before it fires automatically.

This temporal structure is very similar to that of Merlin's *Time Petri nets* [Mer74, MF76], with a few differences. The two time values for each transition here are integers, whereas Merlin used reals. We also have a need for the maximum latency to possibly be unbounded, using the special designation ∞ which is not in Merlin's model. Finally, times are not thought of as durations for transition firing in Trellis. Transitions are still abstractly considered to fire instantaneously, like the clicking of a button in a hypertext interface. Time values in Trellis are thought of as defining ranges for the *availability* of an event.

2.3 COLLABORATION PROTOCOL STRUCTURE (CPS)

The timed Trellis model of hypertext uses the structure and execution rules of timed Petri nets to specify both the linked form and the browsing semantics of a hypertext. This logical structure then is interpreted through a layer of indirection to arrive at a displayed form for reader consumption and interaction. Hypertext content and linked structure are thus effectively separated by the timed Trellis model.

Definition 2 *Collaboration protocol structure*

A collaboration protocol structure *is* $CPS = < CTN, \ M_0, \ C, \ W, \ B, \ P_l, \ P_d > in \ which$

$CTN = < S, \ T, \ F, \ \kappa, \ \tau > is \ a \ colored \ timed \ net,$

$M_0 : S \rightarrow < c_1, c_2, ..., c_r > is \ an$ initial marking *(or* initial state*) for* CTN, *where*
$r = |\kappa| \ and \ \forall p \in S, M_0(p)_i = c_i = \kappa_i(p),$

$C \ is \ a \ set \ of \ document$ contents,

$W \ is \ a \ set \ of$ windows,

$B \ is \ a \ set \ of$ buttons,

$P_l \ is \ a \ logical \ projection \ for \ the \ document,$

$P_d \ is \ a \ display \ projection \ for \ the \ document.$

A CPS consists of a CTN representing the document's linked structure, a marking to tell how many tokens of each color start in each net place, several sets of human-consumable components (*contents, windows,* and *buttons*), and two collections of mappings, termed *projections*, between the CTN, the human-consumables, and the display mechanisms. A window from W is a logically distinct locus of information. A button from B is an action that causes the current display to change in a specified way. Content elements from C can be many things: text, graphics, tables, bit maps, executable code, sound, or, most importantly, *another CPS.*

A logical projection P_l provides mappings from components of a CTN to the human-consumable portions of a group work environment as mentioned above. Each place in the CTN has a content element from C mapped to it, as well as an element of W for the abstract display of the content. Each transition in the net has a logical button from B associated with it. The display projection P_d is a set of mappings that take the logical components and

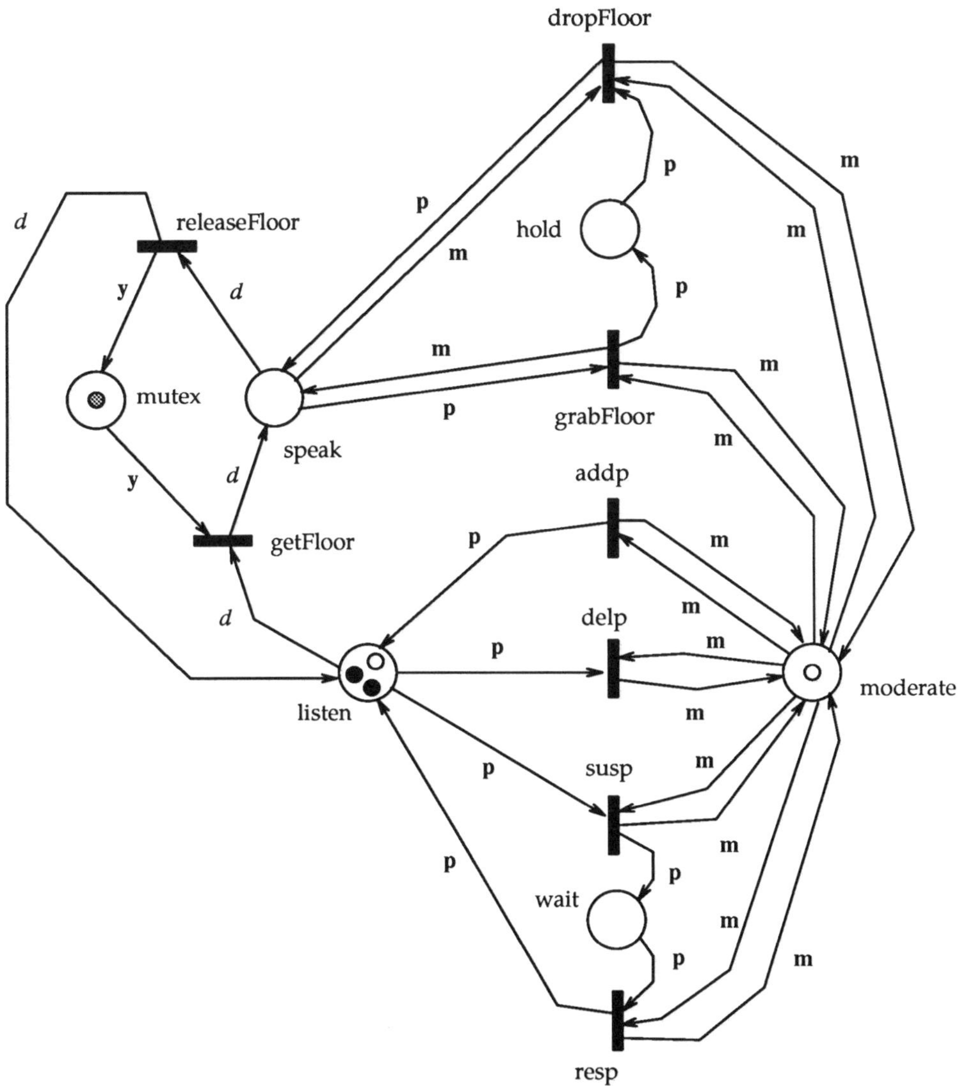

Figure 1: CPS for a simple meeting protocol.

produce tangible representations, such as screen layouts, sound generation, video, etc. P_d determines how things like text and buttons are visibly displayed, *e.g.*, whether a user selects a link from a side menu or from highlighted words (or icons) within the content display.

The net marking M_0 enables a CPS to represent both the logical structure of a collaboration and the current state of inter-activity within it. Together with the execution rules of the CTN, every marking is a characterization of the possible paths in a collaboration from the browsing point it represents. Different browsing patterns (for, say, different classes of reader) can then be enforced on a single CPS simply by choosing appropriate different initial markings.

2.4 EXECUTION RULES FOR A CPS

The execution behavior of a CTN provides an interpretation of the collaborators' experiences when interacting under the control of a CPS. As in the original Trellis model, a token in a place p indicates that the contents of the place $C_l(p)$ are displayed for viewing (or editing, or some other interaction). Content elements come into and go out of view (or begin and end execution) as tokens move through the net. Transitions are fired by selecting logical hypertext buttons. When a transition t is fireable in the timed Petri net, its logical button $B_l(t)$ is displayed in some selectable area of the screen, such as on a (highlighted) word in a text section, or in a separate button menu.

The general execution behavior of the CTN in a CPS requires pattern matching to be done on all arc expressions that are inputs to a transition. The transition is enabled if there is one or more consistent color substitutions for the expressions. When the transition fires, one of the valid substitutions is chosen, the proper color tokens are removed from the input places, and output tokens are produced according to the substitution and the expressions on the output arcs.

Rather than being excessively formal, we will explain CTN execution behavior informally thorough the examples in the next section. We will explain the projections and the interpretation of net annotations during execution in section 5 on prototyping.

3   CTN Examples

In the next few sections we present the basic functions of a CPS through an extended example. Following this illustration, we describe the methods we are using to analyze and verify the behavior encoded in a CPS. After analysis, we explain a major application for CPS—enacting and improving the process of software system development.

3.1 EXAMPLE: SIMPLE MODERATED MEETING

Figure 1 shows a CPS that encodes a simple moderated meeting. To enhance the clarity of this example, we have made some simplifying assumptions about the actions in such a meeting; we discuss more realistic complexities following an initial explanation.

We envision a meeting with two classes of agent: *participants*, and a *moderator* (who may also act as a participant). Participants can be in either of two states: listening, or speaking. When listening, they can request and possibly obtain the floor to speak; when speaking, they can release the floor, to return to listening. The moderator has more extensive abilities. In addition to acting as a participant, the moderator can: add or delete participants in the meeting; suspend participants for a time, and return them to a meeting (we presume that suspension is different from being deleted, as something like a history would be kept for suspended participants); grab the floor, preempting the current speaker, and drop the floor, returning the preempted participant to speaking.

In the CPS shown, we have represented the participants all with one color; that is, we have used color to represent the entire class rather than individuals. Consequently, the net is simpler for an initial discussion, but no participant can be distinguished from another. We will remedy this shortly. We have assigned a second color for the single moderator, and we have used a third color for a token providing mutual exclusion of potential speakers.

Color constants

In this simple protocol, the moderator is fixed for the duration of the meeting (we will explain a more complicated alternative to this, as well, following). To understand the notation on the net, consider the action "add participant" that the moderator can perform. This is represented in the net as the transition labeled "addp". There is one input arc to this transition, labeled **m** coming from place "moderator". The label **m** in boldface indicates a color *constant* which we have selected for the moderator token. The "addp" transition has two output arcs: one labeled **p** to place "listen", and another labeled **m** back to place "moderator". As before, **p** is a color constant representing the participant class.

When a token of color **m** is present in place "moderator", the operation can be invoked (*i.e.*, the moderator can invoke it whenever desired... no other preconditions exist). Firing the transition consumes the **m** colored token, but it also places one back into the moderator place (*i.e.*, the moderator does not give up his role by adding a new participant). Firing also places a new **p** colored token into place "listen", thereby increasing the number of participants by one.

Color variables

So far we have seen behavior that is accomplished with color constants indicated on arcs. However, the real power of the CPS notation comes in allowing color *variables* to appear on arcs. Such a structure appears in figure 1 on the left side, in the net region containing the "getFloor" and "releaseFloor" operations. Note that an **m** colored token in located in place "listen" along with all the **p** colored tokens. This, along with color variables on arcs, implements our claim that the moderator should be able to act as a participant also.

The arc leading from place "listen" into transition "getFloor" is labeled with the expression d, where the italics indicates a color variable. The arc leading out of "getFloor" to place "speak" is also labeled with d. Note that the arc leading to "getFloor" from place "mutex" is annotated with the color constant **y**. The meaning of this net fragment, then, allows "getFloor" to fire with some variability in its input token colors—not just with specific input colors, as in the previous example. Transition "getFloor" may fire if there is specifically a **y** color token in place "mutex" (*i.e.*, if there is no one currently speaking), and if there is some token of *any* color (call it d) in place "listen". When it fires, the **y** color token is consumed from place "mutex"; in addition, a token of whatever color d stands for is removed from

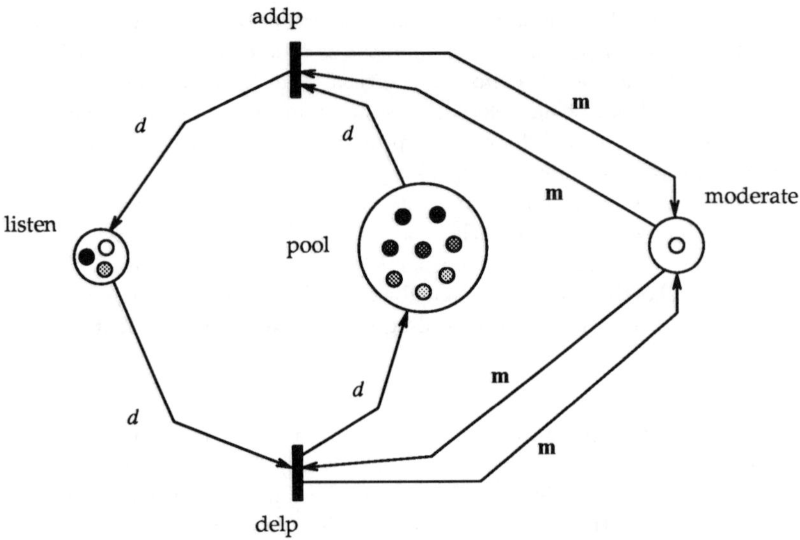

Figure 2: Detail for participant allocation.

"listen", and a token *of that same color* is deposited into place "speak". This means that the single operation "getFloor" may be used to move either an **m** color token or a **p** color token into place "speak".

The same sort of color variable behavior controls the firing of transition "releaseFloor" when someone wished to stop speaking.

3.2 Example: Distinguishing participants in a meeting

Let's consider other CPS structures that add more detail to the simple protocol previously discussed.

One reasonable change is to allow a different color for every meeting participant. This can be done by creating a finite pool of differently colored tokens that is held in reserve. When a new participant is to be added, a "new" color is allocated from the pool and added to the meeting; when a participant is removed from the meeting, the color is returned to the pool.

This alteration is depicted in the CPS fragment of figure 2. In addition to the pool of colored tokens, the net shown in figure 1 has been changed to include varying token colors in place "listen". Also, arcs leading from the moderator operations to place "listen" are now labeled with the variable expression *d* instead of with the constant **p**.

It is important for analysis purposes (explained later) that the pool of potential participants be finite. That is, the CPS must specify all colors that might be used by meeting participants, and no truly new color can be injected into the net as a whole during execution. However, the finite number of participants can be arbitrarily large. This limit presents a practical problem only if the meeting protocol to be modeled must allow an unbounded number of participants. Note that the simple example we presented first, in which one

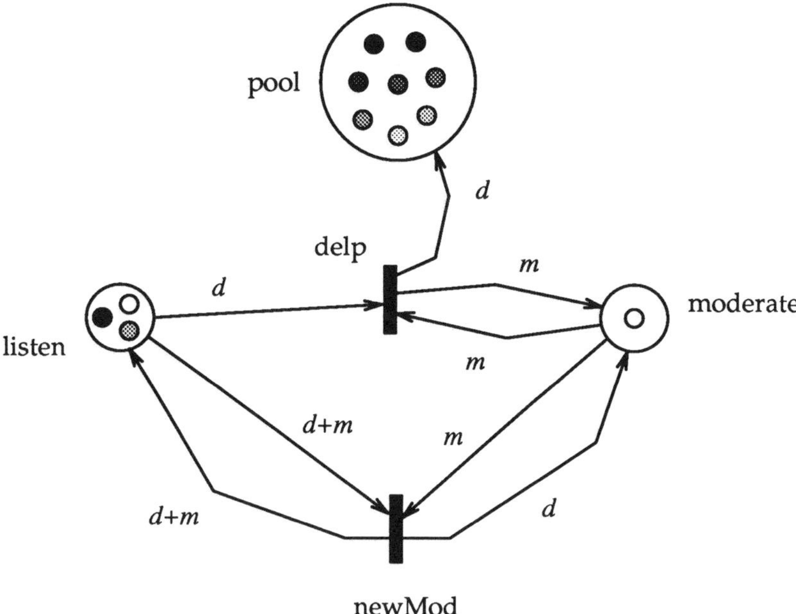

Figure 3: Detail for changing the moderator.

color was used for the entire class, does allows an unbounded number of (indistinguishable) participants. Whether or not a truly unbounded number of participants is a reasonable requirement for a CSCW tool is a point for separate discussion.

3.3 EXAMPLE: CHANGING THE MODERATOR OF A MEETING

Another practical addition to our meeting protocol is the ability to change moderators while the meeting is in progress. For this example, we will build on the one from figure 2 with the finite pool of participants. We continue to assume that each participant, moderator or otherwise, is assigned a unique token color.

Figure 3 shows more CPS details for moderator swapping. In this fragment, we have altered the labels on arcs between the "moderator" place and the previously existing operations (like "addp" and "delp") to have the variable expression m. Labeled in this way, the moderator is not fixed as always being the constant color **m** as before, but instead can be any color; having m on all arcs between place "moderator" and operations like "addp" specifies that execution of such an operation must maintain whatever color m represents (*i.e.*, the moderator cannot change simply by executing "addp" and the other previously discussed meeting control functions).

We have added another operation, "newMod", to specifically perform moderator swapping. The arc leading into "newMod" from place "moderator" is labeled with the expression m, and the arc leading into the transition from place "listen" is labeled $d + m$. This shows

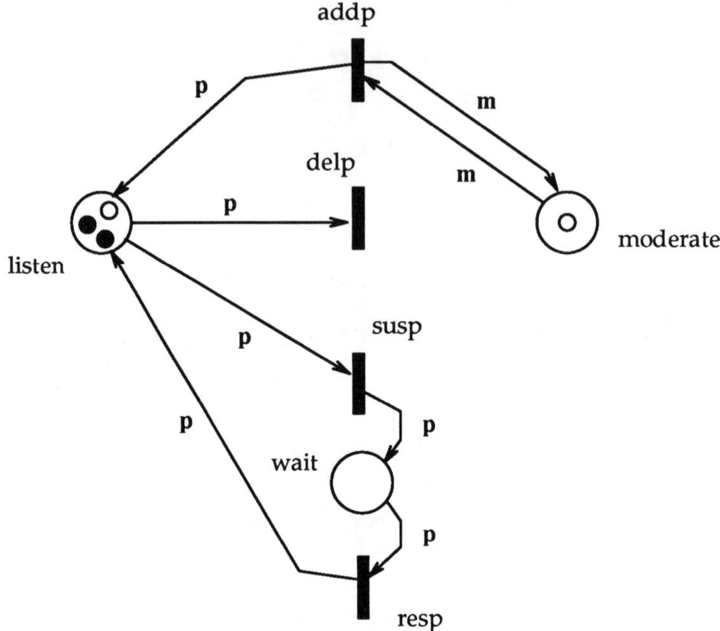

Figure 4: No permission required to leave.

that the "newMod" function can only be invoked if the "listen" place contains both a participant with the same color as the moderator, and another participant with a *different* color from the moderator (we assume no aliasing in color substitutions). When fired, the "new-Mod" transition leaves the token counts in "listen" unchanged, but it takes whatever color was in "moderator" (represented by variable m) and replaces it with a token of whatever color is represented by d. Since we know the value of d is different from the value of m (the "no aliasing" assumption), we know that the moderator has changed. Neither participant leaves the meeting—they just exchange capabilities. Also note that the new moderator color is drawn not from the pool, but from the actual participants found in place "listen". Finally, as written, the CPS allow a moderator to swap only with someone who is listening—a speaker cannot become the new moderator without first releasing the floor.

3.4 EXAMPLE: OTHER MISCELLANEOUS BEHAVIORS

In this section, we return to the simple protocol of figure 1 to illustrate some other behaviors that shed light on the CPS method of specification. Though we present them in the context of the protocol with mostly color constants, the basic interactions will translate into the more complex CPS examples as well.

Note that in the initial protocol, a participant in essence requires the permission of the moderator to leave a meeting, or to be suspended for later rejoining. This condition exists because the moderator place is required to contain a token for all such operations to take

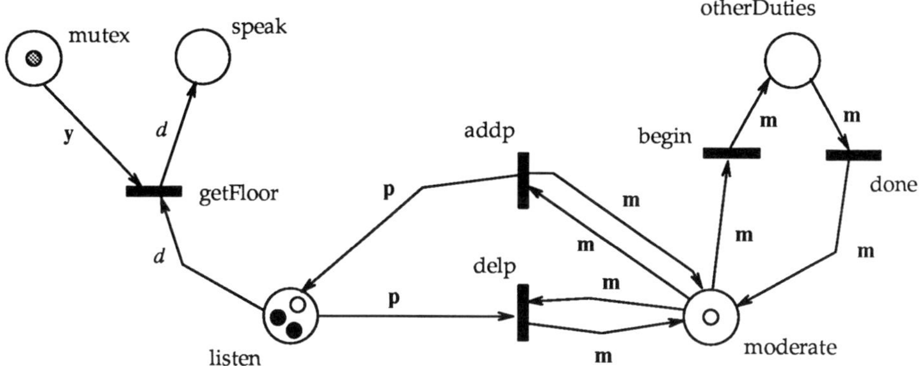

Figure 5: Moderator with other duties, requires permission.

place.

In the example, though, the moderator place *always* contains a token (a condition that can be verified in several ways, including the method we will present following). The moderator is "always home" so to speak, and no net structure is present that would ever cause permission to leave not to be granted. The behavior, then, of our initial simple example is equivalent in one respect to a net in which no moderator permission is required. Figure 4 shows such a fragment; here, moderator permission is required for addition of new participants, but once in a meeting, a participant may leave or suspend (and resume) itself without other permissions.

Of course, the net as originally written might still be preferable as a meeting protocol. Even though the moderator in the original example never denies permission to leave, a designer may well wish the moderator to be involved in such operations if only as a matter of recordkeeping.

Figure 5 shows a further variation on the "permissions" theme. In this CPS fragment, additional duties have been added that may take the moderator away from the main meeting floor for a time. We have indicated this with a place labeled "otherDuties". Firing the "begin" operation will remove the **m** colored token from the "moderate" place, disabling the "addp", "delp", etc. operations until the moderator executes "done" to return from the other duties. It makes good sense to specify that alteration of the makeup of a meeting must be done when the moderator is not busy with other things. However, it does not make sense to specify that the meeting must come to a halt] until the other duties are completed. Note that in our CPS, normal getting and releasing of the floor by participants may still go on while the moderator is otherwise occupied.

We repeat that these behaviors can easily be added into the CPS examples that use color variables too.

3.5 Transition predicates and other high-level nets

As mentioned in section 2.1, there are several functionally equivalent syntaxes for high-level nets. The alternate form provide, in essence, more compactness of expression but do not add modeling power to a CPS. In predicate-transition nets [GL81], for example, tokens carry information (which we have generically called color) and every transition carrys a *predicate* describing how the input tokens may combine to produce output tokens. The use of predicates can allow one transition in a high-level net to represent behavior that what would require several transitions in our simple notation. We will not go into more detail in this report in describing the equivalences. Our current Trellis model does not support transition predicates, but it could easily be extended to do so; either way, the analysis methods we describe below for the current CPS formalism are certainly applicable to alternate high-level net syntaxes as well.

4 Analysis Techniques

The need to analyze a CPS should be apparent to the reader that has spent some time considering the possible behavior of even the simple protocol given in figure 1. As motivation for this section, let us consider for a moment what can happen when the simple meeting CPS is executed.

There are two pairs of operations that are intended to be used in alternation by individual speakers: "getFloor" followed by "releaseFloor", and "grabFloor" followed by "dropFloor" (by the moderator only). If a normal participant executes "getFloor", the "dropFloor" operation cannot be executed thereafter since the arc leading from place "speak" to that transition requires an **m** colored token.

If the moderator executes the "getFloor" operation, as a normal participant would, it might appear that the moderator could then execute the "dropFloor" operation, in violation of the informal expectation. In fact, the net structure prevents this by requiring a **p** colored token to be in place "hold" for firing transition "dropFloor". In other words, "dropFloor" can only be executed if the "grabFloor" operations has first placed a participant on hold. It would appear, then, after a quick informal analysis that the net maintains our intentions.

However, more careful reasoning about the protocol uncovers this interesting behavior. If a moderator first executes "grabFloor" and puts a speaker on hold, there is no requirement in the net that the next operation be "dropFloor". Once an **m** colored token is in place "speak", the "releaseFloor" operation can be executed, no matter how the **m** token got there. In essence, if a moderator grabs the floor, it can then behave as if it obtained the floor through the normal channel. If such a moderator follows "grabFloor" with "releaseFloor", a second **m** colored token will be deposited into place "listen".

If a participant does "getFloor" to begin speaking, this scenario can then be repeated. The moderator can again execute "grabFloor" followed by "releaseFloor", putting a second participant on hold and putting a third **m** colored token in place "listen". This behavior can continue until all participants are put on hold, and "listen" contains a number of moderator tokens equal to one greater than the number of participants on hold.

This behavior can also be undone. While participants are on hold, the moderator can execute "getFloor" with one of the **m** colored tokens in place "listen", and then (against the alternation assumption) follow that with "dropFloor", releasing one of the held **p** colored tokens and eliminating one of the extra **m** tokens. The participant, now speaking again, can execute "releaseFloor" to rejoin the "listen" pool. The moderator can repeat this cycle, releasing in turn all held participants.

Several points should be made about this situation. First, even simple protocols can exhibit complex behavior. Secondly, complex or not, the behavior of a CPS easily can be unexpected. We did not intend for the example protocol to have the behavior described; the "covert" operations were discovered well after its design as other aspects of the CPS structure were being discussed. This surprise, though small, illustrates our point about analysis quite well. In this case, the CPS behavior is not particularly harmful; however, its operation does not match the specifications we had in mind, and its extra behavior does not map well onto the natural and expected actions of a meeting.

Thirdly, informal reasoning cannot be counted on to reliably uncover all the possible behaviors of a CPS. We draw an analogy to program testing vs. program verification; testing (sampling) is necessary, but not sufficient for full confidence. In our example, we first concluded that a moderator could not execute "getFloor" followed by "releaseFloor", arguing that a **p** token was needed in place "hold". We then went on to contradict this conclusion, discovering another vector by which that precondition could in fact be obtained. With such informal reasoning, one cannot be sure all important behavior has been deduced. When is it safe to stop reasoning?

In the next section we present a formal analysis method we have developed for a version of Trellis that is based on a non-colored PT net. Following that, we discuss how this analysis is being extended to the colored nets in a CPS.

4.1 Model checking for Trellis hyperdocuments

Trellis and its implementations provide a formal structure for hyperprograms, and net analysis techniques have been developed for exploiting this formalism. One very promising approach involves our adaptation of automated verification techniques called *model checking* [CES86] from the domain of concurrent programs. This approach allows verification of browsing properties of Trellis hyperprograms expressed in a temporal logic notation called CTL. An author can state a property such as "no matter how a document is browsed, if Node X is visited, Node Y must have been visited within 10 steps in the past." The model checker efficiently verifies that the PT net structure maintains the validity of the formula denoting the property.

In model checking, a state machine (the model) is annotated with atomic properties that hold at each state (such as "content is visible" or "button is selectable"), and then search algorithms are applied to the graph of the state machine to see if the subcomponents of a formulae hold at each state. By composing the truth values of these subformulae, one obtains a truth value for the entire formula. For PT nets, we obtain a useful state machine from the *coverability graph* explained in an earlier Trellis paper [SF89].

The details of our use of CTL are discussed elsewhere [SFR92]. For this rationale, it is sufficient to give an idea of how the method is applied to Trellis models. The Trellis document shown in Figure 6 is a small net that expresses the browsing behavior found in some hypertext systems, namely that when a link is followed out of a node, the source

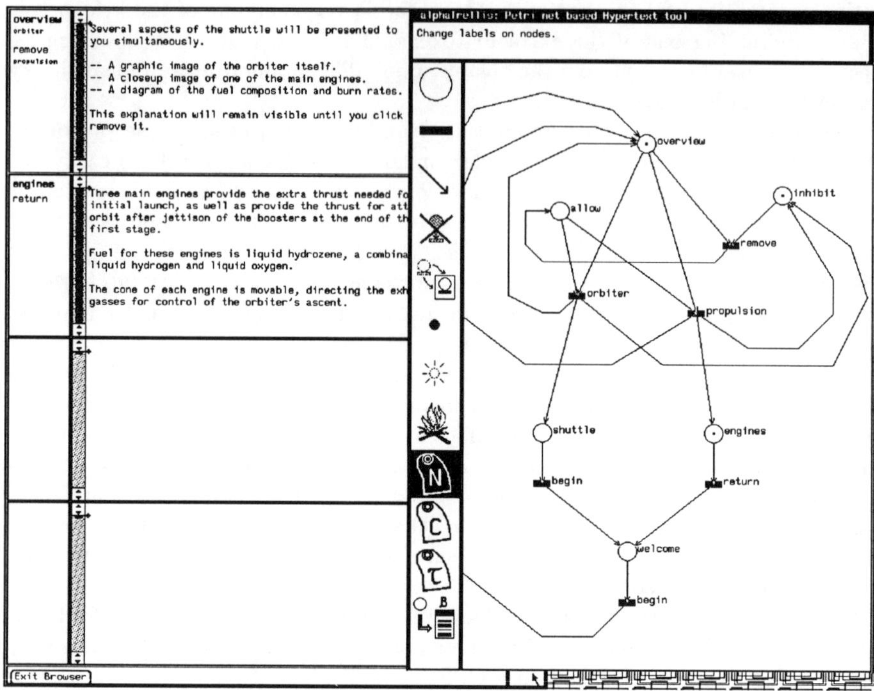

Figure 6: Trellis CPS with programmed browsing behavior.

content stays visible and the target content is added to the screen. The source must later be explicitly disposed of by clicking a "remove" button.

After computing the coverability graph and translating it into the input format required by the checking tool, the model can be queried for desired browsing properties. These examples use the syntax of Clarke's CTL model checker, and show its output:

- Is there some browsing path such that at some point both the "orbiter" and "propulsion" buttons are selectable on one screen?

  ```
  |= EF(B_orbiter & B_propulsion).
  The formula is TRUE.
  ```

- Is it impossible for both the "shuttle" text and the "engines" text to be concurrently visible?

  ```
  |= AG( ~C_shuttle | ~C_engines ).
  The formula is TRUE.
  ```

- Can both the "allow" access control and the "inhibit" access control ever be in force at the same time?

  ```
  |= EF(C_inhibit & C_allow).
  The formula is FALSE.
  ```

- Is it possible to select the "orbiter" button twice on some browsing path without selecting the "remove" button in between?

  ```
  |= EF(B_orbiter & AX(A[B_remove U B_orbiter])).
  The formula is FALSE.
  ```

This particular Trellis model is very small compared to those encountered in realistic applications. Our checker has also been tested on larger Trellis documents—for example, the one we built to represent a CSP parallel program [SF90a] contains about 50 places and transitions, and generates a state machine of over six thousand states. Using a DECstation 5000/25, the performance of the model checker on formulae like those above is mostly on the order of a few seconds each, with the most complicated query we tried (not shown) requiring about 15 seconds to answer. We suspect that authors of Trellis models will find such performance not at all unreasonable for establishing the presence or absence of critical browsing properties, and we also expect that future implementations will exhibit improved performance.

4.2 EXTENSION TO CPS ANALYSIS

We are currently building tools to adapt the basic model checking form of net analysis to the colored PT nets used in Trellis CPS models. The basic approach depends on the well-known result from PT net theory that high-level nets provide more compact, more expressive, modeling notations but do not extend the basic power of classical PT nets. In essence, a high-level net can represent a net fragment that would require several structurally-similar net fragments in classical notation. Correspondingly, techniques are known for "unfolding" a high-level net into an equivalent non-colored PT net.

Figure 7 shows such an unfolding of a portion of the colored net used in our simple meeting protocol. We analyze a CPS by generating a state space automaton for the equivalent unfolded net and applying the model checker as just described. Our current research efforts are concentrated on a tool for helping a developer to interpret the unfolded net and CTL queries in terms of the original colored net.

We have included timing on transitions as part of the Trellis model, but in this paper we have not dealt with that aspect in modeling or analysis. We should note that if the untimed subset of Trellis is used (that is, if all transitions are $(0,\infty)$), then the complete analysis we have described here is possible. Analysis in the presence of timing is a subject for other papers.

5   Trellis: Prototyping and enacting a CPS

In this section we explain the basic architecture of a Trellis-based implementation and show how it can be used for prototyping and enactment of a CPS. This should illustrate more clearly the earlier observation that a hyperprogram integrates task with information.

Recall that a CPS is a colored timed PT net (CTN) that is annotated with fragments of information (text, graphics, video, audio, executable code, other hyperprograms). The CTN encodes the basic individual actions and group interations of a CSCW application, but appropriate visual interfaces are needed to provide users with a tangible interpretation

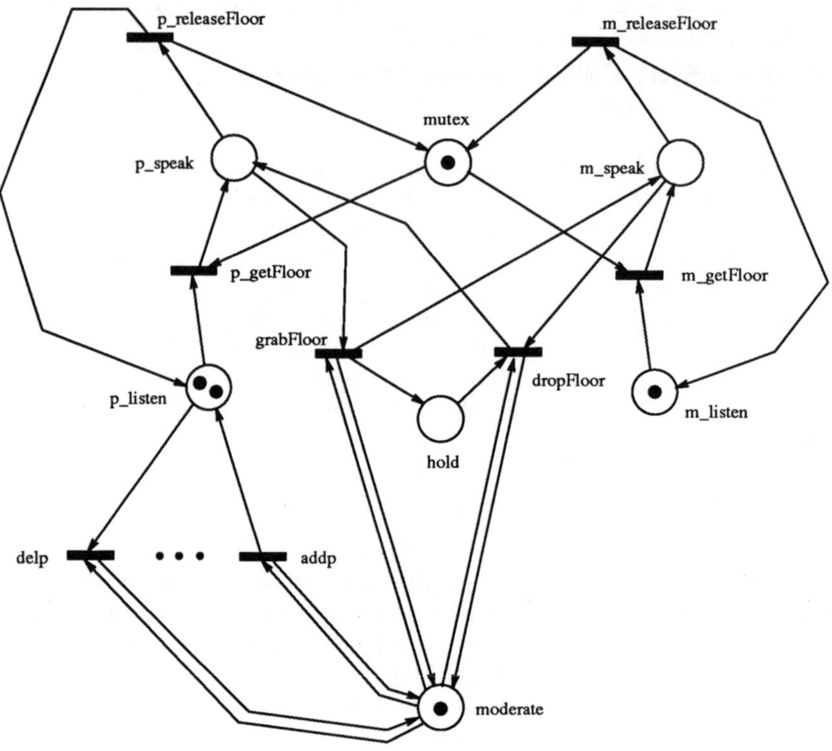

Figure 7: Expansion of colored net to Petri net for analysis.

of the net and its annotations. For example, annotations on net places might be Unix file names, with display names attached to transitions. When a token enters a place during net execution, the file for that place would be presented for viewing. The names of enabled transitions leading out of the place would be shown as a menu of selectable *buttons* next to the file. Selecting a button (with a mouse, usually) would cause the net to fire the associated transition, moving the tokens around and changing which content elements would then be active.

In a Trellis implementation, this cooperative separation between net and interpretation is realized by a distributed *client/server* network, as shown in figure 8. Every Trellis model is an information server—an engine that accepts remote procedure call (RPC) requests for its services. The engine has no visible user interface, but does have an API that allows other remote processes to invoke its functions for building, editing, annotating and executing a PT net. Interface clients are separate processes that have visible user interfaces and communicate with one or more engines via RPC. Clients collectively provide the necessary views, interactions, and analyses of a net for some specific application domain. Simply put,

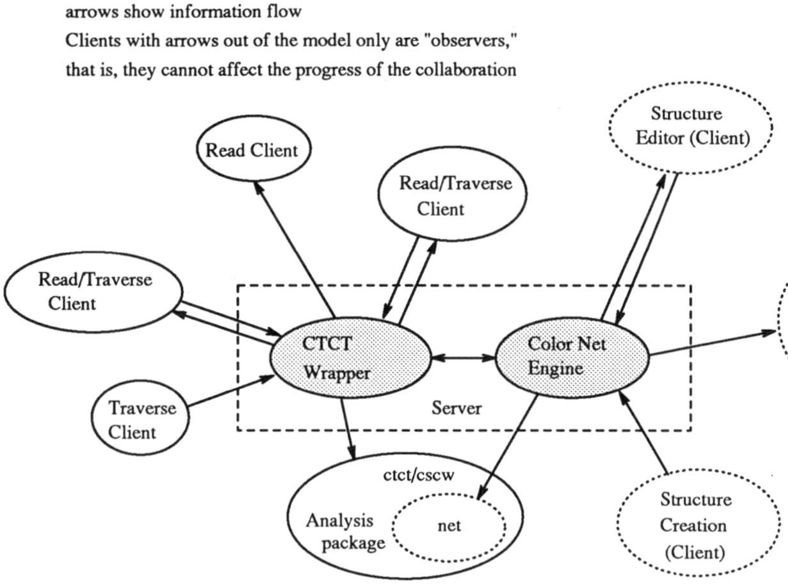

Figure 8: System architecture of a Trellis implementation.

Trellis clients are the syntax of an application, whereas Trellis engines are the dynamic semantics; clients and servers provide an application's look and feel respectively.

Figure 6 shows two clients for the original Trellis system, αTrellis (based on non-colored PT nets). Here, a graphical editor client (on the right) allows construction and execution of a net, and provides a visual representation of the structure of the automaton. A text browsing client is show on the left; it renders the annotations on marked net places as visible text when the net is active. Each client is executing as a separate process, and both are communicating via RPC with the engine process. Figure 9 shows the colored net editor xTed that has been written to interact with the new Trellis engine for CPS construction.

In the early stages of a collaborative tool design, a Trellis document is built that encodes the desired interactions, as in the meeting protocol discussed in section 3. Text and graphics (or video, if needed) are created to explain each portion of the CPS; as part of the authoring process for the hyperdocument, the net components are annotated with the names of the files containing these explanatory content elements. Testing and analysis can then be done on the CPS using the CTN structure; usage trials can be done through browsing with the collective Trellis client interface. At this stage, a CPS prototype is an "active hyperdocument" with interactions simulated, but with actual information exchange and data manipulation simply explained with text and graphics stubs.

Once a CPS design is stable, a more substantial implementation can proceed by replacing the text and graphics that specify behavior of each part of the CPS with executable components that realize the actual behavior and provide the designated services. In the curent

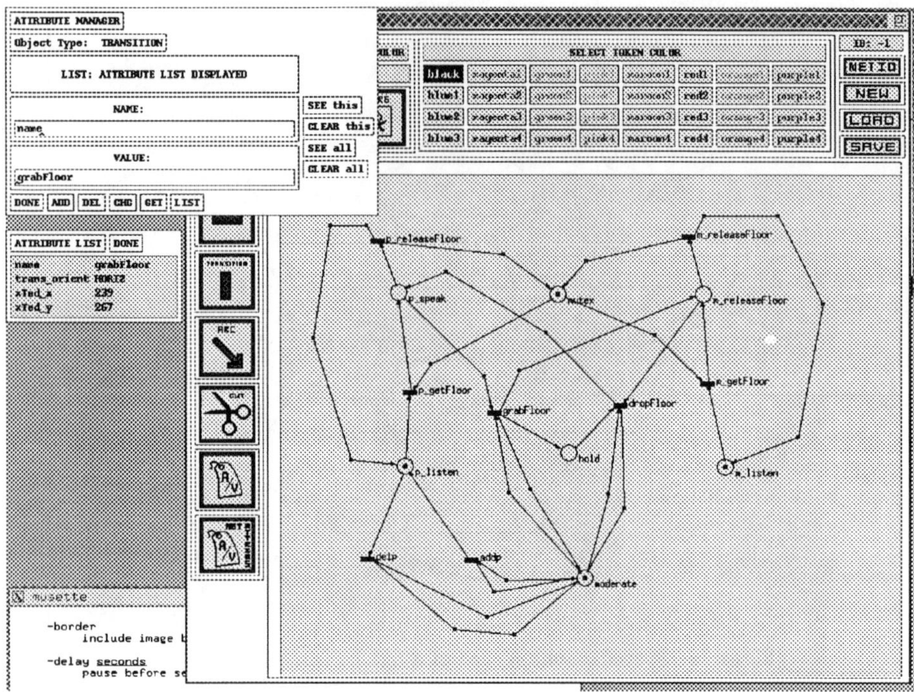

Figure 9: xTed CPS editor client for Trellis.

Trellis engine a Lisp interpreter is provided to assist in this stage of prototpying. The result is a program with a hypertextual interface that controls group use of some information base. An example of a Trellis hyperprogram using the Lisp interpreter is given in [SF92].

6   CPS Application: Software process modeling

In addition to the meeting structure we have shown, and in addition to CSCW tools in general, many different forms of software can be based on CPS models. We have investigated several and are currently constructing others. The common thread in all these applications is multiple users that need to be aware of their mutual existence, and an associative link structure relating together the components of the applications (agents, activities, information, etc.).

Our domains of Trellis investigation include: hypertext and hypermedia, in which multiple readers can share and interact with the linked information elements [SF91]; image browsing indexes, where images are classified according to common characteristics, and the manner of filtering and sharing images among collaborating astronomers is encoded in the net structure defining the index; parallel program browsers, where the coordination of a CPS is applied to program processes rather than to people [SF90a, SF92].

One application we will discuss is some detail here is modeling the process of developing software systems, from requirements to product, from initial design and specification through maintenance. This project, called σTrellis, is a specific version of the Trellis engine, and a specific collection of interface clients tailored to the capture, analysis, and management of information crucial to the software development process, and tailored to representation of the behaviors of the agents involved in the process. In section 6.1, we outline the basic requirements and resulting components of a software process model. In section 6.2 we demonstrate how a Trellis CPS satisfies these requirements. In section 6.3 we conclude with an outline of the main structure, capabilities, and uses of the σTrellis system and methodology.

6.1　BASIC CAPABILITIES AND REQUIREMENTS

Our process models are being designed by integrating ideas drawn from the materials on process from the SEI [Hum88]; the MVP project of Rombach [LR90]; the material contained in the proceedings of the past software process workshops; and from personal communications with representatives of the 14 industrial affiliates of the Purdue/Florida/NSF Software Engineering Research Center.

Process models represent document, design, and code descriptions, relationships among these components, constraints on their temporal and logical creation points, data flow paths among them, and pre/post conditions describing their effects on the overall information content of a development effort. Any process modeling mechanism must assist these areas of software development (from Rombach [LR90]):

- understanding and communication
- measurement
- analysis of captured data
- planning
- execution (actual development)
- process improvement

To produce models meeting these goals, a formalism must satisfy these requirements:

- express relationships among product, process, quality, and resource components;
- support instrumentation mechanisms (data capture);
- support analysis (data reduction);
- support composition of existing models;
- support tailoring to specific projects or sites;
- support execution of models as guidance to the process.

6.2　TRELLIS CPS MEETS THESE REQUIREMENTS

The software process is a mixing of humans, information objects, programs as data, and programs as computations. This mixing with humans is the key to a good modeling formalism. Our detailed discussion of the Trellis technology was meant to emphasize the fact that, while Trellis allows information structures with programmable behavior, it supports and even emphasizes user control of execution ("browsing") over these structures. Since Trellis is basically a man/machine interaction model, it already provides many of the facilities identified above as requirements for a software process modeling formalism.

Human interaction is where Trellis differs strongly with other efforts at representing and enacting process. Most other efforts have concentrated on programming-language-like features and implementations for a model; the σTrellis methodology will emphasize the man-machine interaction and will make "notation" of secondary importance. σTrellis conceives a process model as being mostly human-driven, and therefore presents a largely hypertextual interface rather than a programming-language-like interface.

After several discussions with Rombach's MVP group, we now believe that two major missing aspects of other process modeling efforts are support for human-directed model enactment, and the personal annotation facilities of a hypertext system. We believe that Trellis is uniquely suited for rectifying these problems. Trellis has been well-explored as a hypertext vehicle; it also has inherent concurrency semantics, collaborative multi-user formalism, and a client-server architecture–all of which make it process-oriented, unlike other hypertext systems.

The process of software development is inherently a coordinated activity of (often) many people (engineers, customers, managers). Any system that purports to coordinate the activities of multiple agents, be they computers or humans, must provide several basic services, as does the new Trellis model:

- *activities:* the basic units of collaborative work;
- *agents:* the effectors of activity;
- *information:* the data to be acted upon during collaborative work;
- *relations*: for expressing how agents and information components depend on, or affect, each other;
- *exchange mechanism:* for sharing information, data or parameters among agents;
- *threads:* the collections of possible agent behaviors;
- *synchronization mechanism:* for coordination of parallel activity threads.

A good model should provide these fundamentals with as little other structure as possible, in order to maximize the analytical power one can bring to bear on the model. With extra baggage, designed with good intentions to provide extra "expressiveness" or more "flexibility," one often ends up with a model that become analytically intractable.

Our previous work with Trellis shows it to be an effective and analyzable model for hypermedia documentation. The early parts of this report show how Trellis can encode the behavior of group interactions. As mentioned earlier, this unique blending of both *task* and *information* in one formal structure makes Trellis CPS hyperprograms singularly appropriate for both specification and enactment of CSCW applications.

6.3　Past, present, and future information

In this section we outline the main components of σTrellis, in which the past, present, and future of a development effort is represented in a unified framework: *past* means that historical information about the progress of a system effort is captured and made available for browsing; *present* means that the model always reflects the current state of development; *future* means that an important goal of modeling is to improve the processes that are represented as σTrellis models.

σTrellis consists of:

- a new Trellis engine, adapted from the current colored engine and tailored with capabilities specific to the needs of a software development process;

- a model editing client for X windows, derived directly from the current Trellis editor xTed;

- a model browsing client for X windows, allowing exploration of the linked information database of a model; this client will be hypertextual, allowing the display of documents associated with the process, annotations provided by participants in a process, video/audio data gathered from meetings, data gathered during development, etc;

- a model enactment client for X windows, executing the colored timed PT net in a CPS under human direction, in Trellis fashion; enactment may be *simulated*, for training, or it may be *actual*, for control of a real development effort;

- several model analysis clients, allowing measurement, examination, and reasoning about both the static (links) and dynamic (net execution) aspects of a model and its process; one of these clients is a *model checker* that adapts the method discussed in the technical rationale to the new colored timed PT nets;

- a *process capture* client, allowing a model to be built indirectly by measurement and observation of an actual development (as opposed to direct model construction with an editor); we think of this as a *descriptive* technique, since the resulting model will indicate what *was done* rather than what should be done.

The model browsing and enactment clients offer different views of models from different perspectives (user, manager, engineer; functional, structural, temporal dependencies; past, present, future form of the process; etc.).

The editing client is augmented with other forms of prescriptive model construction; for example, translators will be written as appropriate to convert information in machine readable source notations directly into model format. Thus we do not expect that all parts of a realistic model will be hand constructed with a graphical editor.

Process improvement

The last item above, the process capture client, is especially interesting. We think it is important in our early work to create a *prescriptive* capability in σTrellis (which the model editor gives), so that ideal or experimental processes can be unambiguously defined and studied. However, it is the process capture client that will given real improvement leverage to an organization that develops large software. The capture client gives σTrellis a *descriptive* capability—that is, the ability to attach to a development project, unobtrusively gather data, measurements, observations about what activities happen, when they happen, where, by whom, etc., and then to construct a model from this data of the actual process. The capture client obviously will not be of the same nature as the other, highly visual clients;

rather, it will monitor development in the background, adding structure to an σTrellis model as development progresses by making RPC invocations Trellis engine services.

An important characteristic of an effective software process is *predictability* [Hum88]. This attribute requires the ability to measure the crucial aspects of development, and to apply statistical (repeatable) controls to the procedures involved with the measured quantities. Trellis models serve as a basis for implementing various measurements and evaluations of system development practices.

Improvement will then come from comparison of ideal or defined models with the captured models. Ideal models will give property measures that are expected or desired; captured models will be subjected to the same measures, and where differences are found from the ideals, engineers can be put to work on specific refinements.

Other sources of improvement we expect from using σTrellis will be in training of new engineers by "replay" of past developments as simulated process enactments; from unambiguous and accessible definition and documentation of a process for those participating in it; and from the ability to incorporate more directly information from past development into new efforts (reuse).

Product models, too

A final point to make is that σTrellis provides a vehicle for broader modeling than we have emphasized here. Though we are concentrating initially on the development process, the basic Trellis engine we are building to support σTrellis can be applied in other aspects of software systems as well; we will call these other aspects *product models*. For example, we mentioned that Trellis can be used to express and browse the control structure of program source code [SF90a, SF92]. As a related example, module designs can be expressed as Trellis models, too; σTrellis clients can then be written to apply, say, Zage's design quality metrics to the design models. As another example, a high-level system structure derived from requirements can be expressed as a Trellis model; an interface client can then be written to apply, say, the COCOMO estimation method to the model.

No special techniques are required to integrate product models with process models in σTrellis . Since the same Trellis engine is used for representing each, the hierarchy in Trellis will directly allow product models to be components of a development process model, and *vice versa*. The interface clients will operate on both types of model for construction, browsing, and analysis.

7 Comparison with related research

In general, the previously cited papers defining the various forms of high-level PT nets all mention the appropriateness of the model for representing interactions among users and computations. Our project goes beyond such recognition by providing a modeling framework that includes unique analysis methods, as well as a system design for prototyping and simulation of collaborative tools.

Other research projects have looked at various aspects of the domain we are studying. Fischer is using IO automata [Fis91] to model human/computer interactions in CSCW settings. The Suite project [DC92] is system for construction of CSCW tools; its prototyping facilities are more sophisticated than those of Trellis, but no emphasis is given in Suite on formal methods or analysis of the underlying protocol.

A commercial package, *Design/CPN*, is available from MetaSoftware providing extensive editing capabilities for building hierarchical colored Petri nets. Temporal logic has been used to describe structural aspects of hypertext [BK90], but the goal in that work is to define subgraphs of a structure rather than the dynamics of browsing, as in Trellis. The only project we know of other than Trellis that uses temporal logic for PT net analysis is by Sinachopoulos [Sin89]; the emphasis in this work is on timing in a timed net model. No other project we know of uses model checking for PT net state space analysis.

The use of PT nets as a specification medium for man-machine interaction appears previously in the literature. For example, van Biljon [vB88] has described a special grammar-based notation for designing man-machine dialogues as languages, which are then realized with a hierarchy of PT nets as recognition automata. Another example is the work of Holt [Hol88], who has designed a PT-net-based graphical specification language for coordination of multiple cooperating agents in an information processing organization. Trellis is a more complicated model than these previous proposals, because it encompasses more than just the control aspects of man-machine interactions. It contains an inherent notion of information presentation (text, graphics, executable code), has timing for events, and in later versions includes a Lisp interpreter as a attribute processing facility.

The underlying Trellis information engine supporting σTrellis is related to other hypertext engines that have been used in experimental software support systems. The HAM (hypertext abstract machine) [CG88] was developed in 1986 by Textronix, and was used as the basis for a hypertextual software support system called Neptune. The uniqueness of Trellis is the basis on a parallel collaborative computation model—colored timed PT nets. This gives the model an elegant structure that can be both programmed and analyzed. Scacchi and Garg have also used a hypertext mechanism in a software engineering context [GS87]. Their project, though, concentrated on the object-base aspects of a software project and did not have a formal model for representing process and enactment.

Intellectual leadership in the field of software process modeling comes from SEI, with its process assessment procedures, and with the writings of Watts Humphrey [Hum88].

In terms of experimental projects, Kellner has described a study of how StateMate can be used to present several different views of a development process [Kel89]. This work describes an experiment at SEI, and most closely parallels the approach of σTrellis; we feel the basic idea of unified static and dynamic properties in one model deserves an industrial trial.

The MVP-L project [Rom91] has looked as an Ada-like syntax for expressing salient process properties in a form that can be machine translated into plans, code, documents, etc. Marvel [KFP88] has looked at using rule-based systems for assistance in the software development process.

There are numerous other process projects documented in the proceedings of the annual software process workshop. We have mentioned only a few to give an idea of their nature. These efforts have taught us useful views of processes, but they have not been comprehensive in their support for the human/computer and human/human interaction that is central to a collaborative effort. They have been mostly language oriented, or have looked at applying some particular technological area (like expert systems) to process, while retaining a flavor of traditional computing research.

σTrellis differs by offering hypertextual interaction with a model (i.e., associative linking and retrieval of its components) and by having a direct formal representation of collaborative interaction among agents in a process. The Trellis CPS model directly integrates the dynamics of process with the information entities and relationships of software development.

σTrellis also differs from existing projects in that the Trellis CPS implementation framework allows RPC interaction with the model. Any new interface a customer needs can be constructed fairly easily and will communicate with existing models. Thus σTrellis is an open system.

References

[BK90] C. Beeri and Y. Kornatzky. A logical query language for hypertext systems. In A. Rizk, N. Streitz, and J. André, editors, *Hypertext: Concepts, Systems, and Applications*, pages 67–80. Cambridge University Press, November 1990. Proceedings of the European Conference on Hypertext.

[CES86] E. M. Clarke, E. A. Emerson, and A. P. Sistla. Automatic verification of finite-state concurrent systems using temporal logic specifications. *ACM Transactions on Programming Languages and Systems*, 8:244–263, 1986.

[CG88] Brad Campbell and Joseph M. Goodman. HAM: A general purpose hypertext abstract machine. *Communications of the ACM*, 31(7):856–861, July 1988.

[DC92] P. Dewan and R. Choudhary. A high-level and flexible framework for implementing multi-user user interfaces. *ACM Transactions on Information Systems*, 10(4):345–380, October 1992.

[Fis91] M. Fischer. Decision making based on practical knowledge. In *Proc. of the 1991 Coordination Theory and Collaboration Technology Workshop*, pages 89–97. National Science Foundation, June 1991.

[GL81] H. J. Genrich and K. Lautenbach. System modeling with high-level Petri nets. *Theoretical Computer Science*, 13:109–136, 1981.

[GS87] P. Garg and W. Scacchi. On designing intelligent hypertext systems for information management in software engineering. In *Proceedings of Hypertext '87 (Chapel Hill, NC, November 1987)*, pages 409–432, 1987.

[Hol88] Anatol W. Holt. Diplans: A new language for the study and implementation of coordination. *ACM Transactions on Office Information Systems*, 6(2):109–125, January 1988.

[Hum88] W. S. Humphrey. Characterizing the software proess: A maturity framework. *IEEE Software*, 5(2):73–79, March 1988.

[Jen81] Kurt Jensen. Coloured Petri nets and the invariant-method. *Theoretical Computer Science*, 14:317–336, 1981.

[Kel89] M. I. Kellner. Software process modeling: Value and improvement. *Technical Review 1989*, pages 23–54, 1989.

[KFP88] G. Kaiser, P. H. Feiler, and S. S. Popovich. Intelligent assistance for software development and maintenance. *IEEE Software*, May 1988.

[LR90] C. M. Lott and H. D. Rombach. A mvp-l1 solution for the software-process modeling problem. In *Collected Solutions from the 6th International Software Process Workshop* (Hakodate, Japan), October 1990.

[Mer74] Philip M. Merlin. *A Study of the Recoverability of Computing Systems.* Ph.D. dissertation, University of California at Irvine, Department of Information and Computer Science, Irvine, CA, 1974. Also available as Technical Report 58, Department of Information and Computer Science, University of California at Irvine (1974).

[MF76] Philip M. Merlin and David J. Farber. Recoverability of communication protocols–implications of a theoretical study. *IEEE Transactions on Communications*, COM-24(9):1036–1043, 1976.

[Mur89] Tadao Murata. Petri nets: Properties, analysis and applications. *Proceedings of the IEEE*, 77(4):541–580, April 1989.

[Rei83] W. Reisig. Petri nets with individual tokens. *Informatik-Fachberichte*, 66(21):229–249, 1983.

[Rei85] Wolfgang Reisig. *Petri Nets: An Introduction.* Springer-Verlag, 1985.

[Rom91] H. D. Rombach. Mvp-l: A language for process modeling in-the-large. Technical Report CS-TR-2709, Department of Computer Science, University of Maryland, College Park, MD, June 1991.

[SF89] P. David Stotts and Richard Furuta. Petri-net-based hypertext: Document structure with browsing semantics. *ACM Transactions on Information Systems*, 7(1):3–29, January 1989.

[SF90a] P. David Stotts and Richard Furuta. Browsing parallel process networks. *Journal of Parallel and Distributed Computing*, 9:224–235, 1990.

[SF90b] P. David Stotts and Richard Furuta. Temporal hyperprogramming. *Journal of Visual Languages and Computing*, 1(3):237–253, 1990.

[SF91] P. David Stotts and Richard Furuta. Dynamic adaptation of hypertext structure. In *Proceedings of Hypertext 91*, pages 219–231, December 1991.

[SF92] P. D. Stotts and R. Furuta. Hypertextual concurrent control of a lisp kernel. *Journal of Visual Languages and Computing*, 3(2):221–236, June 1992.

390

[SFR92]   P. D. Stotts, R. Furuta, and J. C. Ruiz. Hyperdocuments as automata: Trace-based
          browsing property verification. In *Proceedings of the 1992 European Conference
          on Hypertext (ECHT92: November 30–December 4, Milan, Italy)*, pages 272–281.
          ACM Press, New York, 1992.

[Sin89]   A. Sinachopoulos. Logics for petri-nets: Partial order logics, branching time logics
          and how to distinguish between them. *Petri Net Newsletter*, pages 9–14, 8 1989.

[vB88]   Willem R. van Biljon. Extending Petri nets for specifying man-machine dialogues.
          International Journal of Man-Machine Studies, 28:437–455, 1988.

BLACKBOARD BASED COORDINATION IN COOPERATIVE PROBLEM SOLVING

Selahattin KURU and H.Levent AKIN
Department of Computer Engineering
Bogazici University
80815 Bebek, Istanbul
Turkey

ABSTRACT. This article introduces a control architecture for blackboard based coordination in cooperative problem solving. The basic elements of the architecture are goals, policies, strategies, methods, and knowledge sources. The basic control loop employs a bidding mechanism to determine the knowledge source to be executed at the current cycle. The architecture employs separate control and domain blackboards, and separate knowledge sources for the control problem and for representing the domain knowledge. It has a simple and uniform structure, and it is based on a formal basis, namely, extending a partially complete general goal tree. The architecture is implemented in Smalltalk and tested on a multiple-task planning problem.

1. Introduction

A distributed problem solving system is a system that is composed of a set of problem solving agents collaborating in solving a problem. Collaboration means the sharing of raw or processed data. Collaboration is necessary when no single agent can solve the entire problem. The problem solving agents must often work in parallel for reasons of speed and feasibility.

Having a distributed problem solving system has the following benefits[1]: (1)It is possible to build modular systems in an incremental fashion, (2)Portions of the knowledge base can be developed by different people using different domain models, (3)By using only the knowledge that is necessary for the task at hand the efficiency of the system increases with the rational utilization of the resources of a distributed problem solving system, (4)It is possible to use different reasoning techniques which most suit the subproblem considered, (5)Inherently distributed problems can be solved in a natural way, (6)It is possible to implement a distributed problem solving system on a parallel architecture leading to an increase in computational speedup, and, (7)The reliability of the system increases since there are always agents to perform problem solving in the case of the failure of a problem solving agent.

There are a number of issues in designing a distributed problem solving system: (1)How to manage coordination and cooperation between problem solving agents (this is called the control problem), (2)What communication medium to use, and (3)What communication techniques to use. These issues are discussed below.

1.1. COORDINATION AND COOPERATION

The two extremes of the spectrum of possible control modes in coordinating the activities of problem solving agents are: (1)Centralized control, and, (2)distributed control. In centralized control only and only one agent is allowed autonomy and it tells every other agent what to do and when to do it. In other words, this is a system with one master and several slaves. In distributed control, on the other hand, there are several masters and several slaves. Each master is completely autonomons and acts on its own initiative. Each

S. Y. Nof (ed.), Information and Collaboration Models of Integration, 391–399.
© 1994 *Kluwer Academic Publishers. Printed in the Netherlands.*

slave may belong to several masters and a slave only acts when it is told to.

The centralized control mode is simple but narrow. Growth and performance are limited by the master's capabilities. In decentralized control mode there is greater potential for growth, high performance and robustness; but there is also greater potential for instability.

There are two possibilities for avoiding instabilities in decentralized control systems: (1)Endow each agent with enough civic sense, social responsibility, and, if necessary, intelligence to ensure that it interacts productively with its neighbors, (2)Introduce a managerial force of expertise (leaders, moderators, critics, repairmen etc.) to encourage desirable behaviors among the agents and limit the effect of their bad behaviors.

1.2. COMMUNICATION

Problem-solving agents collaborate by sharing data. This sharing can be realized through the exchange of messages containing new or processed data. Messages can be exchanged if a link is provided between every pair of agents; but this is expensive. A cheaper way is to provide a central facility for collecting and distributing messages. A blackboard is a database that serves as a central facility for messages. Agents are allowed to post messages on the blackboard and read messages from it.

2. Blackboard Systems

A blackboard system treats problem-solving as an incremental, opportunistic process of assembling a satisfactory configuration of solution elements. The standard blackboard architecture employs a sophisticated scheduler [2], and it entails three basic assumptions: (1)Solution elements generated during problem-solving are recorded in a global database called the blackboard, (2)Solution elements are generated and recorded on the blackboard by independent processes called knowledge sources (KSes), and, (3)On each problem-solving cycle, a scheduling mechanism chooses a single KS to execute its action. KSes have a condition-action format. The condition specifies the situation in which the KS can contribute to the solution of the problem. The action specifies the behavior of the KS in terms of the creation or modification of the elements on the blackboard. Note that the standard blackboard architecture closely resembles the structure of the classical production systems. The blackboard corresponds to the working memory and the scheduler plays the role of the conflict-resolution strategy.

The are two fundamental issues related to blackboard systems: The direction of search and the control problem. The direction of search may be data-driven (forward-chaining) or goal-driven (backward-chaining). In the data-driven search, actions are triggered by data to generate new data or to modify existing data, and the solution proceeds form the initial state to the final state. In the goal-driven search, on the other hand, the solution proceeds from the final state towards the initial state, and each action is an attempt to prove that a subproblem of the original problem is solved. At any point in the problem-solving process there are typically more than one potential action that can be performed. The control problem is to come up with a mechanism to decide on what action to take at each point. Early blackboard systems did not separate the control problem completely from the solution mechanism. Later, special architectures are proposed to solve the control problem. These architectures may be categorized as event-based architectures, meta-level architectures, and control architectures.

In event-based architectures [3], each change to the blackboard constitutes an event. A particular blackboard event can trigger one or more KSes, and when a KS is executed it typically produces new blackboard events. The architecture employs a sophisticated scheduling mechanism to decide on which KS to consider first on each cycle.

Meta-level architectures [4] distinguish domain and control actions. Meta-level actions and domain actions form a hierarchy in the sense that meta-level actions dictate which domain action to execute. Actually, there is a hierarchy of meta-level actions as well, i.e. meta-meta-level actions, etc. Meta-level actions treat the actions at lower levels in the hierarchy as domain actions. Hence, meta-level actions at some level may determine the ideal meta-level action at one level down in the hierarchy. The built-in hierarchy is the major drawback of meta-level architectures for the inflexibility it introduces into the system.

Control architectures [5,6] explicitly represent domain and control problem-solving in terms of KSes designed specifically for these purposes, integrated in a basic control loop. The architecture is expected to interpret and modify its own knowledge and behavior, and thus adopt itself to dynamic problem-solving situations.

This article discusses a blackboard architecture for the goal-driven approach that uses the control architecture paradigm. It may be viewed as a refinement of Wehe et al.'s AKORN G architecture [6]. The major elements and the basic control loop of the two architectures are essentially the same. On the other hand, the architecture of AKORN G is not really clearly defined in that it employs some seemingly useless elements and it has no formal basis. The architecture proposed in this paper has a simple and uniform structure, and its basic control loop is based on a formal basis, namely, extending a partially complete general goal tree.

3. Elements of the Architecture

The basic control loop of the blackboard control architecture employs the following three steps:(1)Update the set of pending goals, (2)Select a pending goal, and, (3)Execute the owner KS of the goal selected. Thus, we express the basic control loop in terms goals rather than KSes. The set of all goals relevant to a problem form a general goal tree, which we introduce as a generalisation of a goal tree. For a formal definition of a general goal tree see the appendix.

Speaking in terms of the terminology of searching general goal trees, the basic control loop finds an immediate extension of a partially complete general goal tree in each cycle. As discussed in the appendix, a partially complete general goal tree may be extended at an or node, or at an and node. We will make three assumptions for the sake of simplicity: 1)And nodes are extended immediately when a particular and node is generated, 2)The generation of an and node implies an execution order for its children, and 3)There is a strategy which specifies which of the or nodes is to be extended first. These three assumptions define a global strategy completely for a general goal tree if we have a mechanism for making a choice among the children of or nodes. We will first develop a basic architecture that addresses this question.

3.1. THE BASIC ARCHITECTURE

The basic architecture incorporates a bidding mechanism for making a choice among the children of an or node. The mechanism asks the owner KSes of the pending goals for their bids for the goals they have posted. The bid is a measure of how strongly the owner KS supports the goal it has posted. The bid is expressed in terms of a set of parameter values. An integration rule states how the parameter values are used in determining the winning bid, and hence the KS to be executed at the current cycle. Typically it is a static rule which simply forms a linear combination of the parameter values. The integration rule plays the role of a local evaluation function in the classical terminology in the sense that it effects the choice of the new branches to be generated in the search process. Unlike in the classical terminology, an integration rule is used locally, and it is subject to change at every cycle of the basic control loop.

The integration rule of a node is called the policy of that node. Policies of different nodes need not to be the same. There are several integration rules in use at any problem-solving step, one for each or node. Policies change dynamically as the solution proceeds. Policies are specified as part of the problem definition.

The basic architecture of the blackboard system incorporates a domain blackboard and a control blackboard. The domain blackboard stores goals and policies, and the control blackboard stores bids and the current policy. Execution of the basic control loop is controlled by the Control KS. The Control KS selects a subset of the goals posted onto the domain blackboard by domain KSes and asks the owner domain KSes for their bids. After domain KSes make their bids, the Control KS selects the best bid according to the current policy and triggers the owner KS for executing the goal selected. Execution of a domain KS causes new goals to be posted onto the domain blackboard.

For illustration purposes let us consider a simple problem defined by the set of domain KSes given in Table 1. The KSes are defined in terms of the goals they bid for and the subgoals they produce. Also specified in the definition of KSes are the values for policy parameters. The general goal tree implied by the search space of this problem is the same as the one in Figure A.1. Assume that the three assumptions introduced earlier hold, i.e. and nodes are extended immediately when a particular and node is generated, the generation of an and node implies an execution order for its children, and there is a strategy which specifies which of the or nodes is to be extended first. Also assume that this strategy chooses or nodes created earliest first, and that the policy specified for or nodes is an integration rule of the form of a weighted sum with the following weight values for the parameters relevant to the particular goal

$$
\begin{array}{ll}
\text{for a2 :} & w(p1)=2,\ w(p2)=10,\ w(p3)=10 \\
\text{for a13:} & w(p1)=2,\ w(p4)=2,\ w(p5)=10
\end{array}
$$

where $w(pi)$ represents the weight of policy parameter pi. Note that the values for these parameters are given in Table 1.

Table 1. KSes for the Illustrative Example

KS no	rule	values for policy parameters
1	a:-a1,a2,a3	v(r1,a1)=10,v(r1,a2)=8,v(r1,a3)=4
2	a1:-a11,a12,a13	v(r1,a11)=10,v(r1,a12)=6,v(r1,a13)=5
3	a12:-a121,a122	v(r1,a121)=8,v(r2,a121)=6,v(r3,a121)=8 v(r1,a122)=5,v(r2,a122)=10,v(r3,a122)=2
4	a13:-a13a	v(p1)=10,v(p4)=8,v(p5)=8
5	a13:-a13b	v(p1)=8,v(p4)=8,v(p5)=10
6	a2:-a2a	v(p1)=10,v(p2)=8,v(p3)=10
7	a2:-a2b	v(p1)=6,v(p2)=6,v(p3)=10

Assuming that all the goals at the leaves of the tree can be satisfied by the facts in the domain blackboard and that the constraints are handled properly, the trace of the execution of the system can be given as in Table 2. Note that the goals on the domain blackboard is represented as a list of lists, where the elements of a list are separated with a comma for and nodes, and with a semicolon for or nodes. In the third and the last columns a colon separates KS number from the bid values or from the goals that are created or satisfied.

Let us consider Cycle 6. At this cycle, there are 3 pending goals on the domain blackboard. The strategy requires KS 4 and KS 5 to be selected for bidding. Their bids, as computed according to the policy specified for goal a13, are 116 and 132, respectively. Hence KS 5 is selected for execution.

3.2. EXTENDING THE BASIC ARCHITECTURE

At this point we will relax the three assumptions made earlier and extend the basic blackboard control architecture in several directions: (1)Determination of an order of execution for the children of and nodes, (2)Specification of a global strategy, and (3) Specification of a solution method for the problem.

Table 2. Trace of Execution of the Algorithm of the Control Loop

cycle	goals on Domain BB	KSes bidding	KS selected
0	a	1	1:(a1,a2,a3)
1	(a1,a2,a3)	2	2:(a11,a12,a13)
2	((a11,a12,a13),a2,a3)	none	:a11
3	((a12,a13),a2,a3)	3	3:(a121,a122)
4	(((a121,a122),a13),a2,a3)	none	:a121
5	(((a122),a13),a2,a3)	none	:a122
6	((a13),a2,a3)	4:116,5:132	5
7	(a13b,a2,a3)	none	:a13b
8	(a2,a3)	6:200,7:176	6
9	(a2a),a3)	none	:a2a
10	a3	none	:a3
11	none	none	none

Determination of an order of execution for the children of an and node employs a similar bidding mechanism. The difference is that there is now a single domain KS bidding for an order of its own subgoals. Again the Control KS determines the order according to the policy specified for the and node.

Let us use the same example problem to illustrate bidding in the existence of a policy for and nodes. Let us consider KS 3, which is bidding for the and node a12. Parameter values for the local policy for this and node as given in Table 1 are

for a121: $v(r1)=8$, $v(r2)=6$, $v(r3)=8$
for a122: $v(r1)=5$, $v(r2)=10$, $v(r3)=2$.

Let us also assume that the weights for these parameters as specified by the policy for this node are

$w(r1)=5$, $w(r2)=10$, $w(r3)=7$.

Then the scores computed as the weighted sum of the parameter values for goals a121 and a122 will be 156 and 139, respectively. This means that the goals will be ordered as a121 followed by a122.

A global strategy specifies the search direction. The strategy may be depth-first, breath-first, or best-first. A partially complete general goal tree is extended first at the children of the most recently extended node in the case of depth-first strategy, and at the children of the node that was extended the earliest in the case of breath-first strategy. The best-first strategy employs a bidding mechanism for choosing among the children of different nodes according to a global policy.

Methods are used to express domain-dependent high level knowledge on how to solve a type of problem. A method is a partially complete general goal tree. Methods are interpreted by the Control KS either to reduce the number of children to be considered for or nodes or to order the children of and nodes. In the first case, the size of the problem is reduced by pruning the tree that defines the search space of the problem. This is done as follows: When a domain KS is executed at the final step of the basic control loop, Control KS checks if there is a restriction on the domain KSes that may bid for a goal. If there is, it asks only those that are specified by the method for their bids. In the second case, it is expected that consideration of critical branches first may result in a saving in search time.

Again using the same example problem we may illustrate the use of a method as follows. Let us assume that a method is specified for a particular instance of the problem that names goal a13a as one of the goals that should be in the solution if possible. Eventhough goal a13b wins the bidding over a13 a to satisfy goal a13 according to the policy specified for this goal, then the tree will be extended at goal a13a, rather than at goal a13b.

When all these fatures are included in the architecture the basic control loop works as follows to extend a parial solution: At any problem-solving step we have a set of pending goals stored in a queue in the order implied by the global strategy so that the goal to be processed at the current cycle is in front. The goals generated at the end of the current cycle are merged into the queue while maintaining the ordering. If the goal processed is at an and node, then an ordering is determined for the children of the node immediately using the policy for the node. If the goal processed is at an or node, then one of the goals corresponding to the children of the or node is selected using the policy for the node. The ordering of the queue is maintained by placing the newly generated goals in front of the queue if the strategy is depth-first, and at the rear of the queue if the strategy is breath-first. The best-first strategy requires a score to be computed for the newly generated goals and the queue to be sorted in decreasing order of scores. The scores are computed according to the global policy. The algorithm for the control loop is given in Figure 1.

Take the goal that is in front of the queue
If the goal is at an or node
    then
        identify the owner KSes for the subgoals of the goal
        if any of the subgoal appear in the method specified
            then select the owner of the goal as the winning KS
            else
                identify the policy for the goal
                get the bids of the KSes
                select the winning KS according to the policy
        execute the KS to generate the goal to be posted
If the goal is at an and node
    then
        execute the owner KS to generate the subgoals to be posted
        if its subgoals appear in the method specified
            then sort the subgoals as specified by the method
            else
                identify the policy for the goal
                sort the subgoals according to the policy
Merge the subgoals into the queue according to the global strategy

---

Figure 1. Algorithm for the Control Loop

Methods, policies, and strategies are specified at the problem definition level. There is a single Control KS that is responsible for governing the bidding mechanism and for updating the current policy. Methods, policies, and goals are stored in the domain blackboard. The current policy, the strategy and the bids are stored on Control Blackboard.

4. Implementation

The proposed architecture is implemented in the object-oriented language Smalltalk [7]. There is a natural parallelism between the object-oriented paradigm and the blackboard paradigm. Both are based on the view that the world can be modelled in terms of objects communicating through the use of messages. Objects are actually instances of classes and the object-oriented approach has a built in mechanism to support a hierarchy between classes. Classes implement the algorithms that define their responses to different messages. Each such algorithm is called a method in this terminology. The object-oriented approach makes it very easy to introduce new classes of objects by adding more features (i.e. methods) to existing objects. One may create as many instances (i.e. objects) of a class as he wants. All these features makes the object-oriented approach very attractive for implementing blackboard systems.

The Smalltalk implementation is based on the identification of all major elements of the architecture as Smalltalk classes. Thus, domain and control blackboards, control KS and domain KSes constitute the main classes. The class corresponding to Control KS implements the basic control loop as its main method. Other mechanisms such as goal posting and bid passing are the methods of corresponding classes.

The current implementation of the proposed architecture is tested by solving a multiple-task planning problem [5]. Details are given in [8,9].

5. Evaluation and Conclusion

The control architecture introduced for blackboard based coordination is based on searching a general goal tree. The basic elements of the architecture are goals, policies, strategies, methods, and KSes. It employs a basic control loop that uses a bidding mechanism in choosing the knowledge source to be executed at the current cycle. The bidding mechanism is guided by a local scheduling criteria, called a policy. A strategy, on the other hand, is a global scheduling criteria such as depth-first, breath-first, or best-first. Strategies and policies together determine how a partial solution is to be extended in each cycle of the basic control loop. The search space may be reduced by pruning a general goal tree using high level knowledge on how to solve a type of problem. This type of knowledge is specified in terms of methods.

The architecture employs separate control and domain blackboards, and separate KSes for the control problem and for representing the domain knowledge. The basic control loop is simple. As a matter of fact, the skeleton of the basic control loop stays in paralel with the three-step control loop of the classical production system architecture. It is able to adopt itself to dynamic problem-solving situations by interpreting and modifying its own knowledge and behavior with the help of policies, strategies, and methods. It realizes all these features in a simple and uniform mechanism. With these features, the control architecture achieves the behavioral goals set for an intelligent control architecture in [5].

The architecture is implemented in the object-oriented language Smalltalk, and tested on a multiple-task problem.

Acknowledgements

The presentation of this work in the NATO ARW was supported in part by the Bogazici University Research Fund.

References

1 Rich, E. and K. Knight, *Artificial Intelligence*, 2nd ed., McGraw-Hill Co.(1992).

2 Lesser, V.R. and D.D. Corkill, "Functionally Accurate Cooperative Distributed Systems," *IEEE Transactions on Systems, Man and Cybernetics*, No 1 (1981) pp 81-96.

3 Nii, H.P., E.A. Feigenbaum, J.J. Anton and A.J. Rockmore, "Signal-to-Signal Transformation: HASP/SIAP Case Study," *AI Magazine*, No 3 (1982) pp 23-35.

4 Rychener, M.D., R. Banares-Alcantara and A.W. Westerberg, *A Rule-Based Blackboard Kernel System: Some Principles in Design*, Research Report DRC-05-04-84, Design Research Center, Carnegie-Mellon University (1984).

5 Hayes-Roth, B., "A Blackboard Architecture for Control," *Artificial Intelligence*, No 26 (1985) pp 251-321.

6 Wehe, R., K. Lien and A.W. Westerberg, *Control Architecture Considerations for a Separation Systems Design Expert*, Research Report EDRC-06-29-87, Engineering Design Research Center, Carnegie-Mellon University (1987).

7 Goldberg, A., *Smalltalk: The Interactive Programming Environment*, Addison-Wesley Publishing Co. (1984).

8 Kuru, S. and F.Bek, "Goal-Driven Blackboard Control Architecture Based on Extending Partially

Complete General Goal Trees," *Knowledge Based Systems*, Vol 3, No 44 (1990) pp 236-244.
9 Bek, F., *A Goal-Driven Control Architecture for Blackboard Systems*, M.S. Thesis, Bogazici University, Istanbul, Turkey (1989).

Appendix: General Goal Trees

Before we introduce the concept of a general goal tree we give the definition of a goal tree for the sake of completeness.

A goal tree [1] is an and/or tree. And nodes represent choices, and or nodes represent simultaneous goals that must have compatible solutions. The children of and nodes are or nodes, and the children of or nodes are and nodes. A partial solution to a node n is a subtree rooted at n, with the property that if it contains an or node m, then it contains exactly one child of m. A complete solution to a node n is a partial solution to n with the property that if it contains an and node m, then it contains every child of m, and the solutions to the children satisfy C(m), the constraint at m. A solution to the entire goal tree is a complete solution to its root. Partial solution s2 is an extension of partial solution s1 if s1 is contained in s2.

We introduce the concept of a general goal tree by allowing a goal tree to violate the requirement that the children of and nodes are or nodes and the children of or nodes are and nodes. A partial solution to a node n is a subtree rooted at n, with the property that if it contains an or node m, then it contains either exactly one child of m or no child of m at all, and if it contains an and node u, then it contains either every child of u or no children of u at all, and the solutions to the children satisfy C(u), the constraint at u. A complete solution to a node n is a partial solution to n, with the property that if it contains an or node m, then it contains exactly one child of m, and if it contains an and node u, then it contains every child of u, and the solutions to the children satisfy C(u), the constraint at u. A solution to the entire general goal tree is a complete solution to its root. Partial solution s2 is an extension of partial solution s1 if they are both partial solutions to the same node and if s1 is a subtree of s2. An extension s2 of s1 is an immediate extension of s1 if it is an extension of s1 with the property that, in addition to the nodes of s1, it contains either a child of a leaf node u of s1 if u is an or node, or every child of a leaf node v of s1 if v is an and node.

Consider the general goal tree in Figure A.1. Letters denote and nodes, numerals denote or nodes, and these are concatenated in the obvious way, as shown in the figure. The root has name 0 or a depending on whether it is an or node or an and node. We use the notation n:{l1,l2,...,lk} to refer to the tree rooted at node n and terminating at nodes {l1,l2,...,lk}. Note that the terminating nodes may be and nodes, or nodes or a mix of and and or nodes. One partial solution to node a is a:{a11,a12,a13,a2,a3}. A complete solution might be a:{a11,a121,a122,a13a,a2a,a3} provided that the solution a12:{a121,a122} satisfies the constraint C(a12), the solution a1:{a11,a12,a13} satisfies the constraint C(a1), and the entire solution satisfies C(a). The partial solution a:{a1,a2,a3} may be extended to a:{a11,a12,a13,a2,a3} or to a:{a1,a2a,a3} by finding one of its immediate extensions. In other words, we may extend the partial solution at an and node or at an or node. While every child of an and node are included in the extension, we include only one child of an or node. The situation is illustrated in Figure A.2.

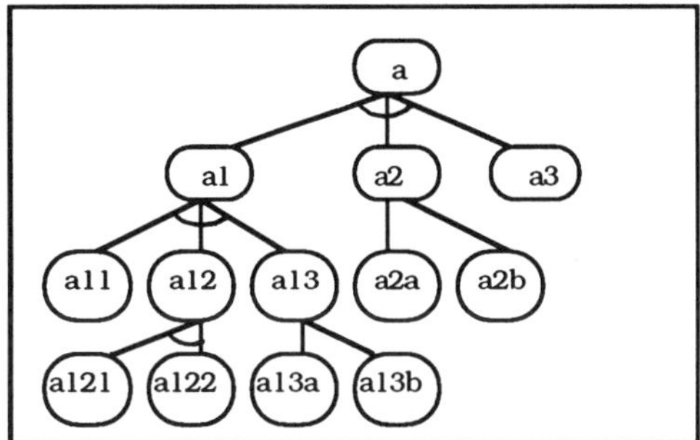

Figure A1. A General Goal Tree

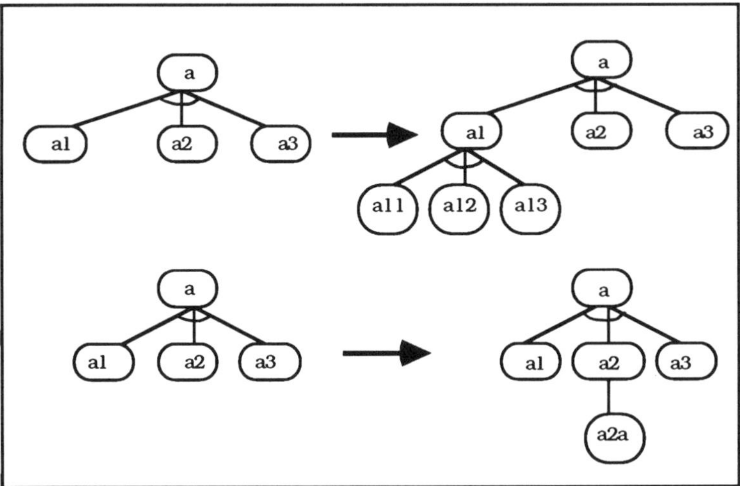

Figure A2. Extension of a Partially Complete General Goal Tree

A GENERIC ARCHITECTURE WITH NEUTRAL INTERFACES FOR OFF-LINE ROBOT PROGRAMMING AND SIMULATION

E.TROSTMANN, L.F.NIELSEN, S.TROSTMANN, F.CONRAD
Instituttet for Styreteknik, bygn. 424
Danmarks Tekniske Højskole
DK-2800 Lyngby
Danmark

ABSTRACT. A generic architecture for a new robot off-line programming and simulation system, ROPSIM, has been synthesized. The generic architecture is based upon the open systems concept and identifies the robot program interface, the necessary robot system models, simulation algorithms, and their interconnections. Neutral interfaces, based on the proposals ISO STEP and ISO ICR, allow for the exchange of robot models and robot program models with other systems making ROPSIM a true CIME subsystem. A simulation in ROPSIM includes the total robot system, i.e. the manipulator geometry and kinetics *and* the robot controller.

As an example a simulation of a hydraulic two link test robot, driven by a STEP model and an ICR program, is presented and compared with experimental results.

Keywords: CIME, Robot Models, Simulation, Neutral Interfaces, Generic Architecture, ICR, STEP, Control.

1. Introduction

Robots offer a high degree of automation, flexibility, dexterity, programmability and repeatability. This makes robots highly suitable as real-time production components in computer integrated manufacturing and engineering (CIME) systems. The programmability allows for making changes or optimizations in the production. The repeatability secures the quality control of the produced products.

Today the task of generating robot programs is a bottleneck for obtaining a better performance and a more effective use of robots. Only when cost effective programming tools become available robots will be attractive for many more production applications.

The ambition to produce at a high speed by using robots, has brought up critical problems in the timing of the robot arm geometrical movements and the control of the dynamic behaviour of the robot. Further the robustness of the control system with respect to changes due to payload, wear and other time varying properties has become an important issue. Also vibrations (stability) of the robot mechanical structure and

401

S. Y. Nof (ed.), Information and Collaboration Models of Integration, 401–433.
© 1994 *Kluwer Academic Publishers. Printed in the Netherlands.*

collision avoidance become significant problems. To overcome such problems the robots may be provided with possibilities through sensors to measure not only the state of the robot itself but also the state of the environment. The sensor signals are fed back to the robot controller and used for corrections or generation of updated control references.

The most common method used for robot programming today is on-line programming. The on-line programming procedure is based upon using the physical robot itself in a socalled "teach-in" mode. In this mode the robot task is programmed by guiding the robot through the required positions. During the teach-in mode the robot positions are recorded and after some manual editing the robot program can be "played back" as many times as needed.

As a consequence, however, the robot itself and the equipment to be served by the robot are idling during the programming phase. This idling period can be costly especially in companies whose production is mainly based on a small series of frequently changing products. And it explains why robots still today are mostly used in large series and mass production.

The on-line programming is carried out using a relatively low robot velocity during the programming phase. Normally the velocity has to be increased considerably when the robot is executing the program in production. Consequently the developed program is only valid for operating the robot as long as the dynamic properties of the robot does not effect the precision of the robot movement too much. Corrections to counteract dynamic errors can by manual programming only be introduced by a trial and error procedure.

Newer methods based on off-line programming seem to be the best proposal for solving the programming problem taking into account the total behaviour of the robot system. In preceding studies of the information flow in CIME systems [Trostmann,E. (1987)] including off-line programming of robots have shown that much of the information needed for off-line programming is already residing in the data bases of other CIME subsystems.

In off-line programming, see Fig. 1, the robot program is generated in a software system by using mathematical models of the robot system, the robot process, the tools, the production scene etc. This means that the programming of the robot is decoupled from the real-time production system.

Although several commercially available off-line programming systems exist on the market today, the task of generating programs is still a bottleneck for obtaining better performance and more effective use of robots. The main problem is that the mathematical models used during the programming phase may be too inaccurate.

In the programming and simulation of the robots the modeling of the robot system is still a critical issue. In fact the accuracy and effectiveness of the programming, simulation, and control of the robot depend on the robot model. The modeling facilities of most programming and simulation systems are not capable of describing neither the full dynamic behaviour of the total robot system, i.e. the robot manipulator and the robot controller, nor the behaviour of external sensor feed-back for generation of data from robot, process state and environment.

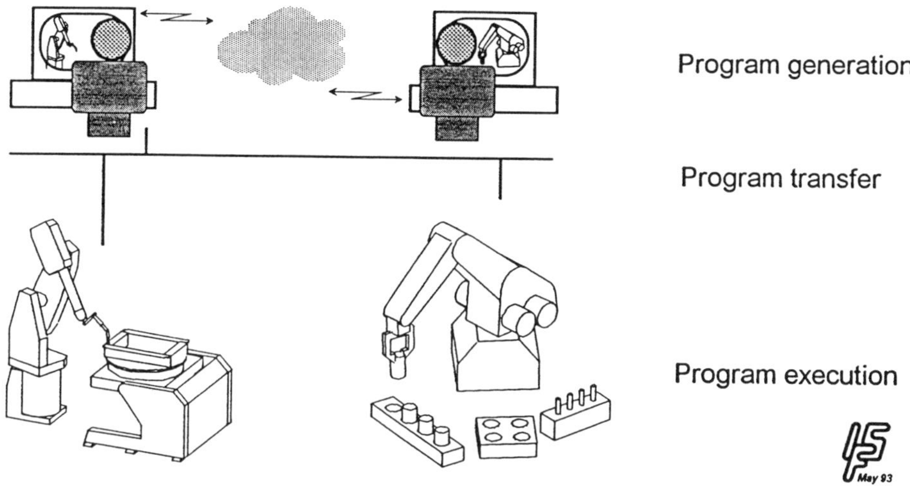

Figure 1. Off-line programming

In conclusion therefore, there is a need for model driven programming and simulation systems for robots illustrating the true real-time behaviour of the total robot system, manipulator, controller and sensors. Such systems should in order to be true CIME subsystems facilitate the exchange and reuse of different robot model definition data and different robot program definition data between systems of other origin or different functionality.

A robot off-line programming and simulation system based on a generic architecture with neutral interfaces is described in the following paragraphs.

2. Neutral Interfaces

Today it is widely believed and accepted that manufacturing companies who achieve large-scale integration of their operations will be competitive to much higher degree than those who only integrate and automate in disparate areas.

One of the most difficult problems to overcome in advancing the concept of integrated manufacturing and engineering systems is the integration process itself. Generally speaking integration is here understood as the linking of different system components together into a complete system, which fulfils the objectives and specifications of the integrated manufacturing system. The basic function of the system components is to transform inputs to desired outputs by certain mechanisms (transfer functions) and controlled by constraints specifying defined goals, specifications, plans etc. to be fulfilled.

In Fig. 2 a schematic illustration of the functional integration of the various subfunctions in an industrial company is illustrated.

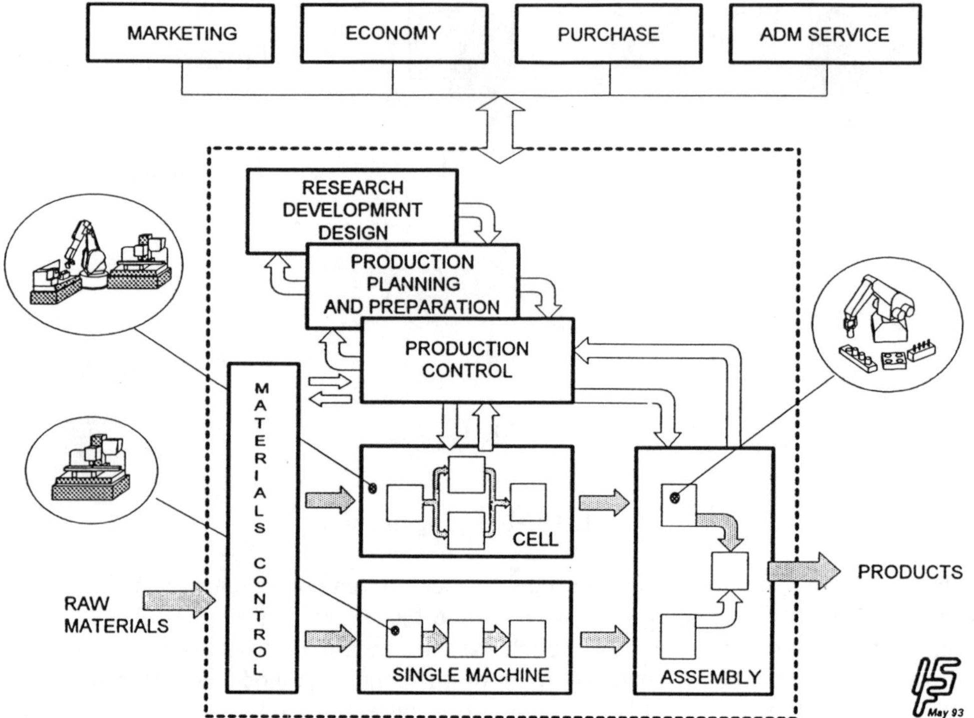

Figure 2. Schematic illustration of the functional
integration in an industrial company.

The general tools applied in this integration process are computers and use of a long series of software aids like CAD, CAE, CAPP, CAM, FMS, CNC, Robots etc. The integration of a set of some or all of these tools in industry has coined the terms "Computer Integrated Manufacturing, CIM" and "Computer Integrated Manufacturing and Engineering, CIME".

The abovementioned information systems have been developed over the latest 3-4 decades by many different people. More systems are being developed every day. Because these systems have evolved independently from one other, they tend to use unique representations for product definition data. Unfortunately, each system is only able to use data that has been defined in the particular representation that it was programmed to understand. Getting several of these systems to work together is like trying to run an organization where everyone speaks a different foreign language. Computer support of isolated engineering and production activities leads to the wellknown state of "islands of automation".

One of the main targets for research in the domain of integrated manufacturing and engineering systems is therefore addressing the properties, conditions, possibilities and means by which different system components can be integrated and how the integrated system can be modified, updated and controlled.

An integration should allow for the building of integrated manufacturing and engineering systems of a type, size, order and implementation speed that fits into the practical needs of a given user company - whether it is a small, medium-sized or large company. This philosophy leads one to the adaptation of standardization and modularity concepts in the manufacturing system design. It leads further to the application of the open system's concept [Trostmann,E.(1987)] in accordance to which one can add, exchange subsystems of different origin when needed without loosing functionality and operationality of the total system. In other words the open system concept guaranties the robustness of the system's architecture enabling the use of multivendor system components.

The key issue in the open system's concept is the application of intelligent, neutral interfaces between subsystems. The proper design of such interfaces requires deep insight into the manufacturing system's functionality and its mapping on to the needed dataprocesses, -models and -structures. Examples of research and development activities of this kind are the PDES project in US the CAD*I, NIRO and INTERROB projects within the ESPRIT programme of the CEC.

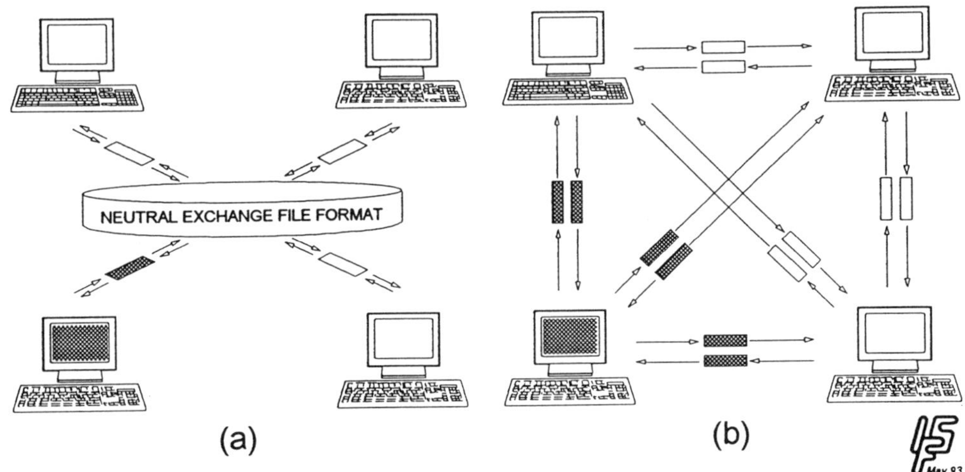

Figure 3. (a) Open systems concept for sub-system communication
(b) Monolithic system concept for sub-system communication

The most promising principle for communication of product definition data between CIME systems or subsystems of different origin is the use of intelligent, neutral data formats and interfaces. By an intelligent interface is here meant that the receiving system is not only receiving the data files correctly but also interpreting the meaning of the file exactly in the way the sender wants it. By neutral interface is here meant an interface (a data format) that is universally accepted, i.e. standardized.

The principle of using a neutral file format for communication between different systems is illustrated in Fig. 3, [Schlechtendahl(1989)]. As can be seen the use of native formats require ½·n·(n-1) pair of processors for translation. Whereas the use of a shared neutral format only requires n pair of processors. Also the architecture based on a neutral file format becomes much more robust and less costly when system components are replaced or new ones added.

The application of neutral file formats has led to a new, general concept for the architecture of a CIME-system, see Fig. 4, [Trostmann,E.(1987)]. The fundamental scheme necessary for an effective communication of product definition data between systems of different functionality and origin is based upon the definition and use of neutral file formats, whereby a native system format is transferred to and from the neutral file format by applying pre- and postprocessor software modules.

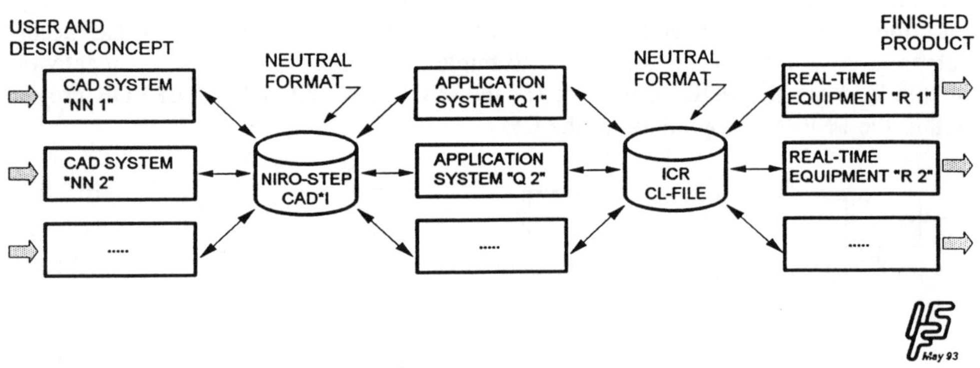

Figure 4. CIME system conceptual schema

In the concept in Fig. 4, [Trostmann,E.(1987)], three main levels of subsystems are identified: CAD Systems, Application Systems, and Real-time Production Systems. Two different neutral formats, the NIRO-STEP format [Kroszynski(1987)] and the ICR/CL-FILE formats [ISO/DP 10562-1.2(1989);Pressmann(1977)], are illustrated and refer to the international standardization work within the CIME area.

The fundamental components or tools necessary for communication between two different systems through a neutral interface/neutral file require as mentioned above two

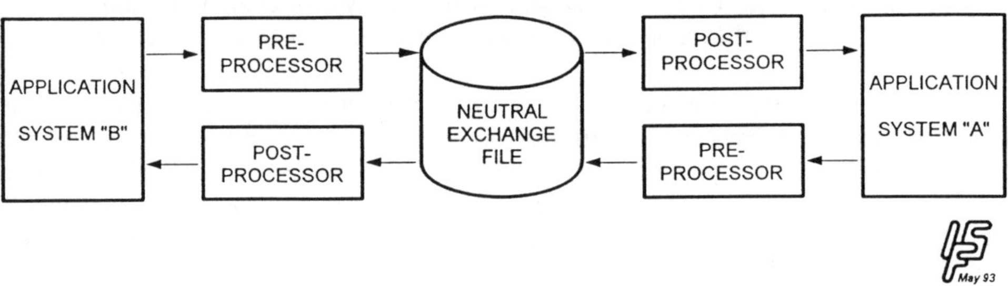

Figure 5. Principle of pre- and post-processors

dedicated software processors called a preprocessor and a postprocessor depending on the communication direction, see Fig. 5. A processor is here defined as a translator whose input and output are in high level language form. The processors translates native files to neutral files and vice versa.

An actual implementation of a subset of the proposed CIME system conceptual schema, Fig. 4, for the off-line programming of robots based upon CAD-geometry has been carried out at the Control Engineering Institute, the Technical University of Denmark, see Fig. 6. It has been demonstrated that the interconnection of existing commercially available subsystems and specific tailored subsystems through a systematic use of neutral, standard interfaces is feasible.

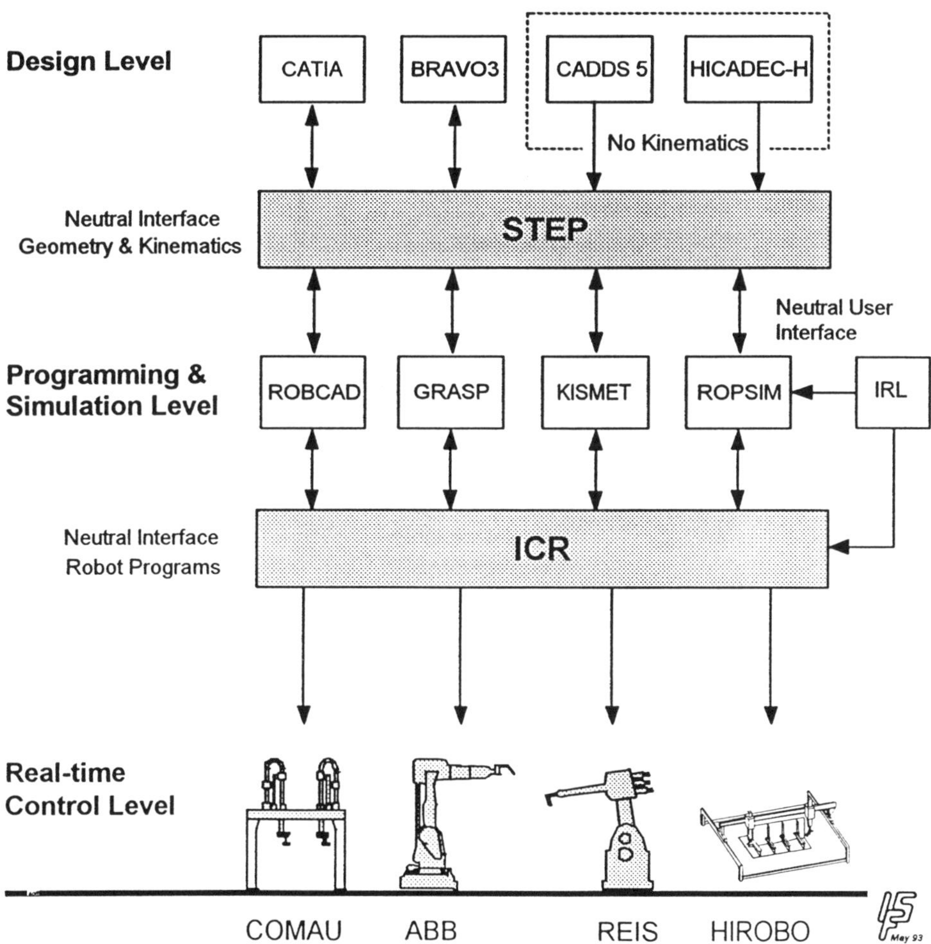

Figure 6. An actual implementation of the conceptual
schema in Fig. 4.

3. Generic architecture

3.1. GENERAL

As a basis for the design of a new off-line programming and simulation system for robotic systems a generic architecture has been developed, [Trostmann,S.(1992)]. This development has focused upon two fundamental challenges:

* The functionality of the system of the system should enable a model driven real-time simulation of robotic systems consisting of robot manipulators (geometry, kinematics, dynamics, drives, loading forces/torques) *and* robot controllers (control functions, dynamics and sensors).
* The implementation of the system should be based on the open system's concept allowing for the exchange of robot models and robot program models with other software systems via neutral interfaces.

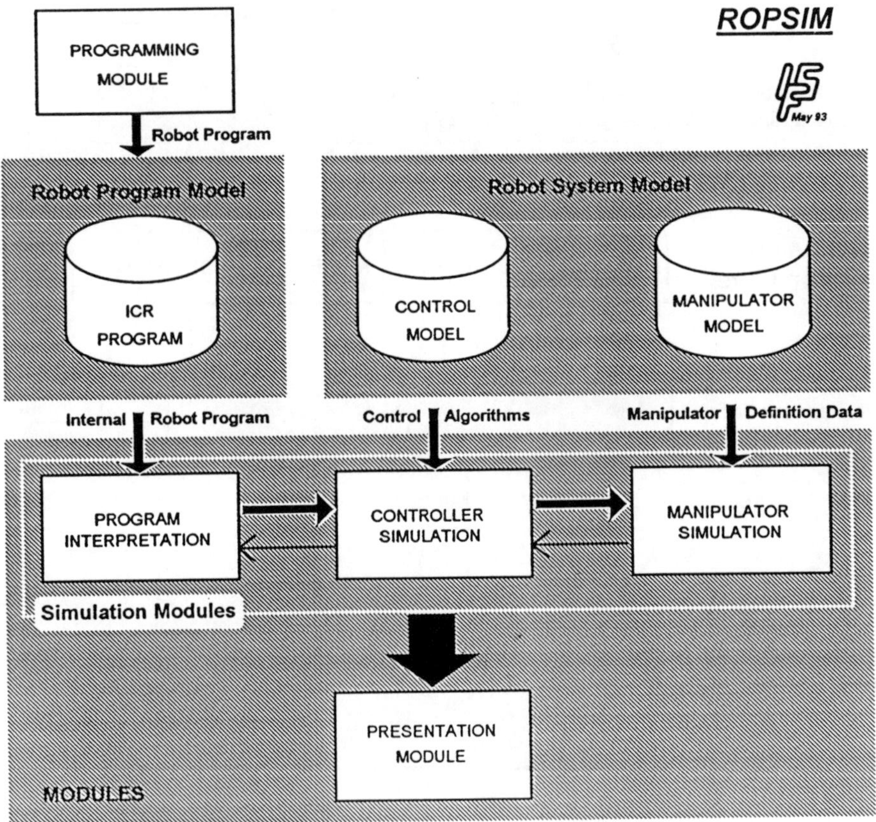

Figure 7. Generic architecture for robot off-line programming and simulation systems.

Through an analysis partly based on experience with existing off-line programming and simulation systems (such as ROBCAD, GRASP, KISMET) and robot systems (such as ABB IRB 2000, Hitachi PW 10, Hirobo) and partly based on control engineering systems' theory essential characteristics of robot models and algorithms (models) have been identified and inspired the synthesis of a generic architecture as illustrated in Fig. 7.

In this figure the models (represented as "cylinders"), the modules (represented as "rectangles"), the interfaces between the modules (represented as horizontal arrows), and the information flow between the models and modules (represented as vertical arrows) are shown.

The thick vertical arrow represents the information flow from the three simulation modules (program interpreter, controller simulation, and manipulator simulation) to the presentation module. Each of the components may be found in commercially available robot off-line programming and simulation systems at a different level of implementation. The components can be categorised into models and modules. The models carry essential robot system information and the modules carry out the necessary data processing.

The application of the generic architecture is used for the development of a new **Robot Off-line Programming** and real-time **SIM**ulation system called ROPSIM. The general architecture will ensure that ROPSIM has the desired functionality and that it will be a true CIME subsystem in the sense that it supports the neutral interfaces for application systems defined in the conceptual schema for CIME systems, refer to Fig. 4.

In the next sections, the requirements to the identified models and to the functional properties of the individual modules are defined and described.

3.2. MODELS

According to the focus in the development of the generic architecture the description of the three models, illustrated in Fig. 7, is presented in the following two sections. The first describes the essential information content of the models. The second describes requirements to the representation of the models with respect to their ability to support the open systems integration concept.

3.2.1. *Information content of the Models*. The models must carry the information relevant for obtaining the functional objective of the modules. The actual information content of the individual models depends on the complexity of the robot off-line programming and simulation system.

Robot program model. Generally speaking the information in a *robot program* is a description of actions the robot must perform in order to accomplish a task [Trostmann,S.(1989)]. The robot program forms the description and is expressed in the given robot language. The instructions of a robot language can be divided into two categories. One part deals with the data manipulation and another part deals with the robot control. The data manipulation instructions contain the logical structure of the program. The robot related instructions contain the control statements of the robot. Also instructions for external sensor feed-back are included in the latter. A suitable robot programming language should include facilities for implementing both categories of instructions. The robot program serves the information needed in the program interpretation module.

Robot system model. As illustrated in Fig. 7, the *robot system model* consists of a *control model* and a *manipulator model*.

The *control model* holds a description of the robot controller, which may include functions for interpolation, inverse kinematic transformation, joint control algorithms, etc. Also the interrelationship between these functions and possibilities for external sensor feed-back should be defined in the control model. The control model provides information for the control simulation module.

The *manipulator model* must carry information about the kinetics of the robot manipulator. Kinetics is the study of all aspects of moving objects, and therefore the manipulator model may include information about the kinematic structure, the dynamic relationship and the geometric shape of the manipulator. The kinematic structure describes the relationship between the links and the joints in the manipulator, and defines the realisable manipulator movements. The kinematic structure is the essential core for simulation of manipulator movements. The dynamic relationship may be described by mass properties (concentrated parameters defining the mass distribution) and elasticity of the manipulator links and the behaviour of the servo motor drives. The dynamic relationship is essential for simulation of manipulator movements under the action of torque/force. The geometric shape describes the shape of the individual links in the manipulator. This geometric information of the manipulator is essential for performing collision tests and for visualisation via a 3D animation of the robot movements. The description of manipulator kinetics provides the information for the manipulator simulation module.

So far only a model of a robot system has been discussed. Often, a robot work scene includes physical objects that are not able to move (the parts to be handled, fixing tools, air ventilation equipment etc.). During the manipulator simulation, information concerning the geometric shape, mass properties and elasticity of such objects in the work scene may be relevant. The geometrical information is applicable in collision tests. The mass properties are necessary for identifying changing constraints in the kinetic manipulator simulation (e.g. the manipulator has varying payload). Such non-moving objects can also be included in the manipulator model. The only difference from the representation of the robot manipulator itself, is that the models of non-moving objects do not include kinematic entities.

Summarising the generic robot system model consists of submodels describing:

- A model of the robot controller
- A model of the moving robot manipulator
- A model of the non-moving environment.

3.2.2. Models with respect to open systems integration concept. It has been decided to use the ISO proposal ICR [ISO/DP 10562-1.2(1989)] as a standardised neutral interface for communication of *robot programs*.

Robot program model. The features of ICR comply with the requirements for such a neutral interface described section 1. The communication of robot programs through the ICR neutral interface is relevant for two purposes.

Firstly the neutral interface is useful when verified and optimised robot programs have to be transferred to dedicated and maybe various control units of the robot system selected to carry out the task without reprogramming. The communication of robot programs from the off-line programming and simulation system to the physical robot systems is traditionally carried out by a translation from the robot programming language in the programming system (e.g. TDL [Robcad(1991)] in the case of ROBCAD) to the native programming language representation of the robot controller (e.g. ARLA [ABB Robotics(1988)] of the ABB IRB2000 robot). Such an integration approach requires development of processors handling the translation of the robot program in the off-line programming and simulation system into the native programming language of *each* of the applied robot systems. The limitation in such an integration strategy lies primarily in the low capability of the native programming languages of the robot systems. Even though the programming language in the off-line programming and simulation system is capable of expressing almost any conceivable task, it is not always possible to transfer a task description that can be expressed in the native programming language of the dedicated robot controller. In such cases the robot task needs not to be reprogrammed but can be transferred through the ICR interface to a more suitable robot controller, assuming that the robot belongs to the family of kinematic systems that may be able to execute the program.

As a result the application of the standardised neutral interface ICR for robot programs can ensure the required information flow between the robot off-line programming and simulation system and the control units of the robot systems in the CIME system architecture illustrated in Fig. 7.

Secondly the neutral interface ICR is useful when robot programs are communicated between two robot off-line programming and simulation systems. For the system sending the robot program this situation is exactly the same as if the receiving system was a robot control unit. The communication of robot programs between two application systems does therefore not introduce new requirements to the sending system. The receiving system, on the other hand, must satisfy certain requirements.

The ICR robot programming language is at the intermediate code level, meaning that the language instructions is at the lowest possible abstraction level and still hardware independent. As it is not possible to translate from a program representation with a low abstraction level, to a program representation with a higher abstraction level, and the ICR program representation is at the intermediate code level, the receiving system must be able to apply the robot program in ICR representation. In practice this means that the application system must include an ICR interpreter or translator.

ICR is close to the German standard, DIN 66313, and it may be expected that ICR will also become an ISO standard. Therefore, it is not unrealistic to require that robot off-line programming and simulation systems support ICR programs.

Summarising, the application of the ICR neutral interface for communication of robot programs from the application system to the robot controller requires that the information in the application system robot program can be exported (e.g. exist on a file format), and that the information can be translated into ICR. The application of the ICR neutral interface for communication of robot programs between application systems requires that the receiving system can read the ICR programming language.

Robot system model. The development and implementation of a *generic robot system model* is described in the next section. The generic robot system model provides the neutral interface for integration of CAD systems with application systems with respect to exchange of robotic models. The generic model is developed using several neutral interfaces defined in STEP (**ST**andard for the **E**xchange of **P**roduct model definition data) [ISO/CD 10303-42(1991);ISO/CD 10303-105(1990)]. As a conclusion the generic robot system model is an extremely helpful tool for sharing robot model data within software systems of different origin and functionality and it has been selected as the interface for robot model data in the generic architecture.

The application of the generic robot system model requires that information concerning robotic models can be exported from the sending system, and that it can be imported in the receiving system.

3.3. MODULES

In this section, the modules in the generic architecture will be described using a functional approach. The modules contain model driven algorithms and can be viewed as a functional unit that include the necessary models, functions and interfaces. The input to and the output from the module, together with a definition of the function will form the description.

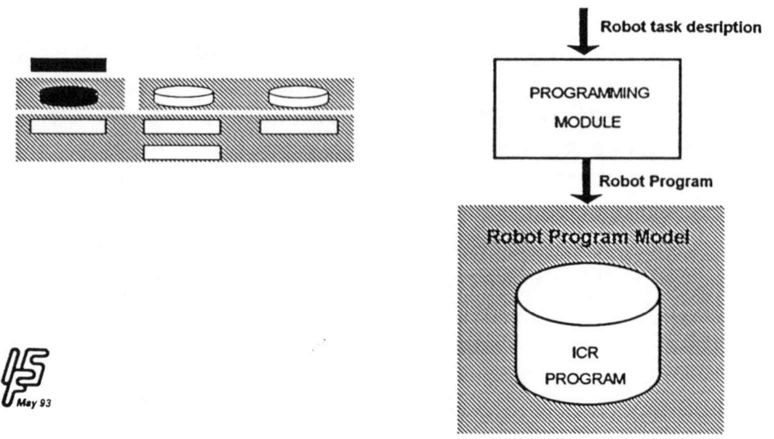

Figure 8. The input and output model of a programming module.

3.3.1. *The programming module.* The functions of the programming module is used to transform a given task description into a robot program. There are many methods for the generation of robot programs. Depending on the methods used in the generation of the robot program various information are needed. The input and output relations of the programming function are illustrated in Fig. 8.

For example, a geometrical description of a part is often needed in order to let the robot manipulate the part. The resulting robot manipulation program must include a definition of geometrical trajectories, which the robot tool has to follow when manipulating the part.

3.3.2. *The program interpretation module.* The program interpretation module can be viewed as a functional unit. The input is a robot program. The robot program is a description of the actions the robot system should perform in order to accomplish a robot task. The program interpretation module perform an interpretation of the robot program on basis of a syntactic and semantic model of the applied programming language. The interpretation takes one statement at the time and performs the semantic actions of the statement (e.g. changes the value of variable or send a move instruction to the controller). The actions will depend on the internal state of the interpreter (e.g. the values of variables). The output is a flow of robot instructions passed to the robot control simulation module.

As some robot tasks need information about the robot system in order to perform the succeeding actions, various kind of information may be fed back from the controller simulation module on requests. These input/output relations and the program interpretation are illustrated in Fig. 9.

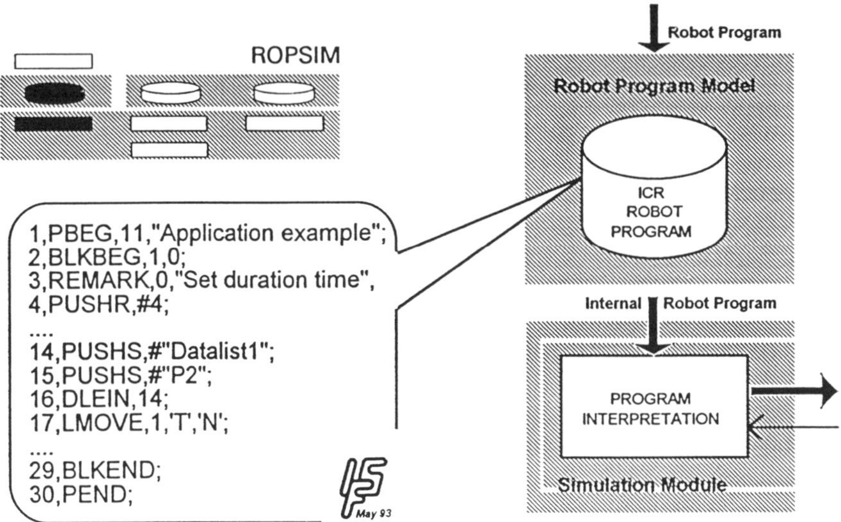

Figure 9. The input and output model of an interpretation module.

3.3.3. *The controller simulation module.* The controller simulation module calculates the control signals on basis of the control model. The input to the controller is the flow of

414

robot instruction generated by the interpretation module. The controller model includes algorithms that can effectuate the instructions and generate actuator control signals as output.

As most robots use sensors in the manipulator the sensor information must be fed back to the controller. The controller algorithms derive internal control states which are updated during the simulation.

The accuracy of the control simulation is heavily depending on the controller model. Therefore as the control simulation should be as close as possible to the controller working in the real-world, the *exact same algorithms* should be used in both situations if possible. These input/output relations and the controller simulation are illustrated in Fig. 10.

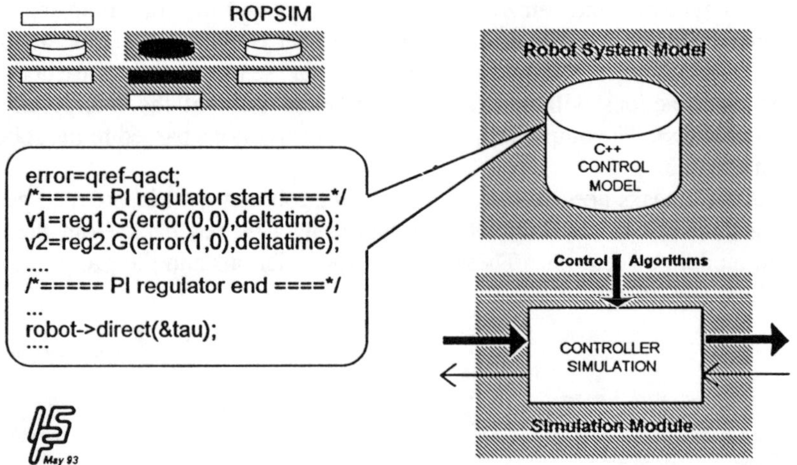

Figure 10. The input and output model of a control simulation module.

3.3.4. *The manipulator simulation module.* The manipulator simulation module calculates the kinetics relationships of the manipulator on basis of the manipulator model. The accuracy of such simulation depends on the manipulator model. The dynamic behaviour can only be simulated when the manipulator model includes the pertinent dynamic information.

The input to the manipulator simulation are the actuator control signals generated by the controller simulation. These control signals will effect the manipulator through the actuators (e.g. a torque) and the internal states (e.g. the position, velocity and acceleration of the robot joints) will be updated. The result of the manipulator simulation is an updated manipulator model. These input/output relations and the manipulator simulation are illustrated in Fig. 11.

3.3.5. *The presentation module.* The presentation module is the module where selected states of the system model are presented. The states of the total system are generated and

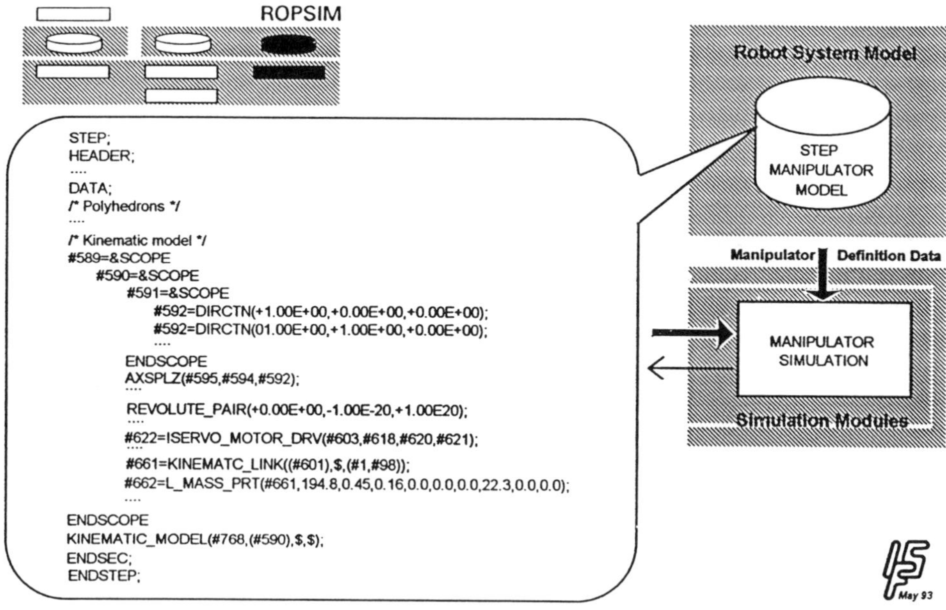

Figure 11. The input and output model of a manipulator simulation module.

updated by the program interpretation, the controller simulation and the manipulator simulation modules. The presentation may have many forms depending of the application. Some examples are illustrated in Fig. 12.

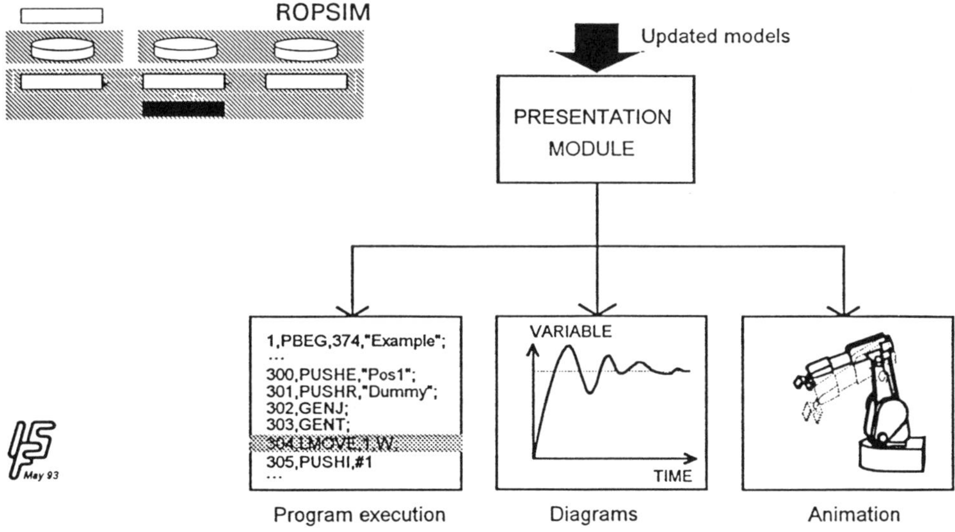

Figure 12. The input and output model of a presentation module.

Visualisationthrough animation has shown to be a very effective way to illustrate the geometric relations between objects. Animation is very useful to illustrate the movements of a robot in surroundings, which might include other moving objects. In Fig. 12, the animation is illustrated by the right rectangle with the "moving" robot.

Some information are better understood by audio-visual aids. Audio signals can be used as a warning when some critical situation is appearing, e.g. a collision or a heavy loaded motor. The audio presentation can be anything from a simple tone (a beep) to computer generated voice.

Diagrams have been applied for presentation of relationships between significant variables in many domains. Transient response diagrams can be used to evaluate state variables of the robot system over time. This is illustrated by the lower middle rectangle in Fig. 12.

To indicate the status of variables or program status the program execution text can be applied. A pointer to the actual program instruction is applied showing the current program execution, see the lower left rectangle in Fig. 12.

4. Generic robot system model

The objective of the generic robot system model is two-fold. The first objective is to enable a generic description of the robot system inclusive effects like the dynamics of the manipulator movement and the use of external sensor feed-back in the controller. Secondly the intention of the generic model is to use it as the fundamental information carrier between different CIME subsystems when a robotic model is exchanged between such subsystems. The generic model is expressed using the neutral file concept and as a result the generic model is an extremely useful tool for the sharing of robot model data within software systems of different origin and functionality.

The generic model of a robot system can be divided into two logical but independent components, a model of the robot controller and a model of the manipulator. These two models are described in the following sections.

4.1. GENERIC MANIPULATOR MODEL

In order to make the generic model of the robot manipulator portable, it is developed on basis of STEP neutral representation of product model definition data. The intention of STEP is to develop one unified information model capable of describing all aspects covering the whole life cycle of a universal product ranging from initial product design to disposal, and to make this an ISO standard. To ensure a formal and robust formulation of such an information model, a new information modelling language EXPRESS, [ISO/CD 10303-11(1990)] has been defined within STEP. Obviously it is of great concern in the STEP approach that the representation of the information model is consistent. As a result a large amount of work has been dedicated to the definition of EXPRESS. In order to structure the development of the large information model, it has been broken up into smaller information models called resources. A resource is an information model of the product, which deals with a certain topic or perspective of the product.

Within STEP, resources concerning product geometry [ISO/CD 10303-42(1990)] and kinematics [ISO/CD 10303-105(1990)] have been proposed. These resources are currently under evaluation at the ISO level. Within the Esprit project NIRO [NIRO (1990)], these original resources have been reduced, using EXPRESS, into an information model applicable to robotic models [Kroszynski(1991]. The same approach is used here, but extended to meet the objectives of the generic manipulator model in ROPSIM. As yet none of the resources in STEP deal with dynamic properties of rigid bodies nor the definition of servo motor drives, the derived STEP information model of kinematics and geometry has therefore been further extended to describe rigid body dynamics and servo motor drives [Trostmann,S.(1992)].

The resulting general information model is a convenient tool in the development of a neutral file format representing the generic model of a robot manipulator. STEP provides a standardized procedure for the development of such neutral file formats [ISO/CD 10303-21(1991)]. The resulting neutral file format is a handy way of communicating an information model between different software systems. The neutral file format can be thought of as a language with syntax and semantics capable of describing a wide range of different robot manipulators.

The different parts of the generic model of a robot manipulator is described in the following sections.

4.1.1. *Kinematic model.* Kinematics deals with the analytical study of motion excluding the influences of masses of the involved objects and excluding the forces causing the motion. Traditionally robot kinematics deals with analytical description of the spatial displacement of the robot manipulator as a function of time, in particular the relations between joint-variable space and the position and orientation of the robot end-effector.

Basically two problems of both theoretical and practical interest are fundamental in robot kinematics. The first problem is usually referred to as *the direct kinematic problem* and it deals with the coordinate transformation from joint space to cartesian space. The second problem is referred to as *the inverse kinematic problem* and it deals with the coordinate transformation from cartesian space to joint space. Since the independent variables in a robot manipulator are the joint variables, and the robot task is typically stated in terms of cartesian coordinate frames, the inverse kinematic problem is most often dealt with.

A robot manipulator can be modelled as a chain (serial or closed) of several rigid bodies (links) connected by kinematic pairs (joints). One end of the chain is attached to the ground and the other end is free and carrying an end-effector (tool). The relative motion of the joints results in the motion of the links that positions and orientates the end-effector at a certain location.

In order to deal with the complex kinematics of a robot manipulator a notation for kinematics must provide rules for affixing coordinate systems to the various parts of the robot manipulator and to describe the transformations between these coordinate systems.

In dealing with the kinematics of rigid links, the geometry (shape) of the links remain constant as the robot moves. The variables are indicative of joint motions. Hence it is natural to select a notation for kinematics that consist of a constant part defining the geometry of the rigid link and a distinct variable part representing the joint motion. If the links are not considered rigid and the joints are not ideal such a separation also opens up for description of effects like elasticity in the links and friction in the joints. Several notations for kinematic modelling are available.

Due to the obvious advantage of a detachment of link geometry and joint motion, the kinematic part of the generic model is based on such a detachment. In STEP the kinematic topology is stated in [ISO/CD 10303-105(1990)]. This kinematic resource contains much more information than needed in this context. In collaboration with the ESPRIT project NIRO [Kroszynski(1991)] the STEP kinematic model has been reduced to the for robotics significant kinematic information.

4.1.2. *Geometric model.* The geometry model of the robot describes the physical dimensions of all the parts in the robot manipulator. In CAD terms the geometry model consists of a series of solid models representing all the parts in the robot manipulator. The geometry model is necessary for computer animation of robot manipulator movements, off-line programming, collision tests, etc.

In mechanical CAD systems the role of topology has traditionally been limited to the well-known Boundary Representation (B-rep) of solid models. As a consequence the topology in the geometric part of the generic model presented here is based on a facetted B-rep representation defined in the STEP resource for product geometry [ISO/CD 10303-42(1990)].

Because of the separation of link shape and joint motion in the kinematic model the geometry model of links can easily be hooked on the link entity in the kinematic model.

4.1.3. *Dynamic model.* Dynamics of moving objects is the study of the relationships between the motion and the forces affecting the motion. Apart from the kinematic model, information about the mass properties of the individual links is central in such a study. Also additional information concerning the joints in the kinematic model is necessary. This information involve which joints are driven by motors, dynamic characteristics of these motors, dynamic characteristics of gears, etc.

Because STEP does not so far deal with this type of information it has been necessary to state the entities of the dynamic model in EXPRESS and to fit them into the overall structure of the information model, [Trostmann,S.(1992)].

4.2. GENERIC CONTROLLER MODEL

The model of the robot controller is just as significant as the model of the robot manipulator when the total behaviour of a given robot is simulated. The situation today is that the development of the total robot system model cannot be accomplished using one system alone. Since there exists no standardised neutral interface, which can be applied for exchange of control models, such sub-models of a robot system must be designed, developed, simulated, and evaluated using different software systems. The partial results must then be evaluated, adjusted, and assembled *manually*, for the intended application (implementation). This is, of course not a desirable situation for future applications.

Based on the above observations, it is clear that there is a need for a general (generic) model which defines the true behaviour of a robot controller, and which can be a foundation for definition of a neutral interface. The generic control model presented here, is a first step towards the definition of such a generalisation. Hopefully, the generic model can contribute to future definition of a complete information model leading to a neutral interface for control model communication between different categories of software systems for robotics.

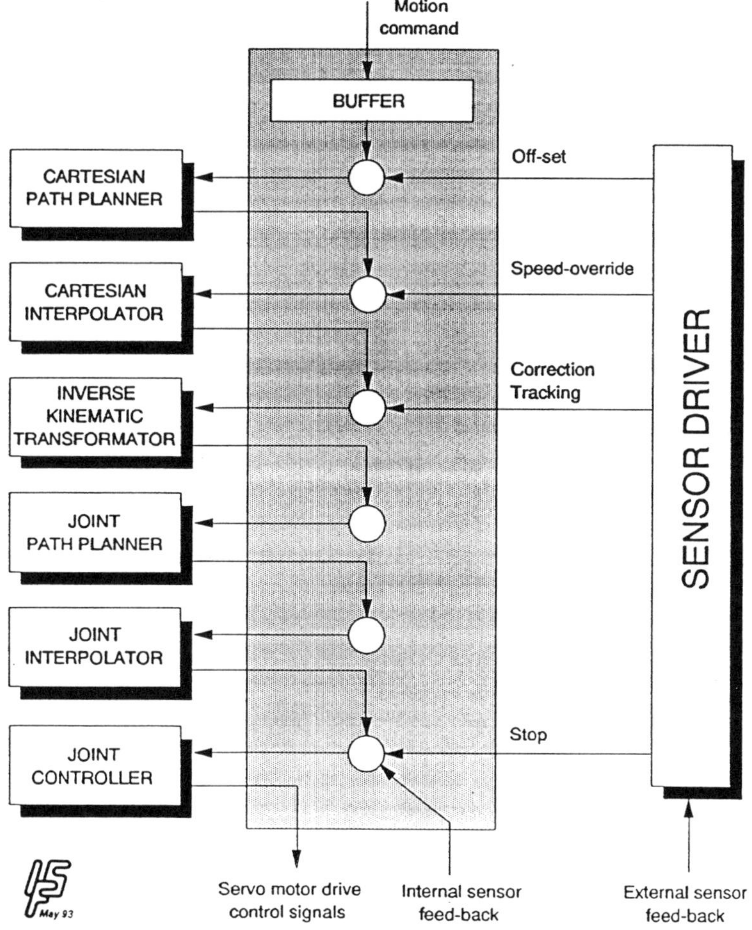

Figure 13. The state-structure model for a generic
motion controller.

The generic control model must be able to describe the wide range of control schemes applied in robotics such as motion control, force control, etc. The generic controller model can be described at two levels, the state-structure level and the behavioural level.

4.2.1. *State-structure level.* At the state-structure level the generic controller model represents entities and states mapped from the real-world robot system. With a state-structure model a programming and simulation system, is capable of handling a wide range of different controllers including motion control, force control, adaptive control, etc. Because almost all industrial robot control schemes are focused on motion control, this presentation is also focused on motion control.

The state-structure model of a generic motion controller is illustrated in fig. 13. In this model the important aspect of external sensor integration are covered. Sensors in the servo loops are considered as internal sensors.

The generic motion control model classifies seven entities normally found in an industrial robot controller. Six of these are functions or algorithms: Cartesian path planner, cartesian interpolator, inverse kinematic transformator, joint path planner, joint interpolator and joint controller. The last entity is an interface for external sensors. The generic motion control model allows a wide spectrum for the actual implementation of the individual functions. That is handled at the behavioural level.

4.2.2. *Behavioural level.* At the behavioural level the function of the individual entities are expressed as mathematical functions or algorithms. This level can be regarded as the lowest level in the generic controller model. The state-structure model define how these functions are interconnected.

The behavioural level can for instance be formalised in matrix algebra and programmed in an ordinary programming language. In this way it is possible to represent a wide range of control functions.

4.3. ICR, INTERMEDIATE CODE FOR ROBOTS

The programmability of robots makes them capable of handling many different tasks. As robot simulation and programming systems as well as robot systems, normally are components in a larger CIME system the neutral interface concept are used together with a neutral robot programming language. The demands to such a neutral robot programming language are [Trostmann,S.(1989)]:

- The robot language must be robot independent, because the language may be used on different robots.
- The robot programming language must be at the intermediate code level, because compilation from different high level languages probably originating from multiple sources are needed.
- The robot language must be general in the sense that every task the robot shall perform must be represented in the robot programming language. Otherwise the robot programming language will be a bottle-neck in the CIME system.

    - The robot programming language must be easy to read and interpret in computers.

These requirements are met by the ISO proposal ICR [ISO/DP 10562-1.2(1989)] for a neutral interface for robot programs. Experience with ICR has shown that ICR is capable of handling complex programs and is useful as a Neutral Interface [Trostmann,S. (1989)].

The instructions in an ICR robot program fall in two categories. One category is dealing with data manipulation instructions and another category is dealing with robot related instructions.

The data manipulation instructions of ICR comply with the assembler level in ordinary computer programming. For this reason ICR can represent the information content of a high level programming language like IRL [ISO/TC 184(1991)], Pascal, C, etc.

The robot related instructions are implementing the control of a robot. The important aspect here is that the information in the individual ICR instructions is robot independent. No constraints introduced from actual robots are enclosed in the ICR program. This is a significant feature of a neutral file format for robot programs for maintaining the robustness of the CIME system architecture. It should be noted that the task represented by the ICR program can implicity require that the robot belongs to a family with a specific kinematic configuration, e.g. 6 degrees of freedom.

The neutral interface concept gives the programmer the flexibility to select the best programming module for accomplishing a specific planning/programming task. For instance one programming system can be used to generate robot independent paths to be followed by the robot end-effector. These are translated into ICR. Another programming system can then be used to create the overall program structure (loops, jumps, etc). Also communication with external equipment (e.g. sensors) can be defined here. The necessary robot paths are references to the paths created in the first programming system. The complete program describing the robot task is then merged at the ICR level.

5. Model Driven Simulation Modules

It is in the simulation modules, section 3.3, Program Interpretation Module, Controller Simulation Module and Manipulator Simulation Module that the simulation computations are performed. The different simulation modules are structured so different program models and robot manipulator and controller models can be imported whereby the simulation becomes *Model Driven*.

5.1. PROGRAM INTERPRETATION MODULE

The Program Interpretation Module is in principle an interpreter for the ICR code. It can be viewed as a function which takes an ICR program [ISO/DP 10562-1.2(1989)] as argument and produces a flow of control commands (e.g. for linear movements) as output.

An ICR interpreter consists of two domains. The first one describes the actual state of the interpreter and the second one incorporates instructions for state changes. A program written in ICR is a list of statements in which one can jump. When starting an execution of a program one starts out from an initial state. Stepping through the programming one statement is executed at a time and each of the statements results in a state change of the interpretation.

5.2. CONTROLLER SIMULATION MODULE

When the objective is to investigate the total dynamic behaviour of the robot system by simulation of a dynamic model of the controller must be included. As described in section 4.2 a generic controller model has been formulated. So far a neutral interface at STEP-file level has not as yet been developed. However this problem a.o. is being addressed in the CEC ESPRIT project No. 6457 INTERROB - Interoperability of Standards for Robotics in CIME. The project was initiated Jan. 1, 1993.

In a model driven simulation the import of a specific controller model to the controller simulation module therefore for the moment cannot be based on a neutral interface description (file). In stead control algorithms and parameters specific for an individual simulation can be "imported" as executable program code.

5.3. MANIPULATOR SIMULATION MODULE

Over the last ten years, several approaches for general dynamic simulation of robot mechanisms have been published [Walker(1982); Featherstone(1987)]. In the present work the main objective has been to develop a model-driven kinetics simulation of industrial robot systems, i.e. the simulation must include the robot systems' behaviour as moving geometrical objects and their kinematic and dynamic properties.

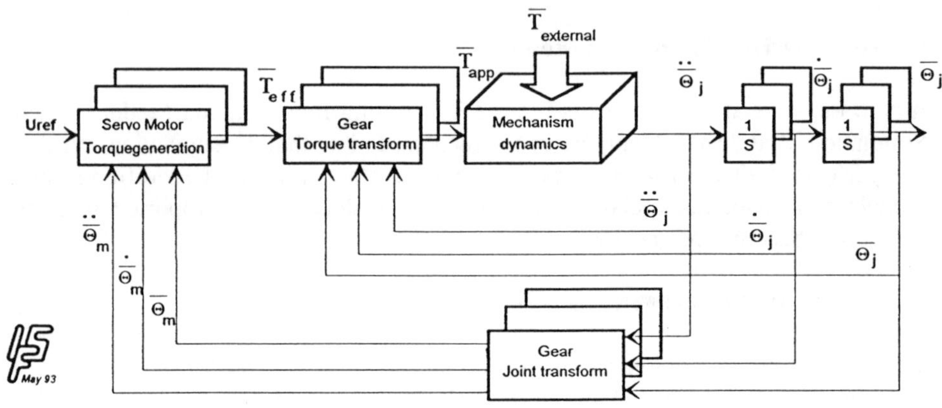

Figure 14. The block diagram of the decoupled simulation schema.

The model driven simulation requires that the submodels of the manipulator model are exchangeable, i.e. these models must be decoupled. The submodels are:

- The servo motor model
- The gear model
- The mechanism model

As a consequence the decoupling of the models requires decoupled simulation algorithms as well.

The decoupling of the models has lead to the following simulation schema illustrated by the block diagram shown in Fig. 14. The block diagram reflects the *model-driven* simulation approach as the algorithms are decoupled into functions related to their respectively models (Servo motor, Gear, and Mechanism).

The block diagram in Fig. 14 illustrates the interconnections of the servo motor, gear and mechanism dynamics. For simplicity reasons the block diagram illustrates only one servo motor drive. At the left the control signals U_{ref} is applied to the servo motor. The servo motor generates a torque T_{eff} which effectuates the gear. The generated torque depends on the status of the servo motor joint, $\ddot{\Theta}_m, \dot{\Theta}_m, \Theta_m$. The gear transforms the effective torque T_{eff} to the torque T_{app} applied in the joint of the mechanism. The applied torque T_{app} is driving the mechanism in parallel with torques from the other servo motors. This is illustrated by the several levels of "rectangles" representing the motors and gears in Fig. 14. The mechanism dynamics is illustrated by a block. This shall indicate that the mechanism dynamics depends on the applied torques from all the servo motors and gears. The thick arrow into the mechanism dynamics function represents the influence from external torques $T_{external}$ on the mechanism.

The mechanism dynamics generates the joint acceleration $\ddot{\Theta}_j$. When the acceleration of the joints are derived, the velocities and the positions are computed by integration. Different numerical integration methods can be applied. The integration method and step time (sampling time) are critical choices and may lead to nonstability. The sampling time is determined by the frequency of the dynamics. A typical sampling frequency is 1000-10000 Hz.

The joint state of the mechanism influence the force transformation of the gear. The mechanism joint state is via the gear transformed to the servo motor joint state $\ddot{\Theta}_m, \dot{\Theta}_m, \Theta_m$.

6. Simulation of robot systems

6.1. GENERAL DESCRIPTION OF ROPSIM

A new **R**obot **O**ff-line **P**rogramming and real-time **SIM**ulation system, called ROPSIM, has been designed, developed and implemented as background for the work presented here, see Fig. 15, [Trostmann,S.(1992)]. ROPSIM has been developed on basis of the generic architecture for off-line programming and simulation systems presented in section 3, see Fig. 7.

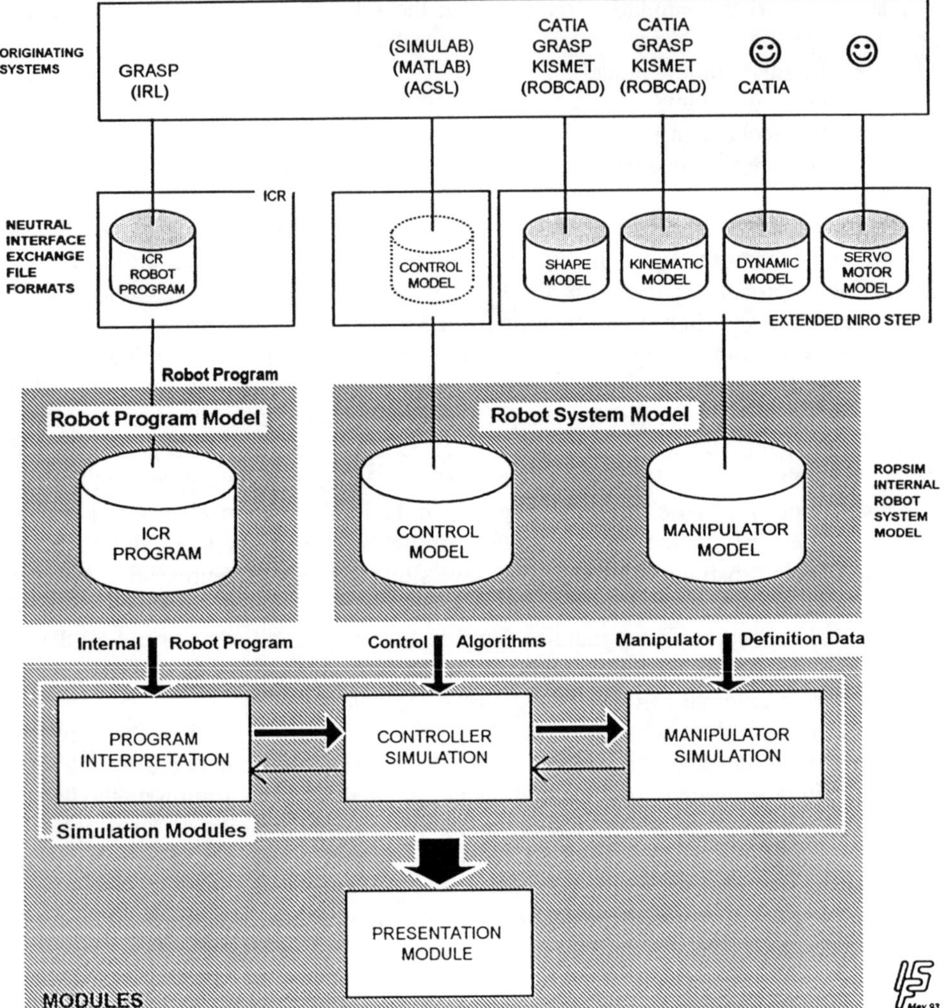

Figure 15. The integration of ROPSIM in the CIME system presently operative at the Control Engineering Institute, the Technical University of Denmark.

The purpose of ROPSIM is to provide a new model driven simulation system for the simulation of real-time, motion-controlled objects including their geometry, kinetics, and control. The robot system model, illustrated in Fig. 7 by the two right-hand side cylinders, is formulated as described in section 4. The simulation modules illustrated by the rectangles in Fig. 7 are described in section 5.

Real-time simulation in this context means that there exists an exact defined coherence between the representation of time applied in the simulation and the actual time in the robot task process under consideration.

Because of the generic architecture, ROPSIM is a transparent software system in the sense that the interaction and interplay between the models and modules are clearly identifiable and well-defined.

As described in section 3, ROPSIM in summary offers three new important facilities over existing commercially available systems:

- ROPSIM features a real-time simulation of a complete robot system (controller *and* manipulator).
- ROPSIM features a model-driven simulation approach.
- ROPSIM is a true CIME sub-system.

Firstly, it is of great importance that a simulation system can handle models of the necessary complexity. Using a simple model (e.g. a pure kinematic manipulator model) may in many cases degrade the correctness of the simulation. To overcome this problem ROPSIM is designed to simulate the dynamic behaviour of *both the robot control algorithms and the robot manipulator.*

Secondly, the simulation algorithms in ROPSIM are designed and implemented by using a *model-driven approach* (section 5). Because ROPSIM is based upon a generic architecture and model driven algorithms dependency on dedicated models is avoided. As a result, only the models representing the robot system must be developed and described in a neutral format when a robot system is simulated in ROPSIM.

Finally, in view of modern technology, a simulation system like ROPSIM must be prepared for an easy integration into a more comprehensive CIME environment. To achieve this, ROPSIM is designed with respect to the open systems concept for information exchange using neutral interfaces. Two different neutral interfaces are presently used in ROPSIM, ICR [ISO/DP 10562-1.2(1989)] for communication of robot programs and EXTENDED NIRO-STEP [Kroszynski(1991);Trostmann,S.(1992)] for communication of robot manipulator models. Further, the modular structure of the models and modules defined by the generic architecture has prepared ROPSIM for a future application of a neutral interface for control model exchange.

The use of such standardised neutral interfaces ensures a straight forward and well-defined procedure for the integration of the ROPSIM system into a larger CIME system architecture so robot programs and robot system models can be automatically exchanged. As a result the best suited sub-system can be used for definition of the individual models.

Further, the components in ROPSIM is designed and implemented in a way such that the architecture of the simulation system remains robust when a component is changed or replaced. This is an inherited characteristic of the object-oriented modelling and programming techniques used for the implementation of the software.

ROPSIM is integrated in a CIME environment presently operative at the Control Engineering Institute, the Technical University of Denmark.

Each of the levels illustrated in Fig. 15 are explained below.

6.1.1. *Originating systems.* The originating systems group contains the commercially available software systems which can be used for development of the neutral exchange files (models). The CIME sub-systems listed in this group are all available at the Control Engineering Institute. The systems without parenthesis are presently integrated with ROPSIM. The integration is implemented by means of pre-processors. The systems shown in parenthesis represent planned future extensions of sub-systems in the illustrated CIME system. The originating systems group can be extended with other new sub-systems. Because of the opens systems' concept for information exchange (neutral interfaces) the integration of new sub-systems with the existing CIME system is straight forward and well-defined requiring only a pre-processor.

6.1.2. *ROPSIM internal models.* The data structure represented in the neutral exchange files are defined with the focus on data (model) exchange. As a result the exchange file data structure are not directly applicable for simulation purposes. Consequently ROPSIM includes an internal data structure of the three essential models (robot program, robot control, and robot manipulator), which are modelled for simulation purposes. The internal models are all implemented in C++.

The internal model of the robot program is constructed by the interpreter module. The internal controller model must be implemented within the ROPSIM system, presumably using the generic control model developed and described in section 4. The internal manipulator model is constructed from the EXTENDED NIRO-STEP exchange files by means of a post-processor.

6.1.3. *ROPSIM modules.* In the light of the internal models, the ROPSIM modules perform the simulation and presents the results. Each of the illustrated modules contains the model-driven algorithms and can be viewed as a functional unit.

6.2. SIMULATION OF ROBOT SYSTEMS USING ROPSIM

With the implemented models and modules the ROPSIM system can be used to investigate a wide range of robot system capabilities.

Basically the three models illustrated in Fig. 15 (robot program model, control model, and manipulator model) can be viewed as parameters in a simulation.

6.2.1. *Robot manipulator.* The robot manipulator model can be imported from other CIME sub-systems using the EXTENDED NIRO-STEP neutral interface, refer to Fig. 16. An example of a manipulator model expressed in EXTENDED NIRO-STEP file format is described in Fig. 16. The manipulator is shown in Fig. 18.

The EXTENDED NIRO-STEP as well as the internal ROPSIM manipulator model is capable of describing the following aspects:
- Kinematic structure including closed chains of the manipulator.
- Shape of the manipulator and its environment.
- Mass distribution of the manipulator links.
- Dynamic behaviour of the manipulator servo motor drives.

The simulation algorithms implemented in the ROPSIM manipulator simulation module is driven by the internal manipulator model, refer to Fig. 15. As a result the manipulator

```
 1  STEP;
 2  HEADER;
 3    FILE_IDENTIFIER('hyd robot',
 4    FILE_DESCRIPTION(('Test file'),
 5  ENDSEC;
 6  DATA;
 7    /* polyhedrons */
 8    ...
 9    /* kinematic model */
10    #589=&SCOPE
11    /* mechanism */
12    #590=&SCOPE
13      /* base frame */
14      #591=&SCOPE
15        #592=DIRCTN(+1.0000E+00,+0.0000E+00,+0.0000E+00);
16        #594=DIRCTN(+0.0000E+00,+0.0000E+00,+1.0000E+00);
17        #595=CRTPNT(+0.0000E+00,+0.0000E+00,+0.0000E+00);
18      ENDSCOPE
19      AXSPLZ(#595,#594,#592);
20      #596=REVOLUTE_PAIR(+0.0000E+00,-1.0000E+20,+1.0000E+20);
21      #901=!SECORD(4.64E-05,1193.81,0.63);
22      #901=!SERVAL(0.6,1.33E-02,850.0,#901);
23      #902=!PISTON(340.0,7.0,9.456E-04,1.667E-12,
24                   7.0E+08,1.6E-04,1.6E-04);
25      ...
26      #615=PAIR_PLACEMENT_STRUCTURE(#609,#610);
27      ...
28      #661=KINEMATIC_LINK((#602),$,(#1));
29      #951=!L_MASS_PRT(#661,194.8,0.4512,0.1630,0.0,
30                   0.0,0.0,22.35,0.0,0.0,0.0);
31      ...
32      #666=KINEMATIC_JOINT(#661,#662,#596);
33      ...
34      #671=KINEMATIC_STRUCTURE((#666,#667,#668,#669,#670),$);
35    ENDSCOPE
36    MECHANISM(#671,#661,#591);
37    /* the ground */
38    #672=&SCOPE
39      #673=&SCOPE
40        #674=DIRCTN(+1.0000E+00,+0.0000E+00,+0.0000E+00);
41        #676=DIRCTN(+0.0000E+00,+0.0000E+00,+1.0000E+00);
42        #677=CRTPNT(+0.0000E+00,+0.0000E+00,+0.0000E+00);
43      ENDSCOPE
44      AXSPLZ(#677,#676,#674);
45    ENDSCOPE
46    GROUND(#673,(#1));
47  ENDSCOPE
48  KINEMATIC_MODEL(#672,(#590),$,$);
49  ENDSEC;
50  ENDSTEP;
```

Figure 16. Extract of the EXTENDED NIRO-STEP exchange file for
the Multidos robot (see Fig. 18).

simulation performed in ROPSIM features a simulation, which makes it possible to study all aspects of the manipulator motion comprising both manipulator kinematics, manipulator dynamics, and servo motor drive dynamics.

6.2.2. Robot controller. As described above at present time no neutral interface for control model definition has been developed and standardised. As a result the control model used for simulation cannot be imported automatically from other CIME subsystems, but must be defined within ROPSIM. For this purpose the generic motion control model described in section 4 can be used. Definition of the control model of a specific robot controller using the generic control model to program in an executable

ICR program of a square task.

```
1,PBEG,11,"Application example";
2,BLKBEG,1,0;
3,REMARK,0,"Set duration time";
4,PUSHR,#4;
5,W_MOVTIM;
6,REMARK,0,"datalist handling follows";
7,PUSHS,#"datalist1";
8,DLOPEN;
9,PUSHS,#"datalist1";
10,PUSHS,#"P1";
11,DLEIN,14;
12,REMARK,0,"Linear move command";
13,LMOVE,1,'T','N';
14,PUSHS,#"datalist1";
15,PUSHS,#"P2";
16,DLEIN,14;
17,LMOVE,1,'T','N';
18,PUSHS,#"datalist1";
19,PUSHS,#"P3";
20,DLEIN,14;
21,LMOVE,1,'T','N';
22,PUSHS,#"datalist1";
23,PUSHS,#"P4";
24,DLEIN,14;
25,LMOVE,1,'T','W';
26,PUSHS,#"datalist1";
27,DLCLS;
28,GOTONR,7;
29,BLKEND;
30,PEND;
```

ICR datalist example.

```
1,DLHEAD,"datalist1",4;
2,DLDAT,"P1",14,#(1.6,1.4,0,1,0,0,0,0,0,0,0,0,0,0,0,0,0);
3,DLDAT,"P2",14,#(1.4,1.4,0,1,0,0,0,0,0,0,0,0,0,0,0,0,0);
4,DLDAT,"P3",14,#(1.4,1.2,0,1,0,0,0,0,0,0,0,0,0,0,0,0,0);
5,DLDAT,"P4",14,#(1.6,1.2,0,1,0,0,0,0,0,0,0,0,0,0,0,0,0);
6,DLEND;
```


Figure 17. An ICR program used for simulation.

code a set of control algorithms mapping the behaviour of the real control algorithms of the robot controller.

When the control model is implemented, ROPSIM can be used to investigate the behaviour of the control algorithms. The algorithms can be evaluated with respect to their performance: Are they correct? Do they behave satisfactorily? etc.

As ROPSIM features a real-time simulation of the manipulator movement, also the cooperation between the control algorithms and the controlled manipulator can be evaluated: Can the algorithms control the manipulator satisfactory? etc.

6.2.3. Robot program. The last parameter to be discussed in the simulation is the robot program. The robot program carries a description of the robot task to be performed.

Robot systems differentiate themselves from "fixed" automation by being "flexible", which in this context a.o. means re-programmable. Not only are the movements of the robot manipulator re-programmable, but through the use of sensors and via communication with other factory automation, the robot systems can adapt to variations as the task proceeds. To benefit from this flexibility it is important that the programming language used for the programming is advanced enough to utilise such abilities.

Considering the state of the art in robot programming languages of commercially available robot systems today, it is questionable if these programming languages can fully

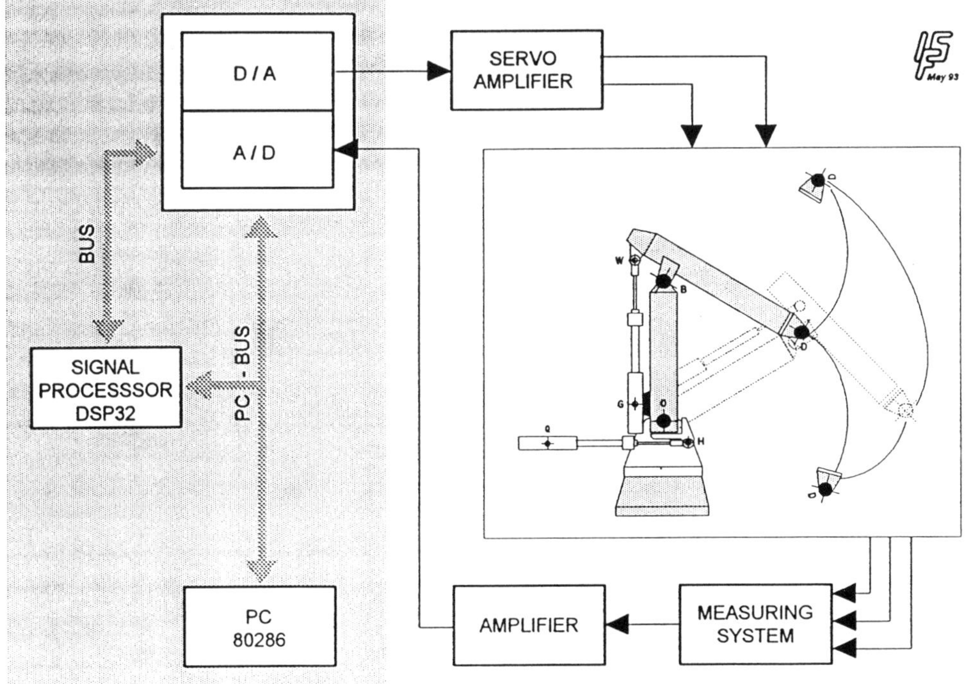

Fig. 18. The architecture of the Multidos robot system

430

support the above mentioned flexibility of the robot system. Usually the robot programming languages of available robot systems do not support for example simple arithmetic operations.

To overcome this problem the ISO standard proposal ICR for a robot programming language has been implemented as the format for robot programs in ROPSIM. An example ICR program is presented in Fig. 17.

6.2.4. *Example, Real-time simulation of the Multidos robot system.* The hydraulic Multidos test robot [Conrad(1992)] has been designed, constructed and implemented at the Control Engineering Institute. As a result, all the detailed information needed for implementation of the control model as well as the manipulator model is immediately available. Because the validity of the robot system model is crucial for the correctness of the simulation in ROPSIM, it was decided to use the Multidos robot as a typical example to demonstrate some of the capabilities of ROPSIM.

The architecture of the real-time control of the Multidos robot system is illustrated in Fig. 18.

The manipulator model of the Multidos robot was developed and the EXTENDED NIRO-STEP exchange file is presented in Fig. 16.

The manipulator is equipped with a measuring and instrumentation system including devices for measurement of various dynamic variables.

In the present Multidos control system the amplified transducer signals are sent to an analog-to-digital converter card (ADC/DAC) in an IBM PC-AT, where the transducer signals are converted into a 12-bit digital representation. The digital words are transferred via a fast data bus to a processor which uses the control algorithms to calculate the control signal. The number representing the control signal is then transferred to ADC/DAC converter and is supplied to the servo valve amplifiers.

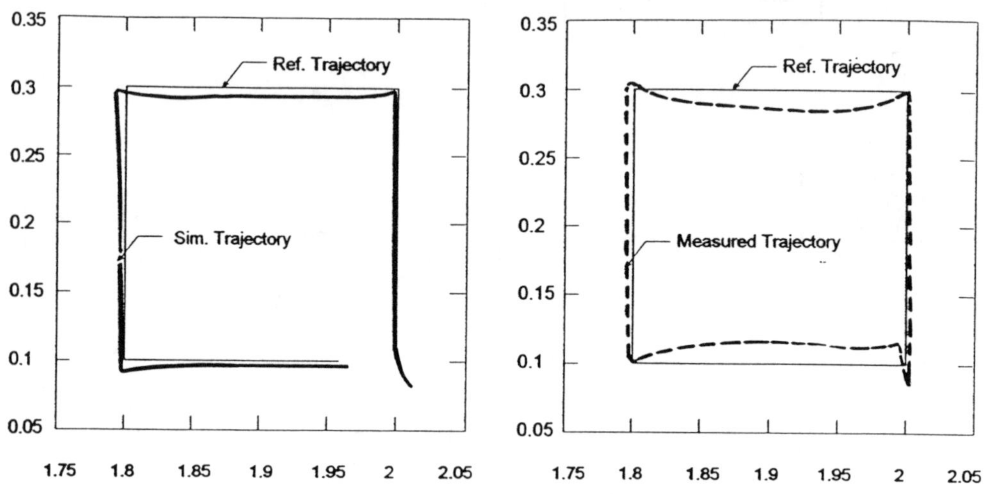

Figure 19. Simulated and experimental obtained results.

The control processor is a single chip AT&T WE DSP 32-bit floating point signal processor with a cycle time of 160 nsec. The AT&T signal processor is implemented on an extension card which complies with the standard slots of a PC-AT. In this way the control algorithms are implemented by using standard programming languages which run on PC-computers. The Multidos control algorithms considered here are implemented in Turbo C ver. 2.0. [Siggaard(1992);Trostmann,S.(1992)].

The control algorithms implementing the real-time control of the Multidos robot is transferred and implemented within the generic control model. As a result, *the same controller used for real-time control of the robot was applied for real-time simulation in ROPSIM*. This means that experimental results can be compared directly with simulated results and used for validation of the robot behaviour via simulation.

A set of simulation results are shown in Fig. 19. Please note that the results shown are not considered to be optimal from a control point of view.

7. Conclusions

The first important steps towards a general robot programming and simulation system have been taken. It has been demonstrated that manipulator and control model designs carried out in other software systems can be imported and reused in ROPSIM via the application of standardized, neutral interfaces. By using a generic architecture ROPSIM becomes a model driven simulation system applicable to many different robot systems without reprogramming. The concept of applying neutral interfaces makes ROPSIM a true CIME subsystem.

The future research and development related to the work presented here is identified to the following areas:

- *Firstly*, further *conceptual research* and development includes the development of a neutral interface for control models. The work presented here has contributed to the objective concerning definition of a neutral control model interface in the CEC ESPRIT project No. 6457, INTERROB started Jan. 1, 1993.
- *Secondly*, extension of the *capabilities* of ROPSIM includes extensions to the kind of behaviours simulated by the ROPSIM system. This includes extensions to the manipulator simulation regarding flexible links, friction in the joints, and movement in media given resistance (e.g. movement in fluids). Also extensions of the control must be considered, such that force control can be simulated. A programming interface (CLDATA) for CNC-machines may be implemented, enabling simulation of CNC machine behaviour in parallel with robotic systems.
- *Thirdly*, the *computer science* related developments are important. Code optimisation and a professional man-machine interface are of utmost importance. Further, the implementation of simulation of mechanisms with more than one closed kinematic chain, concurrent simulation of more robot system, simulation of Automated Guided Vehicles (AGV) are interesting future topics.
- *At last*, the *modelling work* must be continued. Full operative models of industrial robots are of great relevance. Such models must include both the

controller and the manipulator dynamics. Furthermore, a broad library of models of servo motor drives are convenient. On the controller side many controller strategies are relevant for implementation.

8. Acknowledgement

The development and implementation of the **R**obot **O**ff-line **P**rogramming and **SIM**ulation system, ROPSIM, at the Control Engineering Institute, the Technical University of Denmark is primarily sponsored by the Danish Technical Council (STVF). This support is gratefully acknowledged. The results of the CEC ESPRIT projects No. 322 CAD*I and No. 5109 NIRO have been fundamental to the development of ROPSIM.

9. References

[1]   **ABB Robotics (1988).**
ARLA reference manual 2.0
ASEA, 1988.

[2]   **Conrad, F., Zhou, J., Siggaard, M. (1992).**
Experimental Evaluation of Two Feed-back Linearisation Control Schemes for Hydraulic Robot Manipulator.
ASME, Fluid Power System Technology 1992.

[3]   **Featherstone, R. (1987).**
Robot dynamics algorithms.
Kluwer Academic Publishers.

[4]   **ISO/CD 10303-105 (1990).**
Industrial Automation Systems. Part 105: Kinematics

[5]   **ISO/CD 10303-11 (1990).**
Industrial Automation Systems. Part 11: Express Language Reference Manual.

[6]   **ISO/CD 10303-21 (1991).**
Industrial Automation Systems. Part 21: Specification for the Exchange File Implementations of the Standard for Exchange of Product Data (STEP).

[7]   **ISO/CD 10303-42 (1990).**
Industrial Automation Systems. Part 42: Shape representation.

[8]   **ISO/DP 10562-1.2 (1989).**
Manipulating Industrial Robots - Intermediate Code for Robots (ICR).

[9]   **ISO/TC 184 (1991).**
Industrial Robot Language (IRL), Language Description.

[10]   **Kroszynski, U., Sørensen, T., Schlechtendahl, E. G. (1991).**
NIRO-STEP Specification for the Transfer of Robotic Models. Version LO1VO3 "February 91".
NIRO Report: NIRO.WGA.DTH.002.91, Rap.Nr. S91.25, 1991.

[11]   **NIRO. (1990).**
Technical Annex for the ESPRIT project: Neutral Interfaces for Robotics (NIRO).

[12]   **Presmann, R.S., Williams, J.E. (1977).**
Numerical Control and Computer Aided Manufacturing.
John Wiley & Sons. 1977. New York.

[13]   **Robcad (1991).**
Basic Training 2.2.
Technomatix 1991.

[14]   **Schlechtendahl, E. G. (1989).**
Intelligent Communication of Product Definition Data.
Proc. of the IFIP 11th World Computer Conference, San Francisco, USA.
Aug. 28 - Sep. 1, 1989, pp. 1029-1032.

[15]   **Siggaard, M. (1992).**
Model-based Control and Implementations to Hydraulic Robot Manipulator.
M.Sc. Thesis. Rap.Nr. S92.27, 1992. Control Engineering Institute, Technical
University of Denmark.

[16]   **Trostmann, E. (1987).**
Cad Data Interfaces for Robot Control.
2. Duisburger Kolloquium, Automation Und Robotik.
Duisburg, 15.-17. July 1987. Proc. pp. 73-96.

[17]   **Trostmann, E. (1991).**
Intelligent Interfaces in Robotics.
Proc. of 1st Euro-CIM-Conference "Megatronics and Robotics, Aachen,
Germany, July 8-10, 1991, pp. 231-239.

[18]   **Trostmann, S., Nielsen, L. F., Palstrøm, B. (1989).**
Irdata implementations at DTH/IFS - status and overview.
Rap. Nr. S89.30, 1989. Control Engineering Institute, Technical University of
Denmark.

[19]   **Trostmann, S., Nielsen, L. F., Trostmann, E., Conrad, F. (1992).**
Real-time Robot Simulation Driven by Geometric and Kinetic Models.
24th CIRP international seminar in manufacturing systems 1992.

[20]   **Trostmann, S., Nielsen, L.F. (1992).**
Model Driven Simulation of Robot Systems. Models and Algorithms.
Ph.D. Thesis. IFS Rap. nr. S92.66, 1992. Control Engineering Institute,
Technical University of Denmark.

INTEGRATED SUPPORT FOR COOPERATIVE DESIGN COORDINATION:
MANAGING PROCESSES, CONFLICTS AND MEMORIES

M. Klein
Boeing Computer Services
Building 33-07, MS 7L-44
2760 160th Ave SE
Bellevue WA 98008 USA

ABSTRACT. Design has increasingly become a cooperative endeavor that requires effective coordination in the face of highly complex inter-dependencies exposed by distribution over participants, time and perspectives. Distinct coordination support technologies have emerged for each of these kinds of distribution, but all face important limitations. This paper presents an integrated model of cooperative design coordination that synergistically combines existing coordination approaches in a way that avoids many of their individual limitations and combines their strengths. An initial implementation of this model is presented and challenges for future evolution of this technology are identified.

1. Why We Need Coordination Technology

Increasingly, complex artifacts are designed[1] in large manufacturing concerns by complex processes distributed across time, participants and functional perspectives. The design of a commercial jet, for example, requires the integrated contribution of thousands of individuals spread over several continents and a span of decades. Effective coordination is critical to the success of this cooperative process since the distributed activities are typically highly interdependent e.g. due to shared resources, input-output relationships and so on. The sheer complexity of these inter-dependencies, however, has come to overwhelm traditional manual organizational schemes and paper-based coordination techniques, resulting in huge unnecessary rework costs, slowed schedules and reduced product quality. As a result, while individual productivity may be high, failures of existing coordination support practices and technologies have severe impacts on the bottom line.

But what actually do we mean by coordination? Support for cooperative design can be viewed as being divided into three layers, each built on top of the ones beneath it:

---

[1] In this paper, the word "design" is taken to describe all aspects of an artifacts' description, including the requirements, geometric definition, materials, manufacturing plan, supporting documentation and so on. "Collaborative" design thus refers to collaboration both between disciplines at a given stage (e.g. between various systems and structures groups in design engineering) as well as between stages over the product life cycle (e.g. between design and manufacturing engineering).

S. Y. Nof (ed.), Information and Collaboration Models of Integration, 435–459.

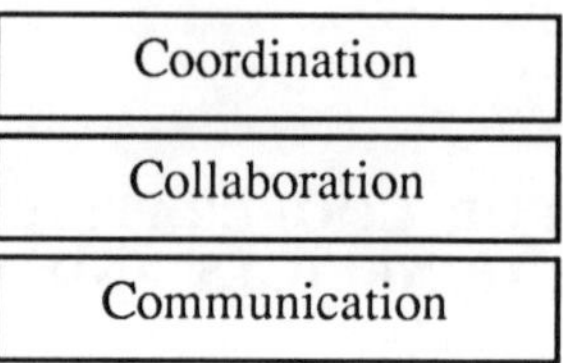

Figure 1. Layers of Support for Cooperative Design.

- <u>communication:</u> allowing participants in the design process to share information (this involves networking infrastructures)

- <u>collaboration:</u> allowing participants to collaboratively update some shared design description (this involves support for tele-conferencing etc.)

- <u>coordination:</u> ensuring the collaborative actions of multiple participants working on a shared design are coordinated to achieve the desired result as efficiently as possible

This paper focuses on the topmost layer, with the understanding the significant challenges exist in providing mature effective technology for the supporting layers. In the conclusion to this paper the requirements that integrated coordination technology presents for the underlying collaboration and communication layers is discussed.

Each of the three kinds of distribution in cooperative design have been addressed to date by three distinct coordination support technologies (Figure 2):

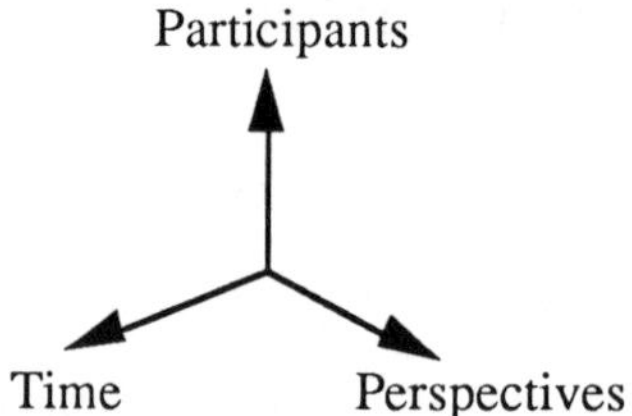

Figure 2. Types of Distribution in Cooperative Design

- <u>distribution across participants</u> requires support for the sequencing of tasks and flow of information among participants; this is addressed by process management (i.e. workflow, planning/scheduling) technologies.

- <u>distribution across perspectives</u> arises because the participants involved work on different interacting aspects of the design and/or have different often-incompatible goals; consistency among their design actions can be maintained with the support of conflict management (i.e. constraint or dependency management) technologies.

- <u>distribution across time</u> arises because the nature and rationale for design decisions made at an earlier stage often need to be available later on, for example to support product changes as well as the design of other similar products; this is addressed by memory management (i.e. rationale capture, organizational memory) technologies.

Each of these technologies face different challenges that have limited their effectiveness in different ways. The central thesis of this paper is that truly effective coordination in cooperative design requires an integrated approach wherein dependencies across all the dimensions of design distribution are modelled and managed in a single computational framework. This integrated approach, I believe, synergistically combines the strengths and avoids the weaknesses of the contributing coordination support technologies. This paper will review the state of the art of existing coordination technologies, identify their current limitations and present an implemented model of integrated cooperative design coordination called iDCSS. The paper concludes with a discussion of directions for future work.

2. Current Coordination Support Technology

In this section we will review the state of the art of the three types of coordination technology currently available for cooperative design support. For each technology we will consider the coordination problem addressed, the current state of the art as well as the technical challenges we must meet to provide improved support given current limitations.

2.1. PROCESS MANAGEMENT

2.1.1. Problem Addressed. The distribution of cooperative design across multiple participants requires that we be able to control the flow of tasks and information among them, i.e. be able to define and perform effective business processes. Generally such processes are documented in difficult-to-access and quickly obsolete paper documents if they are formally modelled at all. As a result, it can be time-consuming to train new people in how one's business works, valuable expertise about effective business processes can be lost as employees become unavailable, and existing processes can become increasingly ineffective in the absence of a good global understanding of what they actually are. The execution of business processes, in addition, is typically manual and paper-based. Bottlenecks (e.g. due to uneven workload distribution), long flow times (e.g. due to slow work package distribution) and errors (e.g. because the preferred process was not followed) can occur all too often. There is also very little information available on the state of the work flowing through a business process, making tracking of work items and the analysis of business process effectiveness a time-consuming and error-prone task.

2.1.2. State of the Art. A wide range of process management technologies have become available to address these concerns. These technologies can be placed on a continuum according to the extent to which they impose limits on how a business process can be performed (Figure 3)

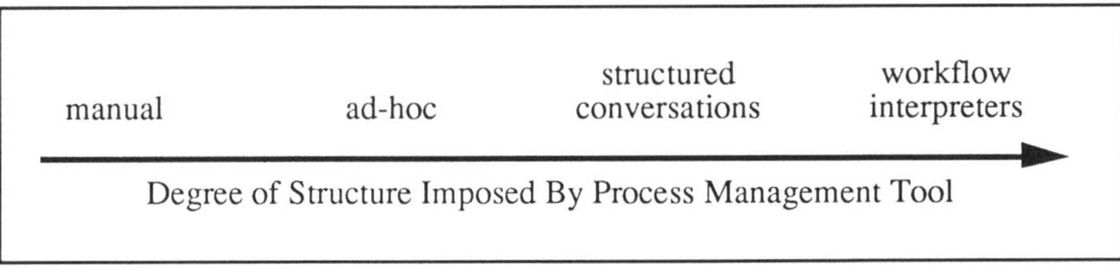

Figure 3. Types of Process Management Systems

Manual systems impose essentially no constraints on how a process is performed, allowing work packages for example to be lost, be routed to obsolete mail addresses and so on.

Ad-hoc systems, typically based on electronic mail, depend on process participants to decide to whom a work package should go next but ensure that the work package is routed correctly once this decision is made. They also typically provide some degree of tracking.

Structured conversation systems such as those based on the contract-net [27] or speech act [34] approaches provide a system-imposed protocol that structures how tasks can be distributed. These typically allow one to track entire "conversations" (series of task assignment negotiations and progress updates) as a unit.

Workflow systems interpret a process model produced by a process capture tool[2]. Processes are modelled as sequences of tasks, each assigned to a given individual or group, interspersed with decision points; these tools typically provide a graphical display of the process like the following (Figure 4):

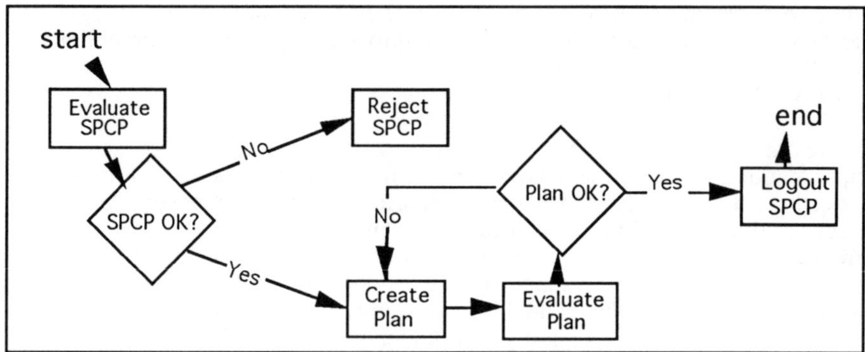

Figure 4.*A Graphical Process Model*

Once a process has been modelled in this way, participants in a given process simply perform their individual tasks and rely on the workflow system to understand the process and ensure it is followed and tracked correctly.

2.1.3. Challenges. The key limitation of current process management technology is that the process execution structure it enforces is related indirectly at best to the underlying realities that shape these processes. Processes arise when a given task is too large to be performed effectively by a single individual, and thus must be decomposed (often recursively) into subtasks that taken together achieve the top-level goal. These tasks have interdependencies that may require some serialization in task execution. Some tasks for example may require as input the outputs produced by other tasks. Tasks that both use some limited organizational resource (e.g. a piece of equipment) may be serialized to avoid contention for that resource. Tasks can be merged if it turns out that one task, perhaps expanded slightly in scope, can achieve the requirements for two or more tasks more efficiently. New

---

[2] Such tools usually allow a human user to define a business process using text-based or drawing tools, but in some cases can provide automated process definition (i.e. planning) support, wherein a process model is derived from a description of the requirements it must satisfy.

tasks may be generated dynamically to resolve conflicts between design actions taken by participants performing different tasks in the process. Finally, tasks will be assigned to individuals or groups based on how skills and responsibilities are distributed throughout the organization. Since there are often many different alternatives available for how we do any of these things, processes are shaped by (implicit or explicit) meta-level control strategies[3] that determine which options will be selected. We can thus summarize the factors that produce a given business process as follows (Figure 5):

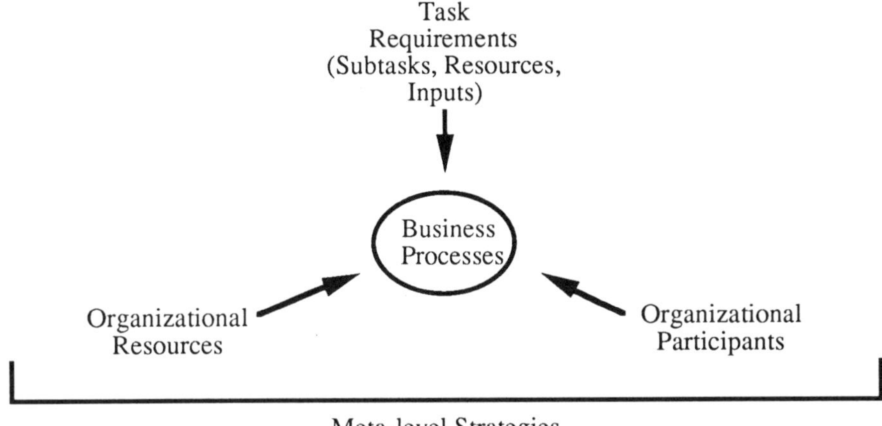

Figure 5. The Factors That Shape Processes

How well do current process management technologies account for the constraints that shape processes? They fail in two ways. Most process management technologies (i.e. the manual, ad hoc and structured conversation approaches) include no model of any of the above-described constraints and thus offer no way of ensuring that the constraints are enforced. They include, for example, no model of how tasks should be sequenced, what resources are available to perform them and so on. We have to rely on the human participants to make the correct decisions, but it is the very difficulty of making and enforcing these decisions in complex processes that led to the development of process management technology in the first place.

Workflow systems, by contrast, are capable of enforcing constraints on the detailed nature of a process. We can specify precisely what tasks will be done in what sequence by whom and using what resources. The problem with this approach is that any given process model represents a "frozen" or "compiled" picture of what an appropriate process was for a set of task, resource and organizational constraints and policies that applied at some time in the past. This model can become inefficient or even harmful should these change, i.e. should an *exception* occur. We can consider an exception to be any departure from an "ideal" process that perfectly utilizes the current organizational resources, satisfies all extant policies and encounters no problems during process execution. Exceptions can thus include any significant change in resources, organizational structure, company policy, task

---

3 An example of a meta-level policy for task assignment is "support cross-training by routing tasks to people who need experience with that kind of task" or "maximize throughput by routing tasks to acknowledged experts".

requirements or task priority. They can also include incorrectly performed tasks, missed due dates, resource contentions between two or more distinct processes, opportunities to merge tasks between different processes, conflicts between actions taken in different process steps and so on. Process models typically compile in conditional branches to deal with common anticipated exceptions. If unanticipated exceptions occur, however, which they do frequently in modern organizations, we are faced with "patching" the compiled process model without any computer support to ensure that we are not violating the relevant organizational constraints and policies. As a result, even organizations that make extensive use of workflow technology can find it difficult to respond quickly and appropriately to changes in their requirements, policies and resources.

More effective support for process management in cooperative design, then, requires that we have a rich model of the task, resource and organizational constraints as well as policies that apply for a given organization, and be able to dynamically define the most effective process as exceptions occur.

2.2. CONFLICT MANAGEMENT

2.2.1. Problem Addressed. Cooperative design is distributed across multiple functional perspectives addressing inter-related aspects of a single design. When many perspectives are involved, maintaining consistency (i.e. avoiding conflicts) among these distributed design activities becomes a significant challenge with major potential impacts on product cost, quality and timeliness. The design of an airplane, for example, involves groups dedicated to many different structures (e.g. wings, doors, struts), systems (e.g. fuel, electrical, hydraulic), analysis disciplines (e.g. noise propagation, stress analysis) and manufacturing functions (e.g. major assembly planning, detailed part planning, tool design). The functions/groups working on these different perspectives are highly inter-dependent. For example, a change in the geometry of a wheel-well structure requires corresponding changes in the routing of the hydraulic lines since they are typically attached to major structural members. Changes in the hydraulic line routing necessitates in turn changes in the electrical line routing because of safety-related requirements on the physical separation between hydraulic and electrical systems. If the hydraulics group is not made aware of the relevant structural changes in a timely manner, it will waste valuable time designing hydraulic structures based on an obsolete view of the structural members and be forced to rework the design once the conflict is finally discovered. This is problem is exacerbated by the ripple effect of such changes: the electrical group and all those groups affected by its design decisions will be forced to abandon and rework the results of their efforts. Since maintaining scheduled commitments is often a priority in large manufacturing concerns, the group doing the rework may have to resort to short flow-time solutions that typically cost much more and may offer reduced product quality.

Current conflict management approaches in large manufacturing concerns are almost entirely manual. Functional groups within a given life cycle stage typically maintain consistency by the use of pre-defined interface conventions (e.g. hydraulic lines are always routed before electrical lines) as well as by more-or-less formal communication techniques such as cross-functional meetings and coordination memos that inform all potentially affected groups of a change made to some portion of the design. Consistency across life-cycle stages (e.g. between design engineering and manufacturing engineering) is maintained by the use of design-for-X (e.g. design for manufacturability) rules, concurrent engineering team meetings and "liaison" organizations that communicate conflicts detected by later life cycle perspectives (e.g. manufacturing) to earlier stages (e.g. design engineering) and request changes to resolve the conflicts.

Such manual approaches to conflict management have become increasingly costly and ineffective as the sheer scale of cooperative design activities have grown. Pre-defined design-for-X and interface rules can be needlessly restrictive and lead to non-optimal designs. Engineers typically find themselves inundated with high volumes of coordination memos and cross-functional team meetings, many of which turn out to be irrelevant to their particular job. Errors of omission can occur, wherein a conflict may not be detected and addressed until, for example, the product is actually being assembled. Once a conflict has been detected, it may take weeks for a change request to propagate back to the responsible design group and result in the appropriate design modification. All of this conflict management activity is mediated by large liaison and change management organizations that tie up significant amounts of human and other resources.

2.2.2. State of the Art. Conflict management technology is emerging to address these issues by providing computer-supported propagation of the consequences of design decisions. This involves notifying affected perspectives of (proposed or actual) design decisions and their potential impact. This can include simply defining the constraints some design decision places on another (so the affected designers can pro-actively avoid conflicts) as well as indicating when a design change made elsewhere conflicts with design decisions one has already made. Systems based on least-commitment design approaches [29] allow preliminary information about design decision impacts to be received even before a detailed design has been completed by another perspective.

There are relatively few kinds of design inter-dependencies. A partial taxonomy is shown below (Figure 6):

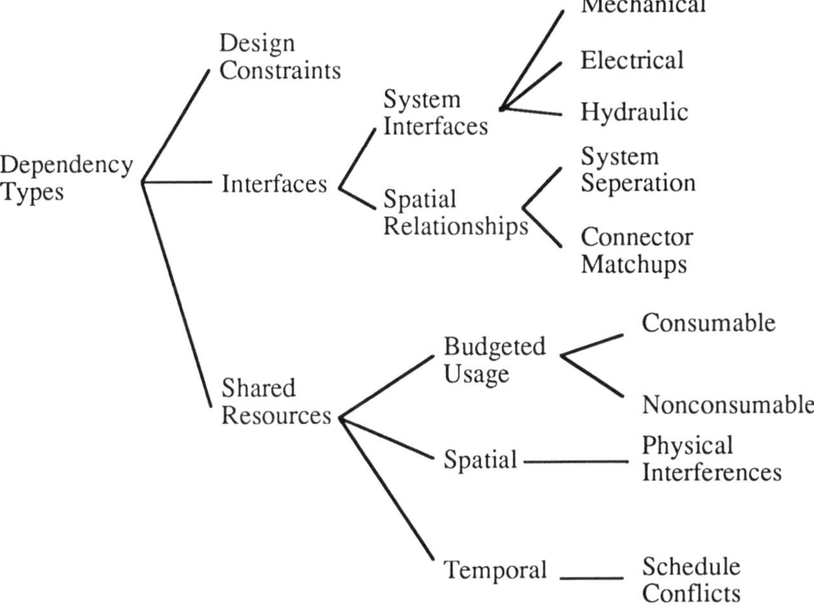

Figure 6. A Partial Taxonomy of Design Dependencies.

Design decisions can be inter-dependent because they share a limited resource (e.g. a power supply, available RAM), because they affect components connected by some kind of interface (e.g. connected components in a hydraulic or electrical system) or because they are explicitly constrained to be so by the designer (e.g. the volume and density of a part specified to have a given weight). Different technologies are available to support impact detection for each kind of dependency. Geometric conflict detection technology (i.e. checking for physical interference between adjoining parts) is widely available, as are systems for detecting temporal conflicts (e.g. scheduling conflicts) and resource budget violations for most kinds of resources[4]. Spatial relationship constraint violation detection (e.g. for checking if two connectors match up properly, or that fuel and electrical systems are separated sufficiently to meet safety requirements) is also becoming available, based on geometric feature detection technology . System interface and design constraint dependencies can be propagated using a constraint propagation infrastructure. A number of constraint propagation systems are available including commercial products such as PECOS as well as research systems [20, 28, 3].

Support for conflict resolution is an area of active study in research settings [30, 10, 8, 4, 22, 11, 12, 33, 18] but technology for this purpose is currently unavailable in commercial settings. This technology relies in general on the fact that there are relatively few abstract classes of conflict and associated general strategies for resolving them. Conflict resolution can then be achieved by heuristically classifying the conflict [7] and then instantiating an associated strategy to generate a suggested conflict resolution approach [14].

2.2.3. Challenges. Determining the cross-perspective impact of design decisions require design representation standards capable of representing design information throughout the product life cycle, including requirements, functional architecture, product geometry, manufacturing plans and so on. These standards must support high-level features (including components, their functions, resource uses and so on) as well as abstract descriptions to allow least-committment design. Existing standards do not provide adequate coverage or expressiveness yet.

Many kinds of design impacts can be inferred using basic physical principles. For example, we can use Ohm's law to infer how a change in the voltage at one end of a resistor will affect the voltage at the other. In many cases, however, the inter-relationships between design decisions will be idiosyncratic to a design and must be expressed explicitly by the designers. For this purpose the design representation language must be able to represent decision inter-relationships. Interfaces to computer aided engineering (CAE) tools[5] need to be extended to allow one to describe in this language the inter-dependencies between design decisions captured using different systems. Currently, such systems stand

---

[4] Resource limit violation is detected differently depending on the kind of resource involved. For example, over-utilization of a monetary budget can be found simply by summing individual expenditures and comparing them with the budget. Detecting space over-utilization is somewhat more complicated; even in the two-dimensional case (i.e. checking for adequate floor space); equipment whose total area does not exceed the available floor space may not fit in a given area due to the particular shapes of the equipment involved. A functional resource that is not used up over time but can only support a finite number of simultaneous users (e.g. a computer terminal) has resource over-utilization detected in yet a third way.

[5] Computer aided engineering tools include CAD (computer aided design), CAM (computer aided manufacturing) and CAPP (computer-aided process planning) systems.

alone, making it impossible to describe for example how some part geometry motivated a particular manufacturing plan decision.

Another key challenge is improving the scalability of the design decision impact assessment process; untrammeled dependency propagation can quickly become computationally intractable and overwhelm designers with floods of information on the impact of multitudes of design changes made elsewhere. Abstract or qualitative representations of design decisions and dependencies are needed to allow meaningful if approximate assessments of design decision impacts at significantly reduced computational cost. We also need to be able to specify context-sensitive policies concerning what kind of dependency impact detection should be done when.

Finally, better support needs to be provided for computer-supported conflict resolution. The first step in resolving a conflict is typically to understand how it occurred and why. Current conflict management technology records what other design decisions impinged on this one but doesn't keep track of the intent or history (e.g. rejected options) behind the decisions. We need to be able to access a rich representation of the rationale behind conflicting design decisions. In addition, the resolution of a conflict is typically reached through a multi-step process (e.g. negotiation over resource assignments) involving the people who produced the conflicting design decisions. These processes need to be integrated into a general and robust process management approach.

2.3. MEMORY MANAGEMENT

2.3.1. Problem Addressed. The distribution of cooperative design across time implies the need to remember the reasoning underlying design decisions made throughout the product life cycle. When an artifact is designed the typical output includes blueprints, CAD files, manufacturing plans and other documents describing the final result of a long series of deliberations and tradeoffs. The underlying history, intent and logical support (i.e.the rationale) for the decisions captured therein, however, is usually lost, or is represented at best as a scattered collection of paper documents, project and personal notebook entries as well as the recollections of the product's designers. This design rationale information can be very difficult to access by humans and is represented such that computers can provide little support for managing and utilizing it.

The potential benefits of more effectively capturing such rationale are manifold. The reasoning behind decisions becomes available for all team members to critique and augment [19]. Participants affected by design changes can be identified readily [21]. Existing designs which addressed similar requirements can be retrieved, understood and modified to meet current needs [26]. As we noted above, the causes and potential resolutions for conflicts between designers can be identified [14]. Design decisions can be easily documented for new team members, new designers and product users [1].

2.3.2. State of the Art. Many decision rationale capture approaches have been developed (e.g. [19], [35], [23], [21], [9]). While the details differ, all represent rationale as graph structures, where nodes represent entities such as issues and criteria, and links between nodes represent relationships between the connected entities. A typical example is given below (Figure 7):

444

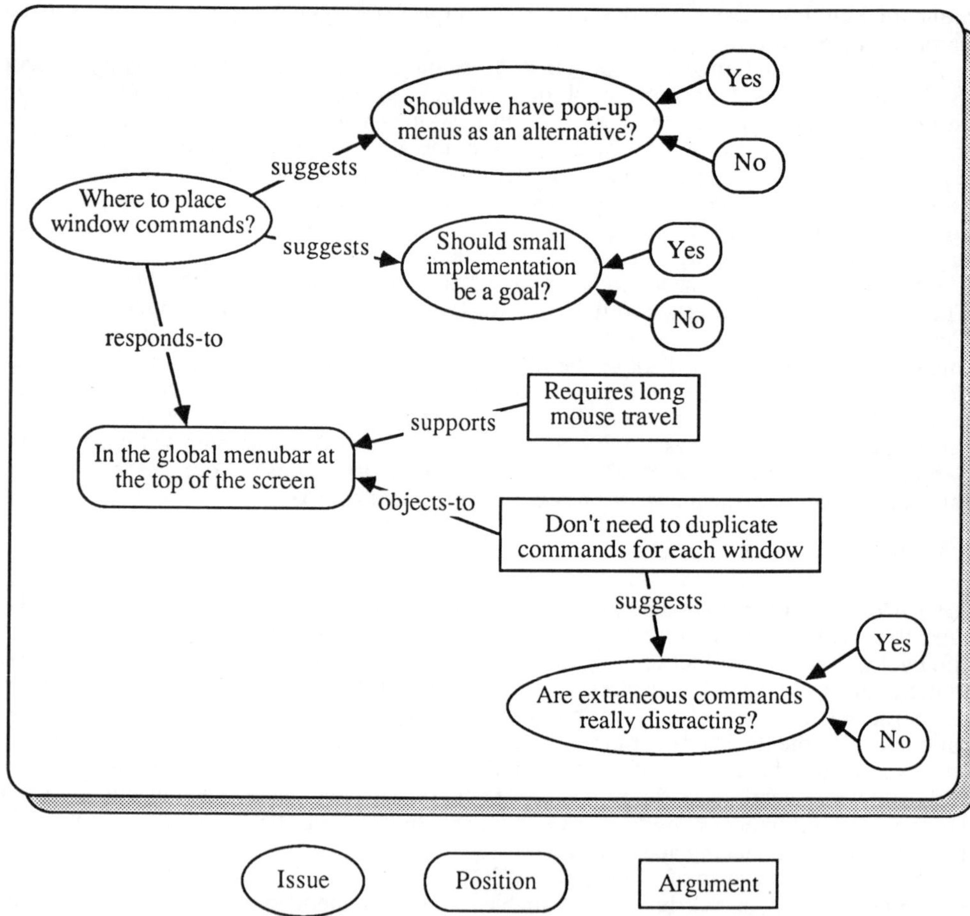

Figure 7. An Example of Rationale Representation

Existing rationale languages differ in the kinds of entities that are used as well as how they can be interrelated. While they may differ in expressiveness they overlap substantially, sometimes simply using different terminology for entity and relationship types with substantially the same semantics.

These approaches typically capture the rationale for decision-making in general but not design in particular; in design settings they add in effect another document to the set produced by existing design tools (Figure 8).

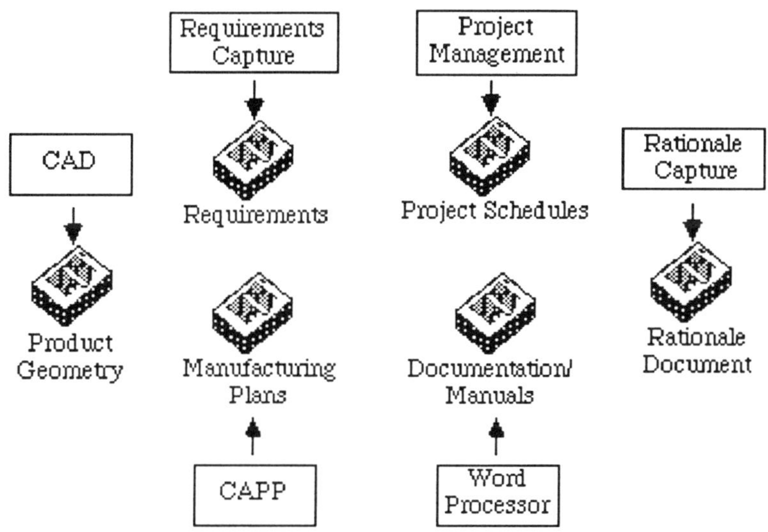

Figure 8. Rationale Captured as a Distinct Document.

While one system that integrates design and rationale representation does exist [9], it uses a domain-specific design representation (for kitchen design) not easily generalized to other domains.

2.3.3. Challenges. Existing rationale capture languages face limited expressiveness and therefore limited computational usefulness due to lack of integration with generic design representations. Since the associated decision rationale capture tools are not integrated with design tools, they face the potential for inconsistency between the rationale and design descriptions, spotty capture of design rationale and the tendency to waste time on issues that later prove unimportant [9]. What we *really* need are systems that allow cooperative design participants to conveniently describe the dependencies between decisions captured by existing design tools (Figure 9).

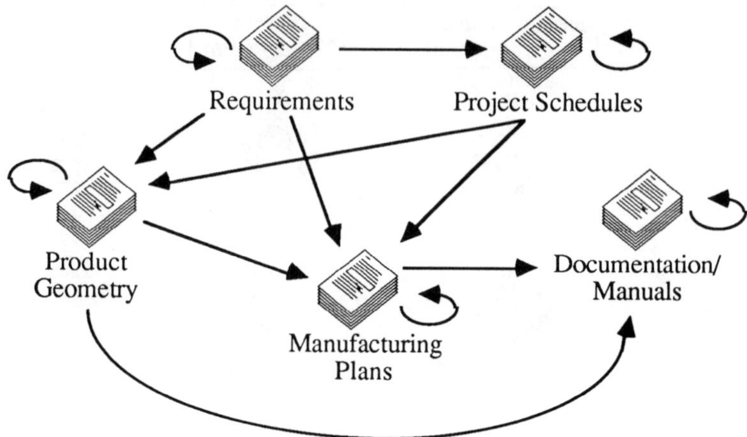

Figure 9. Rationale as Decision Interdependencies.

The rationale for a product geometry decision, after all, consists of such things as the requirements it attempts to satisfy and the other geometry decisions on which it logically depends. A manufacturing plan decision, similarly, is justified by supporting decisions such as project schedules and product geometry.

Existing rationale language technology has been used mainly to capture the pros and cons concerning alternative solutions to a given problem, but not design intent, the relationships between design decisions or their history, all of which, of course, can be extremely important.

Rationale capture can be burdensome, especially given that the person who benefits is generally different than the person required to describe it. This is exacerbated by the fact that it is often unclear where and in how much detail rationale should be described, since the person who enters the rationale is unlikely to know how it will be used or even by whom. We should be able to describe preferred rationale capture processes to delimit what and how detailed rationale should be captured. If possible, these preferred processes should provide default rationale templates that can be quickly customized to describe the rationale for a given decision.

3. iDCSS: An Integrated Approach to Cooperative Design Coordination

Coordination support technology has evolved to support effective coordination in the face of dependencies between different aspects of cooperative design. The different technologies each focus on one set of dependency types but fail to account for others; this is the source of many of their individual limitations. I describe below an integrated approach that attempts to model and account for *all* the kinds of dependencies that occur in cooperative design and as a result synergistically combines the strengths and avoids the weaknesses of the contributing technologies. A preliminary implementation of this approach has been created called iDCSS (the Integrated Design Collaboration Support System; pronounced "IDEX"). iDCSS is the latest result of five years of work including studies of cooperative design [13] as well as several previous systems including the DCSS conflict management system [14], the DRCS design rationale capture system [17] and a prototype workflow system[6]. In the sections below we consider iDCSS's architecture, underlying design model and how it supports the three kinds of cooperative design coordination identified above.

3.1. ARCHITECTURE

iDCSS has a distributed client-server architecture (Figure 10). Shared product data and services are made available by servers to all clients via a network. Clients are design agent/assistant pairs where agents can be either human or machine-based and assistants are computer programs that help coordinate the design activities of their associated agents. Human designers work in front of dedicated workstations executing their assistant, while machine-based design agents are independent processes communicating with distinct assistant processes. The code currently runs on networked Symbolics Lisp Machines and is written in Common Lisp.

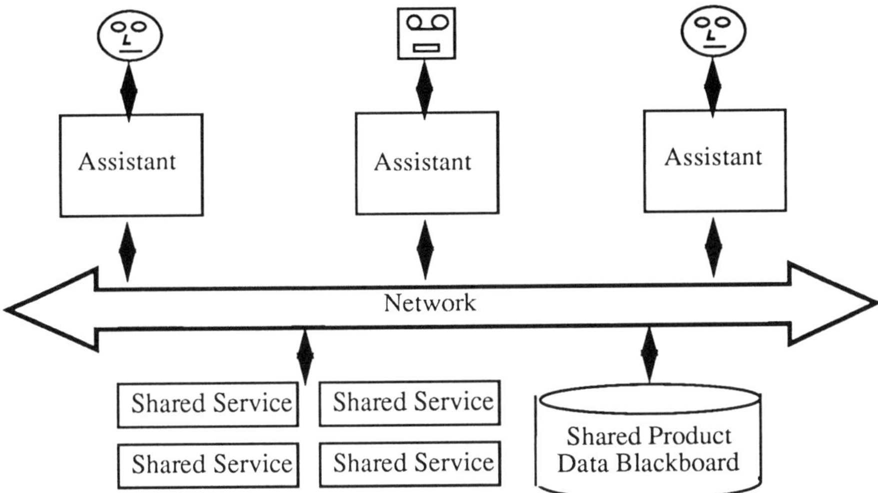

Figure 10. The iDCSS Architecture.

3.2. DESIGN MODEL

iDCSS is based on a model of cooperative design derived from classical Systems Engineering work [2] as well as AI models of artifact design [24, 22, 5, 25, 31, 15] and planning [29, 6].. These models have been applied successfully to many domains including electrical, electronic, hydraulic and mechanical systems as well as software.

In this model, physical artifacts are viewed as collections of modules, which can represent entire systems, subsystems or their components, each with characteristic attributes and whose interfaces (with their own attributes) are connected by typed connections. Artifact designs are refined using an iterative least-commitment synthesize-and-evaluate process. An artifact description starts as one or more abstract modules representing the desired artifact with specifications represented as desired values on module

---

6 Many of the process management ideas in this paper were developed for the TCAPS workflow management system described in [36].

attributes. This is refined into a more detailed description by constraining the value of module attributes, connecting module interfaces (to represent module interactions), decomposing modules into sub-modules and specializing modules by refining their class (Figure 11):

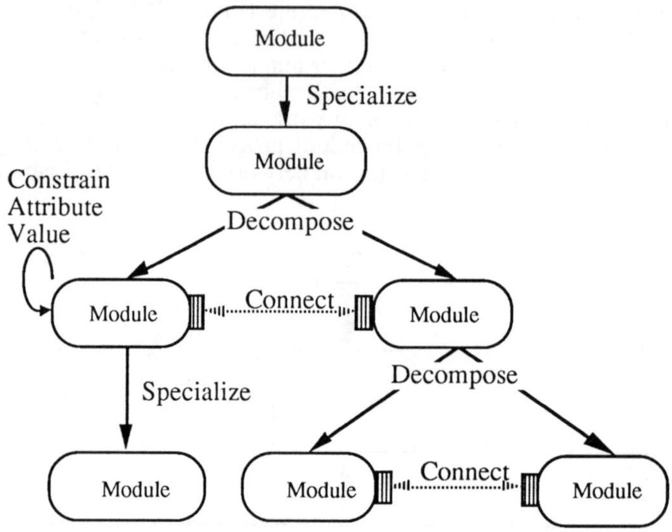

Figure 11. The design refinement process.

Plans are viewed as (perhaps partially) ordered collections of tasks, and are also defined in an iterative least-commitment manner.

In parallel with the iterative refinement of the design description is evaluation of the design with respect to how well it achieves the design specifications. Based on this analysis we may choose to select one design option over another or modify a given option in order to address an identified deficiency. The stages of specification identification, design option definition, evaluation and selection/modification can be interleaved arbitrarily and thus often occur in an iterative fashion throughout the design process.

Designers, in addition to reasoning about the design itself (i.e. at the domain level) , also reason about the *process* they use to define the design (at the *meta*-level) [32]. A designer may have a plan, for example, for how the design itself will be created. If several design options are available, a designer may reflect on which option to select. If a conflict between two or more design goals and actions occurs, a fix for the conflict needs to be found. The design reasoning process is generally goal-driven, in the sense that actions are taken as part of strategies intended to achieve goals such as meeting a specification, refining a design option, making a control choice, resolving a design conflict and so on.

3.3. PROCESS MANAGEMENT

From the perspective of a designer, the design process starts when a goal shows up in his or her ToDo list; at the very beginning of the product life cycle it will be the goal to design a product that meets a given set of requirements:

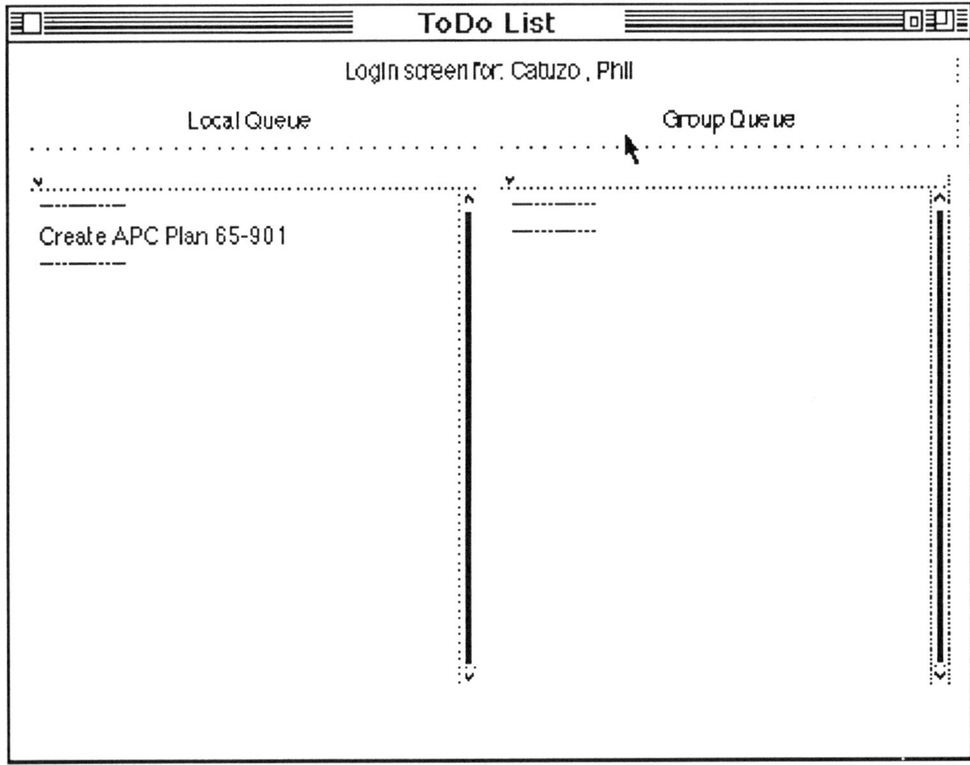

Figure 12. A ToDo List.

The local queue lists goals assigned to this designer specifically, while the group queue lists goals that can be performed by any of a group of people that includes the current designer. In the latter case, the designer must volunteer to take on that goal, at which point it moves off of all the group queues and onto the designer's private queue. Both domain-level goals (e.g. to find the voltage for a power supply) and meta-level goals (e.g. decide which of several alternative solutions to select for a given goal) appear on this ToDo list.

To start performing the task specified by the entry in the ToDo list, the designer double-clicks on the goal to create a "task performance plan" with an associated window like the following:

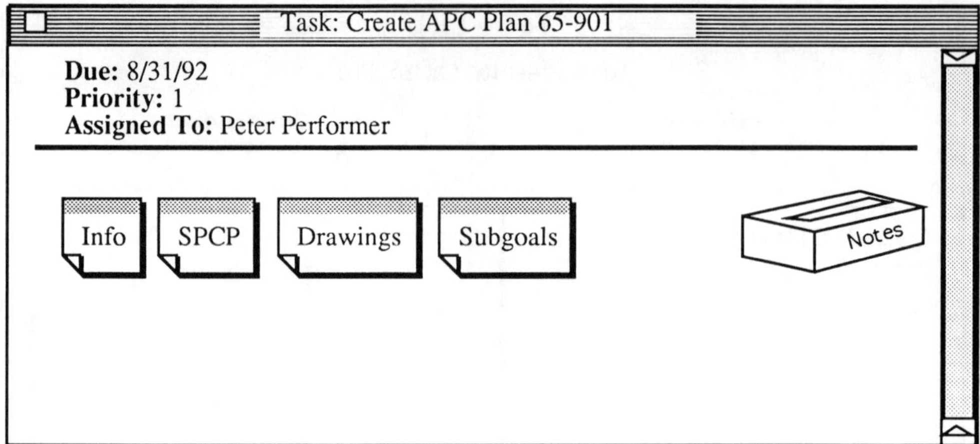

Figure 13. A Task Performance Window for a Goal.

A task performance plan collects all the information, tools and sub-processes needed by a designer to achieve a given goal. Using the associated task performance window, a designer can bring up views allowing one to look at and update different subsets of the complete set of product data from different perspectives. Every icon in the figure above represents one view. A view can correspond to a CAE application and may display product requirements, geometry, manufacturing plans, design process plans, supporting design documentation or even the rationale (underlying argument structure) for a given design decision. In the following example, views display the design version history, related functional components, attributes and manufacturing plan for a component:

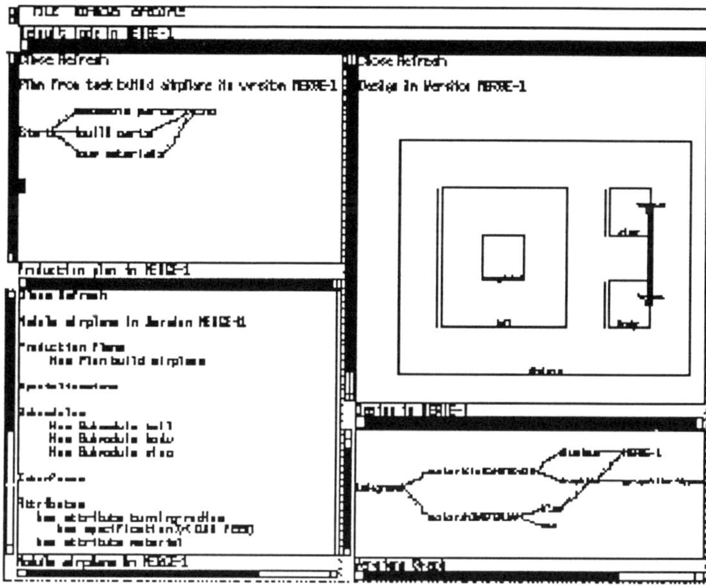

Figure 14. Task Window With Several Open Views.

The rationale for any design decision can be captured using the iDCSS rationale language. This is a typed link language designed to match the iDCSS design model and built from previous work in decision rationale capture ([19], [35], [23], [21]). In this language, design rationale is represented as sets of claims. Any claim can serve as part of the rationale for another claim, so we can make claims about the design, claims describing the rationale for design decisions, claims describing why we should believe this rationale (or not) and so on. The iDCSS rationale language is summarized in Figure 15 below: the plain font items represent types of design decisions while the italic labelled arrows represent allowable kinds of typed links between them, pointing from the sources to the targets.

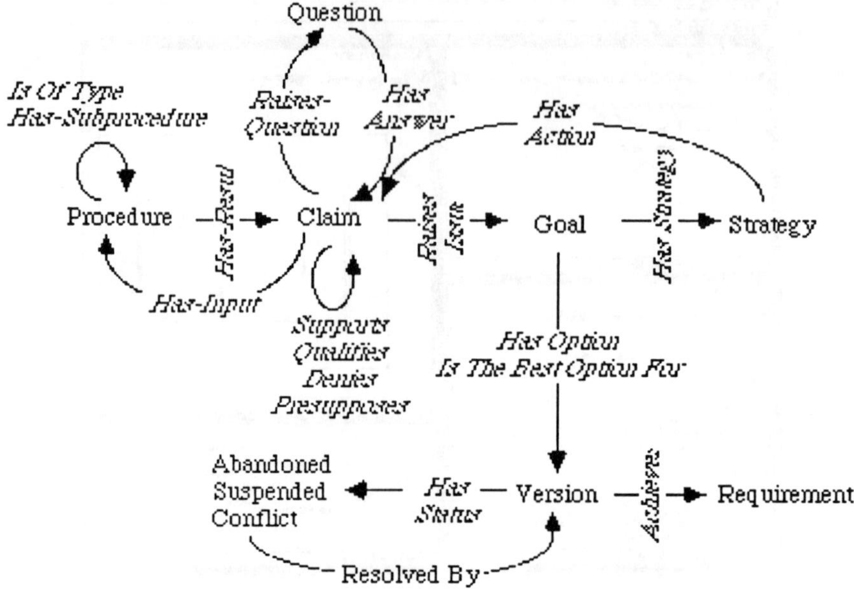

Figure 15. iDCSS Design Rationale Language.

A designer can build up a rationale description using a simple extension of the user interface provided for describing the designs themselves. The menus brought up by selecting an assertion with the pointing device include context-sensitive lists of the types of links allowable starting from that assertion. To create a link, one simply selects the link type and the assertion that is the target of the link. The net result of describing design rationale in this way is a graph of decisions inter-related by typed rationale links. A detailed description of the iDCSS rationale language and user interface scheme is given in [16].

A designer will often need to create subgoals to handle different tasks whose results are needed to achieve the current goal. The subgoals for a task performance plan appear in the "subgoals" view for that plan's window.

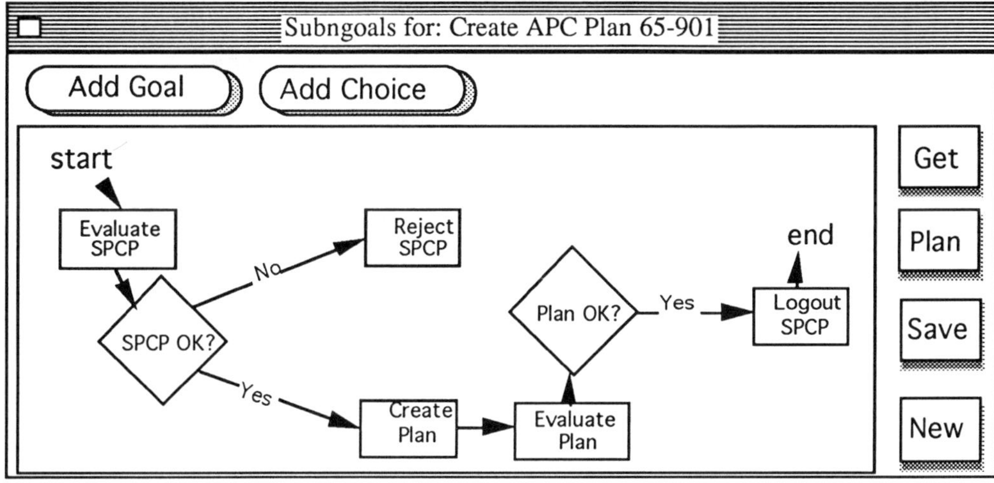

Figure 16. The Subgoals View for a Task Performance Plan.

The preferred process for performing subgoals can be defined graphically or by using an automated planning tool that defines a process based on the subgoal requirements, organizational resources and current policies.

By default, subgoals are assigned to the person who generated them and appear in his or her ToDo list nested under their parent goal. A designer may need to assign these goals to someone else, e.g. because they require specialized expertise or the designer is simply too busy. To decide who the task should be assigned to the user defines task assignment constraints (in terms of desired skills, experience, organizational position, post etc.) and allows the system to identify the one or more candidates who meet these job descriptions. iDCSS uses an organizational model for this purpose; this model stores the organization chart, the members, managers, support people and other resources for each organizational unit as well as the job descriptions that each person can fulfill. An organization model can be displayed and edited using the following view (Figure 17):

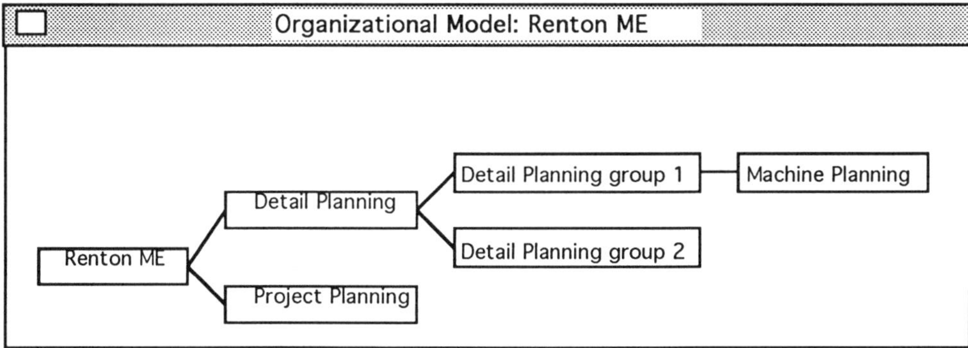

Figure 17. The Organization Model View.

Job descriptions for an individual are viewed and updated using the following view:

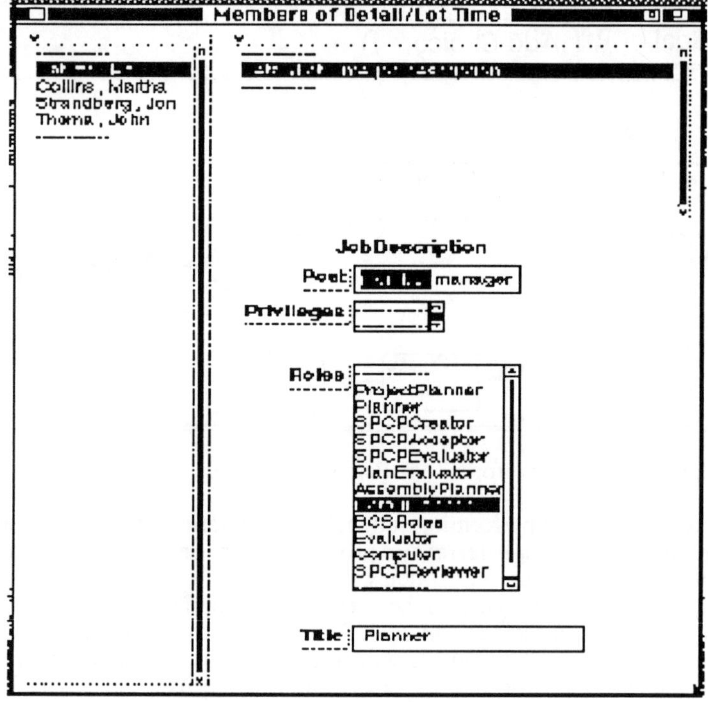

Figure 18. The Job Description View.

The rationale behind task assignment and sequencing decisions can be captured, just like all other design decisions in iDCSS, using the link language. One can justify assigning a task to a given individual, for example, by a link to a job description for that individual in the organization model which shows he or she has the necessary skills. Similarly, one can justify serializing two tasks by a link from the resource conflict that is shown to result if the two tasks are performed simultaneously.

A user's assistant can store predefined task performance plans for accomplishing different goals; these can be retrieved by the assistant and provided as starting points that the users may then modify to provide the desired task performance environment. This can be used to represent the company's preferred processes for achieving given goals; the preferred task performance window is then automatically opened when the designer opens the task for the first time. These predefined windows can be defined manually or may be "learned" by recording task performance plan that a user defined for that goal in the past and then storing it as a reusable case. Precompiled task performance plans, however defined, should record the rationale underlying the decisions they store just as user-defined windows do, so we can replan appropriately when exceptions occur.

The iDCSS process management approach is derived directly from existing approaches but adds a critical feature. One can record the dependencies underlying process decisions so that if these decisions change (i.e. an exception occurs) the affected processes can be identified and modified appropriately. In this way, the brittleness of conventional process management approaches can be avoided.

3.4. CONFLICT MANAGEMENT

When design decisions (made about product or plans) interact in some way, the influence of one design decision on another and whether a conflict has resulted, can be detected using the impact detection service. This service can be applied equally well to product or process descriptions. If the service is applied to design process plans, for example, it helps ensure that the multiple sub-processes making up the global design process can be performed without exceeding the organizations' resources. If the service is applied to product decisions, it helps ensure that the design decisions made by different designers are consistent. When a conflict is detected, the meta-level goal of resolving that conflict is asserted, supported (using the link language) by the facts used by the conflict detection procedure to infer the existence of the conflict. This goal appears on the user's ToDo list just like any other goal. A user may manually define a resolution for that conflict using the standard task performance plan window, or else can call on the conflict resolution service to define one or more suggested task performance plans for resolving the conflict.

The integrated iDCSS model offers some important advantages over existing conflict management technology. It provides the beginnings of a least-commitment high-level feature full product life cycle representation so we can maintain consistency across a greater range of the product design life cycle. A uniform and expressive link language/user interface allows the user to define the rationale for any design decision in terms of any other design decision. This can be used by the impact detection service to propagate design decision impacts, and also by the conflict resolution service to help understand the cause and possible resolutions for a conflict. Preferred design processes that specify when and how the impact detection service should be triggered can be defined using task performance plans; this allows us to control the impact detection cost/benefit ratio. Finally, task performance plans provide a mechanism for representing and executing conflict resolution strategies.

3.5. MEMORY MANAGEMENT

iDCSS includes a robust memory management approach as an integral part of its operation. The reasons underlying product and process design decisions are captured in one uniform formalism that combines the argumentation and intent information of conventional rationale languages, the decision dependencies represented in conflict management technology and the history captured by process management technology. This provides greater expressiveness and therefore allows greater computational support than conventional non-integrated approaches. The iDCSS approach also is, I believe, more natural than previous approaches because rationale can be attached directly to the design decision it refers to, rather than to a piece of text that acts as a proxy. This prevents inconsistency between the rationale and design and avoids the tendency faced in conventional rationale systems to waste time on irrelevant issues. Conflict management helps focus rationale capture; it allows us to concentrate on the rationale underlying controversial decisions rather than straightforward ones. Finally, we can use preferred processes defined in task performance plans to describe to what extent the rationale behind given design decisions need to be described; this allows us to focus rationale capture efforts on those design aspects thought most critical. The task performance plans can even specify default rationale templates (link language network fragments) that the user is asked to fill in for a given design goal, thus further reducing the rationale capture burden in routine design settings.

4. Conclusions

The iDCSS system integrates technologies for coordinating cooperative design over agents, perspectives and time in a way that combines the strengths and avoids the weaknesses of the component technologies. We can summarize this synergism using the following figure:

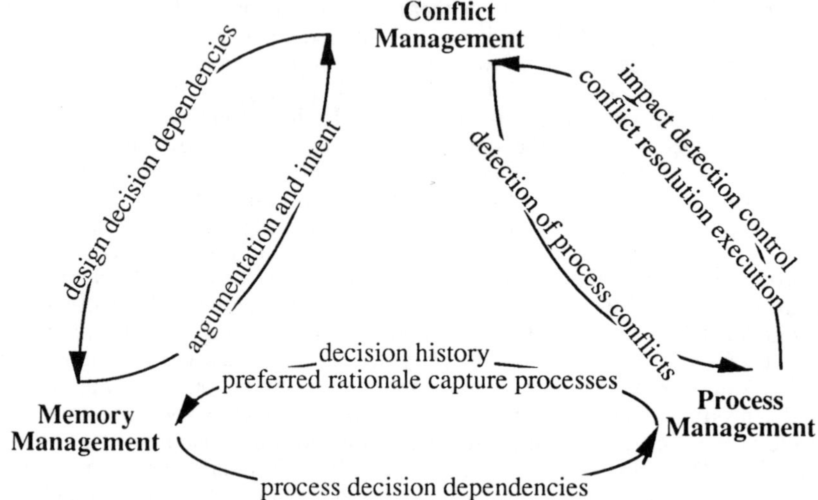

Figure 19. Synergism Among iDCSS Component Technologies.

Process management technology provides processes to control impact detection, implement conflict resolution plans, delimit rationale capture and represent decision history. Conflict management ensures consistency among design processes and provides the design dependency component of the iDCSS rationale language. Finally, the memory management component provides rationale to support conflict management and intelligent process replanning in the face of exceptions. The power of this approach comes from representing all design dependencies in a single uniform way and from leveraging a small set of generic coordination services (for impact detection, rationale capture, process definition and execution as well as conflict resolution) that utilize this dependency information.

While this integrated approach does appear to offer significant advantages over current non-integrated coordination technologies, clearly many challenges remain. Effective design coordination requires a communication infrastructure that provides sufficiently expressive and inclusive data standards (e.g. extending and integrating existing standards like PDES/STEP and IDEF) as well as highly scalable network access to shared heterogeneous data and services (e.g. based on semantic database integration services such as CARNOT). Scalable and effective technology for supporting collaborative editing of shared product datasets is needed, and must address such issues as control of chalk-passing, versioning and release authority control. The existing coordination support services need to be improved (for example, conflict management technology needs to produce more robust generic knowledge bases of impact propagation, conflict classification and conflict resolution expertise) and new services need to be added (e.g. a task merge service that uncovers opportunities to define merged tasks more efficient than their predecessors, or a retrieval service that can fetch previous relevant design experience). Finally, existing CAE applications need to be enhanced so they can take effective advantage of these coordination services. This requires the ability to describe design decisions using high-level features and to express design dependency links across applications as well as making these applications workflow-enabled so they can take advantage of process management technology functionality.

5. Acknowledgements

I'd like to thank my colleagues Roland Faragher-Horwell, Art Murphy, Steve Poltrock, Kish Sharma and Debra Zarley for their substantial contributions to the ideas underlying this paper, especially in the area of process management.

6. References

[1] Balzer, R. Capturing the Design Process in the Machine. In *Rutgers Workshop on Knowledge-Based Design Aids,* 1984.

[2] Blanchard, B.S. and Fabrycky, W.J. *Systems Engineering and Analysis,* Prentice-Hall , Industrial and Systems Engineering(1981).

[3] Bowen, J. and Bahler, D. Constraint-Based Software for Concurrent Engineering. *IEEE Computer 26,* 1 (January 1993) Pps. 66-68.

[4] Brown, D.C. *Failure Handling In A Design Expert System,* Butterworth and Co. (November 1985).

[5] Brown, D.C. Capturing Mechanical Design Knowledge. *American Society of Mechanical Engineers CIME 1985* (February 1985).

[6] Chapman, D. Nonlinear Planning: A Rigorous Reconstruction. *IJCAI-85* 2(1985) Pps. 1022-1024.

[7] Clancey, W.J. Classification Problem Solving. *AAAI* (1984) Pps. 49-55.

[8] Descotte, Y. and Latombe, J.C. Making Compromises Among Antagonist Constraints In A Planner. *Artificial Intelligence* 27(1985) Pps. 183-217.

[9] Fischer, G., Lemke, A.C., McCall, R., and Morch, A.I. Making Argumentation Serve Design. *Journal of Human Computer Interaction* 6, 3-4 (1991) Pps. 393-419.

[10] Fox, M.S. and Smith, S.F. Isis - A Knowledge-Based System For Factory Scheduling. *Expert Systems* (July 1984).

[11] Goldstein, I. Bargaining Between Goals. In *.IJCAI,* 1975, Pps. 175-180.

[12] Hewitt, C. Offices Are Open Systems. *ACM Transactions on Office Information Systems* 4, 3 (July 1986) Pps. 271-287.

[13] Klein, M. and Lu, S.C.Y. Conflict Resolution in Cooperative Design. *International Journal for Artificial Intelligence in Engineering* 4, 4 (1990) Pps. 168-180.

[14] Klein, M. Supporting Conflict Resolution in Cooperative Design Systems. *IEEE Systems Man and Cybernetics* 21, 6 (December 1991).

[15] Klein, M. and Lu, S.C.Y. Insights Into Cooperative Group Design: Experience With the LAN Designer System. In *Proceedings of the Sixth International Conference on Applications of Artificial Intelligence in Engineering (AIENG '91),* Rzevski, G. and Adey, R.A., University of Oxford, UK, July 1991, Pps. 143-162.

[16] Klein, M. DRCS: An Integrated System for Capture of Designs and Their Rationale. In *Proceedings of Second International Conference on Artificial Intelligence in Design,* Carnegie Mellon University, Pittsburgh, Pennsylvania, 1992.

[17] Klein, M. Capturing Design Rationale in Concurrent Engineering Teams. *IEEE Computer* (January 1993).

[18] Lander, S. and Lesser, V.R. Negotiation To Resolve Conflicts Among Design Experts. Tech. Report Dept of Computer and Information Science, August 1988.

[19] Lee, J. and Lai, K.Y. What's In Design Rationale?. *Human-Computer Interaction* 6, 3-4 (1991) Pps. 251-280.

[20] Lu, S.C.Y. Integrated and Cooperative Knowledge Processing Technology for Concurrent Engineering. In *Knowledge-Based Engineering Systems Research Laboratory Annual Report.* University of Illinois, Lu, S.C.Y., April 1991.

[21] MacLean, A., Young, R., Bellotti, V., and Moran, T. Questions, Options and Criteria: Elements of a Design Rationale for User Interfaces. *Journal of Human Computer Interaction: Special Issue on Design Rationale* 6, 3-4 (1991) Pps. 201-250.

[22] Marcus, S., Stout, J., and McDermott, J. VT: An Expert Elevator Designer. *Artificial Intelligence Magazine* 8, 4 (Winter 1987) Pps. 39-58.

[23] McCall, R. PHIBIS: Procedurally Heirarchical Issue-Based Information Systems. In *Proceedings of the Conference on Planning and Design in Architecture,* ASME, Boston MA, 1987.

[24] Mcdermott, J. R1: A Rule-Based Configurer Of Computer Systems. *Artificial Intelligence* 19(1982) Pps. 39-88.

[25] Mittal, S. and Araya, A. A Knowledge-Based Framework For Design. In *American Assocation of Artificial Intelligence,* 1986, Pps. 856-865.

[26] Mostow, J. and Barley, M. Automated Reuse of Design Plans. In *Proceedings ICED,* IEEE, August 1987, Pps. 632-647.

[27] Smith, R.G. The Contract Net Protocol: High-Level Communication And Control In A Distributed Problem Solver. *IEEE Transactions on Computers C-29,* 12 (December 1980) Pps. 1104-1113.

[28] Smith, K., Karandikar, H., Rinderle, J., Navinchandra, D., and Reddy, S. Representing and Managing Constraints for Computer-Based Cooperative Product Development. In *Third Annual Symposium on Concurrent Engineering,* June 1991, Pps. 475-490.

[29] Stefik, M.J. Planning With Constraints (Molgen: Part 1 & 2). *Artificial Intelligence 16*, 2 (1981) Pps. 111-170.

[30] Sussman, G.J. and Steele, G.L. Constraints - A Language For Expressing Almost-Hierachical Descriptions. *Artificial Intelligence 14*(1980) Pps. 1-40.

[31] Tong, C. AI In Engineering Design. *Artificial Intelligence in Engineering 2*, 3 (1987) Pps. 130-166.

[32] Wilensky, R. Meta-Planning. *AAAI* (1980) Pps. 334-336.

[33] Wilensky, R. *Planning And Understanding,* Addison-Wesley (1983).

[34] Winograd, T. A Language/Action Perspective on the Design of Cooperative Work. In *Proceedings of CSCW '86,* 1986.

[35] Yakemovic, K.C.B. and Conklin, E.J. Report on a Development Project Use of an Issue-Based Information System. In *CSCW 90 Proceedings,* 1990, Pps. 105-118.

[36] Faragher-Horwell, R., Klein, M. and Zarley, D. Overview and Functional Specifications for TCAPS Task Coordination And Planning System: A Computer-Supported Workflow Management System. *Tech. Report BCS-G2010-130.* Boeing Computer Services Technical Report, The Boeing Company, December 1992.

PANEL REPORTS

PANEL 1: INTEGRATION OPPORTUNITIES AND REQUIREMENTS

Panel Members: Andrew Whinston (Chair), Agostino Villa (Co-Chair), Allan Hodgson (Provocateur), Thomas Clausen, and Ray Eberts

OBJECTIVE

The aim of this panel was to provide a needs-driven focus. This was achieved by considering the opportunities offered to an enterprise by an "ideal" integrated system and then considering the particular integration-related features or properties required to enable these opportunities. In many cases, these requirements resulted in several further stages of subrequirements. Frequently, similar sub-requirements turned up as "leaves" on different branches of the "requirements tree." It was therefore necessary to go through an additional stage of grouping such sub-requirements to form related issues or research questions.

The major "opportunities" (three in total) are listed below. The major issue areas are then listed, each followed by specific open questions.

OPPORTUNITIES: With an ideal (or optimum) level of integration, opportunities should exist to achieve the following benefits.
1. FUNCTIONAL PERFORMANCE corresponding more closely to business strategies and associated company objectives than at present.
2. COST EFFECTIVENESS at all stages of implementation, not only after integration exercise.
3. FUTURE PROOF SYSTEMS

A. INCORPORATING ENTERPRISE BUSINESS STRATEGIES

Business strategies, where stated, are typically qualitative and aggregate. However, they form (or should form) the basis for the design of appropriately integrated enterprise systems. It is therefore important that they are incorporated formally into the specification/design processes.

Issues
1. Can we bridge the gap between high-level business strategy and (integrated) enterprise systems, to produce models which can be compared and evaluated for consistency?
2. Can we improve the agility of a business strategy model by incorporating appropriate environmental factors?
3. Can any existing or proposed architectural models be extended to include company objectives/business strategies/environmental factors?
4. Can we produce a meta-architecture and toolset to design an enterprise-specific architecture type, populate and implement it?
5. What methodologies and tools are required to enable us efficiently to plan, collect and evaluate appropriate company information prior to integration-related activities?

   Note: Problems associated with collecting data and what to do in its absence appear largely ignored.

S. Y. Nof (ed.), *Information and Collaboration Models of Integration*, 461–470.

B. DEALING WITH LEGACY

A typical enterprise will carry forward into any integration activity a significant amount of aging software, hardware and other components. As this is part of the current operating system, much of it must be included in any integration plans. Some of it may remain for many years. In addition to existing legacy systems, it is equally important to recognize that elements of any proposed integration solution will probably become "legacy systems" at some point in the future. They should therefore be selected or designed with this factor in mind.

Issues
1. How do we deal with current legacy systems when designing integration solutions?
2. How do we minimize the potential problems associated with future legacy systems in a changing environment – including hardware, software, people and standards?
3. (In particular) Can we insulate our expensive functional software code from integration infrastructure obsolescence?
4. Can we "layer" integration infrastructure functionality to extend its life and applicability? (The upper layers would contain, say, industry or function-specific services.)

C. INCORPORATING THE HUMAN ELEMENT

It is now recognized that most "factories of the future" will incorporate human elements at all stages of the life cycle, including specification, design, implementation, fault recovery and change, as well as in normal day-to-day production activities. It is therefore important to ascertain the human-related requirements and aspirations, and enable these to be met by appropriate system design rather than as an afterthought.

Issues
1. Can we create appropriate models of humans and their interactions within integrated organizational systems? In particular, can we model (or cater for) aspects such as culture, context, overload/underload, personal goals and mood? How do we test such models?
2. Can we incorporate (the above) human models into the outputs of on-going CSCW research to improve further the performance of design, planning, management and other software tools?
3. Can we incorporate appropriate human models into software systems in order to detect unreasonable behavior due to, for example, "illegal" untrained usage of a system, malicious intent, illness or influence of drugs?
4. Given the increasing complexity of integrated enterprise systems, can we create adequate human-oriented tools to enable system recovery following failure? This might, for instance, include the innovative combination of intelligent fault diagnosis, human models and virtual reality to allow a directed search of both the software and hardware world.
5. Training needs are increasing. Can we develop an adaptive training methodology which learns both general human responses and those of individuals? This would perhaps run as a training "layer" utilizing the actual information systems. Could such a methodology detect and report "difficult" software areas which cause general confusion due to concepts or presentation?
6. Can we replace the deep knowledge lost to a domain as we remove people, in particular to enable the enterprise information system (1) to check itself for internal consistency and "external reasonability"; and (2) to answer arbitrary queries from humans and other systems?
7. Can we obtain or develop fundamental technology to achieve (6) above, particularly with regard to the creation of an effective conceptual graph-based toolset (with agents)?

PANEL 2: THE DESIGN OF INTEGRATION

Panel Members: Mark Klein (Chair), David Stotts (Co-Chair), Hendrik Van Brussel, Isil Bozma, S. Kuru, Venkat Rajan, Michael Sharpston, Frank Wagner, Richard Westin

The focus of this panel was on identifying research issues relevant to the design of integration processes and computer systems. The process of generating and analyzing enterprise models from different views is the key element in design of integration systems. These views should include specific aspects of an enterprise such as the informational or function issues as well as overall integration issues. Two main underlying themes that need to be considered throughout the design of integration process are:
1. Human and machine elements of an enterprise
2. Legacy and conceptual sub-systems of an enterprise.

An enterprise is typically a system of interactions between humans and machines. Such interaction may be broadly classified to be human-human, human-machine, or machine-machine. The design of integration process is strongly influenced by the existence of one or more of these interactions within the enterprise. There is need for research in the following categories:
1. Developing theoretical models of integration.
2. Developing the actual models of integration at the component and integrated system level.
3. Developing computer based tools, environments and methodologies to support designers.

A. DESIGN MODELS THEORY AND ANALYSIS
Do we have one or more analytic approaches which permit adequate integration of human and organizational issues into an overall formal framework? The traditional "analyze -> solution -> implement" cycle may not work in this context. With people involved it may be impossible or unwise to attempt at the start an exact view of the final outcome; it can be very difficult to foresee exactly how people will use information and communication systems. A high level of user involvement in design and implementation, however, is very beneficial for business outcome. We need a more iterative approach. Rapid prototyping, per se, is not quite the solution. Users must be put in a better power balance with system designers.

The term "semi-formal" system was used in panel discussion to capture the combination of both quantifiable and non-quantifiable information into a single analytical framework. This concept seemed to be what many of the following issues revolve around. There are formal and mathematical aspects that are covered in the respective sections to which they apply. In this section, we concentrate on broad research issues in mathematics that would find application in areas other than integration of information and collaboration models.

Immediate Issues
1. Can one develop a design process, perhaps with semi-formal systems, which appropriately represents the iteration described above between organizational and technological change?
2. Developing new models of interaction (i.e. coordination, conflict resolution, cooperation, etc.) among formal computational agents and more unpredictable behavioral agents (e.g., humans).
3. How to describe distributed decision-making systems (as opposed to decision-tree systems) in a solvable mathematical framework? (Candidate theories are: Markov chains, queuing theory, max. algebra, state space formulation, net modeling).
4. How can the inherently nonformal human components of an enterprise be combined with formal aspects into a "semi-formal" system? What are the fundamental theories of semi formal systems? What analysis techniques can be applied to semi-formal systems?

464

Near-term Issues
1. How is organizational, attitudinal and cultural change mapped against other elements of the integration process (e.g. mechanical, electronic, information and communication systems)?
2. Developing models of change for integrated systems.
3. Developing ontologies for integration modeling (i.e. identification of deep concepts and relations of integration).
4. Development of a theory of flexibility of intermediate solutions maximizing overall system flexibility.
5. Development of a theory of optimal interaction between system elements e.g. reactive scheduler — cell controller (dispatcher).
6. Need for information and communication theories for collaborative multi-agent systems.

Long-term Issues
1. How to educate engineers/computer scientists, etc. in the management of change for humans?
2. How do we control the computational cost of decision impact propagation services so that concurrent engineering participants learn about the relevant external decisions impacting their decisions in a timely fashion?
3. As the scope of an integrated design grows, encompassing more and more of an enterprise, can the number of modeling formalisms be reduced, or must the complexity necessarily increase? Look for unifying formalisms that will reduce the complexity of the intellectual framework while retaining coverage against the question one seeks to answer from the framework.
4. Need for study of effect and levels of local and global objectives of agents (representing enterprise elements).
5. Need to study view (informational, functional, etc.) incompleteness and inconsistency for model creation and analysis.

B. MODEL CREATION
Immediate Issues
1. Evaluating existing representation tools and languages, such as logic, algebra, Petri nets, activity nets, semantic nets. etc. for integration modeling from the point of view of representational power and ease of use.
2. How feasible is it to produce generic models of "machine operators" and "supervisors", both individually and in combination with models of manufacturing machines, which enable their interworking, thereby promoting their collaborative behavior, without placing undue constraint on the environment?

Near-term Issues
1. How do we measure the enterprise vs. a model? Where is the point of diminishing returns with respect to encompassing more and more of an enterprise into a "formalism" framework, or federation of such frameworks?
2. Qualitative vs. quantitative models: When are these different models and hybrid forms _useful_ for the design of enterprise integration? In early stages of design only qualitative information is available. For top level integration (e.g. management) this may be sufficient.
3. Are "competence-checking" models possible? The problem of AI models being "out of competence" is also valid for the design of integration. How can our model (the design of integration) alarm the designer/integrator that it is not valid/sufficient? (Example: Are Petri-Net Models competent for product development processes? What about human factors and other "soft" factors?)

4. What constraint languages or other formalisms effectively capture the dependencies between planning decisions, design decisions, organization resources and policies etc.? Rationale: coordination is required only when our actions are inter-dependent, so computer-supported coordination requires a computer interpretable model of decision dependencies.
5. What are the (preferably generic) approaches people can or should use to deal with exceptions in business processes, and when are they appropriate? Rationale: existing process support technology is brittle in face of exceptions.

Long-term Issues
1. What is the minimally burdensome amount of enterprise and product modeling needed to support the computational coordination services revealed as critical by ethnographic studies (e.g. process analysis including simulation, process definition, organizational memory capture and cross-functional consistency management)?
2. What are the coordination problems (e.g. in concurrent engineering) revealed by ethnographic studies, and what is their relative impact on product quality, cost schedule? Rationale: focus research on the "real" challenges.
3. Development of system element models at different levels based on biological system behavior, genetic algorithms, self-organization.
4. Development of composite designs for generic enterprises in specific market segments such as automobile production, banking, health industry, etc. — similar to generative process planning.

C. INTEGRATION OF MODELS
In this area, our discussion centered on the expectation that no large integrated enterprise model can be employed effectively as the enterprise changes unless there is some interconnection strategy for simpler sub-models, allowing alteration of components without disturbing the whole, and allowing the incorporation of new components from "outside" (i.e., from new modeling methods, from other enterprises, from other tools).

Immediate Issues
1. What are the basic forms of change in an enterprise? How will these changes effect designs? How can model formalisms represent and manage change so that the formal framework can evolve as change occurs in an enterprise?
2. Integration vs. Separation: Which scale of integration is optimal? When is separation better than integration? For what kind of problems which scale of integration is necessary?
3. Assume a set of units with well-defined models are given; for a certain task, what are the candidate models for putting the units together? How do different integration models influence behavior? How can analysis of performance be achieved with these models?
4. Assume a set of units with well-defined models are given; if there are uncertainties in the modeling of the individual units, how does that affect the integrated system's performance?

Near-term Issues
1. I have many designs constructed with system X; you have system Y and are not trained on X, yet you need to use the designs I have created. What technical methods are required to enable the sharing of models/data/designs among different systems?
2. What kind of open interfaces are necessary for integrable sub-models? For the design of integration from heterogeneous elements (e.g. machines, personal, environmental, legal aspect influences, etc.) open and flexible interfaces or the close linkage are necessary.
3. What is the optimal size of integration? From the complexity of enterprise models the design of the integration should be as small and simple as possible. The number of

sub-elements should be as small as possible, but still sufficient for the integration target.

4. Assume a set of units with well-defined models are given; if the units are adaptive, how does that affect the integrated system's performance? How can the restructuring of the overall system occur as the system evolves?

Long-term Issues
1. To what extent and with what benefit can model enactment (re-use of models to drive simulation, emulation and run-time processes) guide the processes of achieving software interoperability in product introduction systems, manufacturing control systems and customer, manufacturer, supplier chains?
2. To what extent and with what benefit can model enactment guide the processes of designing a new generation of highly flexible manufacturing machinery? Connected with this issue will be a study of how abstract models and reality can be merged.

D. TOOLS
The focus here is the framework for information sharing among tools, as well as standardized data formats and interfaces.

Immediate Issues
1. Are there integrated tools for the design of integration? For an effective design of integration an "Integrators Workbench" is necessary. This workbench should consist of a set of tools for all the integration steps of the design. Documentation, storage and release management etc. are necessary features.

Near-term Issues
1. Developing a pilot designers workbench (or a designers workstation) for integration modeling.
2. Develop tool interfaces that appropriately reflect the mixing of humans and computational processes in directing the flow of control and data in an enterprise.

Long-term Issues
1. Can a widely applicable design environment be created to produce a new generation of open and highly configurable control system elements of flexible manufacturing machines, where the machines are to be flexibly integrated into host assembly production line operators in discrete-parts manufacturing industries?
2. How do we enhance existing Computer-Aided Engineering (CAE) tools to be "coordination-enabled," e.g. to support cross-application rationale capture, workflow management, consistency management and design knowledge capture?

PANEL 3: INTEGRATION IMPLEMENTATION

Panel Members: Jack Posey (Chair), Andrew Kusiak (Co-Chair), Ian Coutts (Provocateur), Mario Lucertini, Shimon Nof, Gunther Seliger, and Nicolo Vincenzo.

OBJECTIVE
Implementation bridges the gap between design ideals and physical realities. The panel used a working definition of implementation. It was: *Implementation of integration* is

(1) the acquisition, organization, planning, and scheduling of all the facilities and resources necessary to accomplish the objectives of integration as stated in the integration design specifications, and

(2) the execution of the schedule and control of the implementation activities to ensure that the objectives are attained within stated constraints and limitations.

Seven main topics for implementation research arose from the discussions of the panel members. These were: theory, performance measures, education, institutional and social issues, planning and scheduling, organizational structure, and resources. Outlines and discussion of the questions in each of these groups are presented in the following paragraphs.

The panel members ranked the topics and subtopics in order from highest to lowest research priority. The topics and subtopics are presented in this report in the order that they were ranked. All of the panel members agreed that the theory of implementation and performance measures should be ranked first and second, respectively. The order of the next four topics, education, institutional and social issues, planning and scheduling, and organizational structure, was not as clear. These four should be considered to be tightly grouped with nearly equal priority. There was some disagreement with assigning the lowest ranking to the topic of resources. However, the majority of the panel agreed that the topic of resources was close in importance to the previous four but should be ranked as the lowest priority research topic.

A. THEORY OF INTEGRATION IMPLEMENTATION ISSUES

Theory is the establishment of basic concepts. Following is an outline of five research topics on the theory of implementation. The topics involve preparation for implementation as well as execution of implementation.

    a. Target system model.
1. Develop a target system model based on the design specifications.
2. Define and demonstrate the format of the model.
3. Explain how the design specification serves as a form or guide for the model which in turn serves as the foundation for the implementation process.
4. Describe how the model is used throughout the implementation process.

    b. Existing system model.
1. What is the purpose of a model of the existing system?
2. What resources should be modeled?
3. Define the system behavior rules for the model.
4. How should system effectiveness be measured in the model?
5. Propose methods for model verification and model validation.

    c. How are goals and objectives in the design specifications applied in the implementation?

    d. What are the metrics, tolerances, constraints, and limitations of implementation?

    e. Tactics
1. Define the tactical approaches appropriate for implementation.
2. Develop tools for tactics selection.
3. Describe the tactical methods.
4. Describe tactical contingencies and alternatives.

B. PERFORMANCE MEASURES OF INTEGRATION IMPLEMENTATION ISSUES

Performance measures are used to determine the extent and quality of the integration process.

    a. Define implementation metrics.
    b. Evaluation and testing.
1. Define functional verification and describe the use implementation metrics in the performance of functional verification.
2. Define system verification and describe the use of implementation metrics in the performance of system verification.

C. EDUCATION ISSUES

The education background of the participants in the implementation has a significant impact on what methods can be used for implementation, the schedule of implementation, and the

success of the implementation. Education may be required as part of the implementation plan.
  a. Define the job qualifications for implementation.
  b. Describe methods for taking an inventory of education background and evaluating the job qualifications with respect to the education inventory.
  c. Determine the role of education in motivation for implementation.
  d. Training.
      1. Define the role of training as it pertains to achieving the necessary education inventory.
      2. Determine training profiles and techniques necessary to achieve the required education inventory.
      3. Describe and explain the use of alternative methods for implementation education; e.g., Simulators, Computer-aided instruction, Lectures, etc.
      4. Develop new education and training methods that are specific to integration implementation.

D. INSTITUTIONAL AND SOCIAL ISSUES

Legal requirements and social and cultural considerations must be understood because they affect all aspects and stages of implementation. Key research issues:
  a. Determine labor laws that pertain to the implementation process.
  b. Determine environmental laws that pertain to the implementation process.
  c   Describe the roles of ethics and professional standards.
  d. Explain how safety and occupational health considerations affect the implementation process.
  e. Communication and promotion of project image: Describe the audiences for communication and promotion; What methods should be used? Determine the effect on the implementation process.

E. PLANNING AND SCHEDULING OF INTEGRATION IMPLEMENTATION ISSUES
  a. Verification: Define verification for plans for functions of the system and the whole system; Propose methods for verifying functional and system plans.
  b. Describe the stages of activities to be planned, and explain how to determine schedule duration.
  c. Planning horizon: Define the planning horizon for implementation; Explain how the planning horizon affects scheduling.
  d. Simulation: Explain the role of simulation in planning and scheduling; Describe simulation methods and techniques for planning and scheduling implementation.
  e. Milestones: Define milestones for implementation; Describe how milestones are presented on the schedule; Develop methods for reporting progress along the schedule.
  f. Burn-in: Develop the concept of "burn-in" in the context of integration implementation; Develop methods for including the concept of burn-in within the implementation plan and schedule.
  g. Execution: Describe methods for activating the implementation plan; Plan resource consumption, including how to track it during implementation.

F. ORGANIZATIONAL STRUCTURE ISSUES

Successful integration may require specification of the existing organizational structure or changes in the structure. The changes may be permanent and/or temporary.
  a. What is the role of teams in the implementation process?
  b. Describe how teams are defined, how team members are selected, and sources of team members.
  c. Team structure: Describe different team structures needed for implementation; how team structures change during the implementation process; develop methods for forming, changing, and disbanding teams.

d. Team management.: Describe the hierarchy of teams in the implementation; explain how to motivate teams for implementation; describe methods for communication *between* teams and *among* team members; define team champions and their role in implementation teams.

G. RESOURCES FOR INTEGRATION IMPLEMENTATION ISSUES

a. Define the different types of resources in implementation. Are people, infrastructure, equipment, and capital the only types?
b. Describe how to acquire and distribute resources during implementation.
c. Describe methods for indicating when resources are no longer needed.

Comments

One aspect of implementation that the panel suggested for research is whether an identifiable stage known as *implementation* should even exist in the integration process. The thought was that perhaps the development of the design specification should be the implementation and that the ultimate goal of implementation should be zero cost, zero resource and zero time execution. How reasonable and accurate is this hypothesis? Should it serve as an overarching goal of integration implementation research? Are there variations of this hypothesis that are more practical and useful? Is the goal of the elimination of the implementation stage a useful objective that is never expected to be reached but would promote successful, efficient methods of implementation?

During panel discussions, many different opinions were expressed on many different aspects of implementation. The panel members agreed that integration implementation is a large, important aspect of integration. The panel members believe that the implementation questions and topics presented here will provide a substantial challenge to researchers and will lead to significant results about integration implementation.

PANEL 4: NATURE AND METRICS OF INTEGRATION

Panel Members: Asbjorn Rolstadas (Chair), Bezalel Gavish (Co-Chair), Erik Trostmann, Kurt Kosanke, Mark Fox, and Paul Valckenaers

INTRODUCTION

Definition of Integration: *Enterprise Integration* (an on-going process) can be defined in two parts:

a. Global: Enable organizational units, inside and outside the enterprise, to operate in a coordinated manner to reach the enterprise goal and optimize the results.
b. Operational: Interface functions, information technology systems and resources to act as one entity. Provide information handling transparency. Define standards to enable interpretation consistency.

Some identifiable integration characteristics:
  – Sharing of information between known parties
  – Consistent "information model"
  – Decentralized system
  – Client-server relationship
  – Identification and access to available information
  – Evolutionary process
  – Cutting across enterprise boundaries
  – Visibility and adaptability of organization (flexible!) and operational structures
  – Clearly defined points of control, responsibilities and authorities
  – Predictable (non-deterministic?)
  – "Management of change" support
  – Asset management support
  – Conflict avoidance resolution mechanisms

470

RESEARCH ISSUES

Integration of Legacy Systems: Interfaces; Migration.

Information Technology: Resource management; Network technology; Human-computer interaction; Bench marking; Performance measurements; Security and integrity.

Modelling: Real-time system behavior, feedback co-ops; Re-usability; Model maintenance.

Enterprise: Organizational implications; Flexibility – management of change; Acceptance and education of paradigm shift.

Integration Metrics: Flexibility – adaptation to change (number of changes, efficiency, effectiveness); Operation efficiency/effectiveness (cost, throughput and TA Time, meeting scheduled delivery, productivity, tense) of operational, decision and maintenance processes; Relation to company objectives and goals; Identification of responsibility and authorization; Multidisciplinary communication.

SPECIFIC RESEARCH DIRECTIONS

A. The Role of Models and Modelling in Enterprise Integration.

B. Human Aspects in Integration.

C. Integrating Legacy Systems (Interfaces).

Note: Research directions A, B, and C involve the nature of integration, but they were also discussed by the previous panels and will not be duplicated here.

D. Integration Metrics

Rationale: Enterprise integration will have significant impact on enterprise performance and therefore on the company's bottom line. To monitor and control this impact, relevant metrics have to be applied – some of which may be traditional enterprise performance measurements; e.g., integration metrics on efficiency and effectiveness. Others, such as integration metrics on other terms, may have to be developed and tested for their applicability. Research is required in both areas to identify, develop and test the set of metrics needed to measure the direct and indirect effects of integration on enterprise operations.

Direct integration impact on enterprise operations will come from improvements in asset management, more efficient use of enterprise resources, better control of business processes and their respective information and material flow. Additional influence will be by improved decision support or "management of change" as well as the efforts needed in enterprise engineering to maintain the integration baseline, the enterprise models. Flexibility in enterprise operation to follow changes in demand, supply as well as changes in regulatory legislation, will contribute to the effects of enterprise integration as well.

Quantitative measurements of all these factors will be very helpful to demonstrate the benefits of integration and to promote it in the industry.

Objective: Develop, test and validate a meaningful set of metrics for enterprise integration evaluation and validation.

INDEX